C0 BZC 734

HADRONIC MATTER IN COLLISION

HADRONIC MATTER IN COLLISION

PROCEEDINGS OF THE SECOND INTERNATIONAL WORKSHOP ON LOCAL EQUILIBRIUM IN STRONG INTERACTION PHYSICS

Santa Fe, New Mexico, USA, April 9-12, 1986

Editors: **P. Carruthers and D. Strottman**

Randall Library UNC-W

Published by

World Scientific Publishing Co Pte Ltd.
P. O. Box 128, Farrer Road, Singapore 9128.
242, Cherry Street, Philadelphia PA 19106-1906, USA

Library of Congress Cataloging-in-Publication Data

International Workshop on Local Equilibrium in Strong
Interaction Physics (2nd : 1986 : Santa Fe, New Mexico, USA)
Hadronic matter in collision.

1. Hadrons -- Congresses. 2. Collisions (Nuclear physics)--Congresses. I. Carruthers, Peter, A., 1935- . II. Strottman, D.
QC793.5.H328I584 1986 539.7'54 86-22435
ISBN 9971-50-165-1

Copyright © 1986 by World Scientific Publishing Co Pte Ltd.

All rights reserved. This book, or parts thereof, may not be reproduced in any form or by any means, electronic or mechanical, including photocopying, recording or any information storage and retrieval system now known or to be invented, without written permission from the Publisher.

Printed in Singapore by Fu Loong Lithographer Pte Ltd.

QC
793.5
.H328
I 584
1986

PREFACE

Recent years have witnessed the birth of a new subfield of physics, dealing with the dynamical and statistical properties of all forms of hadronic matter. New discoveries in particle physics experiments have yielded elegant but puzzling results for the strong interactions, while nuclear physicists contemplating facilities in the relativistic domain have begun to realize a common ground with some of the most important problems in traditional high energy physics. Indeed, we now see nuclear physicists learning quantum field theory and particle physicists becoming aware of the power and elegance of the methods of statistical physics. Workshops and conferences such as the present one have resulted in a new conceptual sophistication, exchange of techniques, and, we believe, in rapid progress in understanding. This renewed vigor comes after at least a decade of eclipse of interest in hadronic physics. It is hard to believe that the hydrodynamical description of the gross behavior of hadronic motion, let alone the cncept of quark matter, its phases and collective behavior, were as recently as 1974 held in contempt by otherwise perceptive physicists.

Now that the "standard model" seems safely in place as our current minimal paradigm, we can stand back to contemplate the most fruitful topics for investigation. Surely one of these subjects is the 99.9% of hadronic events whose explication is not to be found in perturbation theory or a Clebsch-Gordan coefficient, nor in the opposite limit of thermodynamic equilibrium. A priori, we cannot say which among all the observed data will lead to basic understanding. Who would have imagined how extraordinarily fruitful would be Debye's idea of quantizing the sound vibrations of a crystal?

The theory of the static, equilibrium properties of hadronic matter has a long, distinguished history, beginning in modern times with the Bethe-Breuckner-Goldstone theory of nuclear matter and lately in the finite temperature lattice gauge theory work of the Bielefeld and Illinois groups, among others. The application of the local thermodynamic (LTE) approximation to selected space time regions of complex nuclear collisions goes back some 50 years. In the early 1950's Fermi and Landau pointed out the relevance of these concepts to the problem of multiparticle production in so-called elementary particle collisions, inventing thereby the first version of the statistical-hydrodynamical model. Not surprisingly, many aspects of such complex collisions were not dealt with in these early investigations. Even now it is not clear how to dissect the different parts of a high energy collision event into parts requiring a coherent quantum mechanical description or a kinetic or even thermalized description. Nevertheless, both experimental advances and conceptual progress suggest that rapid progress is at hand in this field.

The evolution of theoretical expectations following the introduction of the quark model and its persuasive formulation in the form of quantum chromodynamics has led, among other things, to the belief in the existence of a deconfined plasma phase (probably many other phases, were we able to control the external parameters). Despite the continuing absence of a convincing experimental signature for this phase, most scientists accept the basic importance of such a possibility. As an unexpected consequence, it is now possible to propose seriously for hadron physics, topics previously left to condensed matter, traditional nuclear physics and statistical physics, namely the idea of collective coordinates, whose excitations are quasiparticles, and to consider the stochastic behavior of non-equilibrium systems created in hadronic collisions from this point of view.

Experimental investigations of multiparticle probability distributions in hadron-hadron collisions have established the validity of the negative binomial distribution over a vast range of energy. At the same time, the parameter characterizing the negative binomial varies with energy, thereby spoiling what had previously seemed to be a good, though imperfectly understood scaling law due to Koba, Nielsen and Olesen. The understanding of this fact is currently the focus of many theoretical investigations, as shown by some of the papers in this volume. The corresponding distribution of hadrons resulting from electron-positron annihilations is, however, very close to Poisson, pointing to a rather different dynamical framework. Moreover, the study of correlations of various types promises to be a powerful way to distinguish among dynamical schemes equally capable of describing the probability distributions. The results of intensity interferometry, begun in the famous work of Hanbury-Brown and Twiss for electromagnetic radiation from stars, only to reappear in the microscopic domain in the work of Goldhaber, Goldhaber, Lee and Pais on the correlations of pions of like sign, are especially informative in revealing the amount of coherence in the source emissions. Several illuminating papers on this subject were presented at the Santa Fe workshop, and are included here. We expect powerful generalizations of these techniques, many borrowed from the sophisticated repertoire of quantum optics, to be a major topic for detector design and theoretical research. New results continue to appear for hadron-nucleus collisions, the subject of several papers at this workshop. For the present, the hope of a decisive resolution of the space-time properties of the more elementary hadron-hadron collisions event by these measurements has not been achieved.

The theoretical concepts required to describe the data already exist to a large extent, even if not yet adapated to the context of particle and nuclear physics. For the most part, we are dealing with neither a perturbative nor a thermalized regime, but with out-of-eqiulibrium dynamics of some collective variables whose precise identity is not yet clearly understood. The entire set of problems can be cased in the framework of phase space evolutions of suitable distribution functions, whose equations of motion lead in principle to traditional approximations such as Vlasov-Hartree-Fock-Boltzmann descriptions, collective modes or scattering depending on the situation. Or, the lore of stochastic evolutions, associated with many legendary names such as Planck, Einstein, Fokker, and Kolmogorov, can be mentined. As mentioned, many of the methods of modern quantum optics have until recently escaped the attention of people working on strong interactions. The connections to traditinal kinetic and hydrodynamic theory are already being made, as can be seen from many of the papers included in these proceedings. We anticipate a rapid increase in understanding in the near future. The problems are not easy, but the techniques are well delineated, so that the stimulus of experiments can be expected to lead to genuine progress.

We regret that due to the early deadline for manuscript submission, three lectures presented at the Workshop by G. Ekspong, T. Ludlam and D. Scott were not available for inclusion in these proceedings.

This workshop was supported in part by the Los Alamos National

Laboratory and the Physics Division of the National Science Foundation. We also wish to express our appreciation to Mary Ramsey for her considerable aid in the organization of the workshop, and to the Organizing Committee for its help and advice.

ORGANIZING COMMITTEE

G. Baym
University of Ill.
at Urbana-Champaign

W. Busza
Massachusetts Inst.
of Technology

P. Carruthers
Los Alamos National Lab.

A. Goldhaber
State University of NY,
Stony Brook

M. Gyulassy
Lawrence Berkeley Lab.

T. Ludlam
Brookhaven National Lab.

L. McLerran
Fermi National Accel. Lab.

D. Strottman
Los Alamos National Lab.

R. Weiner
University of Marburg

W. Willis
CERN

ADVISORY COMMITTEE

H. Bethe
Cornell University

J. D. Bjorken
Fermi National Accel. Lab.

D. A. Bromley
Yale University

G. Ekspong
University of Stockholm

E. Feinberg
Lebedev University
Moscow

M. Jacob
CERN

K. Kajantie
University of Helsinki

S. Nagamiya
University of Tokyo

H. Pugh
Lawrence Berkeley Lab.

J. Schiffer
Argonne National Lab.

L. Van Hove
CERN

C. N. Yang
State University of NY,
Stony Brook

G. Zinovjev
Inst. for Theoretical Physics
Kiev

EDITORS

P. Carruthers and D. Strottman
(Los Alamos National Laboratory)

CONTENTS

NUCLEAR FRAGMENTATION

NUCLEUS-NUCLEUS REACTIONS

QUANTUM KINETIC THEORY AND HYDRODYNAMICS

PHASE TRANSITIONS

BOSE-EINSTEIN CORRELATIONS

NEW RESULTS ON THE BOSE-EINSTEIN EFFECT IN e^+e^- INTERACTIONS OR THE GGLP EFFECT REVISITED

Gerson Goldhaber and Ivanna Juricic*

Lawrence Berkeley Laboratory and Department of Physics
University of California
Berkeley, California 94720

ABSTRACT

Experimental data on the Bose-Einstein (B.E.) enhancement of pion pairs from e^+e^- colliders is reviewed. The data comes from CLEO, Mark II, and TASSO. A result on KK enhancement from the Axial Field Spectrometer (AFS) is also quoted. Results of new derivations based on the string or Lund model of the inside outside cascade are also discussed.

At the last LESIP conference I prepared a paper dealing with the first 25 years of the GGLP effect.[1] It appears, however, that the field of pion interferometry is just beginning to emerge from the qualitative to the quantitative regime. New data and new insights into the interpretation have appeared, and I want to take this opportunity to survey the progress made in the last two years.

In particular I will primarily concentrate on e^+e^- data. Zajc in his talk at this conference[2] will discuss the relativistic heavy ion data and interpretations.

OUTLINE

The three considerations which were central to the topic of pion interferometry for the first 25 years are still with us:

1. What are the optimal variables?
2. What is the significance of α in the semi empirical expression $1 + \alpha e^{-\beta Q^2}$?
3. What is the most suitable control sample?

To these we can perhaps now add:

4. What is the effect of "long" lived resonances such as ω, η, η', etc. The effect of "particles" such as D^0, D^+, F^+, B^-, B^0, etc., is clear: they decay

*This manuscript is based on a talk given by one of us (G.G.) at LESIP II and is thus in the first person.

weakly and hence are too long lived to contribute a *measurable* interference effect.

5. Can anything new be learnt from *three* or more LIKE pion B.E. interference effects?

The experimental and theoretical results I want to discuss here are:

6. Results from the CLEO collaboration at CESR on B.E. studies in the Υ region.[3]

7. Results from the Mark II experiments at SPEAR and PEP on e^+e^- annihilation and 2γ data. Some of these were quoted in ref 1]. The present results are from a thesis in preparation by Ivanna Juricic.[4]

8. Results from the TASSO experiment[5] at PETRA on e^+e^- annihilation at 34 GeV. A preliminary version of these data was quoted in ref. 1].

9. Papers by M. Bowler[6] and B. Andersson and W. Hofmann[7] who have introduced B.E. correlation into a string model (or Lund model) Monte Carlo calculation. Furthermore, M. Gyulassy and K. Kolehmainen[8] have considered the "inside outside cascade" model of pion production and have obtained explicit analytical solutions for pion interferometry.

10. Results from the AFS experiment[12] at the ISR showing B.E. enhancements for $K^\pm K^\pm$ pairs and the lack of such an enhancement for non identical bosons $K^\pm\pi^\pm$.

1. CONSIDERATIONS ON OPTIMAL VARIABLES

The CLEO collaboration[3] gives a discussion on this topic. They consider both the variables we introduced in our original GGLP paper, as modified empirically by Deutschmann et al.[1] viz.

$$R^L_U = 1 + \alpha e^{-\beta Q^2}$$

where the invariant quantity $Q^2 = -(p_1-p_2)^2 = M^2(12) - 4m_\pi^2$, and R^L_U is the ratio of the experimental distributions for LIKE and UNLIKE pions, as well as the Kopylov and Podgoretsky (KP) variables.[1]

Here α can vary from 0 to 1 and is the deviation from the maximal B.E. effect ($\alpha = 1$), while $\beta = r^2$, the effective radius squared of the pion source distribution. In particular if we define $\vec{q} = \vec{p}_1 - \vec{p}_2 = \vec{q}_T + \vec{q}_L$ and $q_0 = |E_1 - E_2|$ where $\vec{q}_T$ and $\vec{q}_L$ are the transverse and longitudinal components respectively along the $\vec{p} = \vec{p}_1 + \vec{p}_2$ direction, then

$$\begin{aligned} Q^2 &= q^2 - q_0^2 \\ &= q_T^2 + q_L^2 - q_0^2 \end{aligned}$$

where $|\vec{q}_T|$ and q_0 are the KP variables, while $|\vec{q}|$ and q_0 are the Gyulassy, Kauffmann and Wilson (GKW) variables.[1] Here q_T^2 and $(q_L^2 - q_0^2)$ are each relativistic invariants with respect to a boost in the direction of $\vec{p}$. The CLEO group used this feature for the ansatz:

$$R_U^L = 1 + \alpha e^{-\beta q_T^2 - \gamma(q_L^2 - q_0^2)}.$$

They also point out, as have Åkesson et al.[1] that $q_L \approx q_0$ in these high energy reactions. Thus Q^2 can be approximated by q_T^2. This can be illustrated with the Mark II data at the J/ψ. Fig. 1 shows a Q^2 and q_T^2 distribution, the K^0 and ρ^0, which are clearly visible for the Q^2 distribution, appear somewhat washed out for the q_T^2 distribution. The difference between these two distributions gives a qualitative feeling to what extent $q_L \approx q_0$. This approximation improves as the particle energies increase. The CLEO results are quoted as a function of q_T. Furthermore, they showed that to a good approximation the KP distribution, which corresponds to a source distribution on a spherical shell of radius r containing pion radiators of lifetime τ:

$$R_U^L = 1 + \alpha[2J_1(q_T r)/q_T r]/[1 + (q_0 \tau)^2]$$

can be expressed as:

$$R_U^L = 1 + \alpha e^{-\beta(q_T^2 + q_L^2) - q_0^2 \tau^2}$$

which is, however, not an invariant expression. It can thus be concluded that the KP distribution, which works fine at low energies, is not applicable to highly relativistic regions.

2. THE SIGNIFICANCE OF α

Ideally α represents the degree of chaoticity with $\alpha \to 1$ corresponding to a completely chaotic source and $\alpha \to 0$ to a "pion laser" (Fowler and Weiner, Gyulassy, et al.[1]). However, any other effect that tends to dilute the strength of the B.E. enhancement will also reduce the value of α. The result is particularly striking in comparing

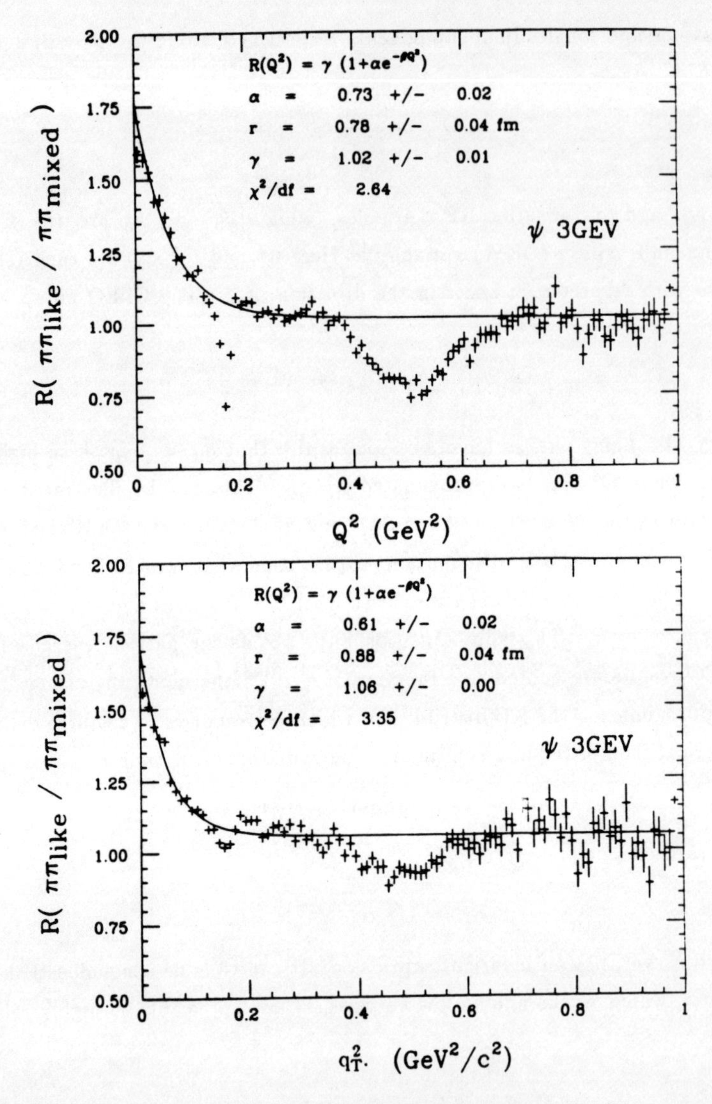

Fig. 1 Mark II data at the J/ψ.

(a) $R_U^L(Q^2)$ distribution.

(b) $R_U^L(q_T^2)$ distribution.

the Mark II SPEAR data at the J/ψ and the 4-7 GeV region. We know of course that the principal new channel that opens up above the ψ is the onset of open charm viz. the D^0 and D^+ mesons. These are weak decays; thus any LIKE pion pair composed of one pion from a D and one directly from the vertex will not give any *measurable* B.E. correlation since these will occur at such low Q^2 values as to be unresolved. The inclusion of such pairs thus *dilutes* the B.E. effect and hence reduces α. It becomes a quantitative question to what extent the reduction in α, in going from the J/ψ to the 4-7 GeV region, is due to the onset of open charm and to what extent to coherence effects in e^+e^- jets as suggested by Giovannini and Veneziano[1].

3. THE CONTROL SAMPLE

Ideally the control sample should have all the features of the like pion pairs except for the Bose-Einstein enhancement. The three e^+e^- experimental results I am discussing here have all attempted various schemes to obtain suitable control samples. All have, however, settled for unlike pion pairs with careful avoidance of known resonance regions K^0, ρ^0, S^{*0}, and f^0. What is not known and could possibly lead to some extraneous effects is the possibility that low $\pi\pi$ mass $I = 0$, or $I = 1$ phase shifts could affect the $\pi^+\pi^-$ system but not the $I = 2$ $\pi^+\pi^+$ system. As I showed earlier (see Fig. 12 of ref. 1]), there is, however, no evidence for such a phase shift effect in the SU(3) analogous systems $K^+\pi^-$ ($I = 1/2$ or $3/2$) and $K^+\pi^+$ ($I = 3/2$), the distribution of non-identical bosons.

4. THE EFFECT OF RESONANCES AND PARTICLES

In the study of the LIKE pion pairs we have to realize that many of the pions come from the decay of resonances and "particles". P. Grassberger[9] and also G.H. Thomas[10] have calculated the consequences of taking resonances into account in the B.E. effect.

Here I give a very simple consideration. The mean path length for resonance (and particle) decay is:

$$L = \beta\gamma c\tau = \frac{p}{M}\, c\tau,$$

where p, M, τ and Γ refer to the resonance. But $\Gamma\tau \simeq \hbar$.

Thus

$$L = \frac{p}{M\Gamma} c\hbar$$

$$= \frac{p}{M\Gamma} 197 \text{ MeV fm.}$$

Hence typical decay distances for pions from ρ, K*, ω, and η decays are 2, 4, 20, and 400 fm respectively.

Typical length dimensions one deduces from the B.E. enhancement in e^+e^- annihilations are ~ 1 fm. Thus one might expect the presence of resonances to have a considerable effect giving rise to enhancements at very small Q^2 values — the regions where corrections due to Coulomb effect and experimental difficulties, e.g. unresolved overlapping tracks, are maximal. We can thus assume that the very low Q^2 region cannot be reliably measured in present day experiments and the effect of these resonances would be to dilute the B.E. enhancement and hence to reduce α.

As I will show for the Mark II detector at SPEAR the surprising feature is that at the J/ψ, $\alpha \rightarrow 1$ and hence *we see no evidence for the dilution factor expected from the resonances* which we know are present in the data!*

The CLEO collaboration have also studied this feature and have explicitly corrected for all resonances and particles present. This results in a distribution for the continuum ~ 10.5 GeV region (see Fig. 2) in which $\alpha_{COR} = 1.45 \pm 0.25$. Again the resonances do not appear to reduce α as much as expected, resulting in an apparent "overcorrection". Of course, within the errors $\alpha_{COR} = 1$ is not ruled out.

5. THREE LIKE PION ENHANCEMENTS

The Mark II and TASSO data have also been analyzed in terms of three LIKE pions.[1] The TASSO group[5] suggest that the information from three LIKE pions can be interpreted entirely as a reflection of the $\pi^\pm\pi^\pm$ case. By neglecting phase factors they suggest that $\alpha_3 \approx 5\alpha$ and $\beta_3 \approx \frac{\beta}{2} \rightarrow \frac{\beta}{3}$ and hence that no useful new information is obtained.

---

*The inclusive branching ratio for $\psi \rightarrow \rho^0 + x$ and $\psi \rightarrow \omega + x$ are $8.3 \pm 0.8\%$ and $13.4 \pm 4.7\%$ respectively (G. Gidal et al., Phys. Lett. **107B**, 153 (1981)). I do not have a good value for $\psi \rightarrow \eta + x$.

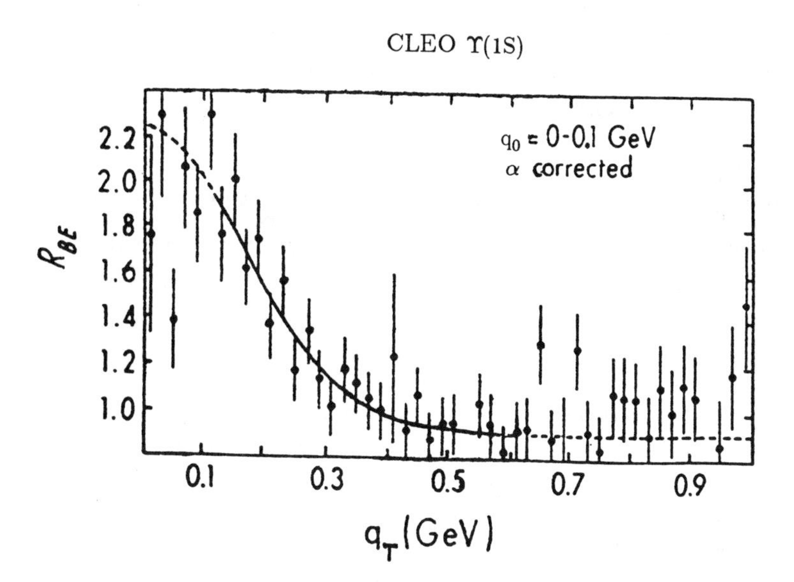

Fig. 2 CLEO data in the continuum near 10.5 GeV. $R^L_U(q_T)$ distribution corrected for all resonances and long lived particles present in the data.

My own opinion is that in view of our many residual uncertainties relating to the interpretation of the *details* of the Bose-Einstein effect, any additional measurement can yield information that will be helpful in the complete understanding of the B.E. enhancement. Thus in Section 7, I quote results from the Mark II data[4] on $\pi^\pm\pi^\pm\pi^\pm$ correlations.

6. THE Υ REGION

A paper has come from the CLEO Collaboration[3] studying the $\Upsilon(1S)$, $E_{CM} \simeq 9.5$ GeV, and nearby continuum regions around $E_{CM} \simeq 10.5$ GeV and 10.8 GeV. In this work the authors found a number of rather interesting features.

6.1. Comparison between $\Upsilon(1S)$ and the Nearby Continuum Regions

In going from the $\Upsilon(1S)$, which is considered a 3 gluon final state, to the continuum region above it — considered primarily a 2-quark jet final state — I had expected[1] an appreciable change in α, where the 3 gluon final state is considered more chaotic (larger α) than the 2 quark jet state.

The CLEO experiment (see Table I) taken at face value observes a slight variation in α, but within the statistical errors the values are not significantly different.

6.2. Dependence on q_0

The CLEO Collaboration present data at the $\Upsilon(1S)$. A distribution of R_U^L as a function of q_T for 3 regions in q_0, is given in Fig. 3. Both α and r increase as $q_0 \to 0$ (see Table II). The experimenters stress furthermore that $R_U^L(q_0)$ is a flat distribution with only a minor enhancement when a cut is made on low q_T values. See Fig. 4.

7. THE MARK II DATA AT SPEAR AND PEP

In the Mark II experiment we have studied four distinct data samples.[4] Here we studied R_U^L vs Q^2.

- the J/ψ, a sample of $\sim$ 1,300,000 events
- the 4-7 GeV region at SPEAR, a sample of $\sim$ 78,000 events

TABLE I

CLEO Results[3]

α and r for different data samples

	α	r(fm)
$\Upsilon(1S)$	0.50 ± 0.09	0.99 ± 0.14
continuum $< \Upsilon(4S)$	0.43 ± 0.07	0.86 ± 0.15
continuum $> \Upsilon(4S)$	0.41 ± 0.04	0.86 ± 0.08

TABLE II

CLEO Results[3] and Mark II Preliminary Results[4]

α and r as functions of q_0

	q_0 (GeV)	α	r(fm)
CLEO	0.0 – 0.1	0.38 ± 0.06	0.84 ± 0.11
$\Upsilon(1S)$	0.1 – 0.2	0.26 ± 0.06	0.74 ± 0.14
	0.2 – 0.3	0.17 ± 0.06	0.77 ± 0.24
Mark II	0.0 – 0.1	1.05 ± 0.04	0.90 ± 0.05
ψ	0.1 – 0.2	0.71 ± 0.03	0.90 ± 0.02
	0.2 – 0.3	0.61 ± 0.04	0.57 ± 0.10
Mark II	0.0 – 0.1	0.63 ± 0.08	0.84 ± 0.17
4-7 GeV	0.1 – 0.2	0.63 ± 0.07	0.71 ± 0.13
	0.2 – 0.3	0.51 ± 0.07	0.58 ± 0.22
Mark II	0.0 – 0.1	1.17 ± 0.17	0.91 ± 0.18
2γ	0.1 – 0.2	1.14 ± 0.15	0.74 ± 0.17
29 GeV	0.2 – 0.3	0.82 ± 0.16	0.55 ± 0.16
Mark II	0.0 – 0.1	0.80 ± 0.23	1.24 ± 0.40
29 GeV	0.1 – 0.2	0.56 ± 0.08	0.79 ± 0.19
	0.2 – 0.3	0.64 ± 0.15	1.03 ± 0.28

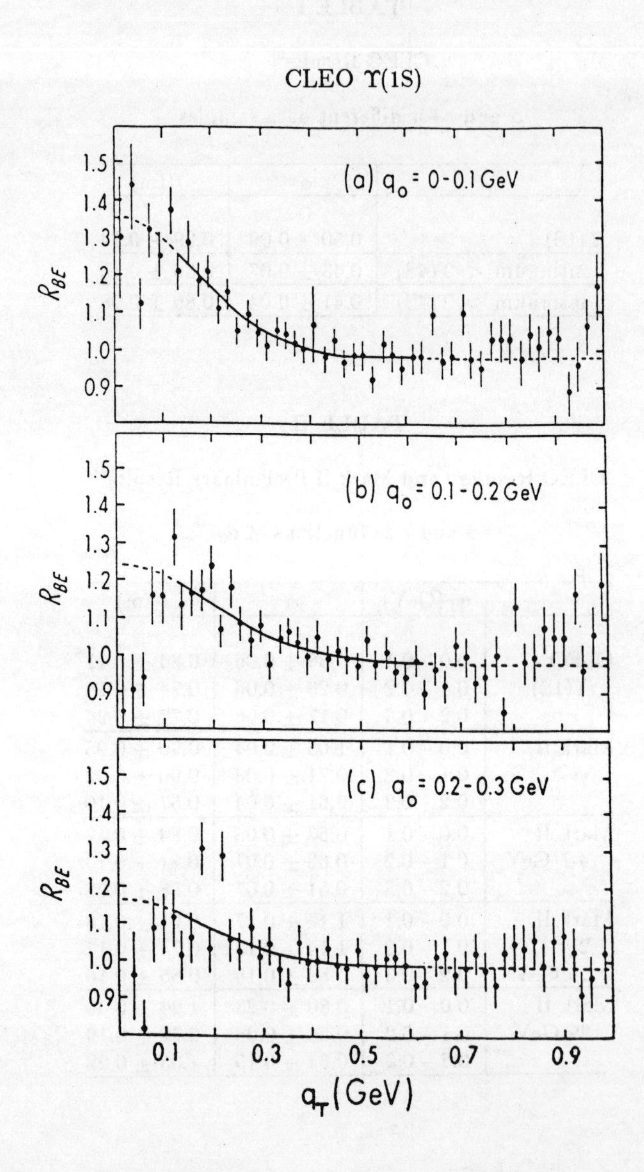

Fig. 3 CLEO data at the $\Upsilon(1S)$. $R_U^L(q_T)$ distributions for 3 q_0 intervals.

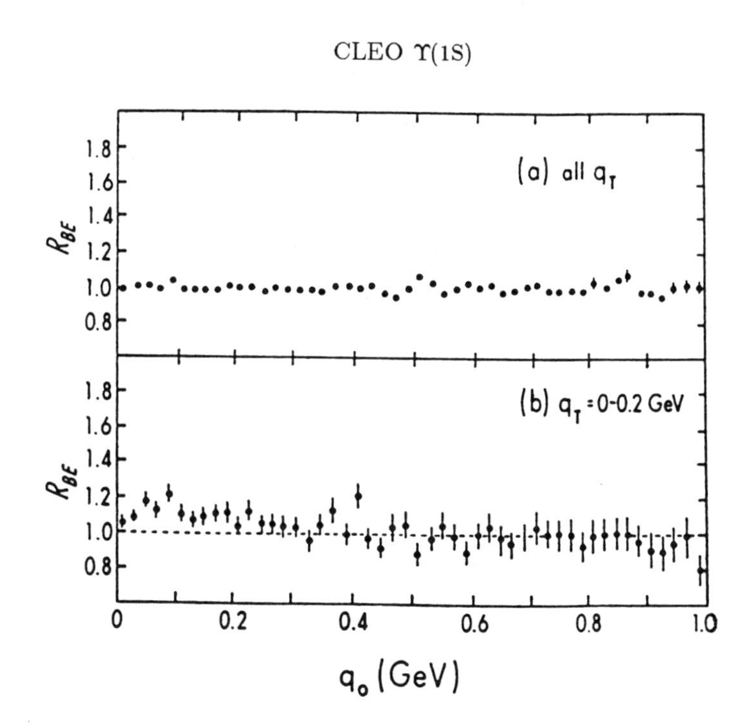

Fig. 4 CLEO data at the $\Upsilon(1S)$. $R_U^L(q_0)$ distributions:
(a) for all the data.
(b) for $q_T < 0.2$ GeV.

- the 29 GeV region at PEP, a sample of $\sim$ 90,000 events
- $\gamma\gamma$ events at 29 GeV, a sample of $\sim$ 42,000 events. These events were tagged by the observation of a single electron at small angles to the beam.

7.1. Corrections and Normalization

The Mark II data was corrected for kaon contamination due to residual K–π misidentification. From Monte Carlo studies[4] we find that $\sim 12\%$ of the LIKE $\pi\pi$ pairs are actually Kπ pairs. The procedure is to subtract a corresponding fraction of the central sample, i.e., the UNLIKE pairs, before R^L_U is calculated.

A further correction is made for the Coulomb repulsion of the LIKE pions and attraction of the UNLIKE pions. The LIKE and UNLIKE distributions are normalized in the regions outside the observed resonances. In a final fit an additional constant γ is introduced which is unity within a few percent. Thus $R^L_U(Q^2) = \gamma(1 + \alpha e^{-\beta Q^2})$ is fitted to the data. The fits for $\pi\pi$ are given in Table III for the four data samples. The corresponding fits for four $\pi\pi\pi$ distributions are given in Table IV. In the three pion case we present the ratio R^L_M where here L corresponds to $\pi^\pm\pi^{\pm\pm}$ and M to $\pi^\pm\pi^\pm\pi^\mp$. Thus the control sample itself contains a B.E. enhancement.[1] The true value for α_3 in this case would thus be larger than the measured value α_3.

7.2. The q_0 Dependence

For each of the four data samples we studied 3 intervals in q_0. Fig. 5 illustrates this for the 29 GeV e^+e^- annihilation events. Table II and Figs. 6 and 7 show α and r for the 4 data samples as well as for the CLEO data at the Υ(1S).

The $R^L_U(q_0^2)$ distribution at the J/ψ is completely flat when no q_T cut is made, similar to the CLEO result. For $q_T < 0.15$ GeV we observe an enhancement with a very steep falloff. This indicates that the enhancement for low q_T is "washed out" by pairs with nearly equal energy (i.e., $q_0 \approx 0$) but with larger q_T values. Thus the e^+e^- results differ from the hadronic data[1] where typically $R^L_U(q_T)$ and $R^L_U(q_0)$ distributions are very similar. In assessing the significance of this result it must of course be remembered that we are dealing with non-invariant variables!

TABLE III

$R_U^L(Q^2)$ for $\pi\pi$

Mark II data[4] — Preliminary

Statistical errors only

DATA	α	r(fm)	γ
SPEAR J/ψ 3.1 GeV	0.77 ± 0.02	0.79 ± 0.03	1.02 ± 0.01
SPEAR 4-7 GeV	0.47 ± 0.04	0.71 ± 0.10	1.00 ± 0.01
PEP $\gamma\gamma$ 29 GeV	0.93 ± 0.08	0.79 ± 0.10	1.04 ± 0.02
PEP 29 GeV	0.50 ± 0.05	0.87 ± 0.12	0.92 ± 0.01

TABLE IV

$R_M^L(Q^2)$ for $\pi\pi\pi$

Mark II data[4] — Preliminary

Statistical errors only

DATA	α_3	r_3(fm)	γ
SPEAR J/ψ 3.1 GeV	3.45 ± 0.20	0.51 ± 0.03	0.97 ± 0.01
SPEAR 4-7 GeV	1.44 ± 0.16	0.43 ± 0.06	0.94 ± 0.02
PEP $\gamma\gamma$ 29 GeV	2.22 ± 0.43	0.47 ± 0.10	0.92 ± 0.04
PEP 29 GeV	1.13 ± 0.21	0.61 ± 0.09	0.99 ± 0.01

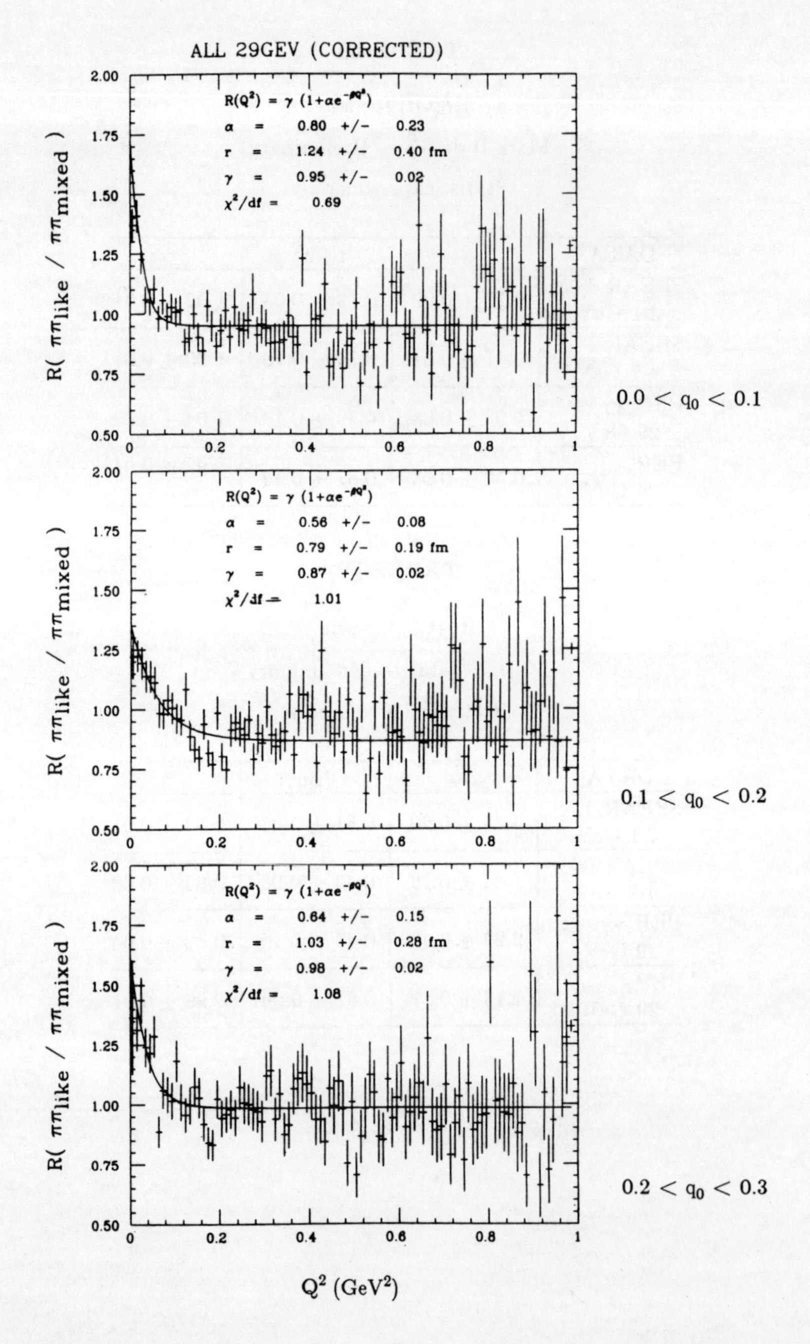

Fig. 5 Mark II data at 29 GeV. $R_U^L(Q^2)$ distribution for 3 q_0 intervals.

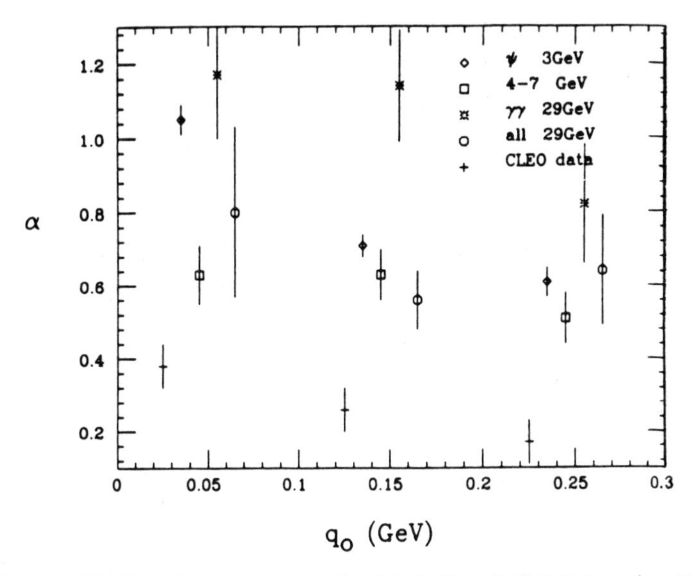

Fig. 6 Display of α values vs q_0 for Mark II and CLEO data (see Table II).

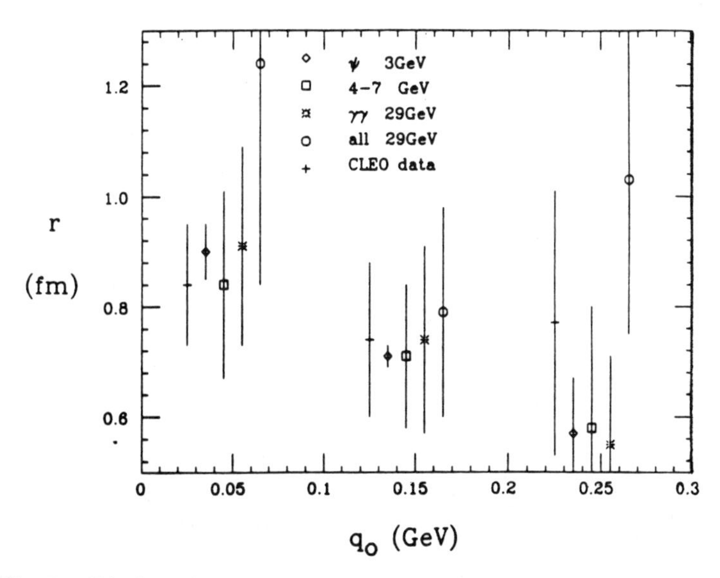

Fig. 7 Display of r values in fm vs q_0 for Mark II and CLEO data (see Table II).

8. THE e^+e^- ANNIHILATION IN THE 34 GeV REGION, TASSO DATA

The TASSO Collaboration at PETRA have studied B.E. effects for some time.[1] The new results are presented in a recent paper[5] together with comparisons with the string (or Lund) model.[6],[7]

8.1. Analysis Method

The TASSO Collaboration present $R_U^L(Q^2)$ distributions in which they remove experimental detector bias and resonance effects by dividing by $[R_U^L(Q^2)]_{MC}$. Here the latter distribution is obtained from Lund Monte Carlo calculations. Fig. 8a gives the experimental $R_U^L(Q^2)$ distribution called r. Fig. 8b gives this same distribution normalized with $[R_U^L(Q^2)]_{MC}$ called R. As may be noted the resonance effects are effectively removed by this procedure. Also the slope in Q^2 which is apparent in (a) is largely removed. The slope in Q^2 has prompted the TASSO experimenters to fit with the empirical form

$$R \text{ or } r = \gamma(1 + \delta Q^2)(1 + \alpha e^{-\beta Q^2}).$$

After MC normalization the slope in Q^2 has essentially gone. However, they leave the slope parameter δ in the fit to R. The results are shown in Table V. The solid curve in Fig. 8a and the dashed curve in Fig. 8b represent these fits.

8.2. Search for Correlation with the Lorenz Factor γ and Also with $\vec{q}$ Either Parallel or Perpendicular to the Sphericity Axis

The TASSO Collaboration has searched for dependence of the parameters α and β $(= r^2)$ on the Lorenz factor γ and also on the direction of $\vec{q} = \vec{p}_1 - \vec{p}_2$ relative to

TABLE V

TASSO data at 34 GeV[5]

Data	α	$\beta(\text{GeV}^{-2})$	γ	δ
Fit to r	0.27 ± 0.03	$21.2^{+3.3}_{-2.8}$	0.78 ± 0.01	0.11 ± 0.01
Fit ro R	0.35 ± 0.03	$16.5^{+2.5}_{-2.1}$	0.96 ± 0.01	0.04 ± 0.01

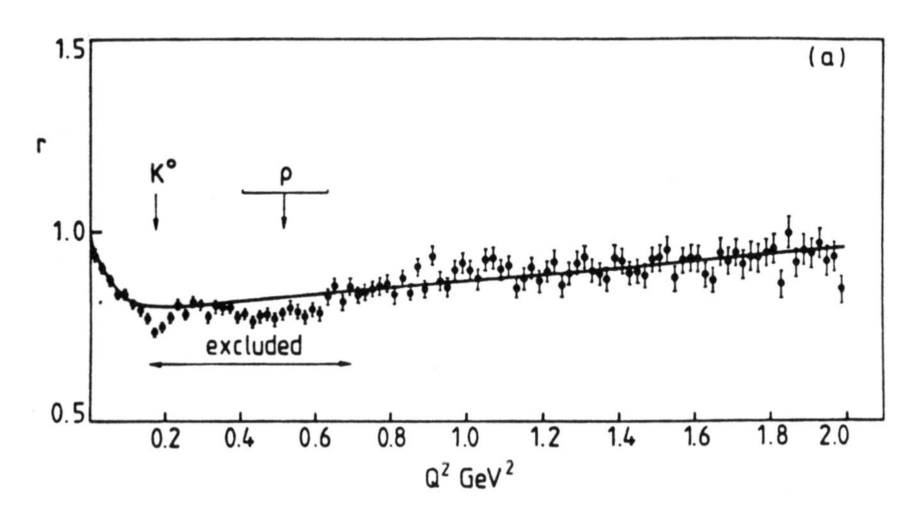

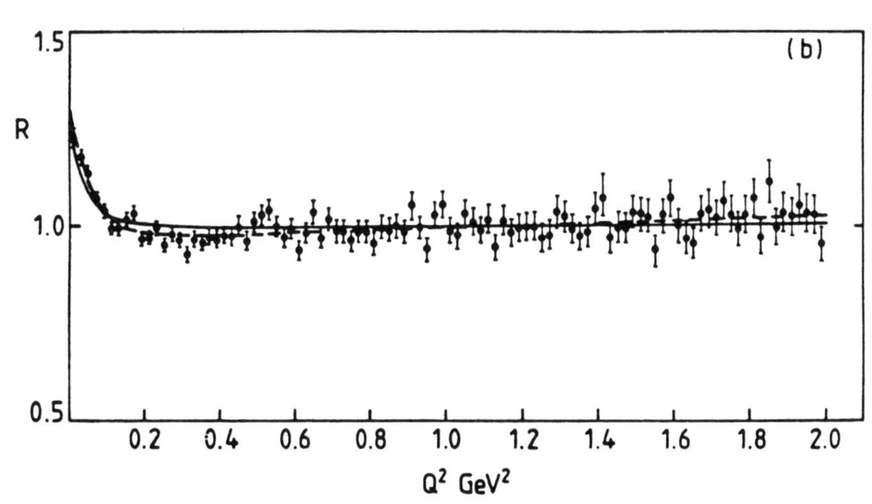

Fig. 8 TASSO data e^+e^- at 34 GeV.

(a) $R_U^L(Q^2)$ experimental distribution (called r here).

(b) $R_U^L(Q^2)/[R_U^L(Q^2)]_{MC}$ distribution (called R here).

the sphericity axis. They define $\hat{S}$ the unit vector along the sphericity axis and $\Delta\hat{k} = \vec{q}/|\vec{q}|$. Figs. 9 and 10 show these R distributions. They find that there is no measurable dependence on these quantities in either case.

9. CALCULATION WITH A STRING AND LUND MODEL INCORPORATING THE B.E. SYMMETRIZATION

Both M. Bowler[6] and B. Andersson and W. Hofmann[7] have found computational methods to incorporate the B.E. enhancement into Monte Carlo models. Bowler's calculations have been compared to the TASSO data, solid curve in Fig. 8b.

Bowler in a very detailed paper[6] has given a series of equations which represent the B.E. effect for pions produced along an "Artru-Mennessier string". He considers both the direct numerical evaluation of B.E. correlation functions based on this ansatz as well as Monte Carlo events weighted with the corresponding enhancements. In an elaborate set of two-dimensional fits between the B.E. symmetrized Monte Carlo calculation and the data good agreement is found except for a small region of phase space where $|\vec{q}|^2 \simeq q_0^2$. A surprising result of the analysis of the TASSO data is that $R_U^L(Q^2)$, which is essentially based on a "fireball" source, still gives the most consistent representation of the data.

The Andersson-Hofmann Lund Monte Carlo symmetrization involves the interchange of two like pions and hence a change in the area of the "Wilson loop". This is illustrated in Fig. 11. Their calculation is compared to $R_U^L(|\vec{q}|)$ from the TPC data[11] in Fig. 12. One noteworthy feature is that when they include the effect of resonances (dotted curve in Fig. 12) the calculations no longer agree with the data. This is another indication that the relation of resonances to the B.E. effect is not fully understood at present.

Gyulassy and Kolehmainen also point out that our current model for pion production is the "inside outside" cascade in which the momentum and production point are highly correlated. (Fig. 11 is a pictorial representation of such a cascade.) They furthermore point out that the traditional expressions for pion interferometry[1] follow only in the ***absence*** of such correlations. They solve analytically for what they consider the correct correlation functions and suggest that one needs to analyze the problem in terms of 3 independent variables: T, the effective source temperature, τ_0, the

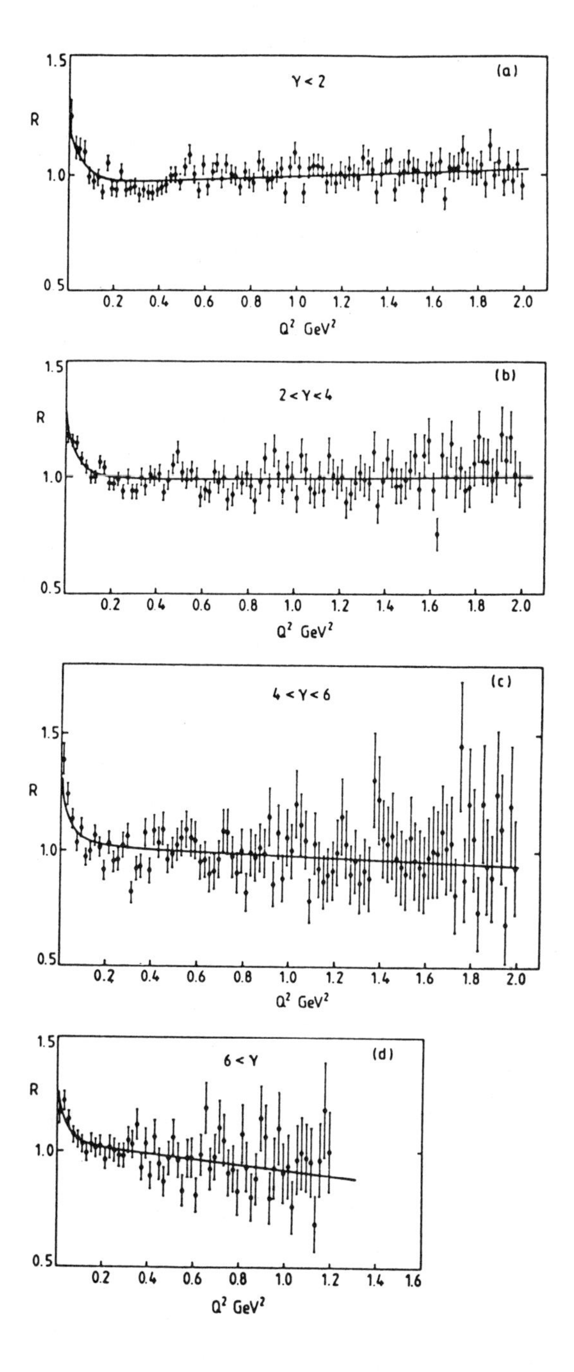

Fig. 9 TASSO data e^+e^- at 34 GeV. Study of correlation with the Lorenz factor γ.

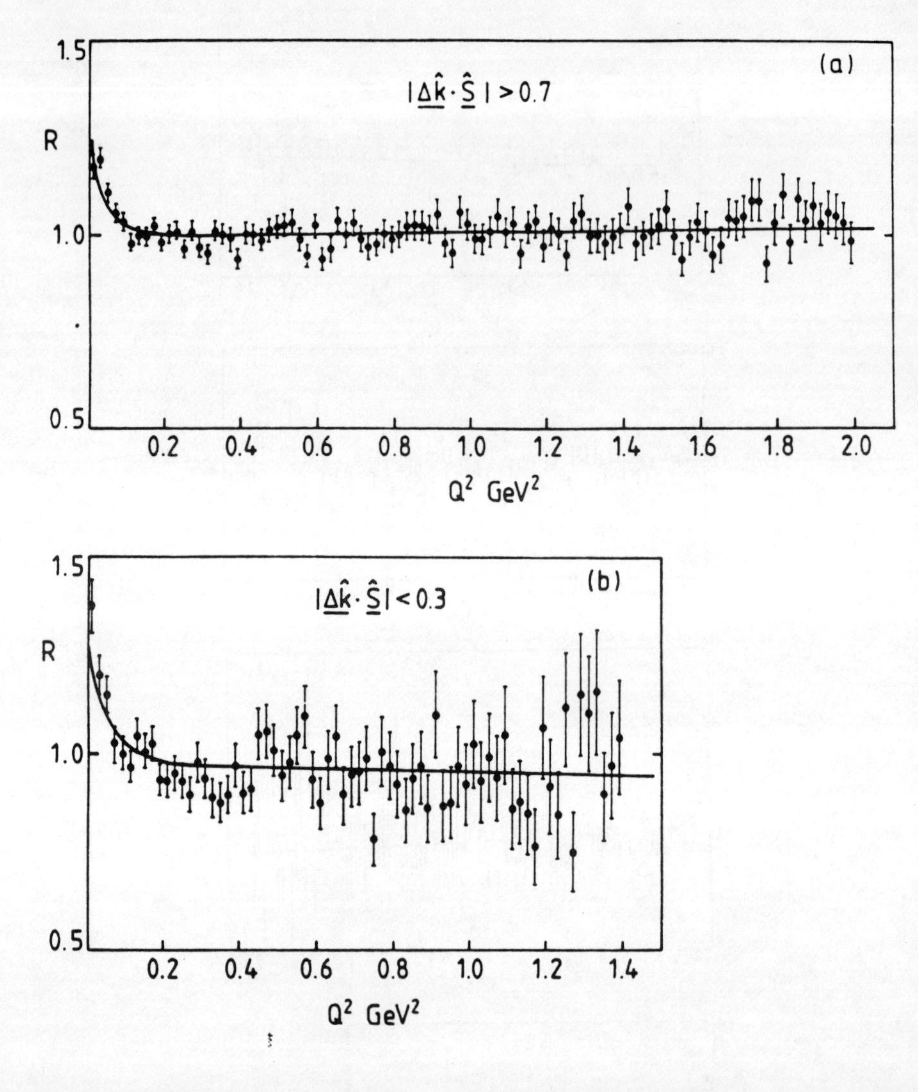

Fig. 10 TASSO data e^+e^- at 34 GeV. Study of the correlation for the direction of $\vec{q}$ (called $\Delta\hat{k}$) relative to the sphericity axis direction $\hat{S}$.

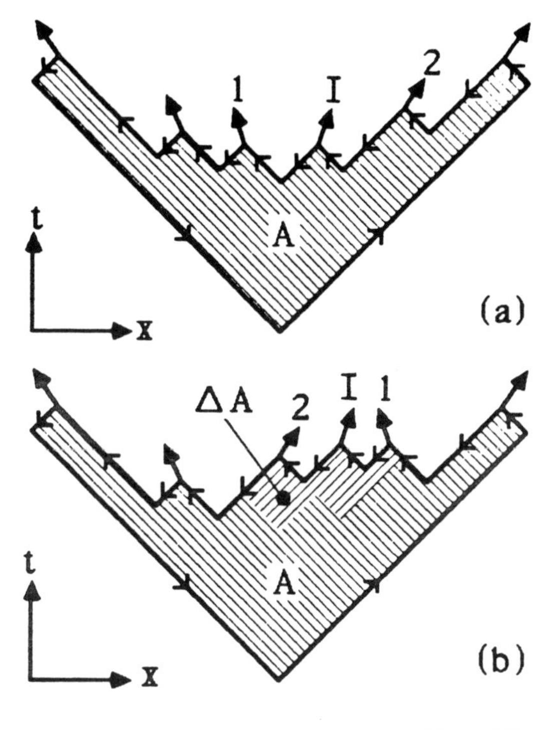

Fig. 11 (a) Space-time diagram for particle production in e^+e^- annihilation, based on the string picture, where energy and momentum of a particle are given by the time and respectively space difference of the production points of its quarks, multiplied by the string tension κ. A denotes the space-time area of the color field enclosed by the quark loop (antiquarks are represented as quarks moving backward in time).

(b) As (a), but particles 1 and 2 are exchanged, resulting in a change of the area A of the color field by ΔA.

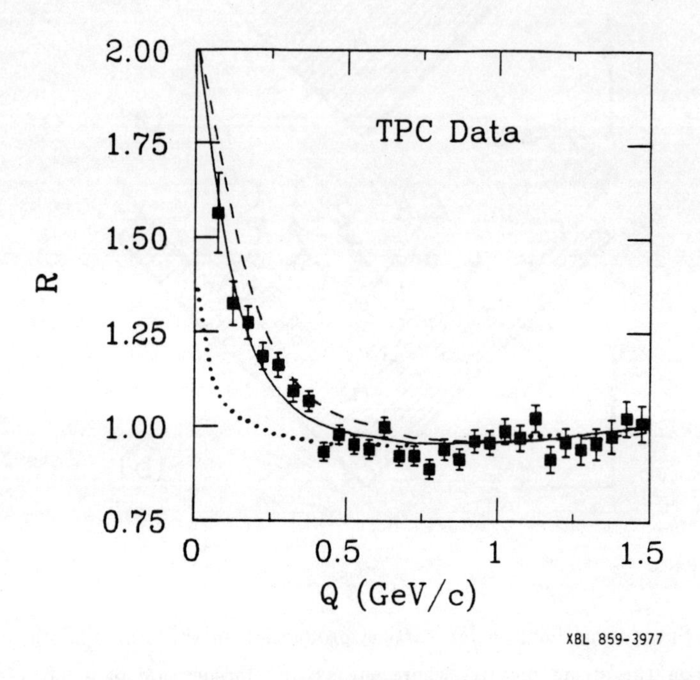

Fig. 12 Comparison of model predictions for R with TPC data,[11] as a function of $Q = |\vec{q}|$. Full line: based on the assumption that resonance production can be ignored for $\kappa = 0.2$ GeV^2 and $b/\kappa^2 = 0.7$ GeV^{-2}. Dashed: using $\kappa = 0.3$ GeV^2. Dotted: including resonances for $\kappa = 0.2$ GeV^2. The curves are very insensitive to the value of b/κ^2. Here b corresponds to the string decay rate.

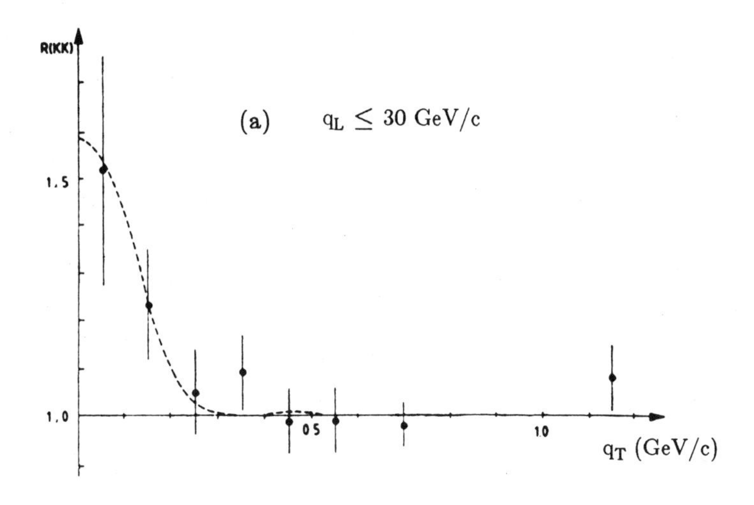

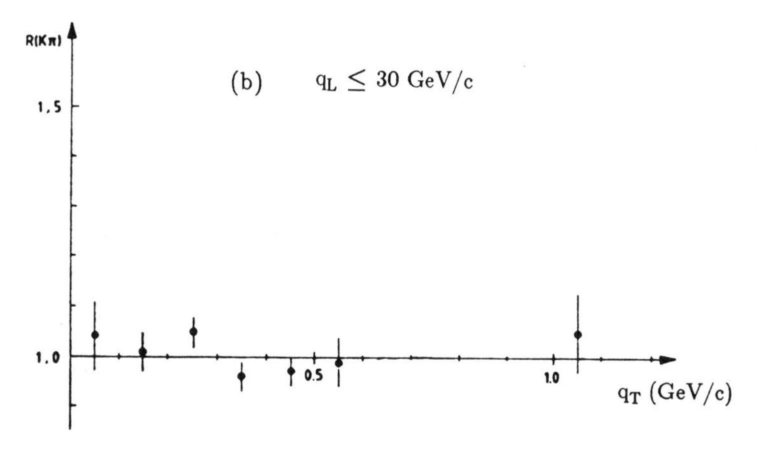

Fig. 13 AFS data at $\sqrt{s} = 63$ GeV.

(a) B.E. enhancement for $K^{\pm}K^{\pm}$ distributions, i.e., for identical bosons.

(b) No B.E. enhancement for $K^{\pm}\pi^{\pm}$ distributions, i.e., for non-identical bosons.

proper time at which the particle interaction ceases, and r_T, the radius of the production region transverse to the cascade. These 3 variables depend on 4 parameters, viz.: m_{1T} and m_{2T}, the transverse masses of the two pions, $y_2 - y_1$, the longitudinal rapidity difference and $\emptyset$, the angle between the two transverse momenta. Here $m_{1T}^2 = (p_{1T}^2 + m_1^2)$, where p_{1T} is the particle momentum transverse to the cascade axis. They present a scheme — which presumably will require very high statistics — for evaluating these parameters and the corresponding variables.

10. BOSE-EINSTEIN CORRELATION BETWEEN CHARGED KAONS

The B.E. enhancement has recently been observed in $\alpha\alpha$, pp and $\bar{p}p$ interactions at the ISR at $\sqrt{s} = 63$ GeV. The work[12] is a continuation of the studies by Åkesson et al.[1] with the Axial Field Spectrometer (AFS). Fig. 13 shows the q_T distribution for LIKE K pairs normalized by K's from different events. In this work this normalization technique was successfully applied. They observe a marked B.E. enhancement for the identical LIKE charge bosons $K^{\pm}K^{\pm}$ but as expected no enhancement for non-identical LIKE charge bosons $K^{\pm}\pi^{\pm}$. This latter observation agrees with the Mark II results presented earlier.[1] The AFS collaboration finds radii of $\sim 2.4 \pm 0.9$ fm for the KK producing region.

ACKNOWLEDGMENT

We wish to thank Valerie Heatlie for the careful typing and assembly of this paper. This work was supported in part by the U.S. Department of Energy under contract DE-AC03-76SF00098.

REFERENCES

1] G. Goldhaber, LESIP I, Bad Honnef, Germany, Editors D.K. Scott and R.M. Weiner, World Scientific Publishing Co., p. 115 (1984). See references quoted therein.

2] W.A. Zajc, paper at this conference, LESIP II.

3] CLEO Collaboration, P. Avery et al., Phys. Rev. **D32,** 2294 (1985).

4] Mark II Collaboration, Ivanna Juricic, Thesis in preparation, 1986, LBL, University of California, Berkeley, CA.

5] TASSO Collaboration, M. Althoff et al. DESY 85-126, OUNP 89/85.

6] M. Bowler, Oxford University OUNP 40/85 revised, Z. für Physik C, in press.

7] B. Andersson and W. Hofmann, LBL-20272 (1985).

8] M. Gyulassy and K. Kolehmainen, Presented at Conference on Ultrarelativistic Nucleus-Nucleus Collisions, Asilomar, Pacific Grove, CA, April 1986.

9] P. Grassberger, Nucl. Phys. **B120,** 231 (1977).

10] G.H. Thomas, Argonne report ANL-HEP-PR 76:33 (1976). Unpublished.

11] H. Aihara et al., Phys. Rev. **D31,** 996 (1985).

12] AFS Collaboration, T. Åkesson et al., Phys. Lett. **155B,** 128 (1985).

MULTIPLICITY DEPENDENCE OF THE TRANSVERSE MOMENTUM AND OF THE PARTICLE SOURCE DIMENSIONS IN HADRON-HADRON INTERACTIONS

R. Campanini

Dipartimento di Fisica, Università di Bologna and
INFN, Sezione di Bologna, Italy

ABSTRACT

In the first part of this report I present preliminary results on the multiplicity dependence of the transverse momentum and of the particle source dimensions obtained by the Ames-Bologna-CERN-Dortmund-Heidelberg-Warsaw (ABCDHW) collaboration in p-p interactions at $\sqrt{s}$= 63,44 and 31 Gev. Then I will comment on results obtained by the UA1 Collaboration on the multiplicity dependence of the transverse momentum in $\bar{p}$-p collisions at $\sqrt{s}$= 200, 540,900 Gev. Finally I will present some considerations on the significance of the presented results for the search of the phase transition from hadronic matter to quark--gluon plasma.

1. INTRODUCTION

Recent experiments at the CERN $\bar{p}p$ collider at various centre of mass energies, $\sqrt{s}$ = 540 GeV [1], $\sqrt{s}$ = 200 and 900 GeV [2] and at the highest ISR energy, $\sqrt{s}$ = 63 GeV [3], have shown that there is an increase of the mean transverse momentum, $\langle p_T \rangle$, with the particle density, $\rho = \Delta n/\Delta y$, in the central region of rapidity.

A number of possible explanations of this increase have been proposed. Among them geometrical models [4], "mini-jets" production [5], and thermodynamical models [6].

An increase with mean charged multiplicity of the particle emitting source radius, r, which may be explained by a geometrical model [7], was measured from a study of Bose-Einstein (B-E) correlations in $\alpha\alpha$ interactions at $\sqrt{s}$ = 126 GeV, $\bar{p}p$ and pp interactions at $\sqrt{s}$ = 53 and 63 GeV [8]. At $\sqrt{s}$ = 31 GeV [3] no clear evidence of $\langle p_T \rangle$ increasing with the charge multiplicity has been observed. Previous measurements of $\langle p_T \rangle$ and r at $\sqrt{s}$ = 19 GeV [9] also showed no dependence on the charge multiplicity.

I recently suggested [10] that in the framework of some thermodynamical models [6] [11] one could find a correlation between the dependence on multiplicity of the pions emitting volume and of the $\langle p_T \rangle$ which can be used to look for a possible phase transition from hadronic matter to a quark-gluon plasma. In such models the p_T distribution of secondaries reflects the temperature of the hadronic system and its evolution in the transverse direction, while the particle density provides a measure of the entropy. The radius, measured via Bose--Einstein correlations, gives information on the final volume fo the particles emitting source after a possible expansion following the early stages of the interaction.

In this talk I will first present preliminary results of a comparative analysis of the dependence of $\langle p_T \rangle$ and r on ρ in pp interactions at $\sqrt{s}$ = 63, 44 and 31 GeV obtained by the ABCDHW Collaboration. Previous results on these subjects, with lower statistics, have been presented [3], [12] [13]. Then I will shortly comment on the SPS COLLIDER's results on the $\langle p_T \rangle$ vs ρ dependence obtained by the UA1 Collaboration at $\sqrt{s}$ = 200,540,900 GeV. Finally I will present some considerations on the possible significance of the presented experimental results for the search of phase transition to quark-gluon plasma.

2. EXPERIMENT ABCDHW AT THE CERN ISR

The experiment was performed by the ABCDHW collaboration at the CERN Intersecting Storage Rings (ISR) using the Split Field Magnet (SFM) detector. The magnetic volume of the detector (with a maximum field strength of 1 Tesla) was filled with Multiwire Proportional Chambers. The momenta of the charged particles were measured on nearly the full solid angle. The performance of the detector is described in previous publications [14].

The experiment used two different triggers. A "minimum bias" trigger, which required the presence of at least one charged track candidate in the detector. It accepted about 95% of the inelastic cross section. An "electron trigger" which selected an electron candidate of positive or negative charge produced at about 90° with respect to the beam direction. A two stage Cerenkov counter selected the electrons. The trigger particles covered a c.m.s. rapidity $|y| < 0.35$ and an azimuthal angle $|\phi| < 9°$ with respect to the beams [15]. Although very much enriched in prompt electrons with respect to the "minumum bias" events, the sample from this trigger still contains events not really associated with prompt electrons. The trigger was in these cases either electrons from a Dalitz pair or from a photon converted near the colli sion vertex, or a hadron accompanied by a delta electron. Events were reconstructed with the standard chain of computer programs for the SFM detector [16]. A good vertex fit, defined by a minimum of two charged tracks, was then required for each event. Globally the present preliminary analysis is based on $\sim$800,000, 600,000 and 1,400,000 "minimum bias" events at $\sqrt{s}$ = 63, 44 and 31 GeV respectively and about 350,000 "electron trigger" events at $\sqrt{s}$ = 63 GeV.

3. ANALYSIS IN THE ABCDWH EXPERIMENT

A number of selection criteria were applied to the data in order to define the characteristics of the events used for the analysis:

(a) The interaction vertex was required to be in the overlap region of the two colliding beams.

(b) Low multiplicity events (less than 5 tracks) which may be contaminated by diffraction were not considered in the analysis.

(c) In order to remove some background from mismeasured tracks at large p_T only particles with $p_T < 2.5$ GeV/c were used in the present analysis. Further only tracks with $p_T > 0.150$ GeV/c have been taken into account. This cut was not applied in the ρ calculation.

(d) Tracks were accepted if the relative precision in the measurement of their momenta, $\Delta p/p$, was less than 0.8. This cut substantially reduces the number of badly measured tracks and gives a $\langle \Delta p/p \rangle \sim$ 0.18, independent of multiplicity.

(e) From the electron trigger sample small angle e^+e^- pairs were removed by energy loss measurements in a proportional chamber close to the interaction region and large angle e^+e^- pairs by a cut on effective mass [15]. With these selections the sample reduces to 170,000 events. The same physical analysis was performed separately for the selected as well as for the full sample.

The charged multiplicity in the central region Δy (defined as $-1.5< \Delta y<1.5$) was not corrected for acceptance losses. With the above cuts, these losses were estimated to be about 20% on average for single tracks (see ch. 3.1). After the above cuts were applied we define the particle density, $\rho = \Delta n/\Delta y$, to be the number of charged tracks per unit rapidity interval.

All tracks were assumed to be pions.

4. AVERAGE TRANSVERSE MOMENTUM DETERMINATION IN THE ABCDHW EXPERIMENT

The average transverse momentum, $\langle p_T \rangle$, for N events having Δn charged tracks in the interval Δy has been computed by the following method [12]. First one computes the arithmetic mean, $\bar{p}_T$, of the Δn tracks observed in the Δy interval, in each event. These $\bar{p}_T$'s are then averaged (to give $\langle p_T \rangle$) over all the N events having Δn tracks in the interval Δy, i.e. with the same particle density ρ.

This method yields overestimated values of $\langle p_T \rangle$ because the preferential loss of tracks at small p_T results in each $\bar{p}_T$ being slightly higher than it should be.

However since our average p_T is actually computed in the interval $.15 < p_T < 2.5$ GeV/c this affects only weakly our results.

In order to check the validity of the average p_T determination, for some events subsamples, a different method was used.

The p_T distributions of the tracks in the range $0.3 < p_T < 0.8$ GeV/c were fitted to the exponential form $d\sigma/dp_T 2 = A \exp(-bp_T)$. The average transverse momentum was then computed from the fitted value of the slope, $\langle p_T \rangle = 2/b$. These checks lead to the conclusion that the $\langle p_T \rangle$ dependence on ρ has the same general shape when computed using either of the two methods.

The reported value of $\langle p_T \rangle$ have not been corrected for acceptance losses. Some comments are needed about this point. Acceptance corrections have been computed via Monte-Carlo simulation over the full space-phase covered by the apparatus.

To analyse the effects of the set-up acceptances we then proceeded following two steps:

(a) first we computed the invariant inclusive cross sections as a funcion of p_T and y in the ranges $0.15 < p_T < 3$ GeV/c, $|y| < 1.5$, using only those regions of the apparatus in which the acceptances are well known and above 90%. This yields to results in complete agreement with published data.

(b) second, in the same mentioned regions, were computed the acceptances corrected values of p_T and ρ. Comparing these corrected values with the non corrected ones computed on the full apparatus, differences were found in the absolute values of the $\langle p_T \rangle$'s and ρ's, which were at maximum around 20%, but the shapes of the $\langle p_T \rangle$ dependence on ρ were essentially unchanged.

This procedure has been repeated for different cuts in $\Delta p/p$, azimuthal angle ϕ and Δy, yielding to the same conclusions.

Since the attention in the present work is mainly pointed to the shape of the $\langle p_T \rangle$ on ρ, we decided to present the results from the full apparatus to gain the maximum statistical significance.

5. MEASUREMENT OF THE PARTICLE SOURCE RADIUS IN THE ABCDHW EXPERIMENT

We use measured Bose-Einstein correlations between like-charged pairs of particles to determine the size of the emitting regions. Results on Bose-Einstein correlations in pp and $\bar{p}p$ interactions at $\sqrt{s}$ = 63 GeV and a detailed discussion on the non-interfering background have been published [13]. The following points should be noted:

(a) Since all tracks were assumed to be pions in the analysis, the small contamination by kaons, protons and antiprotons should result only in the observed interference effect appearing smaller than reality.

(b) For each particle we required, in addition to the cuts mentioned in the previous section, $\Delta p < 0.1$ GeV/c, where Δp is the estimated error in the momentum measurement of the tracks. This results in a relative error $\langle \Delta p/p \rangle \sim 8\%$ which was checked to be independent of particle density.

(c) Events were kept in the analysis only if they contained at least two positive tracks and one negative track or vice versa.

Several parametrisations of the ratio between like and unlike pairs of charged particles, $R = N_L/N_U$, in terms of the variables constructed from the particle momenta were proposed [18]. In this work R was analysed for different central rapidity densities as a function of variable Q^2 as proposed in reference [19]. Q^2 is defined as the negative of the four-momentum transfer squared:

$Q^2 = -(p_1-p_2)^2 = m_{12}^2-2m_1^2-2m_2^2$ (p_1 and p_2 are the four-momenta of the two particles in a pair).

The ratios R for different charged particle densities, ρ, were fitted to a function of the form:

$$R(Q^2) = \gamma\,(1+\alpha e^{-\beta Q^2})(1+\delta.Q) \qquad (1)$$

where the parameter α, which is often called the chaoticity parameter, should lie between zero (no Bose-Einstein effects) and one (maximum chaoticity and thus maximum interference). The parameter β is related to the emission, radius, ($r = 0.197\sqrt{\beta}$ fm), γ is a normalization factor and the term $(1+\delta.Q)$ accounts for the shape of R at large Q, ($Q \equiv \sqrt{Q^2}$). The parametrization (1) gives the space-time dimension of the particle emitting region and leads to values of r somewhat lower than different parametrization used in the literature [20]. All fits had values of x^2/DOF close to one.

6. RESULTS OF THE ABCDHW EXPERIMENT

6.1. Average p_T Versus ρ For Minimum Bias Data

The average transverse momenta of charged particles, computed in the above mentioned p_T interval, as a function of the particle density in the central rapidity region are shown in fig. 1. The plots refer to minimum bias events at $\sqrt{s}$ = 63 GeV. In fig. 1a the $\langle p_T \rangle$'s for all particles are reported, while fig.'s 1b and 1c show the same quantities for negative and positive particles respectively. The rise of $\langle p_T \rangle$ with the multiplicity is clearly seen. Taking the lower edge of the p_T interval to p_T>300 MeV/c makes the rise of $\langle p_T \rangle$ with ρ further enhanced as it was already known from reference [3].

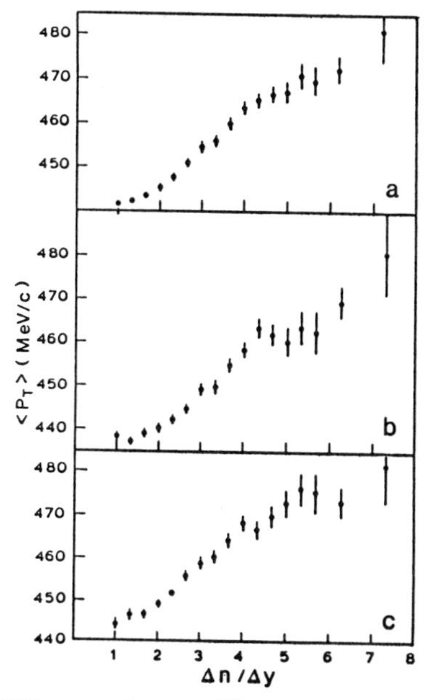

fig. 1 (a) ABCDHW experiment. The average transverse momentum of charged particles, with $0.15 \leq p_T \leq 2.5$ GeV/c, produced in the central region of rapidity ($|y|<1.5$) plotted versus charged particle density, at $\sqrt{s}$ = 63 GeV.

(b) Same as in (a) for negative particles only.

(c) Same as in (a) for positive particles only.

The same analysis of the average p_T as a function of ρ has been done for the minimum bias events at $\sqrt{s}$ = 44 and 31 GeV. The results are reported in fig. 2a for $\sqrt{s}$ = 44 and in fig. 2b for $\sqrt{s}$ = 31 GeV. Comparing with the $\sqrt{s}$ = 63 GeV data one can see that similar dependence of the $\langle p_T \rangle$ on ρ is still present at 44 and 31 GeV but the p_T rise becomes weaker with decreasing $\sqrt{s}$.

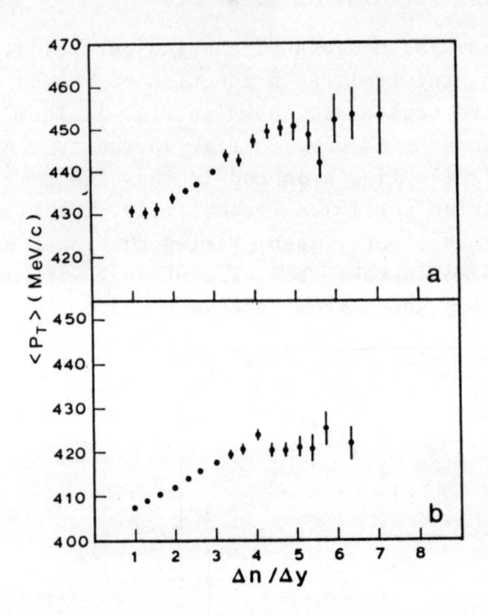

fig. 2 (a) ABCDHW experiment. The average transverse momentum of charged particles with $0.15 \leq p_T \leq 2.5$ GeV/c, produced in the central region of rapidity ($|y|<1.5$), plotted versus charged particle multiplicity density, at $\sqrt{s}$ = 44 GeV.
(b) Same as in (a) for $\sqrt{s}$ = 31 GeV.

6.2. Average p_T Versus ρ For Electron Trigger Data

The mean raw multiplicity in this sample is, in the central region ($-1.5<y<1.5$), about a factor 2.5 larger than that of the minimum bias events. For these data we analized the average p_T dependence on separately for the trigger track and for all the other tracks.

As already stated, this analysis, as well as the studies on the particle source radius, were performed on both the selected and full electron trigger samples. The results obtained from the two sets are in complete agreement and, in the following, only data from the full set of events are presented.

Fig. 3a shows the $\langle p_T \rangle$ of the trigger particle, computed in the range

0.4-1.0 GeV/c. In this range of p_T the hadronic contamination is estimated to be below 30% [15].

The average p_T of all the non trigger particles is shown in fig. 3b computed in the same way as for the minimum bias analysis.

Finally in fig. 3c are reported the average p_T as a function of ρ for non trigger particles that lie in a ϕ interval of $\pm$ 40° around the trigger track.

The $\langle p_T \rangle$ has been computed considering all the tracks with $0.15 < p_T < 2.5$ GeV/c.

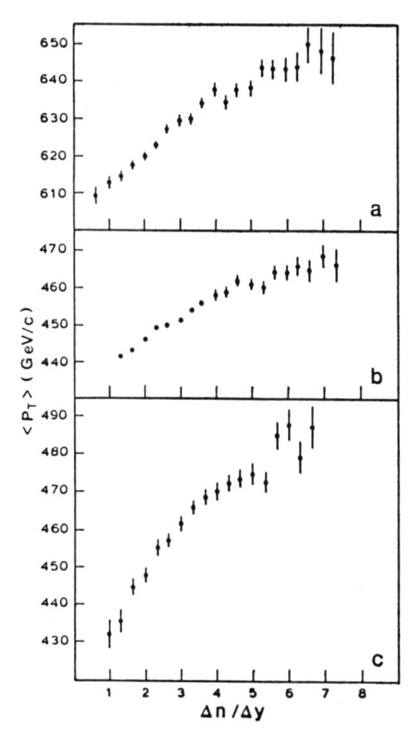

fig. 3 ABCDHW experiment. Electron trigger data; $\sqrt{s}$ = 63 GeV.

(a) The average transverse momentum of the triggering particle with $.4 < p_T < 1$ GeV/c, as a function of = $\Delta n/\Delta y$.

(b) The average transverse momentum of all the non triggering particles having $.15 < p_T < 2.5$ GeV/c as function of ρ .

(c) The average transverse momentum for non triggerint particles, with $.15 < p_T < 2.5$ GeV, lying in a ϕ interval of 40° around the trigger track.

6.3. Average p_T Versus Local Density

Since it is known [17] that the particle density reaches its higher values when $|y|\to 0$, we recompute the average p_T and the particle densities taking into account only the tracks in the y interval $-0.5<y<0.5$. The results for the minimum bias data at $\sqrt{s}$ = 63 GeV are reported in fig. 4a. In fig 4b is shown the $\langle p_T\rangle$ of the trigger track in the electron trigger sample ($0.4<p_T<1$ GeV/c) as a function of ρ, where now ρ is computed taking the tracks lying in an interval $\Delta y = 1$ centered around the trigger particle rapidity.

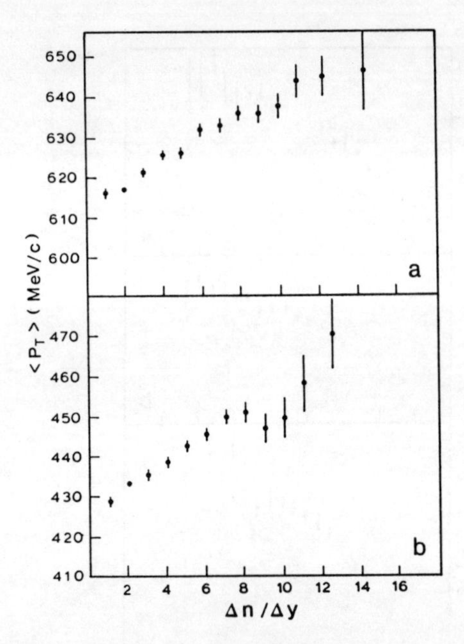

fig. 4 (a) ABCDHW experiment. The average transverse momentum for tracks having $0.15\leq p_T\leq 2.5$ GeV/c and $|y|<0.5$ as a function of the particle density in the same y interval for minimum bias data at $\sqrt{s}$ = 63 GeV.

(b) The average transverse momentum of the triggering track as a function of the particle density computed in a y interval ± 0.5 around the triggering track rapidity.

6.4. Source Dimensions

Fig. 5a, 5c and 5d show the radius r of the particle emitting region as a function of ρ at $\sqrt{s}$ = 63, 44 and 31 GeV respectively. A clear rise of r with increasing particle density is visible in the data at $\sqrt{s}$ = 63 GeV, ranging from ∿ 0.72 fm for ρ<2 up to ∿1.7 fm at ρ ∿ 6. We point out that, due to the finite momentum resolution, the maximum measured value of r can only be an underestimate of the true maximum value [20]. A very weak dependence of r on the particle density in the central region is observed at $\sqrt{s}$ = 31 GeV (see fig. 5d), while a rise of r with ρ is observable at vs = 44 GeV in the ρ region between 1 and 4.

The values of r as a function of ρ for electron trigger events are presented in fig. 5b. They range from ∿ 0.8 at ρ around 2 to about 1.3 at ρ around 6.

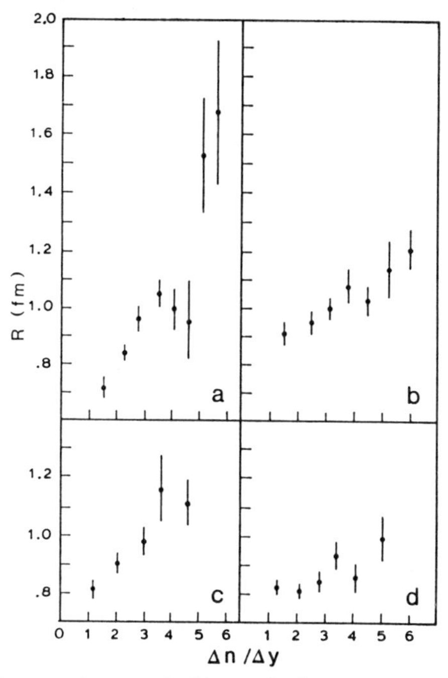

fig. 5 ABCDHW experiment. Radius of the particle emitting source as a function of ρ

(a) Minimum bias trigger at $\sqrt{s}$ = 63 GeV.

(b) Electron trigger at $\sqrt{s}$ = 63 GeV.

(c) Minimum bias trigger at $\sqrt{s}$ = 44 GeV.

(d) Minimum bias trigger at $\sqrt{s}$ = 31 GeV.

The values of the chaoticity parameter α , as shown in figures 6a-6d, lie between 0.50 and 0.30, and slowly decrease with increasing multiplicity, at least up to ρ around 4, for all three energies. The behaviour shown in fig. 6a-6d is in agreement with results reported in ref. [8].

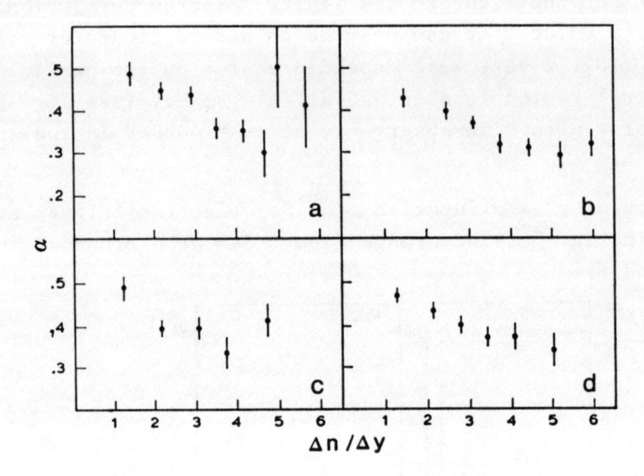

fig. 6 ABCDHW experiment. Chaoticity parameter α as a function of ρ .
(a) Minimum bias trigger at $\sqrt{s}$ = 63 GeV.
(b) Electron trigger at $\sqrt{s}$ = 63 GeV.
(c) Minimum bias trigger at $\sqrt{s}$ = 44 GeV.
(d) Minimum bias trigger at $\sqrt{s}$ = 31 GeV.

7. SUMMARY OF THE PRELIMINARY RESULTS OF THE ABCDHW EXPERIMENT

We have performed an analysis of the behaviour of the average transverse momentum and of the size of the particle emitting region as a function of the variable, $\rho = \Delta n/\Delta y$, at $\sqrt{s}$ = 63,44 and 31 GeV using minimum bias trigger events. An additional sample of electron trigger events at $\sqrt{s}$ = 63 GeV was also analized. The main results are:

i) The mean transverse momentum increases with increasing ρ at all energies. The rise of $\langle p_T \rangle$ becomes higher when $\sqrt{s}$ goes from 31 to 63 GeV. In the electron trigger events an increase of the $\langle p_T \rangle$ of the trigger particle as well as of the non trigger particles,versus ρ is observed. Qualitatively the rise is approximatively linear, in all cases, up to ρ around 4-5. In this region a smoothed flat down is visible which, when going to local density, displaces to ρ around 6-7, which from formulas relating early energy density to final entropy [22] corresponds to initial density energy of about 3 GeV/fm^3.

ii) The particle source size increases with particle density. The rise, very week at 31 GeV, becomes clear at $\sqrt{s}$ = 44 GeV and at $\sqrt{s}$ = 63 GeV both in minimum bias trigger and in the electron trigger data. The increasing of the particle source size slows down at ρ around 4. The chaoticity parameter α slowly decreases with ρ at all energies at least for $\rho < 4$.

8. COMMENTS ON THE UA1 RESULTS

New data on the $\langle p_T \rangle$ vs ρ dependence at $\sqrt{s}$ = 200,540,900 GeV in $\bar{p}p$ collisions have been presented in the 1985 by the UA1 Collaboration [2] [1]. For details on the experimental methods and on the possible interpretation of the results in terms of presence of two kinds of events (jet and no-jet) in the minimum bias sample, see the refs. [1] and [2]. Here I want to stress the following result: in the $\langle p_T \rangle$ vs ρ dependence in the no-jet events, as well in the full sample of minimum bias events, there is a clear slope change at central multiplicity ($|y|<2.5$) around 30, which corresponds to $\rho \sim 6$ (see fig. 7). Taking in account the above mentioned acceptance losses of the ABCDHW experiment, one can think that the slope changes in the UA1 results may correspond to the slowing down in the $\langle p_T \rangle$ increasing seen in the ABCDHW experiment at ρ around 5. In ref. [10] I suggested the hypothesis that the slope change in the UA1 results may be due to a small (in ρ) plateau followed by a new increasing. To test this hypothesis data with binning smaller than the present ones are needed from UA1 experiment. Very preliminary data presented at the International Europhysics Conference on High Energy Physics held in Bari in the 1985 supported the 'small plateau' hypothesis (see ref. [10]).

9. CONCLUSIONS

The results from ABCDWH and UA1 experiments suggest that in the ρ region 4-6, at different energies and in data taken with different triggers, there is a break in the ρ dependence of both the $\langle p_T \rangle$ and particle source dimensions. Recent calculations in statistical QCD [21] showed in the phase transition region a noticeable flattening on the $\langle p_T \rangle$ vs ρ dependence and a stop in the transverse expansion. This suggests that the above presented results may show signals of phase transition from hadronic matter to quark gluon plasma. In the following I suggest measurements which could give additional informations:

1) to study the $\langle p_T \rangle$ vs ρ dependence for particles with different masses and to look at a possible break in the same ρ region.

2) To search for possible differences in the events structure below and above the 'critical' density trough many particle rapidity correlations, backward-forward correlation and transverse energy distribution.

3) To look at possible structure in the p_T dependence vs the central

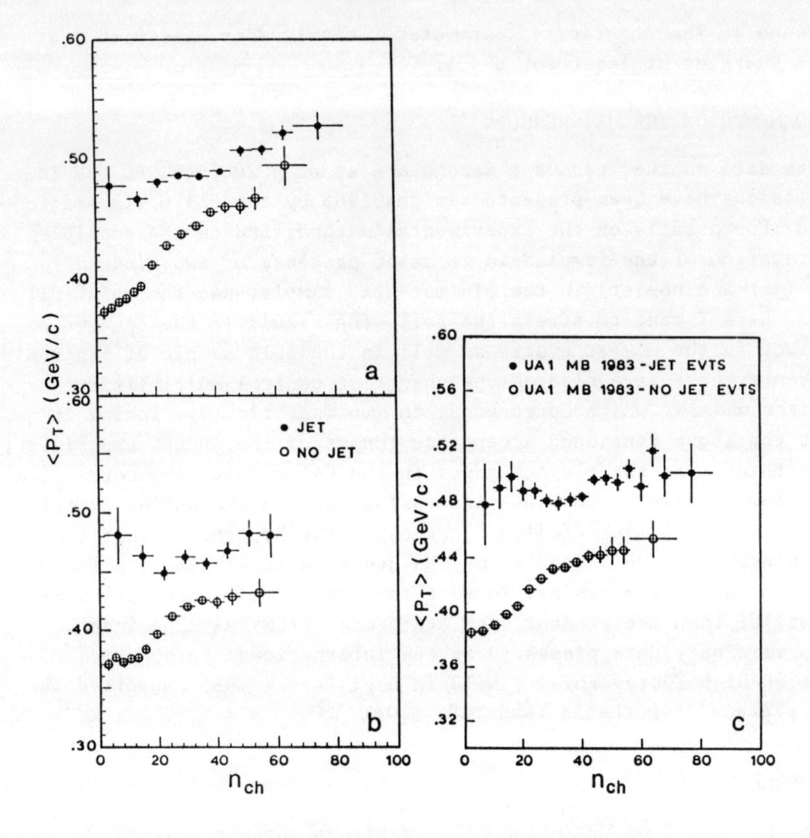

fig. 7 $\langle p_T \rangle$ vs ρ (for $|y|<2.5$) from UA1 experiment
(a) at $\sqrt{s}$ = 900 GeV
(b) at $\sqrt{s}$ = 200 GeV
(c) at $\sqrt{s}$ = 540 GeV

ρ in high p_T events.

4) To study the Kaons/Pions ratio vs ρ both in low and high p_T events.
5) To study the relative amount of high density fluctuations in events below and above the 'critical density'.

10. REFERENCES

[1] G. Arnison et al., Phys. Lett. 118B (1982) 167;
G. Ciapetti, Proceedings of the V Topical Workshop on Proton Antiproton Collider Physics, St. Vincent, 1985.

[2] F. Ceradini, Proceedings of the International Europhysics Conference on High-Energy Physics, Bari, 1985.

[3] A. Breakstone et al., Phys. Lett. 132B (1983) 463.

[4] S. Barshay, Phys. Lett. 127B (1983) 129.

[5] M. Jacob, CERN preprint, CERN/TH 3515 (1983).

[6] L. Van Hove, Phys. Lett. 118B (1982) 138.

[7] S. Barshay, Phys. Lett. 130B (1983) 220.

[8] T. Akesson et al., Phys. Lett. 129B (1983) 269;
T. Akesson et al., Phys. Lett. 155B (1985) 128.

[9] C. De Marzo et al., Phys. Rew. 29D (1984) 363.

[10] R. Campanini, Nuovo Cimento Lett. 44 (1985) 343;
R. Campanini, "Possible evidence of phase transition to quark gluon plasma at CERN ISR and SPS Collider", preprint DFUB 17/85, (1985).

[11] E.V. Shuryak, Phys. Rep. 61 (1980) 71.

[12] A. Breakstone et al., Multiplicity dependence of transverse momentum spectra at ISR energies, EPS, Bari, 1985.

[13] A. Breakstone et al., Phys. Lett. 162B (1985) 400.

[14] M. Della Negra et al., Nucl. Phys. B127 (1977) 1;
W. Bell et al., Nucl. Instr. and Meth. 156 (1978) 111.

[15] D. Drijard et al., Phys. Lett. 108B (1982) 361;
M. Heiden,thesis, Heidelberg (1981).

[16] M. Della Negra, A. Norton: (unpublished), adapted from H. Wind: Function parametrization, Proc. CERN computing and data processing school, Yellow Report CERN 72-21;
D. Drijard: Spline track fit, SFM internal Note 1976, adapted from
H. Wind: CERN/NP/DHG 73-5 (1973).

[17] W. Thomé et al., Nucl. Phys. B219 (1977) 365;
W. Bell et al., Z. Phys. C27 (1985) 191.

[18] G. Goldhaber, "The GGLP effect from 1959 to 1985"; report LBL-19417 (1985).

[19] G. Goldhaber, "Multipion Correlations in e^+e^- Annihilation at SPEAR", presented at the International Conference on High Energy Physics, Lisbon 1981; report LBL-13291.

[20] H. Aihara et al., Phys. Rev. D31 (1985) 996.

[21] K. Redlich and H. Satz, Critical behaviour near deconfinement; BI-TP 85/33.

[22] J.D. Bjorken, Phys. Rev. D27 (1983) 1940.

Prototypical Bose-Einstein Experiments

William A. Zajc
Department of Physics
University of Pennsylvania
Philadelphia, PA 19104

ABSTRACT

The use of Bose-Einstein correlations between like pions to extract geometrical parameters of the pion source is presented from an experimental standpoint. Two quite different experiments are used to illustrate the strengths and weaknesses of such measurements, and to discuss possible systematic errors, the analysis and interpretation of the resulting data, and potential approaches for future experiments. Both relativistic heavy ions and elementary hadron-hadron collisions are considered.

1 INTRODUCTION

In this paper, I intend to explore in some depth the methods by which Bose-Einstein experiments are performed. To do this, I will use two experiments with which I am familiar to illustrate capabilities of typical fixed-target and colliding beam magnetic spectrometers for such work. In addition to being comfortable ground for the author, the two experiments are nicely complementary in that they were performed at vastly different energies, with significantly different target-beam combinations, were analyzed somewhat differently, etc. Thus, some substantial region of the continua of choices in experimental procedure will be explored in discussing them.

A word about the title of this paper: I chose the word "prototypical" in the sense of "illustrative". Much to my chagrin. I have discovered that

Webster's first definition for prototype is "An original on which a thing is modeled." No such claim for priority or originality is intended here. Instead, I hope that the reader will interpret prototypical in the second sense, i.e., "exhibiting the essential features of."

This paper is organized as follows: The next section presents details and results from a (relatively) low energy heavy ion experiment performed at the Berkeley Bevalac. The third section provides a similar description of a high energy experiment at the CERN ISR. Various complications, interpretations, and points of contrast between the two experiments are taken up in Section 4. Future directions and conclusions are presented in the final section. An appendix provides a "minimalist" summary of the formalism and notation; extensive discussion and references to the literature may be found in Ref. [1]. Units will be used with $c = 1$, so that lifetimes will be expressed in fermis. In light of the pioneering work by Goldhaber, Goldhaber, Lee and Pais in this field (Ref. [2]), I will use "Bose-Einstein" and "GGLP" nearly interchangeably, using the latter to emphasize aspects of hadronic physics and the former to include more general properties of bosons.

2 A HEAVY ION EXPERIMENT

The first experiment considered was performed at the Berkeley Bevalac with beams of 1.8 A· GeV ^{20}Ne and ^{40}Ar incident on NaF and KCl targets, respectively. Two pion pairs emitted about $\theta_{LAB} \sim 45^\circ$ were momentum analyzed in the JANUS magnetic spectrometer, shown in Figure 1. This arrangement thereby allowed the study of pion pairs emerging at 90° in the center-of-mass from a collision of roughly symmetric ion-pairs, each with $\gamma_{CM} \sim 0.4$.

As shown in Figure 1, the spectrometer design is simplicity itself: a pair of proportional wire chambers, MWPC1 and MWPC2, define the incoming ray of each pion. The magnet provides a nearly uniform dipole field. with central value of 0.9 Tesla. The outgoing trajectories are measured in MWPC3 and MWPC4. A hodoscope array formed from the A and B scintillation counters, in combination with the timing counters S1 and S2, allows one to trigger on events with two pions within the spectrometer

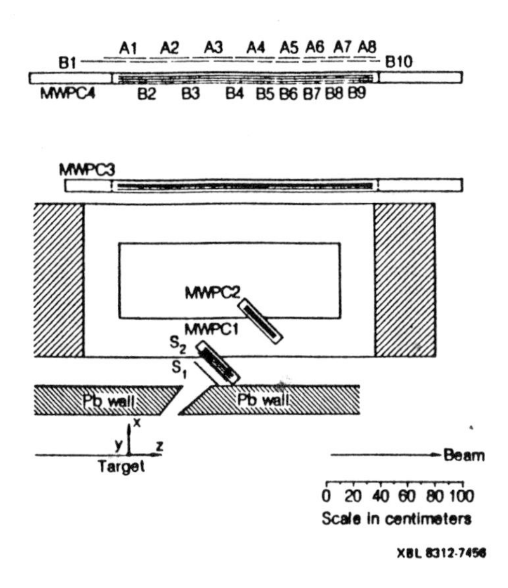

Figure 1. A plan view of the JANUS spectrometer at the Berkeley Bevalac.

acceptance.

Tracks in each event are selected on the basis of geometrical criteria derived from the four wire chambers. Good tracks must trace back to the target region, avoid the lead collimator, and be consistent with charged particle motion in both the (horizontal) bend plane and the (vertical) focussing plane. Two-pion events are obtained by requiring the sample of two track events to have the time-of-flight and pulse-height signature (in the AB counters) expected for pions (as opposed to electrons or protons, the principle sources of background). The end result of these cuts is a set of pion-pair events (contamination is less than 1%), with each pion having lab momentum $250\,\mathrm{MeV/c} < |\vec{p}_{\mathrm{LAB}}| < 600\,\mathrm{MeV/c}$, corresponding to $200\,\mathrm{MeV} < \mathrm{E_{CM}} < 500\,\mathrm{MeV}$.

The analysis of these events is performed in terms of the relative momentum $|\vec{q}|$ and the relative energy $q_0 \equiv |E_1 - E_2|$ (both quantities are in the center-of-mass). If we let $A(|\vec{q}|, q_0)$ denote the actual number of pion pairs observed in a given bin, and write $B(|\vec{q}|, q_0)$ for the background number of pairs expected in the absence of Bose-Einstein correlations, the correlation function is then defined as

$$C_2(|\vec{q}|, q_0) \equiv \frac{A(|\vec{q}|, q_0)}{B(|\vec{q}|, q_0)} \quad . \tag{1}$$

The background pairs are created by combining a pion from one real event with a second pion taken from another real event. In fact, to create a background spectrum with small statistical fluctuations (at least compared to the real events), *all* such possible fake pairs were made from the real events. The essence of this procedure is described in Ref. [3]; further details may be found in Ref. [4].

The correlation function so obtained is fit to the functional form

$$C_2(|\vec{q}|, q_0) = 1 + \lambda e^{-(|\vec{q}|R)^2/2-(q_0\tau)^2/2} \quad , \tag{2}$$

which corresponds to the squared-Fourier transform of a source density function

$$\rho(\vec{r}, t) = \frac{1}{\pi^2 R^3 \tau} e^{-r^2/R^2 - t^2/\tau^2} \quad . \tag{3}$$

Correlation functions averaged over over either $|\vec{q}|$ or q_0 are presented in Figure 2 along with the resulting fit. The results for the various ion pairs and charge states studied are presented in Table 1. The spatial information is summarized in Figure 3, which shows confidence contours obtained by combining data sets in Table 1, assuming that both R and τ vary as $A^{1/3}$. This results in a convenient reduced form for these data: $R \sim A^{1/3} fm$, $\tau \sim \frac{3}{4}R$.

3 A HADRON-HADRON MEASUREMENT

The second experiment was performed with the Axial Field Spectrometer (AFS) at the CERN ISR. Both hadron-hadron and light ion collisions were studied: pp reactions at $\sqrt{s} = 53$ and 63 GeV, $p\,\bar{p}$ events at 53 GeV, and α-α collisions at 31.5 A·GeV. Two-pion pairs emitted in the central rapidity region ($|\eta| < 0.8$) were analyzed using the central drift chamber of the AFS.

This drift chamber, shown in Figure 4, consists of two identical segments of 41 4 ° sectors each, thereby covering nearly the entire azimuth.

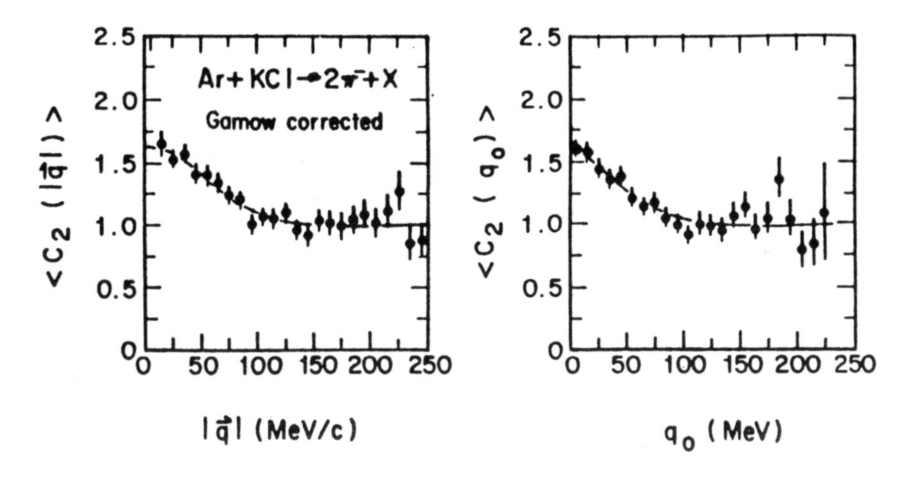

Figure 2: The correlation function for 1.8 A· GeV ^{40}Ar +KCl $\rightarrow 2\pi^- + X$.

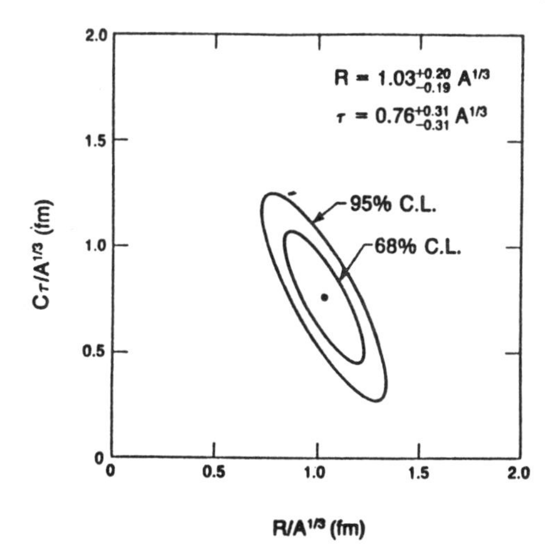

Figure 3: Contours of 65% and 95% confidence levels obtained by combining the three data sets in Table 1 using an assumed $A^{1/3}$ scaling between the Ne and Ar results.

TABLE 1

Fits to $C_2(|\vec{q}|, q_0) = 1 + \lambda e^{-(|\vec{q}|R)^2/2-(q_0\tau)^2/2}$

Reaction	λ	R	τ	χ^2/NDF
Ar+KCl→ $2\pi^- + X$	0.63±0.04	$2.88^{+0.5}_{-0.9}$	$3.29^{+1.4}_{-1.6}$	98.2/80
Ar+KCl→ $2\pi^+ + X$	0.73±0.07	$4.20^{+0.4}_{-0.6}$	$1.54^{+2.5}_{-1.54}$	67.1/81
Ne+NaF→ $2\pi^- + X$	0.59±0.06	$1.83^{+0.8}_{-1.6}$	$2.96^{+0.9}_{-1.0}$	125.7/82

Each sector contains 42 sense wires, providing a spatial resolution in the R-ϕ coordinate of 230 μm. Spatial coordinates along the beam direction were determined via charge division with a resolution of 1.5 cm. Measurement of the pulse height on the sense wires was also used for particle identification based on dE/dx information. An axial magnetic field (i.e., along the beam direction) of 0.5 Tesla gives a momentum resolution of $(\Delta p/p)^2 = (0.025p)^2 + (0.01)^2$ for p in GeV/c. Further details may be found in Ref. [5] and references therein.

Two types of triggers were used in the analysis: The first is a minimum bias trigger. which for these purposes may be regarded simply as the presence of two charged particles in the central region defined by the drift chambers. The second was a high (charged) multiplicity trigger, which provided events with $n_{ch} \sim$ (2-3)$\langle n_{ch}\rangle$. Both of these triggers were implemented based on hits in an 44-element hodoscope of scintillation counters surrounding the beam pipe.

In the off-line analysis, events were required to have a good vertex fit. Individual tracks were selected by goodness-of-fit criteria ($\Delta p/p$ and χ^2 of the fit). Tight cuts based on dE/dx allowed the rejection of p's, $\bar{p}$'s, and

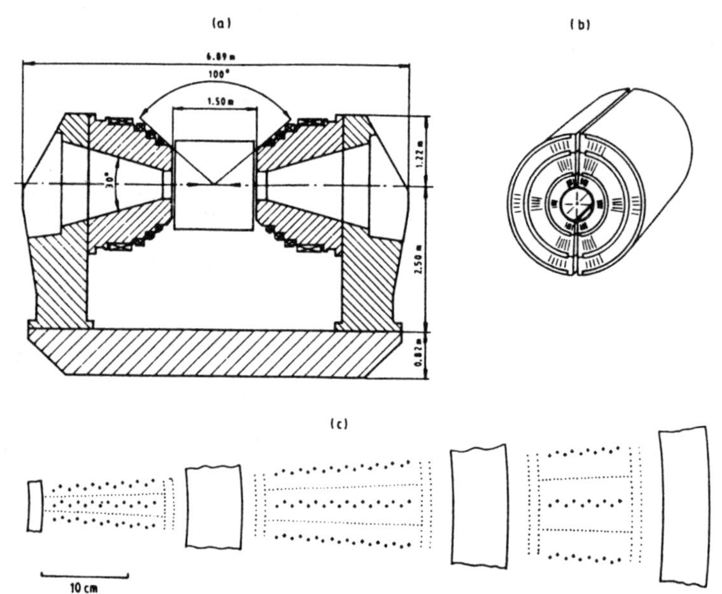

Figure 4: a.) Schematic drawing of the Axial Field Magnet and the drift chamber. b.) End view of the drift chamber. c.) Expanded view of the sector structure. The staggering of the sense wires has been exaggerated by a factor of five.

kaons. Good two-pion events were further required to have relative opening angles for all pairs of particles satisfying $|\Delta\theta| > 2.5°$ and $|\Delta\phi| > 2.5°$ to prevent confusion arising from failure to separate closely spaced tracks.

Analysis proceeds as in the JANUS experiment, in that the ratio of the actual number of events at a given relative momentum is divided by the background number, as per Eq. 1. The number of background pairs is again obtained by mixing pions from different events, but both A and B in Eq .1 are now binned in terms of q_T and q_L, rather than $|\vec{q}|$ and q_0. Here q_T is the component of $\vec{q}$ transverse to the mean momentum of the pion pair (see the appendix for further details). The correlation function in these variables is fit to a function of the form

$$C_2(q_T, q_L) = 1 + \lambda\left[\frac{2J_1(q_T R)}{q_T R}\right]^2 \cdot \frac{1}{1 + (q_L \tau)^2} \quad , \tag{4}$$

corresponding to a uniform spherical source of radius R and exponential lifetime τ.

A typical result for the correlation function is shown in Figure 5. Of some interest is the discovery that extracted value for R depends strongly

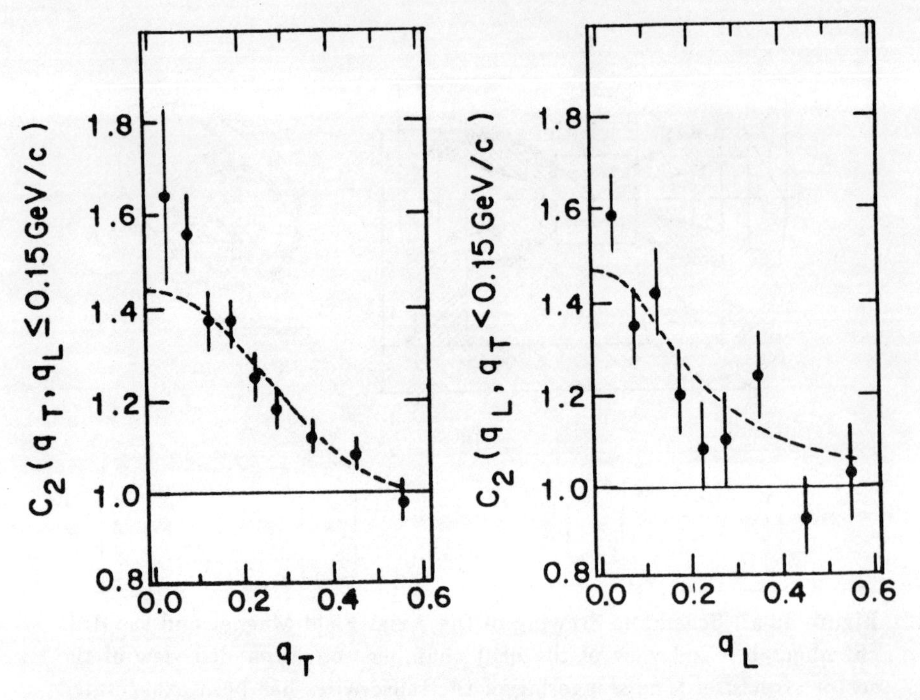

Figure 5: The correlation function measured by the AFS for $p\bar{p} \to 2\pi^+$ at 53 GeV .

on the multiplicity of the event, as shown in Figure 6. (A similar result holds for τ.) Note also the close correspondence between radii measured for $\pi^+\pi^+$ and $\pi^-\pi^-$ pairs; this is a strong check on the systematics of extracting R from different data sets.

4 DISCUSSION

In this section I would like to explore some of the factors that limit our ability to perform more refined GGLP measurements. We begin with the question of obtaining an adequate number of events.

4.1 Event Rates

Naive estimates of the two-pion event rate lead one to expect correlation functions containing hundreds of thousands of events. I will demonstrate this reasoning for both of the experiments considered, and then show the limitations of such a calculation.

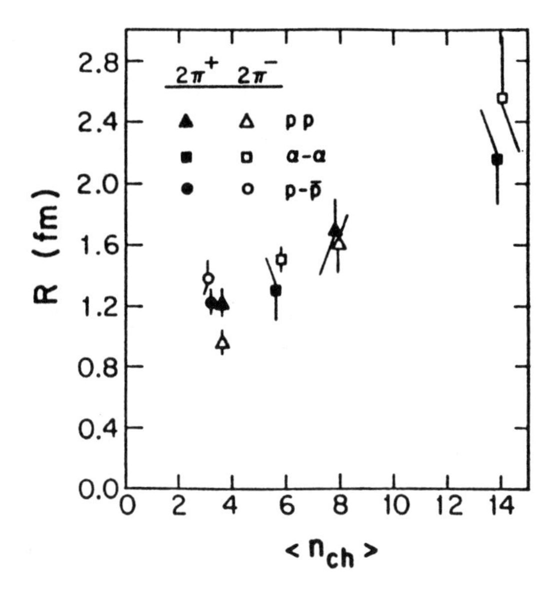

Figure 6: The radius R as a function of average multiplicity $\langle n_{ch} \rangle$ for pp, $p\,\bar{p}$ and α-α collisions.

The solid angle $\Delta\Omega$ for the JANUS experiment is of the order of 0.1% of 4π. Although this is a fixed target experiment, the forward-peaking of the particle flow is sufficiently mild that we make only a small error in assuming an isotropic distribution. The mean number of pairs expected per event is then $\frac{1}{2}\langle n_\pi(n_\pi - 1)\rangle(\frac{\Delta\Omega}{4\pi})^2$. The pion multiplicity distribution is Poisson to a good approximation, so that $\langle n_\pi(n_\pi - 1)\rangle \simeq \langle n_\pi \rangle^2$. For 1.8 A$\cdot$ GeV ^{40}Ar+KCl, $\langle n_\pi \rangle \sim 5$, leading to $\sim 10^{-5}$ pairs per event. (Since this spectrometer can accept only one sign of like-pairs with a given field polarity, we use the multiplicity of π^-'s in this calculation.) Assuming a beam intensity of 10^8 ions/sec, and a 1% interaction length target, we obtain an instantaneous rate of 10 pair events per second. Nearly all of the accepted pairs contribute to the enhanced region of the correlation function, as may be seen by examining the error bars in Fig. 2. Allowing for the beam structure of the Bevalac at this energy (a 1 second spill every 6 seconds), and assuming that the experiment has a 20% live-time, one would expect $\sim$1200 pairs/hr, or about 2×10^5 good events in a one week run.

The estimate for the AFS experiment proceeds in much the same fashion: The rapidity density of charged particles dn_{ch}/dy is about 2 at $\sqrt{s} = 63$ GeV. Assuming that all charged particles are pions, and that the drift chamber covers the region $|\Delta y| < 1$, we have (again for one charge species) $\langle n_\pi \rangle \simeq 2$ per minimum bias (MB) event. The AFS simultaneously accepts $2\pi^-$ and $2\pi^+$ events, so that the number of pairs per (MB) event is now $2 \cdot \frac{1}{2}\langle n_\pi \rangle \sim \langle n_\pi \rangle^2 \sim 4$. (The non-Poisson multiplicity fluctuations increase this number by about 50%, but I ignore such subtleties here.) The principal difference between this calculation and the one for the JANUS spectrometer is that all such pairs are not useful in a GGLP measurement–most have relative momenta much larger than the region of interest for the correlation function.

It is straightforward to estimate the fraction of contributing pairs. Assume a source size of 1 fermi, so that values of $|\vec{q}| \leq 200$ MeV/c contribute. Taking the mean pion momentum $\langle p_\pi \rangle$ to be about 300 MeV/c, this requires pions to have an opening angle $\delta\phi \leq \pi/4$, or about the $\frac{1}{8}$ of the (relative) azimuth. Similarly, using $p_z = m_t \sinh y$ and requiring that δp_z also be small leads to $\delta y \sim \frac{1}{2}$, or about one-fourth the rapidity interval. Taken together. we find that about 3% of all pion pairs will be in the enhancement region. (More sophisticated calculations are consistent with this crude estimate.)

For MB events, the ISR luminosity is sufficiently high that data-taking is limited by tape speed to about 5 events/sec. The above numbers then imply that of order 300 events (of a given charge species) are accumulated in one hour of running, which translates to about 50,000 in a one-week run. It should be emphasized that this is the number in the interesting region of the correlation function; the number of two-pion events in the tails of this function will be about thirty times larger.

A cursory inspection of the data appearing in Figures 2 and 5 indicates that the number of events used to make these figures is far from the number expected from the hypothetical one-week runs estimated above. What has happened? Not surprisingly, a variety of effects contribute. The detection and tracking of two like-sign charged particles is very difficult, both from the standpoint of hardware and software requirements. In the case of the JANUS experiment, there is very little redundancy in the 12 planes of MWPC read-out. A perfect event would require the firing of all 12 planes for each pion. While this is feasible for the chambers operating far from the target, those in the vicinity of the beam suffer severe losses from the high event rate. (The high number of heavily ionizing particles seen by MWPC1 also creates problems in tracking.) The AFS experiment has little problem with redundancy, due to the 42 potential $R - \phi$ digitizings per track. However, in order to determine the particle's identity, good dE/dx information must be obtained, which requires a large number of these digitizings (typically 20 or more). Also, for both experiments, the *unambiguous* identification of tracks that have followed very nearly the same path through the respective spectrometers presents a non-trivial pattern recognition problem to the reconstruction software. These effects lead to high rejection rates (>90%) for recorded events, thereby dramatically decreasing the number of analyzed events available for further study.

Finally, note must be made of two programmatic difficulties. The AFS experiment was largely devoted to the study of events with high transverse momenta at ISR energies. In some sense, the various results for MB events were obtained as a bonus accruing from an elegant detector design. Thus, there was little incentive for extensive MB data-taking at the level necessary for an exhaustive GGLP analysis (say 10^6 events). The JANUS experiment suffered from a an even more mundane malady: there are no one-week runs at the Bevalac. All of the data presented in Ref. [3] were taken in three weekend runs spaced over two years. Such a beam schedule is hardly conducive to the systematic study of a subtle effect.

4.2 On the Interpretation of λ

For a fully chaotic source, the parameter λ appearing in Eq. 2 would be 1. Deviations from this value are often cited as evidence for coherent (i.e., non-chaotic) behavior of the pion source, since the fraction f of pions emitted coherently is given by (Ref. [9]) $\lambda = 1 - f^2$. A completely coherent source would have $\lambda = 0$, i.e., there would be no Bose-Einstein correlations *of the form predicted by Eq.* 2. This is often a source (no pun intended) of confusion, as we normally speak of coherence as resulting from strong correlations. It is perhaps more illuminating to recall that the signal in a Bose-Einstein experiment results from the strong fluctuations caused by

from Bose statistics; these fluctuations are fully developed for a chaotic (e.g., thermal) source and minimal for a coherent (e.g., stabilized laser) source.

Nearly all hadronic experiments that determine λ have found values significantly less than one. The experiments discussed in this paper are no exception. Tempting as it may be to use these results as a probe of the quantum nature of the pion source, some caution must be exercised. While an entire variety of essentially mundane effects can be concocted, I will limit myself to the study of two particularly relevant ones.

The first arises in the analysis of the JANUS experiments. The momentum resolution of this spectrometer is sufficient to resolve the repulsion at very low values of relative momentum induced by the two-pion Coulomb repulsion. This reduces the value of the correlation function, and hence the extracted value of λ according to

$$C_2(|\vec{q}|, q_0) \rightarrow \frac{2\pi\eta}{e^{2\pi\eta} - 1} \cdot C_2(|\vec{q}|, q_0) \quad , \quad \text{where} \quad \eta \equiv \frac{m_\pi e^2}{2\hbar\sqrt{(q \cdot q)}} \quad . \tag{5}$$

The data presented in Section 2 have been corrected for this effect, which typically increases the extracted value of λ by about 20%. Given this dependence, and the fact that such corrections are necessarily approximate, (due to the unsolved multi-body Coulomb effects that are at play in these heavy-ion collisions), it would be premature to ascribe any profound significance to such deviations from unity. Nonetheless, the observed trend to larger values of λ (Ref. [6]) seen in heavier-mass collisions than those described here is most interesting and should provide a quantitative handle on these data.

A second trivial effect reducing the value of λ is that of resonance production. This is largely due to the long lifetime of the vector resonances. Pions originating from the decay of these resonances will have traveled several fermis from the interaction region, thereby increasing the effective source size. Since most hadron-hadron experiments are unable to resolve a source size larger than 2 fermis, the correlation from this larger source goes unresolved, leading to a smaller value of λ. Such an effect cannot be ruled out for the AFS data, given the large number of pions resulting from resonance decays at this energy (Ref. [7]).

To demonstrate the subtleties that such effects may induce, consider the observed increase of R with $\langle n_{ch} \rangle$ measured by the AFS (Fig. 6). A perfectly straightforward explanation of this effect may be had in terms of an impact-parameter model of hadron production (Ref. [8]), in that large multiplicities arise from a larger overlap of the colliding hadronic matter. However, one must regard the resonance production as a systematic correction to this trend: large fluctuations in charged multiplicity could favor events with an above average vector meson content (since they decay into 2-3 charged particles), leading to an increased radius, due to the extended nature of this "source". The fact that the AFS data is at least consistent with a decreasing λ as a function of $\langle n_{ch} \rangle$ indicates that more definitive measurements will be required to disentangle these effects.

5 FUTURE DIRECTIONS

In this section I discuss, in order of increasing speculation, three possible extensions of GGLP methods: simultaneous extraction of three or four-dimensional source parameters, the study of Bose-Einstein correlations in conjunction with multiplicity distributions, and speckle interferometry. Space considerations limit this to essentially the presentation of a wish-list for future measurements.

The first item, determination of the full space-time structure of the source, has already been attempted by several groups, with varying degrees of success. It is clear from the material covered in the Appendix that, in principle, a four-dimensional correlation function will probe all four dimensions of the pion source. In practice, it is often difficult to determine even two such parameters (say, some mean radius R and a lifetime τ) simultaneously. This, of course, results from the limited available statistics, the coupling between lifetimes and source velocities, etc. Nonetheless, both the JANUS experiment (Ref. [10]) and the AFS collaboration (Ref. [11]) have preliminary data indicating the possibility of measuring transverse and longitudinal source dimensions.

Secondly, it would be of some significant utility to measure in the same experiment both the multiplicity distribution of charged particles $P(n)$ (even better would be the distribution for one species of pions) and the correlation function. Such information would not only greatly aid our understanding of the dynamics of the hadronic production mechanism, but would also help to unravel statistical aspects of these reactions. This is particularly true in light of the rediscovery of the negative-binomial distribution and its generalizations as a calculus for the description of such distributions (Ref. [12]). If in fact these distributions result from the statistics of partitioning bosons in phase space, there is then a natural relationship between the "cell parameter" k and the coherence parameter λ. Verification of this prediction in a systematic fashion (say as a function of the rapidity interval) would provide enormous impetus towards this approach.

Finally, we close with the most speculative, yet most intriguing concept–speckle interferometry. As noted by Willis and Chasman (Ref. [13]), in the limit of a large number of pions (say 100 or more), the appropriate measure of Bose-Einstein correlations lies in the statistical occupation of phase space cells. Just as the GGLP effect is the particle physics extension of the Hanbury-Brown–Twiss effect, speckle interferometry is the pionic extension of techniques developed by Labeyrie and his collaborators (Ref. [14]) for imaging of stars under poor seeing conditions.

In our context, we wish to "image" an individual event (e.g., a very relativistic, very high mass heavy ion collision, with $\langle n_\pi \rangle \sim 10^3$) by examining the number and distribution of "lumps", or speckles in phase space. If we restrict ourselves to the angular domain by considering a small interval $\delta p_\pi < 1/R$ about some fixed $|\vec{p}_\pi|$, each pion momentum vector is describable in terms of its angular coordinates $\cos\theta$ and ϕ. The width of speckles, or clumps of pions, on the unit sphere then reflects the dimensions of the source ($\delta\phi_s \sim 1/(|\vec{p}_\pi|R)$) and the number of speckles is of order the number of occupied phase space cells $N_s \sim (pR/2\pi)^2$.

To demonstrate these effects, I have written a program to create a sequence of events with Bose-Einstein correlations between all pions (Ref. [15]). While anticipating the advent of machines such as RHIC, we can develop our speckle technique with such events: Figure 7 shows a sample event for $n_\pi = 500$, $|\vec{p}_\pi| = 300$ MeV/c, and $R = 6$ fm. For reference, an uncorrelated event is shown in Figure 8. While it is clear that the correlated event does have a non-uniform phase space population, it is difficult to make a visual assessment of the statistical significance of one event. (The expected num-

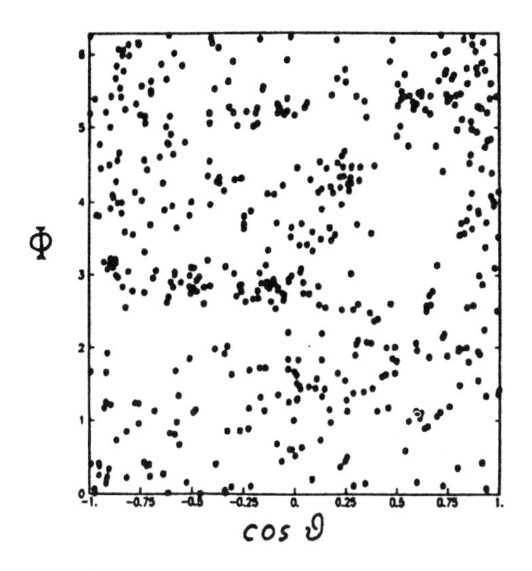

Figure 7: A sample Bose-correlated event with 500 like-pions

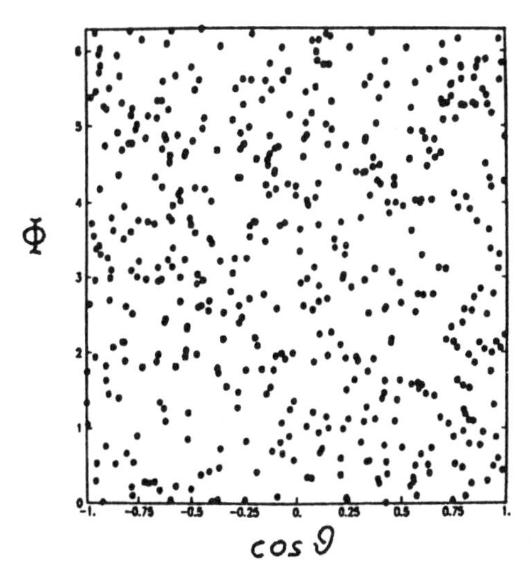

Figure 8: A sample uncorrelated event with 500 like-pions

ber of speckles is 1 or 2, so the event is not inherently wrong.) However, a numerical calculation of the reduced chi-squared $\tilde{\chi}^2$ is quite enlightening: If N_B is the number of bins ($N_B = 900$) in Figure 7, consider

$$\tilde{\chi}^2 \equiv \frac{1}{N_B} \sum_{i=1}^{N_B} \frac{(n_i - \bar{n})^2}{\bar{n}} \quad . \tag{6}$$

Clearly, if the fluctuations are Poisson, we have $\langle \tilde{\chi}^2 \rangle = 1$, with a (near)-Gaussian distribution about this value with width $\sigma = (2/N_B)^{1/2} \simeq 0.047$. For our uncorrelated event, we find $\tilde{\chi}^2 = 0.964$, consistent with unity, while for the Bose-correlated event $\tilde{\chi}^2 = 1.456$, which is roughly a 10σ deviation from the value expected for a random phase distribution. Moreover, the value of $\tilde{\chi}^2$ is very nearly what we would expect for a completely Bose-correlated event, $\tilde{\chi}^2_{BE} = \bar{n} + \bar{n}^2 = 1.56$. (A more sophisticated estimate of the $\tilde{\chi}^2$ distribution (Ref. [18]) leads to $\tilde{\chi}$ being distributed about 1 with standard deviation $(1/2N_B)^{1/2}$, thereby reducing the significance of our event to "only" 8.8σ.)

While $\tilde{\chi}^2$ measures the global significance of each event, much more information may be obtained by an event-by-event correlation analysis. The number of pairs in our sample event is sufficiently large ($\sim 12,000$) to allow a crude extraction of a radius from the usual correlation function. In fact, it is possible to demonstrate that the information content of an event plot is proportional to the number of such relative pairs (Ref. [15]). Nonetheless, the speckle technique provides a powerful tool for visual assessment of such information, as well as dramatizing the subtle interplay between particle and wave behavior that makes this field so fascinating.

6 ACKNOWLEDGEMENTS

It is a pleasure to acknowledge the efforts of the JANUS and AFS collaborations, without whose essential efforts none of this work would have been possible. In particular, the AFS results owe their existence to the careful work of Karin Kulka. The thoughtful reading of this manuscript by B. Callen and M. Andrea is also gratefully acknowledged; all remaining errors are entirely my own.

APPENDIX

This appendix provides a brief summary of the formalism for Bose-Einstein measurements. Commonly used variables will be defined, a schematic derivation of the GGLP effect is presented, and the issue of Lorentz invariance of the corrrelation function is addressed.

We begin by considering two pions with (four)-momentum p_1 and p_2 which are detected at two space-time positions x_1 and x_2, respectively. If these two pions are emitted from an extended source at space-time positions r_1 and r_2 (*not* respectively!), the wave function may be written as

$$\Psi_{p_1p_2}(x_1x_2 \leftarrow r_1r_2) = \frac{1}{\sqrt{2}}\left[e^{ip_1\cdot(x_1-r_1)}e^{ip_2\cdot(x_2-r_2)}+e^{ip_1\cdot(x_1-r_2)}e^{ip_2\cdot(x_2-r_1)}\right] \quad . \quad (7)$$

Assuming that the emission of pions at different space-time points may be considered independent processes, and introducing the distribution of these points $\rho(r)$, the probability P_{12} of the specified detection process is

$$P(p_1 \,\mathrm{at}\, x_1, p_2 \,\mathrm{at}\, x_2) = \int |\Psi_{p_1p_2}(x_1x_2 \leftarrow r_1r_2)|^2\rho(r_1)\rho(r_2)dr_1dr_2 = 1+|\tilde{\rho}(q)|^2 \quad , \quad (8)$$

where $q \equiv p_2 - p_1$ and $\tilde{\rho}(q)$ is the Fourier transform of $\rho(r)$ wrt q.

Expressed in terms of the usually measured experimental quantities,

$$P_{12} = \frac{\langle n_\pi\rangle^2}{\langle n_\pi(n_\pi - 1)\rangle} \cdot \frac{\sigma_\pi \dfrac{d^6\sigma_\pi}{dp_1^3dp_2^3}}{\dfrac{d^3\sigma_\pi}{dp_1^3}\dfrac{d^3\sigma_\pi}{dp_2^3}} \equiv C_2(p_1,p_2) \quad . \quad (9)$$

Assuming that the pions are describable as plane-wave states then allows us to map the above obviously Lorentz invariant form into the usual $C_2(|\vec{q}|,q_0)$ (e.g., see Eq. 2), which is frame-dependent. This observation has caused some authors to state that $C_2(|\vec{q}|,q_0)$ is incapable of describing the data; see, e.g., Ref. [16]. While true, this statement is somewhat akin to stating that $\Phi = e/r$ is incapable of describing the potential of a moving charge. This is only a statement that the expression must be appropriately modified with the invariants at hand, so that the expression becomes $A^\mu = eu^\mu/[-(r\cdot r)^2 + (r\cdot u)^2]^{1/2}$ (the Lienard-Weichert potential, with u the four-velocity of charge).

A similar modification to protect the invariance of $C_2(|\vec{q}|, q_0)$ (as opposed to the covariance of A^μ) was proposed some time ago by Yano and Koonin (Ref. [17]). Their result for a Gaussian source density of the form in Eq. 3, moving with four-velocity u in the frame where q is calculated, may be written as

$$C_2(|\vec{q}|, q_0) = 1 + \lambda e^{-(q\cdot q)R^2 + (q\cdot u)^2[R^2+\tau^2]} \quad . \tag{10}$$

In principle, one could extract the (distribution of?) source velocity by fitting the data to this form, although I am unaware of any successful attempts to do so.

Many authors prefer to analyze their data in terms of the variables q_L and $\vec{q}_T$, which are the components of the relative three momentum $\vec{q}$ longitudinal and transverse to the mean momentum of the pion pair $\vec{P} = \vec{p}_1 + \vec{p}_2$,

$$\vec{q}_T = \vec{q} - (\vec{q}\cdot\vec{n})\vec{n} \quad , \quad q_L = \vec{q}\cdot\vec{n} \quad , \tag{11}$$

where $\vec{n}$ is a unit vector in the direction of $\vec{P}$. These variables decouple the measurement of transverse and longitudinal dimensions (with respect to the line-of-sight) of the source. For example, assuming a long-lived spherical source of uniform surface radiance (e.g., a star), one obtains the q_T dependence of Eq. 4. However, there is a strong coupling between q_L and q_0, as may be seen by noting that

$$q_L = \frac{(\vec{p}_1 - \vec{p}_2)\cdot(\vec{p}_1 + \vec{p}_2)}{|\vec{p}_1 + \vec{p}_2|} = (E_1 - E_2)\cdot\frac{E_1 + E_2}{|\vec{p}_1 + \vec{p}_2|} \simeq q_0 \quad , \tag{12}$$

where the approximation assumes that the average velocity of the pair is relativistic. The physical source of this coupling arises from transit-time effects for pions crossing the volume of the source: If the source is long-lived ($\tau >> R$), we can define a phase-plane for emission perpendicular to the surface and apply Lambert's law, while for a short-lived source significant disassembly occurs by the time pions from the limbs cross the phase-plane. This effect has been periodically rediscovered by various authors; the first reference of which I am aware is that of Kopylov and Podgoretskiǐ (Ref. [19]).

REFERENCES

[1] M. Gyulassy *et al.*, Phys. Rev. **C20**, 2267 (1979).

G. Goldhaber in Proceedings of the First International Workshop on Local Equilibrium in Strong Interaction Physics, D.K. Scott and R.M. Weiner, ed., World Scientific Publishing Co. (1985).

A. Giovannini *et al.*, La Rivista del Nuovo Cimento, **2**, No. 10, 1979.

[2] G. Goldhaber *et al.*, Phys. Rev.. **120**, 300, (1960).

[3] W.A. Zajc *et al.*, Phys. Rev. **C29**, 2173 (1984).

[4] W.A. Zajc, "Two-Pion Correlations in Heavy Ion Collisions", Ph.d Thesis, University of California, (1982). Available as LBL-14864.

[5] T. Åkesson *et al.*, Phys. Lett. **129B**, 269 (1983).

[6] K.M. Crowe, private communication.

[7] T. Åkesson *et al.*, Nucl. Phys. **B203**, 27 (1982).

[8] S. Barshay, Phys. Lett. **130B**, 220 (1983).

[9] G.N. Fowler in Proceedings of the First International Workshop on Local Equilibrium in Strong Interaction Physics, D.K. Scott and R.M. Weiner, ed., World Scientific Publishing Co. (1985).

[10] A.D. Chacon, private communication.

[11] K. Kulka, private communication.

[12] P. Carruthers and C.C. Shih, Phys. Lett. **127B** (1983) 242.

[13] W. Willis and C. Chasman, Nucl. Phys. **A418**, 413 (1984).

[14] A. Labeyrie, Progress in Optics, **14**, 1976, ed. E. Wolf (North-Holland).

[15] W.A. Zajc, Proceedings of the Brookhaven Workshop on Detector Designs for the Proposed RHIC, Brookhaven, NY, April 15-19, 1985 (BNL-51921).

[16] T. Avery *et al.* , Phys. ReV. **D32**, 2294 (1985).

[17] F.B. Yano and S. Koonin, Phys. Lett. **78B**, 556 (1978).

[18] "Review of Particle Properties", Rev. Mod. Phys. **56**, No. 2, Part II, (April 1984).

[19] G.I. Kopylov and M.I. Podgoretskiĭ, Sov. J. Nucl. Phys. **18**, 336 (1974).

BOSE-EINSTEIN CORRELATION IN LANDAU'S MODEL

Y.Hama and Sandra S. Padula

Instituto de Física, Universidade de São Paulo
São Paulo, Brasil

ABSTRACT

Bose-Einstein correlation is studied by taking an expanding fluid given by Landau's model as the source, where each space-time point is considered as an independent and chaotic emitting center with Planck's spectral distribution. As expected, the correlation depends on the relative angular positions as well as on the overall localization of the measuring system and it turns out that the average dimension of the source increases with the multiplicity N_{ch} .

1. INTRODUCTION

The correlation between identical particles produced in a reaction is closely related with the space-time structure of the emitting source of these particles. This is the base of Hanbury-Brown and Twiss method [1], used in radioastronomy to measure stellar dimensions, and also of the GGLP effect[2] in nuclear physics. Indeed, there are correlations of other origins[3], but here we limit ourselves to H-B-T effect.

Several authors have studied this phenomenon[3-7], but so far concrete applications have been restricted to static sources, the ones with the time and the space dependences factorized or field theoretical model with a classical source[3]. While these models clarify several qualitative features of the phenomenon or even are convenient approximations in some cases, several doubts remain when one tries to extract information on a more dynamical processes like hadronic interactions from experimental data. The purpose of the present paper is to discuss the correlation produced by a rapid expanding source as the one given by Landau's hydrodynamic model by giving a preliminary account of such a study.

Several factors appear in this model which must be carefully examined. First, the source is in rapid expansion, so kinematic effect due to Lorentz transformation becomes important. The overall size of the source becomes very large after some interval of time, but only its "surface" emits particles. Also, it is completely asymmetric with large energy-dependent longitudinal dimension and a small more-or-less constant transverse dimension. We have chosen Landau's model, because it gives an explicit space-time development of the source[8] and also a fairly good account of several of the experimentally observed parameters, such as the average multiplicity, the rapidity distribution, the transverse-momentum distribution, etc.. Then, we think natural to verify whether this model reproduces also the experimentally observed properties of the identical-particle correlation[9,10].

2. THE MODEL

We consider a source which is given by the following variant of Landau's model:

a) The constituents are quarks and gluons which form a plasma. Quantitatively, this implies a large increase in the statistical factor which appears in the thermodynamic relations.

b) The fireballmass is not fixed, but it is an event-dependent parameter. Although this hypothesis is not necessary in computing the correlation, it turns out to be essential in comparing the results with some data[10].

In the present version, we completely neglect the transverse expansion, since it is small[11], and for simplicity we use the asymptotic form of Khalatnikov's solution[12] (scale-invariant solution), namely

$$\begin{cases} \xi \equiv \ln \frac{T}{T_0} \simeq - c_0^2 \ln \frac{\tau}{\Delta} , \\ \alpha \simeq \frac{1}{2} \ln \frac{t+x}{t-x} , \end{cases} \tag{1}$$

where

$$\begin{cases} \alpha = \text{rapidity of the plasma,} \\ c_0 = \text{sound velocity } \left(\simeq \frac{1}{\sqrt{3}} \right) , \\ \Delta = \sqrt{\frac{1-c_0^2}{\pi}}\, \ell \ , \ 2\ell = \text{initial thickness of the fireball.} \end{cases}$$

The condition for the validity of this solution is $|\xi| \gg |\alpha|$

Usually, one assumes that the final particles appear when the local temperature reaches a certain critical value, dissociation temperature $T = T_d \simeq m_\pi$, which defines a transition "surface". We will follow this usual version in the present paper, but some remarks should be made. First, there is evidently a component which comes from the evaporation at $T > T_d$, but probably the color neutralization process is damped at high temperatures. The second observation concerns the "surface" where the phase transition occurs. For macroscopic bodies, such a surface exists with a negligible thickness, but for a microscopic system with dimensions comparable to the emitted particles, the thickness of such a "surface" becomes important. Although in this first version we consider $T = T_d$ in the plasma as the emitting source, we are conscious of this ambiguity, so the numerical results presented here may change considerably if the transition is taken into account with more detail. The effects of resonance production, final state interactions[3], as well as of the partial coherence of the source[3,7] should be taken into account, but, as mentioned before, here we simply consider a totally chaotic source, focusing our attention especially on its rapid expansion.

So, each point of the surface $T = T_d$ in (1) is considered to be an independent chaotic source with the momentum spectrum in the rest frame of the fireball with mass M given by

$$f(p) = \frac{1}{(2\pi)^3} \frac{u^\mu p_\mu / E}{\exp(u^\mu p_\mu / T_d) - 1}$$

$$\simeq \frac{1}{(2\pi)^3} \cdot \frac{u^\mu p_\mu}{E} \exp\left(-\frac{u^\mu p_\mu}{T_d}\right), \tag{2}$$

where

$$\begin{cases} u^\mu = (\text{ch}\,\alpha, \text{sh}\,\alpha, 0, 0) \\ \quad \text{is the 4-velocity of the fluid and} \\ p^\mu = (E, p_x, p_y, p_z) \\ \quad \text{is the 4-momentum of the emitted particle.} \end{cases} \tag{3}$$

The amplitude for finding a particle at x and emitted at x' is written

$$J(x, x') = \int d\vec{p} \sqrt{f(p)} \exp\left[-ip_\mu(x^\mu - x'^\mu)\right] \exp\left[i\theta(x')\right], \tag{4}$$

where $\Theta(x')$ is a random phase. Following the notation of Ref.5), the probability of detecting two quanta of momenta p_1 and p_2 in an event is

$$W(p_1, p_2) = \tilde{I}(0, p_1)\tilde{I}(0, p_2) + \left|\tilde{I}\left[p_1 - p_2, \tfrac{1}{2}(p_1 + p_2)\right]\right|^2, \tag{5}$$

where

$$\begin{cases} \tilde{I}(\Delta p, p) = \int d^4x\, d^4\Delta x \exp\left[ix\Delta p + i\Delta x p\right] \int dx'\, I(x, \Delta x, x') \\ \text{and} \\ \left\langle J^*(x - \frac{\Delta x}{2}, x') J(x + \frac{\Delta x}{2}, x'') \right\rangle = \delta(x' - x'') I(x, \Delta x, x'). \end{cases} \tag{6}$$

From eqs. (2) - (4) and (6), we obtain

$$\tilde{I}(\Delta p, p) = (2\pi)^5 \int d^4x' \exp(i\Delta p_\mu x'^\mu) \cdot \exp(-u^\mu p_\mu / T_d)$$

$$\times \left[\frac{u^\mu (p_\mu - \frac{\Delta p_\mu}{2})}{E - \frac{\Delta E}{2}} \cdot \frac{u^\mu (p_\mu + \frac{\Delta p_\mu}{2})}{E + \frac{\Delta E}{2}} \right]^{1/2} . \qquad (7)$$

The integration in the transverse variables y', z' may readily be carried out and gives

$$\int dy' \int dz' \exp\left[-i\Delta p_y y' - i\Delta p_z z'\right] = \frac{2\pi R}{\Delta p_\perp} J_1(R\Delta p_\perp), \qquad (8)$$

where

$$\Delta p_\perp = \sqrt{\Delta p_y^2 + \Delta p_z^2} \qquad (9)$$

and R is the transverse radius which remains fixed in the present version.

In order to calculate the other integrals, first we perform a change of variables

$$\begin{cases} t' = \tau \operatorname{ch} \alpha , \\ x' = \tau \operatorname{sh} \alpha , \end{cases} \qquad (10)$$

use the condition $T = T_d$ to fix $\tau = \tau_d$ and get

$$\tilde{I}(\Delta p, p) = \frac{(2\pi)^6 R \tau_d}{\Delta p_\perp \sqrt{E^2 - \Delta E^2/4}} J_1(R\Delta p_\perp)$$

$$\times \int_{-\infty}^{\infty} d\alpha \sqrt{\left[(E - \frac{\Delta E}{2})\operatorname{ch}\alpha - (p_L - \frac{\Delta p_L}{2})\operatorname{sh}\alpha\right]\left[(E + \frac{\Delta E}{2})\operatorname{ch}\alpha - (p_L + \frac{\Delta p_L}{2})\operatorname{sh}\alpha\right]}$$

$$\times \exp\left[(i\Delta E \tau_d - \frac{E}{T_d})\operatorname{ch}\alpha - (i\Delta p_L \tau_d - \frac{p_L}{T_d})\operatorname{sh}\alpha\right], \qquad (11)$$

with the notation $p_L \equiv p_x$ and $\Delta p_L \equiv \Delta p_x$. Now, we evaluate the last integral by the saddle-point method, obtaining for $\tilde{I}(\Delta p, p)$

$$\tilde{I}(\Delta p, p) = \frac{\sqrt{(2\pi)^{13}}\, R\,\tau_d}{\Delta p_\perp \sqrt{E^2 - \frac{\Delta E^2}{4}}}\, J_1(R\Delta p_\perp)$$

$$\times \left[\left(E - \frac{\Delta E}{2}\right)\left(\frac{E}{T_d} - i\Delta E\,\tau_d\right) - \left(p_L - \frac{\Delta p_L}{2}\right)\left(\frac{p_L}{T_d} - i\Delta p_L\,\tau_d\right)\right]^{1/2}$$

$$\times \left[\left(E + \frac{\Delta E}{2}\right)\left(\frac{E}{T_d} - i\Delta E\,\tau_d\right) - \left(p_L + \frac{\Delta p_L}{2}\right)\left(\frac{p_L}{T_d} - i\Delta p_L\,\tau_d\right)\right]^{1/2}$$

$$\times \frac{\exp\left[-\sqrt{\left(\frac{E}{T_d} - i\Delta E\,\tau_d\right)^2 - \left(\frac{p_L}{T_d} - i\Delta p_L\,\tau_d\right)^2}\,\right]}{\left[\left(\frac{E}{T_d} - i\Delta E\,\tau_d\right)^2 - \left(\frac{p_L}{T_d} - i\Delta p_L\,\tau_d\right)^2\right]^{3/4}} \tag{12}$$

The probability of detecting two identical quanta of momenta p_1 and p_2 in an event may now be computed with the help of (5) and (12). One usually defines a quantity which is the ratio of this probability to the product of the one-particle inclusive probabilities $W(p_1)$ and $W(p_2)$, namely

$$C(p_1, p_2) \equiv \frac{W(p_1, p_2)}{W(p_1)\,W(p_2)}$$

$$= 1 + \frac{\left|\tilde{I}\left[p_1 - p_2, \frac{1}{2}(p_1 + p_2)\right]\right|^2}{\tilde{I}(0, p_1)\;\tilde{I}(0, p_2)} \tag{13}$$

So, from (12) and (13), we finally arrive at

$$C(p_1,p_2) = 1 + \frac{4\left[J_1(R\Delta p_\perp)\right]^2}{T_d R^2 \Delta p_\perp^2}$$

$$\times \frac{\exp\left\{-2\left[\sqrt{\left(\frac{E^2-p_L^2}{2T_d^2} - \frac{\Delta E^2-\Delta p_L^2}{2}\tau_d^2\right)^2 + (E\Delta E - p_L\Delta p_L)^2\frac{\tau_d^2}{T_d^2}} + \frac{E^2-p_L^2}{2T_d^2} - \frac{\Delta E^2-\Delta p_L^2}{2}\tau_d^2\right]^{\frac{1}{2}}\right\}}{\left[\left(\frac{E^2-p_L^2}{T_d^2} - (\Delta E^2-\Delta p_L^2)\tau_d^2\right)^2 + 4(E\Delta E - p_L\Delta p_L)^2\frac{\tau_d^2}{T_d^2}\right]^{3/4}}$$

$$\times \frac{\exp\left\{\frac{1}{T_d}\left[\sqrt{E^2-p_L^2+\frac{\Delta E^2-\Delta p_L^2}{4}+(E\Delta E-p_L\Delta p_L)} + \sqrt{E^2-p_L^2+\frac{\Delta E^2-\Delta p_L^2}{4}-(E\Delta E-p_L\Delta p_L)}\right]\right\}}{\left[\left(E^2-p_\perp^2+\frac{\Delta E^2-\Delta p_L^2}{4}\right)^2 - (E\Delta E - p_L\Delta p_L)^2\right]^{1/4}}$$

$$\times \left\{\left[\left(\frac{E^2-p_L^2}{T_d}\right)^2 + \left(\frac{E\Delta E-p_L\Delta p_L}{2T_d}\right)^2 + (E\Delta E-p_L\Delta p_L)^2\tau_d^2 + \frac{(\Delta E^2-\Delta p_L^2)^2}{4}\tau_d^2\right]^2\right.$$

$$\left. - \left[\frac{E^2-p_L^2}{T_d^2} + (\Delta E^2-\Delta p_L^2)\tau_d^2\right]^2 (E\Delta E - p_L\Delta p_L)^2\right\}^{1/2}, \qquad (14)$$

where

$$\begin{cases} p \equiv \frac{1}{2}(p_1+p_2) = (E, p_L, p_y, p_z), \\ \Delta p \equiv p_1 - p_2 = (\Delta E, \Delta p_L, \Delta p_y, \Delta p_z) \end{cases} \qquad (15)$$

and $\Delta p_{\perp}$ is given by eq. (9). We fix the critical temperatura $T_d = m_{\pi}$ and τ_d shall be determined by imposing the condition $T = T_d$ on the plasma. To doing so, we need the initial temperature T_o, which depends on the mass of the fireball and its initial size. In a series of papers, we have shown that the hypothesis of large-mass-Lorentz-contracted-fireball formation around one of the incident particles provides a nice framework to accounting for several of the experimentally observed quantities[13,14]. In Ref. 14), we show that, if such a fireball is made of quarks and gluons, a very reasonable choice of the initial radius $R \simeq R_{proton}$ leads to the experimentally observed charged multiplicity $\langle N_{ch}\rangle(M)$ as well as the pseudo-rapidity distribution $\frac{d\sigma}{d\eta}(M)$, which have recently been measured in large-mass diffractive dissociation at $\bar{p}p$ collider.[15] In contrast, if such a fireball is made of pions, a very large radius R becomes necessary to fitting the data. Then, following Ref. 14),

$$T_o = \left[\frac{45}{4\pi^3 m_p R^3(g_g + \frac{7}{8}g_q)}\right]^{1/4} \sqrt{M} \simeq 0.118\sqrt{M}, \tag{16}$$

where g_g and g_q are the statistical wights for gluons and quarks respectively and the numerical value has been obtained assuming only the lightest flavours (u,d) and R = 0.75f. By using eqs. (1), the combination $\tau_d^2\,T_d^2$ which appears in eq. (14) turns out to be

$$\tau_d^2 T_d^2 = \Delta^2\left(\frac{T_o}{T_d}\right)^6 T_d^2 \simeq 0.0755\,M, \tag{17}$$

with M given in GeV.

3. RESULTS AND DISCUSSION

In order to better visualize the result expressed by eq. (14), we now consider some particular situations, by fixing the most of the variables which define the measuring apparatus there. Due to the low statistics in the experiments, the data correspond to some average of these particular results and this averaging processes smear out most of more prominent features of the correlation. A purpose of the present report is to show that experimentally more restrictive constraints are necessary in order to learn details

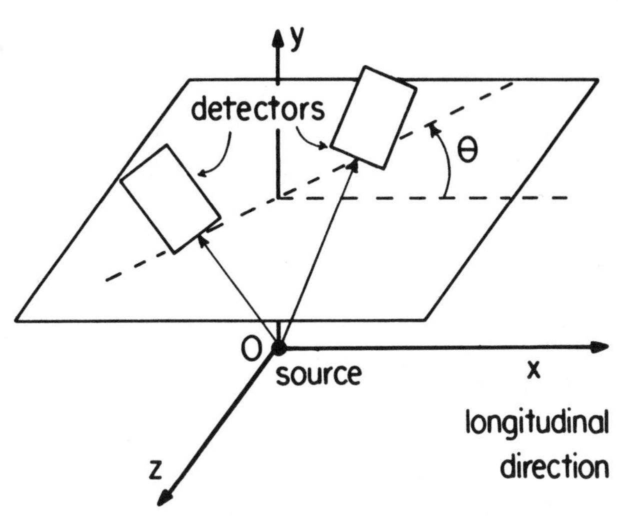

Fig. 1 - Sketch of the experimental situation considered here

of the space-time structure of the pion-emitting source. In the following, we fix the average position of the detectors at 90^0 in the proper frame of the fireball, with respect to the symmetry axis ($p_L = 0$).

We first consider two equal-energy quanta emitted symmetrically with respect to some transverse direction. In a plane perpendicular to this direction, we place the two detectors as indicated in Fig.1. The correlation is then studied for different values of Θ, with M and $p_\perp$ held fixed. Some typical results are shown in Figs. 2, where one sees that, as Θ increases, also the correlation increases, which means that the effective longitudinal size of the source is larger than the transverse one for the particular choices of M and $p_\perp$. Moreover, the comparison of these figures show that the longitudinal-correlation curve becomes narrower as M increases, indicating that the effective source dimension increases with M in that direction as expected. As for the transverse dimension, one sees that the curve for $\Theta=\Pi/2$ remains unchanged as it should be. In Figs. 2, we also display the average value of $C(p_1,p_2)$ when the detectors are rotated around y axis.

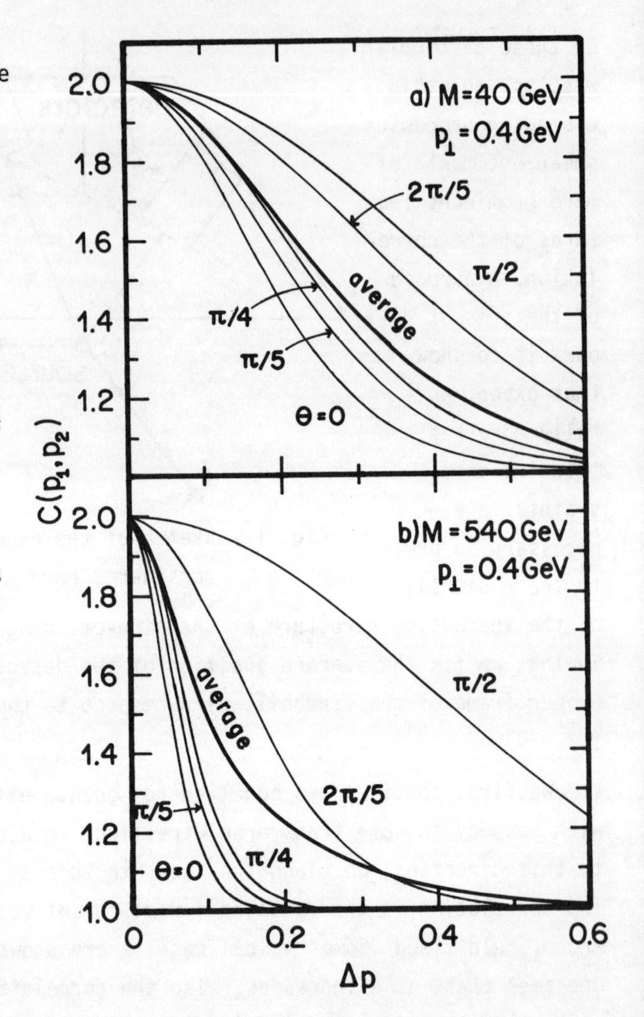

Fig. 2 - Θ-dependence of $C(p_1,p_2)$, with $\Delta E=0$ and $p_L=0$

As mentioned before and discussed in Refs. 13, 14), in our picture the mass M fluctuates and is not fixed for a given incident energy $\sqrt{s}$. When M increases, the average charged multiplicity $\langle N_{ch}\rangle(M)$ increases and so does the central multiplicity. Thus, with increasing M, the multiplicity grows at the same time that the pion emitting source stretches as shown above. However, a quantitative comparison with the data should be done by performing a more complete average over all directions and not only at 90^0.

Next, we vary $p_\perp$, the average momentum of the two quanta. Being the fluid expansion fixed once M is fixed, and the emission occurring from the surface $T = T_d$, one might hastily conclude that $C(p_1,p_2)$ is independent of $p_\perp$. However, it is not so, as seen in Fig. 3, where the effective size of the source decreases as $p_\perp$ increases. If one includes the evaporation, which may exist from the beginning, and tries to study the time - evolution of the source by its means, the effect mentioned above complicates the problem because it acts in the same direction.

Let us now consider a situation in which the two quanta are emitted with different energies but with the same direciton at 90^0. In the notation of Ref.4), the previous case corresponds to $q_\perp$ distribution of $C(p_1,p_2)$, whereas now we are looking at its q_L distribution.

Some of the results are shown in Figs. 4 and 5. Again, contradicting the naive intuition, kinematical effects

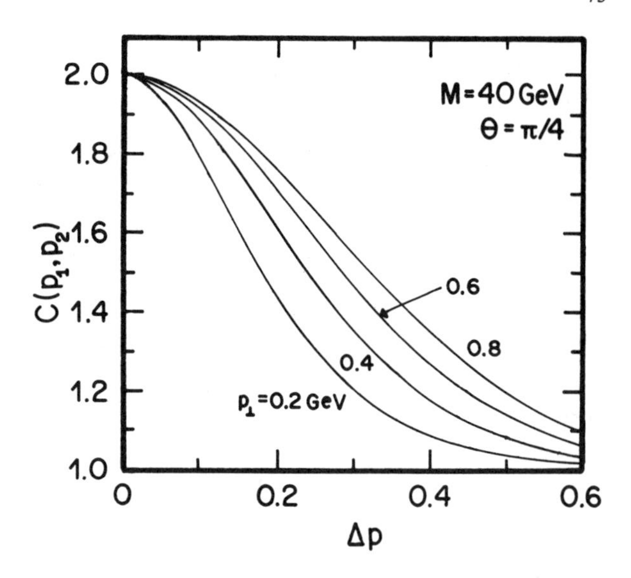

Fig. 3 - $p_\perp$ dependence of $C(p_1,p_2)$, with $\Delta E=0$ and $p_L=0$.

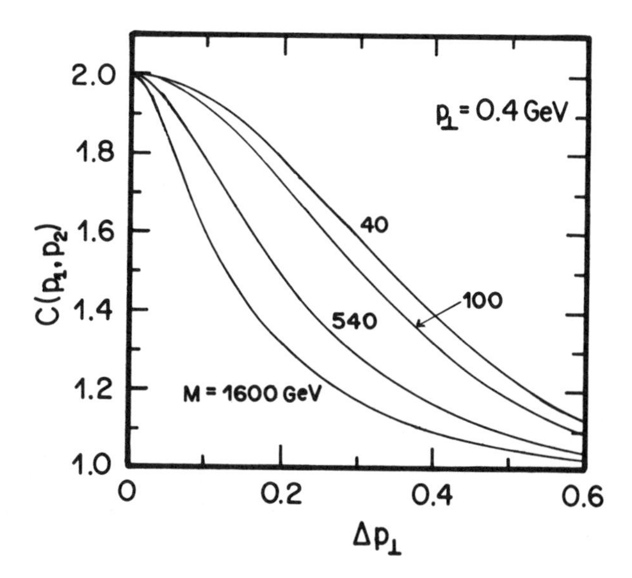

Fig. 4 - M dependence of $C(p_1,p_2)$ with $p_L=0$, $\Delta p_L=0$ and $\Delta p_z=0$

caused by the expansion of the source appear as M-dependence of $C(p_1,p_2)$- Fig. 4 and its $p_\perp$ -dependence - Fig. 5.

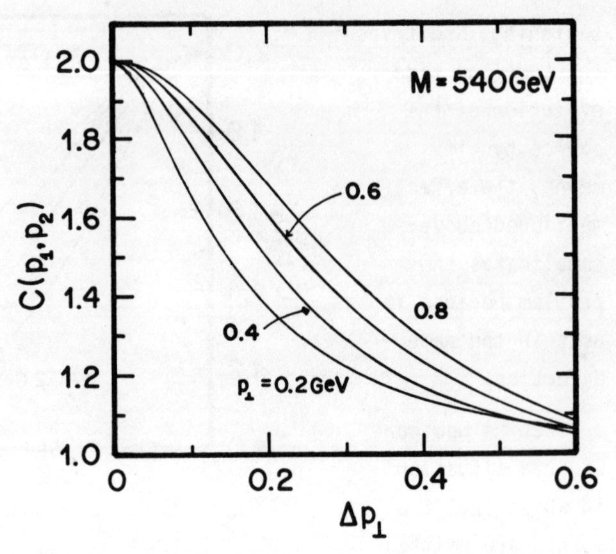

Fig. 5 - $p_\perp$ -dependence of $C(p_1,p_2)$ with $p_L=0$, $\Delta p_L=0$ and $\Delta p_z=0$

Finally, we consider a slightly more general case, in which Δp_y is held constant but $\neq 0$, and study the correlation as a function of $\Delta p = \sqrt{\Delta p_L^2 + \Delta p_z^2}$. Among several cases examined here, this is the closest to the average experimental situation[10]. We show in Fig.6 some of the results. What calls our attention there is the fact that the curve does not start at the maximum-correlation point for chaotic sources but, reflecting our choice $\Delta p_y \neq 0$, it is lower. If the decay time τ_d were slightly larger as expected from the finiteness

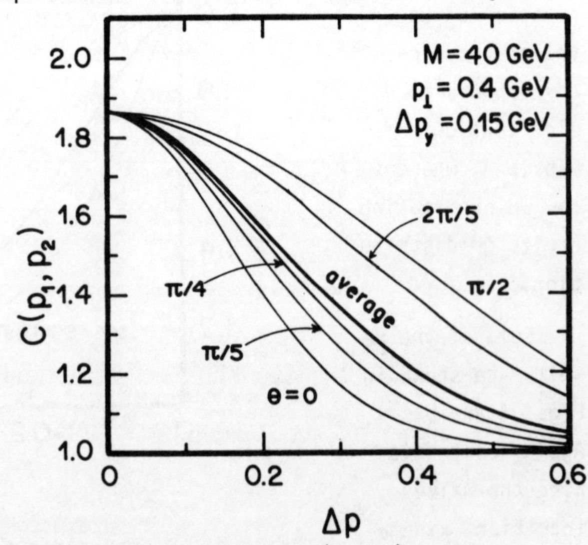

Fig.6 - Θ dependence of $C(p_1,p_2)$, with $p_L=0$, $\Delta E\neq 0$ and fixed Δp_y

of the transition-time interval, this curve could even be lower and also narrower in accordance with the existent data. Usually, the partial coherence of the source is invoked to explain the above mentioned feature of the data, but in our opinion a totally chaotic source is not excluded as far as the correlation data are concerned.

In the present paper, we have presented a preliminary account of a study of the identical-particle correlation, in which particles are emitted from a chaotic and rapidly expanding source. Several of intuitively expected results have been obtained such as the stretching of the longitudinal dimension with M, together with the constant transverse dimension, and some less intuitive ones have emerged, resulting from the kinematics of a rapidly-expanding source. Those related with the depth of the source are not obvious at all. Alghough our curves are not very far from the experimental points, nevertheless we feel that a more complete study, especially regarding the phase transition and so a better estimate of τ_d is needed to achieve a quantitative comparison with the data.

One of us (S.S.P.) acknowledge the Fundação de Amparo ã Pesquisa do Estado de São Paulo (FAPESP), Brazil for the scholarship awarded to her during the present work.

REFERENCES

1) Hanbury-Brwon, R. and Twiss, R.Q., Phil.Mag. 45,663(1954); Nature 178,1046(1956).
2) Goldhaber, G., Goldhaber, S., Lee, W. and Pais, A., Phys.Rev. 120, 300(1960).
3) Gyulassy, M., Kauffmann, S.K. and Wilson, L.W., Phys.Rev. C20,2267 (1979).
4) Kopilov, G.I. and Podgoretzky, M.I., Sov.J.Nucl.Phys. 14,1084(1971); 15,392(1971); 18,336(1974).
Kopilov, G.I., Phys.Lett. 50B,472(1974).
Podgoretzky, M.I., Sov.J.Nucl.Phys. 37,272(1983).

5) Shuryak, E.V. Phys.Lett. 44B,387(1973).
6) Cocconi, G., Phys.Lett. 49B,459(1974).
7) Fowler, G.N. and Weiner, R.M., Phys.Lett. 70B,201(1977).
8) Landau, L.D., Izv.Akad.Nank SSSR Ser. Fiz. 17,51(1953); Collected papers, ed. D.Ter Haar (Pergamon, Oxford, 1965) p.569.
9) Biswas, N.N. et al., Phys.Rev.Lett. 37,175(1976).
ABBCCLVW Collaboration, Detuschmann, M. et al., Nucl.Phys. 204B, 333(1982).
Angelov, N. et al., Sov.J.Nucl.Phys. 35,45(1982).
10) Axial-Field-Spectrometer Collab., Åkesson, T. et al., Phys.Lett. 129B,269(1983); 155B,128(1985).
11) Hama, Y., "Transverse Expansion in Hydrodynamical Models", Proceedings of LESIP-I, eds. Scott, D.K. and Weiner, R.M. (World Scientific, Singapore, 1985), p.60.
Hama, Y. and Pottag, F.W., "Energy Dependence of the Transverse Expansion in Hydrodynamical Models", submitted for publication on Nucl.Phys.B.
12) Khalatnikov, I.M., Zhur Eksp.Teor.Fiz. 27,529(1954).
13) Hama, Y., Phys.Rev. D19,2623(1979);
Hama, Y. and Pottag, F.W., Rev.Bras.Fis. 12,247(1982);
Hama, Y. and Navarra, F.S., Phys.Lett. 129B,251(1983); Z.Phys. C26,465(1984).
14) Hama, Y. and Pottag, F.W., "Diffractive Excitation of the Quark-Gluon Plasma", presented to this Workshop.
15) UA4 Collaboration, Bernard, D. et al., Phys.Lett. 166B,459(1986).

HADRONIC PAIR-PRODUCTION IN HIGH ENERGY COLLISIONS AS A CONSEQUENCE OF SUPERFLUORESCENCE

F.W. Pottag and R.M. Weiner

Fachbereich Physik, Universität Marburg

ABSTRACT

We discuss hadronic pair-production from quark-gluon plasma which is likely to be formed in high energy collisions, as a consequence of a superfluorescent decay of the primary fireball.

The observation of hadronic pairs[1] in high energy cosmic ray physics is an interesting phenomenon which has not gotten yet a satisfactory explanation. We discuss here the implications of the hypothesis that (at least some of) these rare events are due to an early decay of the primary fireball due to superfluorescence following the ideas of a previous paper[2] in which the Centauro events have been interpreted in the same way. In this note we will not discuss the mechanism of superfluorescence[3] but simply assume that it exists. In what follows we point out the main features of this model.

We assume that after the collision (we restrict our study to pp-collisions), seen in the center-of-mass frame, matter is concentrated in a very small Lorentz contracted volume V in which the kinetic energy is converted into thermal energy distributed over quarks and gluons. We have

$$V = \frac{V_0}{\gamma} \; ; \quad V_0 = \frac{4}{3}\pi r_0^3 \; , \; r_0 = 0.7 \text{ fm} \; , \; \gamma = \frac{W}{4m_p} \tag{1}$$

γ is the Lorentz contraction factor and W is the center-of-mass energy. W is given by

$$W = g_q \sum \frac{\varepsilon_i}{e^{\beta\varepsilon_i}+1} + g_g \sum \frac{\varepsilon_j}{e^{\beta\varepsilon_j}-1} \; , \tag{2}$$

where $\beta = 1/T$, T is the temperature, g_q and g_g are the total number of degrees of freedom of quarks and gluons respectively. Considering three flavors for the quarks, we have:

$$g_q = 36 \quad \text{and} \quad g_g = 16. \tag{3}$$

The sums in equation (2) are carried out explicitly over all permitted states in a box of volume V. In this case $\varepsilon = \sqrt{m^2+p^2} \simeq p$ and each component of the momentum is given by $p_i = 2\pi n/L_i$; $n = 0, \pm 1, \pm 2, \ldots$, where L_i is the length of the box in the direction $i = x, y, z$. Then $V = L_x L_y L_z$. Since V is very small for high energies (because of the Lorentz contraction) we can not make the semiclassical approximation:

$$\sum_i = \frac{1}{(2\pi)^3} \int d^3p \, V \tag{4}$$

This was also observed by Nowakowski and Cooper[4]. For a given energy W we obtain V from equation (1) and substituting into equation (2) we get the temperature T of the system. Now we assume that in some rare events the fireball decays immediately at this temperature as a consequence of superfluorescence without cooling down to some dissociation temperature as considered in hydrodynamical models[5].

The number of hadrons for each species is then governed by the statistical function:

$$n_i = g_i \sum_j \frac{1}{e^{\beta\varepsilon_j} \pm 1} , \qquad \varepsilon_j = \sqrt{M_i^2 + p_j^2} \tag{5}$$

where g_i is the statistical weight factor corresponding to species i and M_i the respective mass. Again, the sum is taken over all permitted energy states within the box of volume V. From the form of this equation it follows that the higher the temperature, the more important is the contribution of large masses to the total spectrum of produced particles and resonances. For practical purposes we are interested in the center-of-mass energy near 200 Gev which corresponds to an initial temperature around 1.4 GeV as can be seen in Table I. At these temperatures the contribution of e.g. ρ-resonances is not negligible and becomes more and more important in comparison to direct π-production with increasing temperature. In Table I we give the ratio f of the number of charged pairs $\pi^+\pi^-$, which are decay products of the ρ^0-resonances, to the sum of the number of charged pions coming directly from the fireball and the number of charged pions arising as decay products of the ρ-resonance. Of course, pions can originate also from other resonances, but because of conservation laws their weight is not very important as long as the temperature is not too high.

TABLE I

W (GeV)	T (GeV)	f
0.604	0.1	0.043
2.91	0.2	0.195
13.4	0.4	0.252
32.1	0.6	0.263
59.1	0.8	0.267
94.6	1.0	0.269
138	1.2	0.270
191	1.4	0.271
251	1.6	0.271
320	1.8	0.271
389	2.0	0.272

From this model it follows that if in a few high energy events the system decays immediately after its formation, an excess of number of hadronic pairs will be seen. Also because of the high temperature at which this occurs, the average momentum of the decay products will be much higher than in the "normal" events.

This work was supported in part by the Conselho Nacional de Desenvolvimento Científico e Tecnológico - CNPq, Brazil, and by the Deutsche Forschungsgemeinschaft.

REFERENCES

1) T.H. Burnett et al., the JACEE Collaboration, talk presented by Y. Takahashi to the 19th International Cosmic Ray Conference, San Diego, August/1985.
Y. Takahashi, Invited Review Talk, given at the International Conference "Physics in Collisions 5", Autun, France, July/1985.
2) G.N. Fowler, E.M. Friedlander, and R.M. Weiner, Phys. Lett. 116B,203 (1982).
3) R. Bonifacio and L.A. Lugiato, Phys. Rev. A11, 1507 (1975).
F. Haake, H. King, G. Schröder, J. Haus and R. Glauber, Phys. Rev. 20A, 2047 (1979).
4) J. Nowakowski and F. Cooper, Phys. Rev. D9, 771 (1974).
5) L.D. Landau, Izv. Akad. Nauk SSSR, Ser. Fiz. 17, 51 (1953). Collected Papers, ed. D. Ter Haar (Pergamon, Oxford, 1965) p. 569.

HADRONIC MULTIPARTICLE PRODUCTION

MULTIPLICITY DISTRIBUTIONS IN e^+e^- ANNIHILATION AT 29 GeV

M. Derrick
Argonne National Laboratory
High Energy Physics Division
Argonne, IL 60439

ABSTRACT

This paper presents the charged particle multiplicity distributions for e^+e^- annihilation at $\sqrt{s}$ = 29 GeV measured in the High Resolution Spectrometer. The data, which corresponds to an integrated luminosity of 185 pb^{-1}, were obtained at the e^+e^- storage ring PEP. The multiplicity distribution of the charged particles has a mean value $\langle n \rangle = 12.87 \pm 0.03 \pm 0.30$, a dispersion $D_2 = 3.67 \pm 0.02 \pm 0.18$, and an f_2 momentum of $0.60 \pm 0.02 \pm 0.18$. Results are also presented for a two-jet sample selected with low sphericity and aplanarity. The charged particle distributions are almost Poissonian and narrower than results reported by other e^+e^- experiments in this energy range. The mean multiplicity increases with the event sphericity and for the sample of three-fold-symmetric three-jet events, a value of $\langle n \rangle = 16.3 \pm 0.3 \pm 0.7$ is found. No correlation is observed between the multiplicities in the two hemispheres when the events are divided into two jets by a plane perpendicular to the thrust axis. For the single jets, a mean multiplicity of $\langle n \rangle = 6.43 \pm 0.02 \pm 0.15$ and a dispersion value of $D_2 = 2.55 \pm 0.02 \pm 0.13$ are found. These values give further support to the idea of independent jet fragmentation.

The multiplicity distributions are well represented by the negative binomial distribution. The multiplicity distributions are also studied in restrictive rapidity ranges. As the rapidity span is narrowed, the distributions become broader and approach a constant value of the parameter k.

INTRODUCTION

The measurement of multihadron production in e^+e^- annihilation is the simplest way to study the fragmentation of partons into hadrons since the parton level processes are particularly simple. The center-of-mass energy of the hadronic system is clearly defined, and all of the incident energy contributes to the creation of particles, except for the well-known initial state radiation.

At a phenomenological level, comparisons with other reactions, soft or hard, may reveal which features depend on constraints such as longitudinal phase space and which are intrinsic to simple parton-hadron transitions. In contrast to e^+e^- annihilation, which is a simple process at the parton level, soft hadronic interactions are thought to represent a collective process yielding a forward and backward cone dominated by one leading particle, usually with the same quantum numbers as the initial beam or target, accompanied by a large number of soft particles produced in the collision. Thus, a priori, the mechanism of hadronization in the two reactions is expected to be quite different. The final state in the e^+e^- annihilation process, which results from quark fragmentation, is prototypical, and comparison with the hadronic data may be revealing the characteristics of the partonic process in such reactions.

The results presented are based on data taken with the High Resolution Spectrometer (HRS) operated at the e^+e^- storage ring (PEP) at c.m. energy of 29 GeV. The data sample corresponds to a total integrated luminosity of $(185 \pm 5)\ pb^{-1}$.

EXPERIMENTAL DETAILS

The HRS is a solenoidal spectrometer[1] that measures charged particles and electromagnetic energy over 90% of the solid angle. The tracking system consists of a vertex chamber, a central drift chamber, and an outer drift chamber. The central drift chamber has 15 layers of cylindrical drift planes, eight of which have stereo wires ($\pm$60 mrad) in order to measure the position along the e^+e^- beam direction. The momentum of a charged particle in the 1.62 T magnetic field is measured with a resolution of 3% at 14.5 GeV. The minimum momentum for detecting tracks with good efficiency is about 200 MeV/c. The 40-module barrel shower counter system provides electromagnetic calorimetry over 62% of the solid angle with energy resolution of $\sigma_E/E = 0.16/\sqrt{E\ (\mathrm{GeV})}$.

The beam pipe and the inner wall of the central drift chamber are made of beryllium so as to minimize the conversion of photons into electron-positron pairs; the total material between the interaction point and the central drift chamber is less than 0.02 radiation lengths.

To ensure good tracking efficiency, the thrust axis of the event was selected to be within 30° of the equatorial plane of the detector, and each track had to have an angle with respect to the e^+e^- beam direction of more than 24° and had to register in more than one-half of the drift chamber layers traversed. With these selections, the reconstruction efficiency for isolated tracks was greater than 99%. For a typical annihilation event, with several close neighboring tracks, the reconstruction efficiency was 80% or better. The data sample contains 29649 events.

In addition to the inclusive data set, a sample of two-jet events was selected[2] with sphericity less than 0.25 and an aplanarity less than 0.10. These cuts, which exclude the events with hard gluon radiation, gave a data sample of 24,553 collimated and planar events.

The true multiplicity distribution was determined from the observed data by means of a matrix unfolding technique. If N_m^0 is the number of events observed with m tracks and N_n^T is the true number of events with n tracks (n even), we define M_{nm} such that

$$N_n^T = \sum_m M_{nm} N_m^0 \quad . \tag{1}$$

The matrix M_{nm} was determined from a Monte Carlo (MC) simulation of the experiment which includes the effects of the experimental cuts as well as the tracking inefficiencies.

In the initial data selection, events with $m \leqslant 5$ were removed in order to exclude tau pair events. The numbers of events with n = 0, 2 and 4 charged particles were estimated from the data themselves assuming independent fragmentation of the two jets in the event.[3]

RESULTS

The analyses gave $\langle n \rangle = 12.87 \pm 0.03 \pm 0.30$, where the first error is statistical and the second systematic. The charged particles from heavy meson (K_s^o, D, F, ...) and baryon (Λ, Σ, Λ_c, ...) decays are included in these values. Our measured multiplicities[4] of $\langle n \rangle_{K^o} = 1.44 \pm 0.05 \pm 0.09$ and $\langle n \rangle_\Lambda = 0.220 \pm 0.007 \pm 0.022$ reduces $\langle n \rangle$ to $11.34 \pm 0.03 \pm 0.30$ if the charged particles from these decays are omitted.

The $\langle n \rangle$ value of $12.53 \pm 0.03 \pm 0.30$, measured for the data sample with a two-jet selection, suggests that events with a hard gluon emission have a higher multiplicity. This is shown in more detail in Fig. 1, which shows the variation of $\langle n \rangle$ with event sphericity (S).

The sharp rise at low sphericity comes about because events with low prong numbers give low S values when the sphericity is defined from the charged particles alone. The slower rise, amounting to about 20% in the higher sphericity region, is the effect of hard gluon radiation. The full line, which is in qualitative agreement with the data, shows the prediction of the Lund Monte Carlo program. If hard gluon emission is suppressed in the MC event generation, then no such rise is predicted, as seen by the dashed line. The events at high sphericity are predominantly of the three-jet topology.

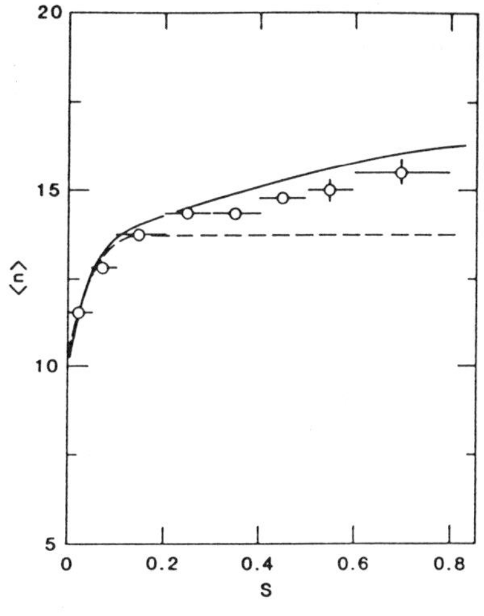

Fig. 1 Mean charged multiplicity as a function of event sphericity, S, for e^+e^- annihilation at 29 GeV. The lines show the Lund Monte Carlo prediction for the complete event sample (full line) and for the events with the hard gluon radiation removed (dashed line).

The rise of $< n >$ with sphericity for $S < 0.2$ agrees with the measurements of the PLUTO collaboration.[5] The increase of $< n >$ for $S > 0.2$, due to hard gluon radiation, was not previously observed.

In a separate study[6] we have isolated a sample of three-fold symmetric three-jet events in which the angle between each pair of jets in the event is between 100° and 140°. This sample, which contains 276 events, has $< n > = 16.3 \pm 0.3 \pm 0.7$ and $D_2 = 4.2 \pm 0.5 \pm 0.3$. If the events result from the symmetric $q\bar{q}g$ partonic state and the gluon splits into a $q'\bar{q}'$ pair then, in the Lund string picture, the mean charged multiplicity should just be given by twice that of the two strings connecting the $q\bar{q}'$ and $q'\bar{q}'$ quarks whose W values can be calculated. Such a model predicts $< n > = 16.6$ in good agreement with the measurement.

These multiplicity distributions are shown in Fig. 2 for the unselected data set (Fig. 2(a)) and for the data with a two-jet selection (Fig. 2(b)). The curves join points calculated from twice the Poisson distribution calculated for the average values, $< n > = 12.87$ and $< n > = 12.53$, respectively. The data are remarkably close to this simple shape, which could be expected for the random emission of single particles. This is certainly an oversimplification since it is known that resonances and heavy quark states are important components in the final hadronic system, and so many of the observed tracks are secondary decay products of such systems.

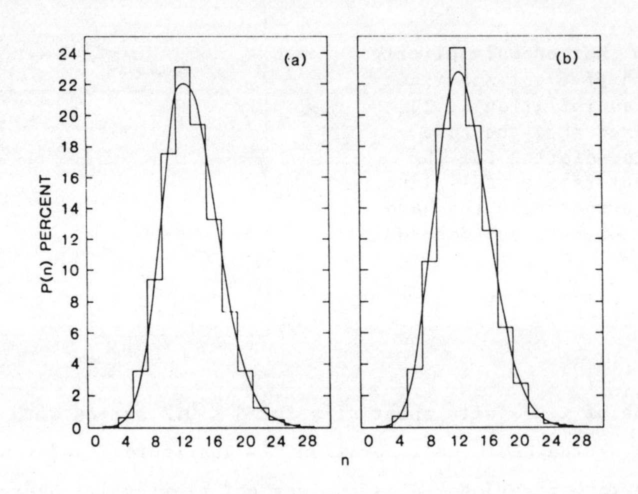

Fig. 2. Charged particle multiplicity measurements (histogram) compared to the line that connects points on a Poisson distribution with the same mean value: (a) inclusive data sample, (b) two-jet data sample.

The properties of the multiplicity distributions are listed in Table I. The f_2 moments are small; positive for the inclusive data and negative for the two-jet sample, consistent with the small deviations seen between the data histograms and the Poisson curve in Fig. 2. The value of the dispersion, D_2, of $3.67 \pm 0.02 \pm 0.18$ is lower than the values previously reported for inclusive e^+e^- annihilation, although not outside of the range defined by the systematic uncertainties of the other experiments.

A more detailed shape comparison, using the KNO form, is given in Fig. 3. If $z = n/< n >$, then $\psi(z)$ is defined by

$\psi(z) = \langle n \rangle \frac{\sigma_n}{\Sigma\sigma_n}$, where σ_n is the cross section for producing n charged particles. The present measurements give a narrower KNO distribution than measured by the TASSO experiment[7] at W values of 14 GeV, 22 GeV, and 34 GeV. The errors shown are statistical.

Table I

Properties of the Multiplicity Distributions

Variable	Inclusive Sample		Two-Jet Sample	
	Whole Event	Single Jet	Whole Event	Single Jet
$\langle n \rangle$	12.87 ± 0.03 ± 0.30	6.43 ± 0.02 ± 0.15	12.53 ± 0.03 ± 0.30	6.26 ± 0.02 ± 0.15
D_2	3.67 ± 0.02 ± 0.18	2.55 ± 0.02 ± 0.13	3.48 ± 0.02 ± 0.17	2.45 ± 0.02 ± 0.12
f_2	0.60 ± 0.02 ± 0.18	-0.07 ± 0.02 ± 0.13	-0.42 ± 0.02 ± 0.17	-0.26 ± 0.03 ± 0.13
$\langle n \rangle/D_2$	3.51 ± 0.18	2.52 ± 0.13	3.60 ± 0.18	2.56 ± 0.13
K GD fit	12.3	7.11	13.3	7.49
1/k NB fit	$3.9 \cdot 10^{-3}$	$1.4 \cdot 10^{-4}$	$4.3 \cdot 10^{-5}$	$1.4 \cdot 10^{-4}$

The HRS data have been fitted to the negative binomial distribution:

$$P(n, \langle n \rangle, k) = \frac{k(k+1)\ \ldots(k+n-1)}{n!} \left(\frac{\langle n \rangle/k}{1+\langle n \rangle/k}\right)^n \left(1+\frac{\langle n \rangle}{k}\right)^k \quad . \tag{2}$$

In this parameterization, the shape of the distribution is determined by the parameter k and the position of the maximum by $\langle n \rangle$. The resulting k values are given in Table I. The variable k is related to the mean multiplicity $\langle n \rangle$ and the dispersion D by

$$\frac{D^2}{\langle n \rangle^2} = \frac{1}{\langle n \rangle} + \frac{1}{k} \quad . \tag{3}$$

Since the HRS data give an almost Poissonian distribution (for which $k = \infty$), the fitted k values are large. The values of the C_q moments for the NB fits are in good agreement with those directly measured.

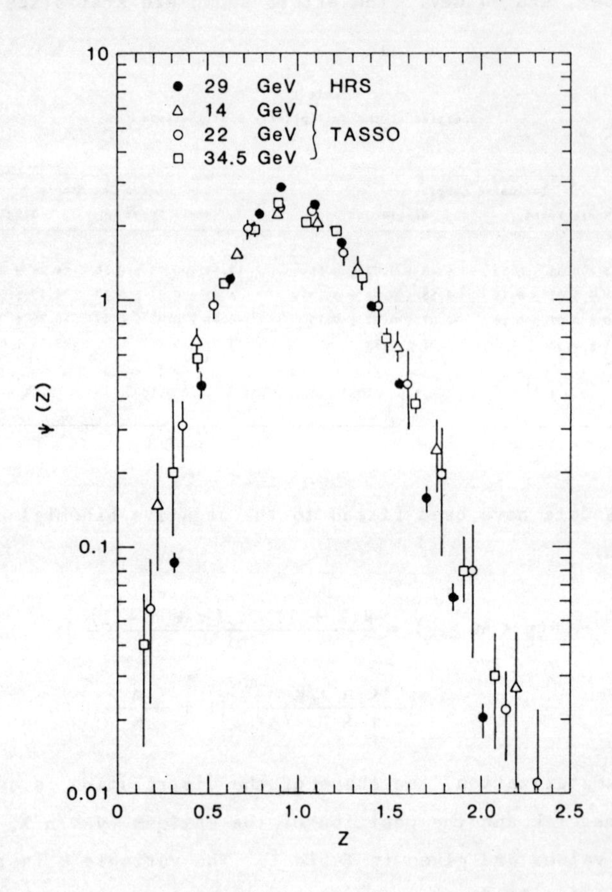

Fig. 3. Multiplicity distribution for e^+e^- annihilation expressed in KNO form. The TASSO data is from Ref. 7.

The data have also been fitted to the gamma distribution, which in KNO form is:

$$\psi(z) = \frac{K^K}{(K - 1)!} z^{K-1} e^{-Kz} \quad . \tag{4}$$

The GD has only one free parameter and the fits, which are less good than the NB, give smaller K values that satisfy the approximate relationship $\frac{1}{K} \approx \frac{1}{k} + \frac{1}{\langle n \rangle}$. Figures 4 and 5 shows the HRS data, together with the best fit curves to the negative binomial and the gamma distribution. In these figures, the errors plotted are the quadratic sum of the statistical and systematic uncertainties.

The multiplicity distributions have also been studied as a function of the rapidity span of the tracks.[8] The KNO distributions for the two-jet data sample and for particles contained in selected rapidity ranges from $|Y| < 0.1$ to $|Y| < 2.5$, corresponding to Y spans from 0.2 to 5.0 units, are shown in Fig. 6. Each successive data set has been displaced lower by a factor of ten for clarity. The two-jet data with no Y selection are also shown with their own ordinate scale. These events are always even prongs so the normalization differs by a factor of two from that of the data with rapidity selection. The distributions clearly widen as the rapidity span is restricted, and for $Y \lesssim 1$ events with z values of 3 to 5 are seen. The results of these fits are given in Table II (statistical errors).

Table II
Fits to the Negative Binomial

Rapidity Range abs(y)	Two-Jet Data k	Two-Jet Data ⟨ n ⟩	Inclusive Data k	Inclusive Data ⟨ n ⟩
< 2.5	57.18 ± 2.62	10.33 ± 0.02	38.50 ± 0.86	11.38 ± 0.01
< 2.0	26.18 ± 0.66	8.97 ± 0.02	17.45 ± 0.36	9.70 ± 0.02
< 1.5	15.95 ± 0.41	6.65 ± 0.02	10.57 ± 0.15	7.39 ± 0.01
< 1.0	10.56 ± 0.27	4.27 ± 0.01	6.99 ± 0.05	4.82 ± 0.02
< 0.5	7.32 ± 0.30	2.12 ± 0.01	5.53 ± 0.19	2.35 ± 0.01
< 0.25	6.14 ± 0.20	1.06 ± 0.01		
< 0.1	4.85 ± 0.50	0.403 ± 0.005	4.42 ± 0.29	0.427 ± 0.002

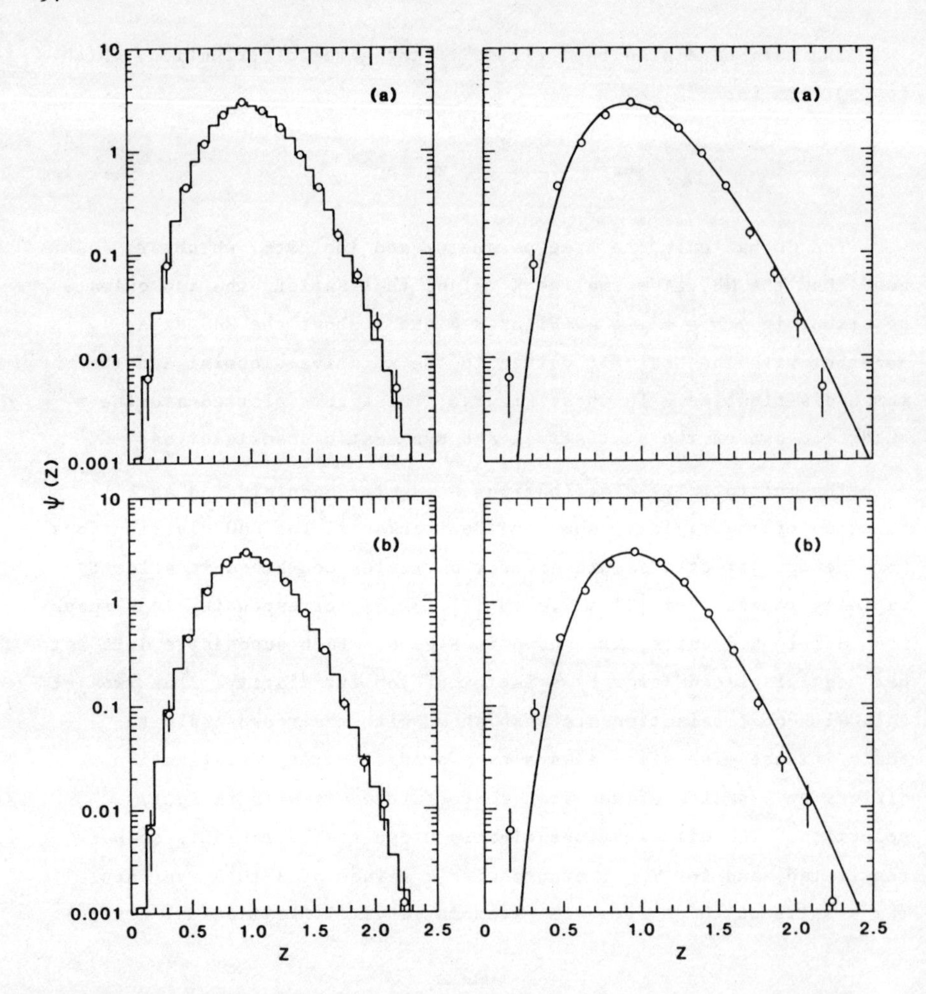

Fig. 4. Charged particle multiplicity distributions for e^+e^- annihilation at 29 GeV compared to the negative binomial distribution: (a) inclusive data set, (b) two-jet data set.

Fig. 5. Charged particle multiplicity distributions for e^+e^- annihilation at 29 GeV compared to the gamma distribution: (a) inclusive data set, (b) two-jet data set.

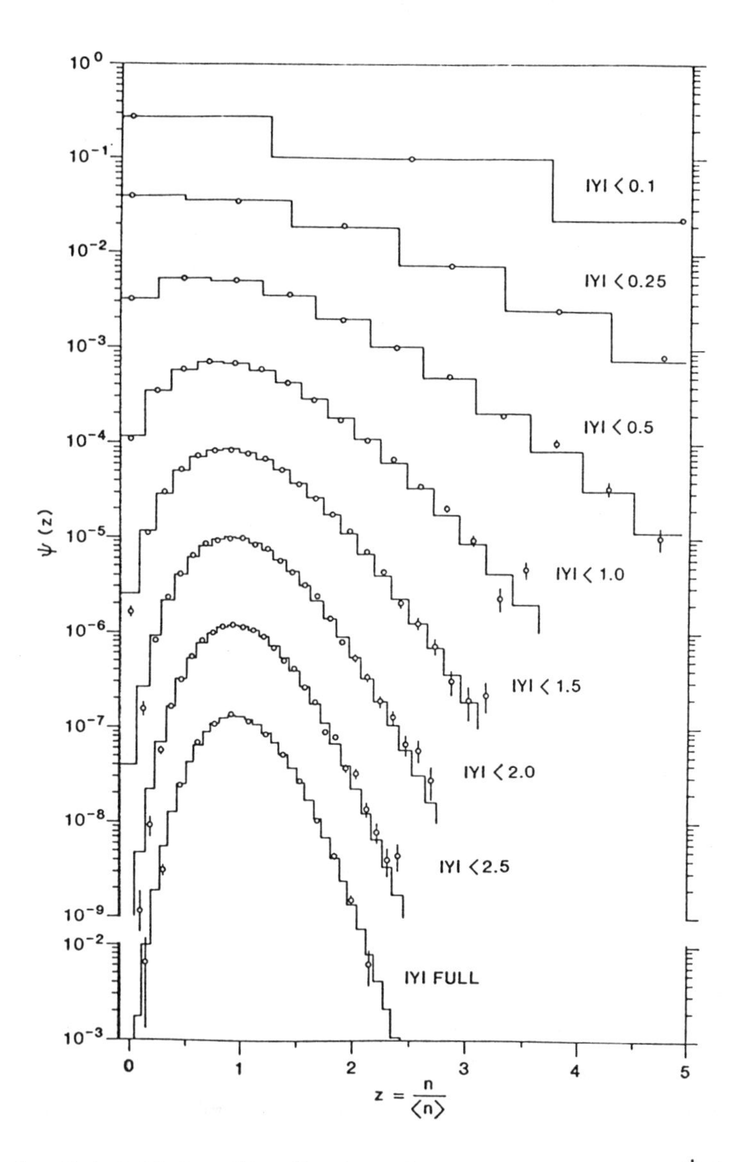

Fig. 6. Multiplicity distributions for two-jet events of e^+e^- annihilation at 29 GeV as a function of the rapidity span selection. The histograms show the best fit to the negative binomial of Eq. (2) expressed in KNO form. Each of the selected distributions has been shifted down by a factor of ten relative to the $|Y| < 0.1$ data. The ordinate scale for the data set with no rapidity selection is shown separately.

The multiplicity distributions for the data in the outer regions of the rapidity span has also been studied. In all cases, the distribution is narrow, even when the Y range selected is small.

Figure 7 shows the inclusive data with $|Y| < 1$ compared to the NB (histogram) and the GD with $K = 3$ (dashed line). In general, the data for $|Y| < 1$ are not well represented by the GD if the events with no tracks in the selected rapidity range are included. If these zero-prong events are discarded, then the gamma distribution provides good fits down to $|Y| < 0.5$ with $K = 2.5$ for this selection.

Figure 8(a) and 8(b) show the single-jet multiplicity distributions for the inclusive and two-jet data samples. The curves again show the Poisson distribution, calculated for the corresponding mean values. As was the case for the prong distributions for the whole event, the simple Poisson shapes give a good representation of the measurements. If the multiplicity distribution for the complete events is an incoherent sum of the individual Poissonian jets, then the resulting distribution will be almost Poissonian, as is observed.

In Fig. 9, the single-jet KNO distribution for the inclusive data set is compared with results from the TASSO experiment[7] at W values of 14 GeV, 22 GeV and 34 GeV. The agreement among the data points is quite good so that approximate KNO scaling also holds for the single-jet multiplicities.

The single jet multiplicity distributions in KNO form are compared to the NB fits in Fig. 10 and to the GD fits in Fig. 11. The errors shown are the quadratic sum of the statistical and systematic uncertainties. Again the NB form represents the data very well, whereas the GD fits are less good. The resulting k (K) values are given in Table I.

The comparison of the widths of the single jet and complete event multiplicity distributions support the idea of independent jet fragmentation. This can be directly checked by measuring the multiplicity correlation between the two-jet hemispheres.[9]

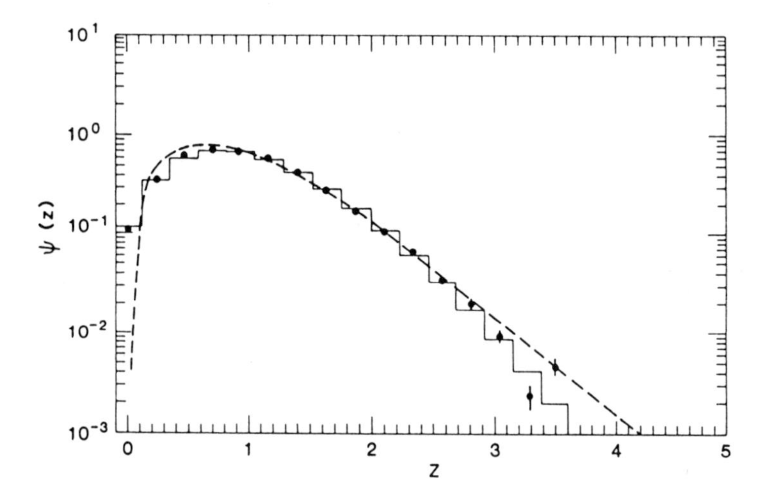

Fig. 7. Multiplicity distribution in KNO form for the inclusive data sample with $|Y| < 1.0$. The histogram shows the best fit of the data to the negative binomial; the dashed line gives the gamma distribution with $K = 3$.

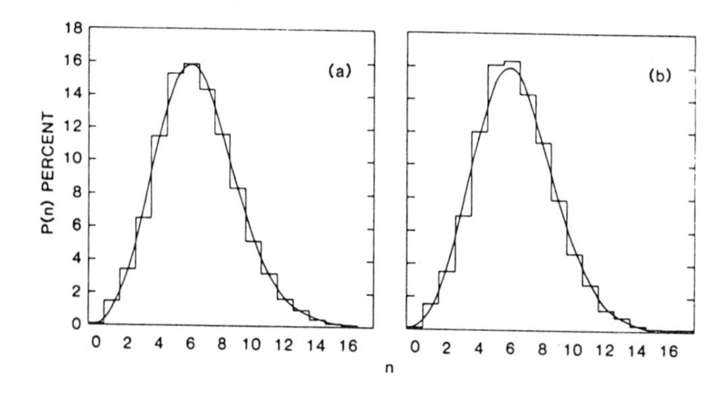

Fig. 8. Charged particle multiplicity measurements (histogram) for single jets compared to the line that connects points on a Poisson distribution with the same mean value for (a) inclusive data sample and (b) two-jet data sample.

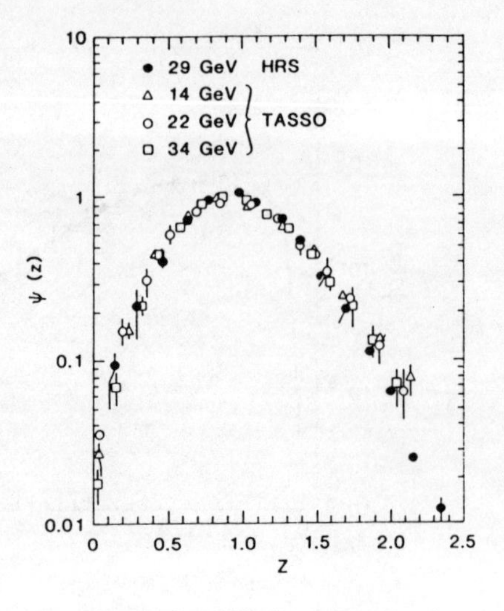

Fig. 9. Single jet multiplicity distribution for the inclusive event sample in KNO form compared to the measurements of the TASSO collaboration.

Figure 12 shows the distribution of the mean multiplicity in one jet (B) compared to the number of charged particles in the second jet (F) for different rapidity selections. There is clearly no correlation between the multiplicities in the two hemispheres. While these results are corrected for inefficiencies and detector smearing effects, they are not corrected for the biases at low multiplicities introduced by the selection criteria ($m \geq 5$), since this correction was made assuming no jet-jet multiplicity correlation, an assertion that we are now testing. This cut is responsible for the small correlation effect observed for low multiplicities in the F hemisphere. The line shows the results of a simple calculation using the measured $\langle n \rangle$ values for (u,d,s), c and b quarks[10], as well as the effect of the cut at $m < 5$.

The analysis has been repeated (Fig. 12(b)) with the central rapidity region removed $|Y| < 1$). This selection is made because of a possible bias coming from particles near $Y = 0$, whose jet assignment could be affected by the uncertainties of the thrust axis direction. Such particles are often of low momentum and so are not reconstructed by the spectrometer with full efficiency. In addition, this selection removes any correlation that could result from the decay of a centrally-produced heavy cluster whose decay products

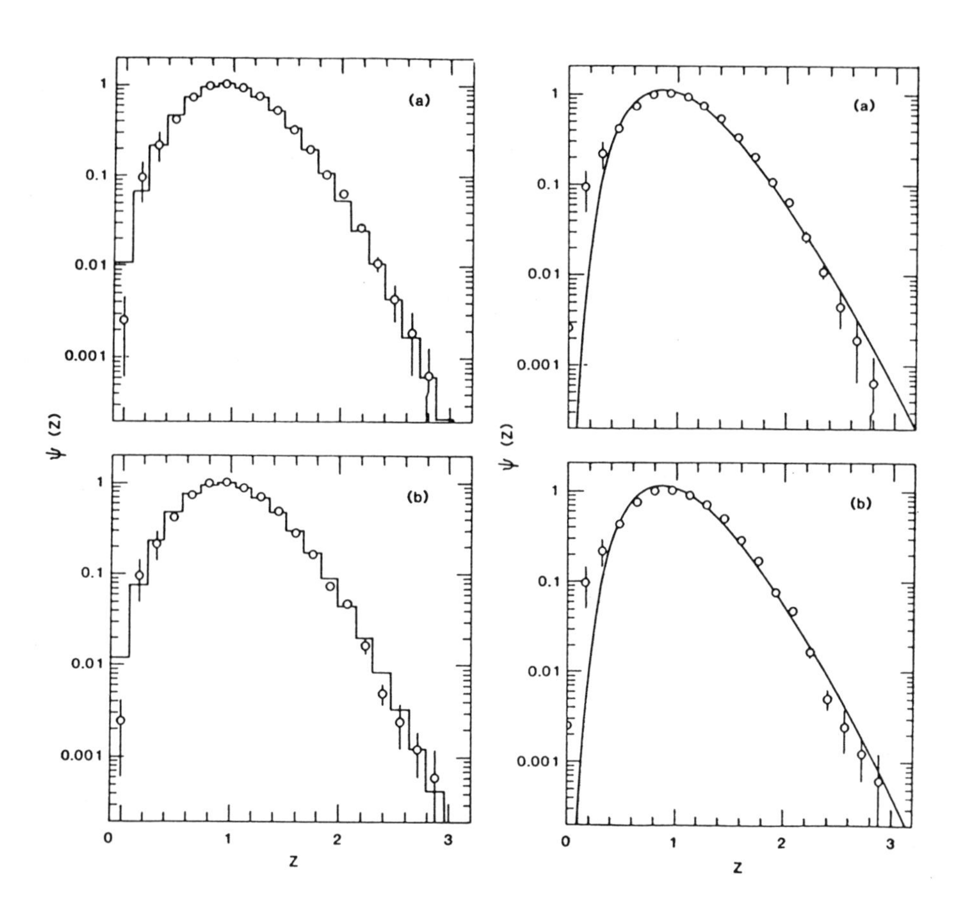

Fig. 10. Charged particle multiplicity distributions for single jets in e^+e^- annihilation at 29 GeV compared to the negative binomial distribution: (a) inclusive data set, (b) two-jet data set.

Fig. 11. Charged particle multiplicity distributions for single jets in e^+e^- annihilation at 29 GeV at 29 GeV compared to the gamma distribution: (a) inclusive data set, (b) two-jet data set.

could fall in both hemispheres, and so give an F:B correlation. Figure 12(c) shows the same correlation for the central region of rapidity.

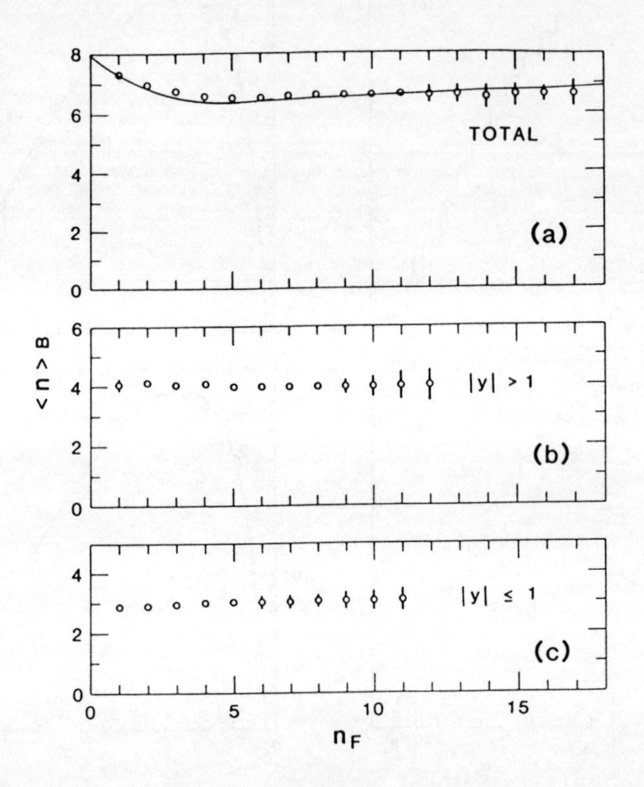

Fig. 12. Forward:backward multiplicity correlation for (a) full rapidity span, (b) tracks with $|Y| < 1.0$ removed, (c) tracks in the region $|Y| < 1.0$.

If the data of Fig. 12 are fit to the linear relationship:

$$< n_F > = a + b\, n_B \quad , \tag{5}$$

typical values of $b = 0.001 \pm 0.001$ are obtained. There is a simple relationship[11] between b and the normalized dispersions of the single jet compared to that for the complete event:

$$\mathbf{D}_2 = \frac{D_2}{\sqrt{2(1+b)}} \quad . \tag{6}$$

The values for the single-jet dispersions ($\mathbf{D}_2$) and those for the whole event (D_2), given in Table I, yield b values of 0.04 $\pm$ 0.02 $\pm$ 0.10 for the inclusive sample and 0.01 $\pm$ 0.2 $\pm$ 0.10 for the two-jet sample, in agreement with the direct measurement.

A further check of the idea that the e^+e^- results show independent emission of single particles may be made by looking at the F:B split for a fixed n value. Specifically, this should be binomially distributed:

$$F_n(\mathbf{n}_B) = \frac{n!}{\mathbf{n}_B!\,\mathbf{n}_F!} \left(\frac{1}{2}\right)^{\mathbf{n}_B} \left(\frac{1}{2}\right)^{\mathbf{n}_F} \quad . \tag{7}$$

In Fig. 13, the F:B multiplicity correlation is shown for several n values, together with the predictions from Eq. (7). Although the data of Fig. 13 is uncorrected, the agreement with the binomial is good.

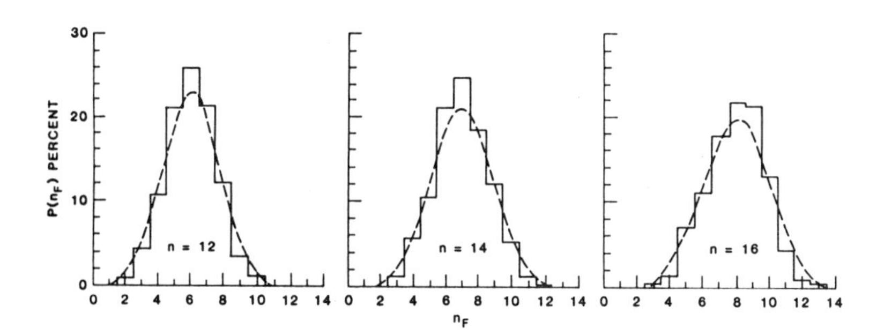

Fig. 13. Forward:backward multiplicity correlations at fixed n compared to the binomial distribution.

DISCUSSION

Many explanations of the origin of the NB and GD have been given.[12] The NB, defined for integral values of n, can be derived assuming that the final state particles obey Bose-Einstein statistics. According to these ideas, k can be thought of as the number of identical emitters or cells in phase space. An alternative interpretation[13] is that the emission of more particles is stimulated by the bosons already present in a given cell, and 1/k represents the average fraction of existing particles that stimulate the emission of additional particles. Expressions giving similar KNO shapes can also be derived from a generating function that describes QCD branching processes at the parton level.[14] The one-parameter GD also gives a reasonable representation of the results, although it falls below the data at the lowest n values. There is, at present, no fundamental understanding of these parametrizations, and so neither expression can be considered to be the more fundamental, although, since the multiplicities are by their very nature discrete, the NB is perhaps more natural.

Giovannini and Van Hove[13] also show the multiplicity distributions, when interpreted with the NB distribution, can be used to estimate the mean multiplicity of intermediate clusters, $\langle n \rangle_c$, assuming that the charged particles result from cluster decay. The result is

$$\langle n \rangle_c = \frac{B}{B-1} \frac{1}{\ln(1-B)} \quad ,$$

where $B = \frac{\langle n \rangle}{\langle n \rangle + k}$. The fits to the HRS data given in Table II result in a cluster multiplicity close to one in agreement with the idea of independent emission of single particles.

Multiplicity distributions have been measured for a number of hadronic interactions and over a large energy range.[15] The pp results are clearly wider than the fits to the (νp, $\bar{\nu} p$) and the $\bar{p}p$ annihilation data. However, our e^+e^- data give a distribution that is still narrower.

These qualitative observations have been interpreted in an appealing geometrical picture[16] in which hadronic reactions are considered to be a superposition of many collisions with different impact parameters. The central collisions give higher multiplicities and a narrow KNO distribution, more akin to the e^+e^- results, whereas the more peripheral collisions populate the lower multiplicities. The $\bar{p}p$ annihilation process is predominantly central and so has a distribution closer to the e^+e^- results. The lepton scattering reactions are point-like, but they give rise to a final state with a quark and a diquark. Since diquark fragmentation gives a higher $\langle n \rangle$ than for the quark, this leads to a result intermediate between the e^+e^- and pp data.

A somewhat related discussions has been given by Chou and Yang[17] who consider that the annihilation process in a given angular momentum state is stochastic, which leads to a Poissonian multiplicity distribution. Since only $\ell = 0$ and $\ell = 1$ is allowed for e^+e^- annihilation through a one-photon intermediate state, the e^+e^- multiplicity distribution should be simple, whereas for the highest energy hadronic collisions ℓ values up to several thousand are possible. These ideas, which are consistent with our data, predict that KNO scaling will not hold for e^+e^- annihilation. These authors specifically consider e^+e^- annihilation leading to two jets, but as is clear from the present results, the differences between the inclusive sample and the two-jet data sample are small. Data of similar precision to the present measurements, but at higher energies, are needed to check this suggestion.

An alternative interpretation has been given by Carruthers and Shih[18] in which the multiplicity distribution is considered to arise from emission by a small number of (k) of cells. As we have discussed, such a picture leads to the negative binomial expression. If there is a coherent component to the emissions, then the KNO distribution can be narrow, whereas an incoherent sum of emitters leads to a broader distribution. This is similar to the situation of a laser operating above or below threshold.

Another explanation of the 540 GeV data[19] involves the suggestion[20] that energy-momentum conservation strongly influences the multiplicity distributions observed at the lower energies when the data are selected in the full rapidity range. According to this idea, the KNO distribution in the region of the rapidity plateau should be nearly free of kinematical constraints, and so give a more direct measure of the production process. The KNO distribution in restricted pseudorapidity intervals for the CERN $\bar{p}p$ experiment[19] is indeed observed to be wider than that for the inclusive sample, as is also the case for the HRS e^+e^- data selected in this same way. This picture, however, does not lead to a natural explanation of the Poissonian multiplicity distributions observed for the unselected e^+e^- data.

CONCLUSIONS

The charged particle multiplicity distributions and forward-backward correlations presented in this paper are in agreement with the simplest picture of the e^+e^- annihilation being dominated by $e^+e^- \rightarrow q\bar{q}$ with the materialization of the q and $\bar{q}$ into hadrons proceeding independently. The detailed results are completely different from the hadronic data. The Poissonian shape of the multiplicity distribution, the lack of correlation between $\langle \mathbf{n}_F \rangle$ and $\mathbf{n}_B$ and the binomial distribution of $\mathbf{n}_F:\mathbf{n}_B$ at fixed $\mathbf{n}$, all suggest the direct and independent production of hadrons. Such a picture must be reconciled with the known strong resonance production in e^+e^- annihilation and with the observed rapidity correlations[9], which are not in agreement with this simple picture. The present e^+e^- multiplicity results also cast doubt on the long-held belief that the clusters needed to interpret the hadronic data are to be identified with resonances.

Data on e^+e^- annihilation at other energies are needed to see if the multiplicity distributions remain Poissonian. At the present energies, more detailed studies, both of the multiplicities in restricted regions of the phase space and of correlations, will be done in order to elucidate the interplay between global measurements such as multiplicity and the detailed nature of the hadronization processes.

REFERENCES

1. Bender, D. et al., Phys. Rev. D30, 515 (1984).
2. Bender, D. et al., Phys. Rev. D31, 1 (1985).
3. Details of the experimental selection, as well as a more complete discussion of the results, are given in: Derrick, M. et al., ANL-HEP-PR-86-03, Phys. Rev. D.
4. Baringer, P. et al., Phys. Rev. Lett. 56, 1346 (1986).
5. Berger, Ch. et al., Z. Phys. C12, 297 (1982).
6. Derrick, M. et al., Phys. Lett. 165B, 449 (1985).
7. Althoff, M. et al., Z. Phys. C22, 307 (1984).
8. Derrick, M. et al., Phys. Lett. 168B, 299 (1986).
9. Valdata-Nappi, M., Proc. of XIV Symposium on Multiparticle Dynamics, Eds. P. Yager and J. F. Gunion, p. 75.; Aihira, M. et al., Z. Phys. C27, 495 (1985); Althoff, M. et al., ibid., C29, 347 (1985).
10. Althoff, M. et al., Phys. Lett. 135B, 243 (1984); Sakuda, M. et al., ibid., 152B, 399 (1985); Kesten, P. et al., ibid., 161B, 412 (1985).
11. Fialkowski, K. and Kotanski, A., Phys. Lett. 115B, 425 (1982); Wroblewski, A., Proc. XV Symposium on Multiparticle Dynamics, Ed. G. Gustafson and C. Peterson, p. 30.
12. Giovannini, A., Nuovo Cimento 15A, 543 (1973); Knox, W. J., Phys. Rev. D10, 65 (1974); Carruthers, P. and Shih, C. C., Phys. Lett. 127B, 242 (1983).
13. Giovannini, A. and Van Hove, L., CERN Th 4230/85, Z. Phys. C.
14. Durand, B. and Ellis, S.D., Proceedings of the 1984 Summer Study on the Design and Utilization of the Superconducting Super Collider, Eds. J. Morfin and R. Donaldson, Snowmass, CO, p. 234.
15. Albini, E. et al., Nuovo Cimento 32A, 101 (1976).
16. Dias de Deus, J., Nucl. Phys. B59, 231 (1973).
    Barshay, S. and Urban, L., Phys. Lett. 150B, 387 (1985).
17. Chou, T. T. and Yang, C. N., Phys. Lett. 135B, 175 (1874); Phys. Rev. Lett. 55, 1359 (1985).
18. Carruthers, P. and Shih, C. C., Phys. Lett. 137B, 425 (1984).
19. Alner, J. G. et al., Phys. Lett. 160B, 193, 199 (1985).
20. Bialas, A. and Hayot, F., Phys. Rev. D33, 39 (1986).

MULTIPARTICLE PHENOMENOLOGY AS A SYNTHESIS OF MICROSCOPIC AND STATISTICAL CONCEPTS*)

by

Richard M. Weiner**)

Physics Department
University of Marburg
F.R. Germany

Physics is an experimental science and any theoretical description has to pass as an ultimate and decisive test, the test of agreement with experimental data. However, this is not the only condition which a satisfactory theory has to fulfill. There are certain constraints like economy of thought or simplicity and uniqueness without which no theoretical understanding is possible. These supplementary criteria are of special importance in the present stage of multiparticle phenomenology, when no complete theory yet exists and when different models try to describe particular aspects of what ultimately must be a multiparticle dynamics.

The history of physics has shown that among the multitude of experimental facts, there are always a few characteristic ones which are reflected in certain fundamental theoretical concepts.

In the following table we enumerate what we believe to be at present experimental cornerstones of multiparticle production processes and the corresponding theoretical concepts.

---

*) Part of this material was presented as an invited review at the High Energy Annual Meeting of the Brasilian Physical Society, San Lorenco, Brazil, Sept. 1985

**) Invited talk presented at the workshop LESIP II , April 1986

TABLE I

Experimental Facts	Theoretical Ideas
1. Average multiplicities $\bar{n}>>1$. $\bar{n}$ increases with cm. energy $\sqrt{s}$.	1. In a quantum field theory (QFT) of strong interactions the coupling constant $g \geqq 1$. The large number of particles produced suggest the necessity of statistical methods.
2. Existence of leading particles and a corresponding inelasticity distribution.	2. The QFT is non-abelian (gluon-gluon interactions in QCD).
3. The average transverse momentum $\bar{p}_T$ of secondaries is a very slow function of c.m. energy $\sqrt{s}$.	3. Local equilibrium - hydrodynamics and statistical physics.
4. Shape and energy dependence of multiplicity distributions P(n); (KNO scaling and its violations); Bose-Einstein correlations are incomplete.	4. Quantum statistics.

Messenger: "Gracious my lord, I should report that which I say I saw, but know not how to do't."..."I look'd toward Birnam, and anon, me thought, the wood began to move."
Macbeth: "Liar and slave!"
Messenger: "Let me endure your wrath if it be not so; within this three mile may you see it coming; I say a moving grove."
Macbeth: "If thou speak'st false, upon the next tree shall thou hang alive..."

Shakespeare, The Tragedy of Macbeth, Act V, Scene 5

Most physicists believe at present that eventually quantum field theory (QFT) and in particular its strong interactions correspondent quantum-chromodynamics (QCD) will become the microscopic theory of strong interactions. So far this goal could not be achieved because the large coupling constant (g) regime could not yet be handled.

However, even if such a QFT would exist, it would not yet suffice for a description of all phenomena of strong interaction physics and of multiparticle production processes in particular. Indeed, multiparticle production processes present certain global features the description of which in terms of microscopic concepts only would be highly uneconomical if not even impossible. These features (1-4 in table 1) demand in their interpretation supplementary theoretical concepts, much in the same way as it does not make sense to reduce condensed matter physics in general and certain specific effects like superconductivity, e.g. to atomic physics only. The very fact 1 that the mean multiplicities are much larger than unity suggests the necessity of a collective-statistical approach. It will be argued that the other 3 fundamental experimental observations indicate the same need. However, as in Shakespeare's play, this message met with resistance and has been accepted only gradually.

In some cases models have been constructed which describe only certain aspects of multiparticle production processes ignoring or postulating away other aspects. (A typical example of this is the constancy of $\bar{p}_T$. This is obviously an unsatisfactory situation and it is the purpose of this review to show that there exists another more fundamental approach which combines the field theoretical and constituent picture with a global description. Essentially this is a quantum statistical approach which exploits the collective nature of multiparticle phenomena.

Historically such a statistical (or hydrodynamical) approach was advocated as a macroscopic description of "hadronic matter". With the discovery of hadronic constituents besides hadronic matter another phase has been introduced, i.e. the quark-gluon plasma (QGP) and now these statistical-hydrodynamical methods have become very popular. However, as will be shown it appears that these methods have to be supplemented by even more powerful methods inspired from quantum optics which is a particular case of quantum statistics or of statistical field theory.

Quantum statistics (QS) of quarks and gluons appears thus as a promising candidate for a theoretical framework of multiparticle production processes and in the following we shall try to illustrate this statement by analyzing features 2-4 of table 1.

The existence in each high energy event of particles which have an appreciable fraction K of the total available energy $\sqrt{s}$ has been known from cosmic rays physics for several decades and has been confirmed by accelerator physics up to ISR energies ($\sqrt{s} \leqq 62$ GeV). From the multiparticle production point of view, this effect has particular importance because only the amount of energy

$$W = K\sqrt{s} \qquad (1)$$

can be used for this production process. Prof. Wilk discusses in detail in his talk (cf. this volume) our present understanding of this remarkable effect and I will limit myself here only to emphasize that this phenomenon is interesting in itself since it is one of the very few if not the only one from which information about the soft gluon-gluon interaction can be obtained and thus the non-abelian nature of QCD tested and studied. Thus this effect is an illustration how statistical methods supplement microscopic tools.

LOCAL EQUILIBRIUM AND ITS CONSEQUENCES

a) Limitation Of $\bar{p}_T$ As A Consequence Of Local Thermodynamical Equilibrium And Of Hydrodynamical Expansion

The well known experimental fact that $\bar{p}_T$ is a very slowly increasing function of energy (cf. fig. 1) is explained by the Landau[1] hydrodynamical model in the following way:

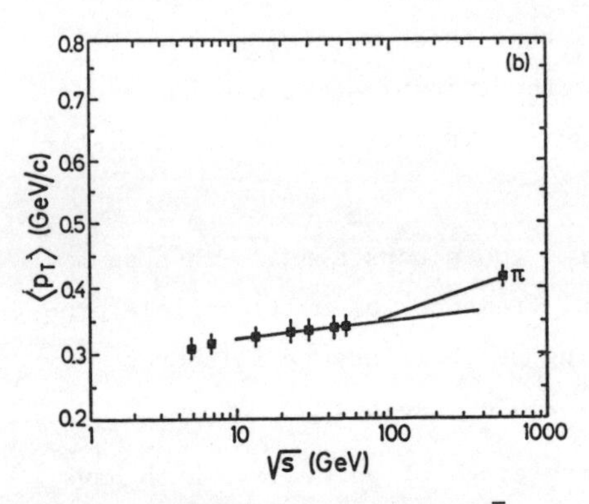

Fig. 1: Energy dependence of $\bar{p}_T$. The accelerator data represented in this figure confirm earlier observations from cosmic rays

The initial system formed in a hadronic reaction is much smaller than the characteristic size m_π^{-1}. In this stage (considered today to be the quark-gluon plasma phase) no free hadrons (pions) can emerge. Only after the system has expanded (mainly longitudinally because of the initial compression) to a size m_π^{-1}, pion production can take place. But this expansion is adiabatic and hence is associated with a cooling of the system to a temperature $T_c \sim m_\pi$. This means that pions are produced at this temperature and hence $p_T \sim 3/2\ T_c \sim 300$ MeV. (Another interpretation of this effect is that due to Hagedorn[2] in which T_c is the maximum temperature of hadronic matter. This leads to an elegant "bootstrap" formulation of the equation of state of hadronic matter in which characteristic properties of strong interactions like resonance production are exploited.) The small increase with s beyond this value is due in the hydrodynamical approach to the correction which the small transversal expansion introduces and to leakage. Hard scattering processes present at large p_T contribute also to this increase. In a first approximation the slow s dependence of p_T can be described by the equation

$$\bar{p}_T \sim (K\sqrt{s})^{\alpha} \tag{2}$$

with $\alpha \sim 1/6$. As far as we can gather there is no alternative explanation for this important experimental observation, and this strongly suggests that at least in a certain stage the system is in local equilibrium, although the details of this process are not yet fully understood (cf. the end of this section). At present T_c is interpreted as the critical temperature at which the phase transition QGP-hadronic matter occurs. The interpretation of T_c as a critical temperature goes back, however, to a time[3] when the QGP had not yet been invented.

Equation (2), if averaged over the inelasticity distribution $\chi(K)$, explains[4] the observed slow increase of $\bar{p}_T$ from ISR to collider energies. Another application of (2) is the calculation of the correlation between $\bar{p}_T$ and the multiplicity n (fig. 2). Suppose $P(n|K)$ is the conditional probability to observe n particles at a given K. Then the average transverse momentum at fixed n can be written as

$$\langle p_T, n\rangle = \frac{\int dK A(K\sqrt{s})^{\alpha}\chi(K,s)P(n|K)}{\int dK\chi(K,s)P(n|K)} \tag{3}$$

$P(n|K)$ can in principle be measured directly or determined from the measured multiplicity distribution at fixed s . The results of this calculation[4] based on eq. (3) are represented in fig. (2) and one sees that the agreement with experiment [5,6] is good. As a matter of fact the apparent plateau in $\langle p_T, n\rangle$ at $\sqrt{s}$ = 540 GeV had been related to a possible phase transition[7] since in a purely thermodynamical approach to $\bar{p}_T$, this quantity is a measure of temperature while $n \sim \Delta n/\Delta y$ is a measure of entropy. The calculation sketched above suggests that the hydrodynamical expansion and the inelasticity distribution can explain so far the observed correlation without invoking a phase transition, a point suggested also in ref. 8. A further increase of $<p_T>$ at large n beyond the plateau could, however, indicate the presence of a phase transition.

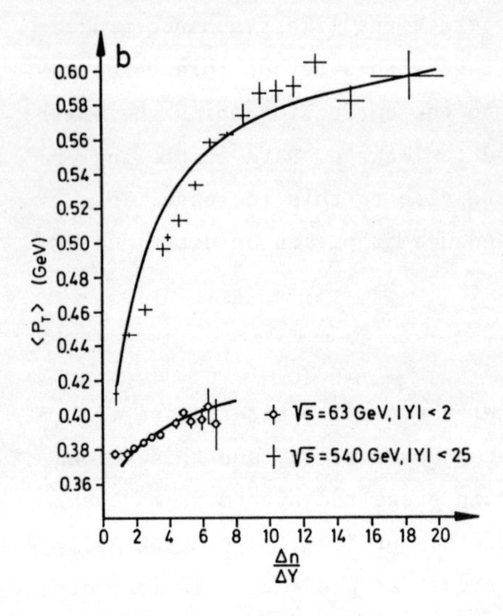

Fig. 2: The correlation between mean transverse momentum $\bar{p}_T$ and multiplicity per unit rapidity interval $\Delta n/\Delta y$. The experimental data are from refs. 5,6 and the theoretical curves are explained in the text.

b) The Rapidity Distribution

This quantity is related to the entropy of the system and was calculated already in 1953 by Landau. The remarkable increase of the plateau with s had been predicted correctly and the later invented Feynman scaling, which contradicts this s dependence, could survive for some time only because of some confusion between rapidity y and pseudorapidity η as well as because of lack of adequate experimental data[9]. It is now well established that this prediction of the Landau model is in agreement with experiment. An illustration of this situation is represented in fig. 3. Here the theoretical curves are calculated for a velocity of sound $c_o = 1/\sqrt{3}$ and by taking into account the change of inelasticity with s (for details cf. ref. 10).

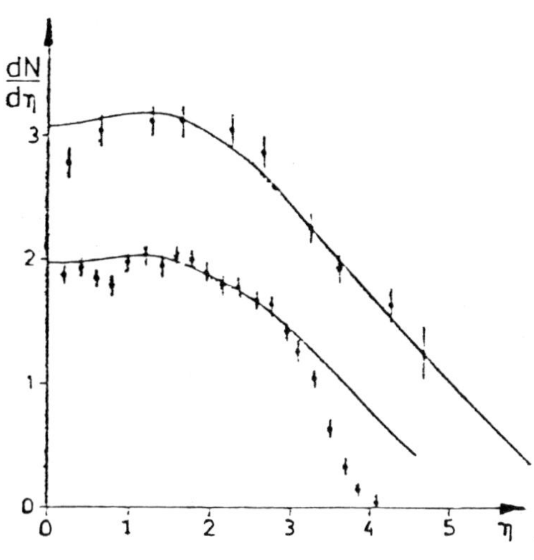

Fig. 3: Pseudo-rapidity distribution at ISR and collider energies (from ref. 10).

The Landau model has been applied successfully also to hadron-nucleus reactions[12] where again new effects could be predicted[13] before their experimental verification (e.g. the ratio

$$\frac{(dN/dy)_{p-A}}{(dN/dy)_{p-p}}$$

at large η) and at present the hydrodynamical model has become an almost indispensable tool in the study of nucleus-nucleus reactions.

c) Initial Conditions In The Landau Model[14]. Energy Dependence Of Mean Multiplicity

In order to solve the equations of hydrodynamics the initial energy ε density has to be specified. We have:

$$\varepsilon = W/V. \tag{4}$$

There are two aspects in ε which have lead to confusion and misunderstanding; one related to the energy W and the other in connection with the initial volume V. Before the discovery of the existence of leading effects and even a long time after, W was identified with the total available energy $\sqrt{s}$. Then the inelasticity was introduced so that $W = K\sqrt{s}$ but the distribution of K was ignored and one limited oneself to putting $K \cong \bar{K} = \frac{1}{2}$. Finally with the introduction of the inelasticity distribution $\chi(K)$ this question has been apparently settled. The situation is not yet as clear what concerns the initial volume V. In the original formulation of Landau V was written

$$V = V_0/\gamma \tag{5}$$

where γ is the Lorentz contraction factor

$$\gamma = \frac{\sqrt{s}}{2m} \tag{6}$$

and V_0 the volume of the hadron at rest and m its mass. This equation however leads to difficulties.

To see how this comes about let us recall that the mean multiplicity of the system is

$$\bar{n} \sim S \sim VT^3 \tag{7}$$

where S is the total entropy. With

$$\varepsilon \sim T^4 \tag{8}$$

we get from (4)

$$\bar{n} \sim V^{1/4} K^{3/4} s^{3/4} .$$

With (5), (6) this leads to

$$\bar{n} = AK^{3/4} s^{1/4} \tag{9}$$

where A is independent of K or s. This last relation could be tested recently when measurements at fixed inelasticity K were performed. It was found by Carruthers and Minh[15], using the data of ref. 16 and 17, that the s dependence of (9) was confirmed but the K dependence not. The experimental data show that

$$\bar{n} \sim W^{1/2} = K^{1/2} s^{1/4} . \tag{10}$$

This can be interpreted as if the Lorentz factor would not be determined by the total available energy $\sqrt{s}$ but by the energy effectively used $K\sqrt{s}$:

$$\gamma = \frac{K\sqrt{s}}{2m} = \frac{W}{2m} \tag{11}$$

Things happen as if only the gluonic part (which is responsable for the inelasticity K) suffers Lorentz contraction and/or the Lorentz contraction takes place after the leading particles (valence quarks) have left the system. This is apparently in contradiction with the parton model which assumes that the valence partons suffer Lorentz contraction while the wee partons (which should be identified with the gluons) not.

That the Lorentz factor for a quark-gluon system can be different from its classical value is after all not so surprising. It was shown by Mueller[18)] that the Lorentz factor depends on the dynamics of the system and can in principle take any value. However, before engaging in so far reaching conclusions, further experimental confirmation is necessary. More information about this very important issue could be obtained by performing measurements at fixed K at higher energies than those of the ISR, e.g. at the CERN-collider ($\sqrt{s} \sim 540$ GeV).

MULTIPLICITY DISTRIBUTIONS, THEIR QUANTUM CONTENT AND RELATED PROBLEMS

As already pointed out, the fact that the first moment of the multiplicity distribution

$$\bar{n} = \Sigma n P(n) dn$$

is much larger than 1, suggests in itself the need and usefulness of quantum statistical (QS) methods in describing the production process. Another hint in this direction are second order correlations. For identical bosons (pions of equal charge e.g.) the "differential" of the second moment, i.e. the second order correlation function

$$C_2 = \frac{\dfrac{\delta^2\sigma}{\delta p_1 \delta p_2}}{\dfrac{1}{\sigma}\dfrac{\delta\sigma}{\delta p_1}\dfrac{\delta\sigma}{\delta p_2}} \qquad (12)$$

in the limit $\vec{p}_1 \to \vec{p}_2$ would reach the value of 2, if only the correlations due to Bose-Einstein statistics would be present. The remarkable experimental fact is that this almost never happens and that C_2 is usually smaller than 2 (Bose-Einstein correlations are "incomplete"), cf. e.g. ref. 19. A natural explanation[19)] for this is the existence of partially coherent fields in the quantum optical sense in analogy to the laser effect. This coherence property reflects a specific collective property of the system characteristic e.g. for condensates. Unfortunately there are other experimental effects, like resonance production, which also contribute to the decrease of C_2 at $|p_1 - p_2| \cong 0$ (cf. G. Goldhaber, this volume).

Finally as will be shown, the entire form of the multiplicity distributions P(n) is also very suggestive for statistical behavior. It has been found that in a first approximation $\bar{n}P(n)$ scales in terms of $n/\bar{n}$, i.e.

$$\psi(z) \equiv \bar{n}P(n) \tag{13}$$

depends only on $z = n/\bar{n}$ (Koba-Nielsen-Oleson (KNO) scaling). In this scaling limit the width of P(n) is constant, i.e. s independent. Expressed in terms of the moments

$$C_j = \frac{\Sigma n^j P(n)dn}{(\Sigma n P(n)dn)^j} \tag{14}$$

KNO scaling is equivalent to the statement that C_j are independent of s. Although this property is not exact (at energies ≧ 540 GeV it fails especially at large n (cf. fig. 4)), the important role which the average multiplicity $\bar{n}$ plays in it might indicate already the existence of some statistical effects.

What is, however, more important, it has been found that the same general QS considerations which explain e.g. the Bose-Einstein correlations, provide a natural framework for the understanding of the form of P(n) and its s dependence including the deviation from KNO scaling.

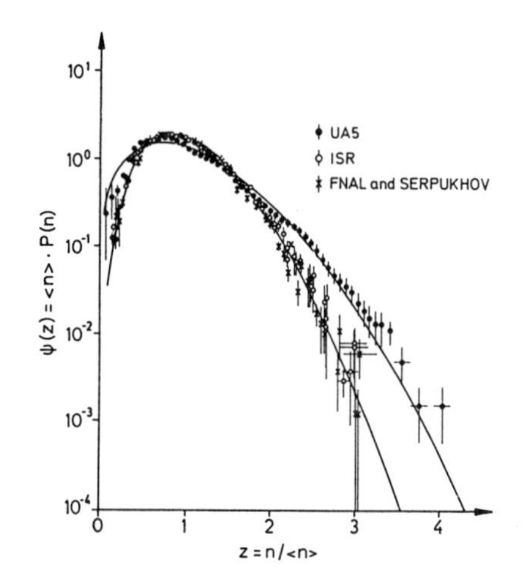

Fig. 4: Deviation from KNO scaling (from ref. 21).

While, as mentioned there is no alternative explanation for the constancy of $\langle p_T \rangle$ but in statistical terms, the situation is yet different in the case of the shape of multiplicity distributions. One can distinguish two big classes of models for P(n): "microscopic" the prototype of which are the dual parton model[22] and the quark-gluon bremsstrahlung model[23], and "macroscopic" which include geometrical (impact parameter) models[24], general statistical[25] and quantum statistical models (ref. 26). Somewhere between are the stochastic evolution models[27]. From a purely experimental point of view it is difficult at present to decide with sufficient confidence which of the models is correct, since most of these models describe rather successfully the data. Therefore, if any preference is to be given at present to some of these models, then this can be done only on heuristic grounds. From the preceding it should be clear where our preference lies. If one asks for a theoretical understanding of all multiparticle phenomena including features 1-4, of table 1, then a quantum statistical approach to the problem of multiplicity distributions appears to be favoured. This approach is quite general since it relies essentially on the density matrix formalism and its heuristic superiority manifests itself also in its suitability to describe collective phenomena, a feature which most of the other models do not posess. It is consistent with thermodynamical and statistical models in the sense that it contains as a special case chaotic-thermal multiplicity distributions. The salient new feature of QS is, as mentioned already, the possibility of including quantum coherence properties.

Before examining multiplicity distributions within the QS context we remind again that in a statistical approach the energy W effectively used for particle production is the relevant variable and therefore QS is expected to define the distribution at fixed $K = W/\sqrt{s}$, i.e. $P(n|K)$. The experimental distribution $P(n,s)$ is related to $P(n|K)$ by

$$P(n,s) = \int P(n|K)\chi(K)dK \ .$$

A quite general, but not the most general, QS expression for $P(n\ K)$ is given by the generalized Glauber-Lachs formula[28)]

$$P(n|K) = \frac{p\,\bar{n}\,k^n}{\{1+(p\bar{n}/k)\}^{k+n}}\, e^{-\frac{q\bar{n}}{1+p\bar{n}/k}}\, L_n^{(k-1)}\left(\frac{-qk/p}{1+p\bar{n}/k}\right) \tag{15}$$

Here k is the number of cells in phase space which can be equated to the number of independent quantum states and/or sources, p the amount of chaoticity $p = \frac{\bar{n}_{ch}}{\bar{n}_{ch}+\bar{n}_c}$, where $\bar{n}_{ch}$ is the average number of particles produced by the chaotic components and $\bar{n}_c$ the corresponding coherent number ($\bar{n}_{ch} + \bar{n}_c = \bar{n}$), q the percentage of coherence = 1-p, and $L_a^{(b)}$ the associated Laguerre polynomials.

There are thus 3 independent parameters which characterize completely $P(n|K)$: p, k, $\bar{n}$. These parameters depend in general on K and s. For the sake of simplicity the K dependence of p and k is neglected and considered only in $\bar{n}$ (cf. eq. (10)). The s dependence of all these parameters, however, is essential, since as will be shown below, it explains among other things the deviation from KNO scaling and has important physical significance.

Eq. (15) has two remarkable limits: For p = 1 corresponding to a completely chaotic distribution it reduces to the negative binomial or generalized Bose-Einstein distribution

$$\lim_{p\to 1} P(n|K) = \frac{(n+k-1)!}{n!(k-1)!}\,\frac{(\bar{n}/k)^n}{(1+\bar{n}/k)^{n+k}} \tag{16}$$

(for k = 1 this reduces to the Bose-Einstein distribution) and for p = 0 corresponding to a completely coherent distribution we obtain the Poisson distribution:

$$\lim_{p\to 0} P(n|K) = e^{-\bar{n}} \frac{(\bar{n})^n}{n!} \tag{17}$$

It is useful to consider the normalized factorial moments ϕ_q of $P(n|K)$ defined by:

$$\phi_j = \frac{1}{\bar{n}^j} \overline{n(n-1) \ldots (n-j+1)} \tag{18}$$

which are linear combinations of the moments C_j (eq. (14)).

They characterize the relative width of the distribution and their upper and lower bounds define the corresponding bounds of $P(n|K)$. The ϕ_q have the property that at fixed p, they decrease monotonically with k and at fixed k, the maximum width of P(n K) corresponds to $p_{max} = 1$ and the minimum width to $p_{min} = 0$. This proves that the negative binomial is the widest distribution and the Poisson form the narrowest distribution compatible with QS. The existence of these bounds[29)] is a characteristic property of QS distributions and is absent in most other approaches.

The experimentally observed broadening of P(n), i.e. the deviation from KNO scaling can be due either to a broadening of $\chi(K)$ with s or to a broadening of $P(n|K)$ with s. The first possibility is apparently excluded (cf. G. Wilk, this volume) so that the only way of explaining the experimental observations is to assume that either k decreases with s or p increases with s or both. The first possibility was considered in a special case by the UA-5 collaboration[30)] who fitted P(n) by a negative binomial. There are several weak points in this attempt among which we mention:

(i) it ignores the role of the inelasticity distribution;

(ii) it ignores the existence of a coherent component;

(iii) the number of cells k decreases with energy while in any intuitive understanding of this phenomenon the opposite effect would be expected.

A different analysis of multiplicity distributions within the QS formalism was performed in ref. 29 where it was shown that a viable alternative to the negative binomial fit can be achieved by considering the increase of chaoticity with energy. An illustration of this possibility is given in fig. 5. Things happen as if with the increase of energy the system eats up its reserves of coherence ("coherophagy").

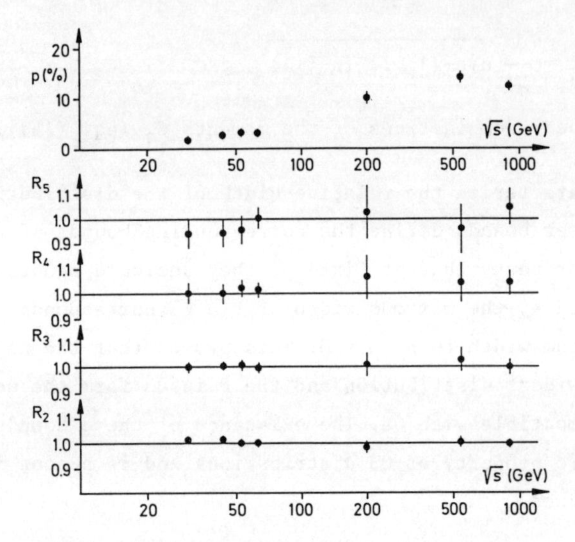

Fig. 5: Ratios R of the first five moments $C_j^{exp}/C_j^{theor.}$ (eq. (14)) of the multiplicity distribution at various energies. The $C_j^{theor.}$ were calculated with a QS distribution (15) including the effect of $\chi(K)$. The upper part of the figure represents the corresponding p values.

From the above considerations another important conclusion follows: The presently observed broadening of P(n) cannot continue for ever. Once the upper bound of $P(n|K)$ is reached the fate of P(n) depends only on $\chi(K)$. If $\chi(K)$ continues then to narrow P(n) can even reverse its trend and start to narrow, too. If and at what energy this happens is an open question to be answered by the new accelerators, the first of which is the Tevatron.

In order to answer the question whether the broadening of P(n) is due to a change of k or to a change of p, the more general formalism[20] of QS which includes the finite correlation length in rapidity has been used[31]. In this formalism the factorial cumulants μ_j which are related by simple algebraic relations to the measured moment C_j of P(n) read:

$$\mu_j = \frac{\langle n\rangle^j}{k^{j-1}} \{(j-1)!p^j B_j + j!p^{j-1}(1-p)\tilde{B}_j\} \tag{19}$$

where p is the chaoticity and B_j, $\tilde{B}_j$ algebraic functions of $\beta \equiv Y/\xi$. Here Y is the total rapidity range and ξ the correlation length in rapidity defined in terms of the fields π through the relation

$$\langle\pi(y_1)\pi(y_2)\rangle = n_{chaotic}\, e^{-\frac{|y-y_2|}{\xi}} \tag{20}$$

k is as before the number of independent quantum states (charges e.g.). As compared with eq. (15) we have now one parameter more, namely β. For $\beta = 0$ corresponding to $\xi=\infty$ we reobtain the Glauber-Lachs distribution.

The dependence of μ on β has as an immediate consequence that with the increase of the rapidity range Y, the distribution P(n) narrows. It turns out that this effect, which corresponds to a new scaling property of P(n) is in quantitative agreement with experiment.

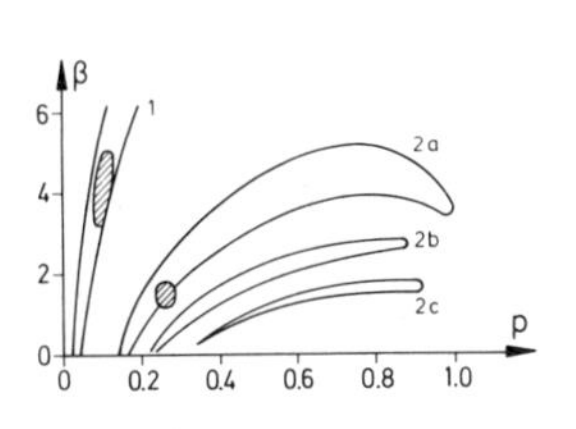

Fig. 6: Approximate one standard deviation (S.D.) contours in the (β,p) plane for fits to the moments of order 2 to 5 from multiplicity distributions at: 1) $\sqrt{s}$ = 30.4 GeV, 2) $\sqrt{s}$ = 540 GeV; 2a) refers to the total multiplicity while 2b) and 2c) refer to pseudo-rapidity cuts at $|\eta|<3$ and $|\eta|<1.5$ respectively. The shaded portions are one equivalent S.D. χ^2-contours for fits to C_2 and b (see text). The data come from: ref. 30 for curve 1,2a,2b and 2c; ref. 32 for the shaded χ^2 contour on curve 2a; ref. 33 for the same on curve 1 (from ref. 31).

In fig. 6 the range of β,p values which are consistent with the values of the moments of P(n) at ISR and collider energies is represented. The existence of solutions for β,p proves that the QS formalism can account for the shape as well as rapidity and energy dependence of P(n).

Another implication of (19) is that QS allows us to predict the moments of P(n) in various rapidity ranges (cf. below). It is important to note that β is effectively the number of cells in phase space and from eq. (19) it follows that the broadening of P(n) or the increase of the moments C_j can be achieved either by an increase of p or a decrease of β (at fixed m). Although we are still faced with an alternative, the situation is nevertheless much more satisfactory, since the decrease of the number of cells β with energy does not contradict any physical expectation. Indeed, the decrease of β with energy can be understood in the sense that the correlation length ξ increases faster with s than Y. Further progress in solving the above mentioned ambiguity could be obtained[31)] by considering the forward-backward correlation of multiplicities.

Experiments at ISR and collider energies show that there exists a linear relationship between the average number of particles $\langle n_F \rangle$ emitted in the forward hemisphere and the number of particles n_B emitted in the backward hemisphere.

$$\langle n_F \rangle = a + bn_B \tag{21}$$

The slope b is an increasing function of s and can be expressed in the QS formalism in terms of the factorial cumulants $\mu_2(\beta)$ of the entire distribution and the factorial cumulant $\mu_{2,F}(\beta/2)$ of the forward hemisphere distribution:

$$b = \frac{\dfrac{2\mu_2(\beta)}{\langle n \rangle^2} - \dfrac{\mu_{2,F}(\frac{\beta}{2})}{\langle n \rangle^2/4}}{\dfrac{\mu_{2,F}(\frac{\beta}{2})}{\langle n \rangle^2/4} + \dfrac{2}{\langle n \rangle}} \tag{22}$$

Using the experimental data for b at different energies as well as the moments of the multiplicity distributions, one is able (after introducing in (22) the inelasticity distribution $\chi(K)$) to narrow very much the range of β,p values (cf. in fig. 6, the dashed areas). One gets

$$0.7 < \xi < 1 \qquad p = 0.1 \ , \sqrt{s} = 30 \text{ GeV}$$

$$3.5 < \xi < 5 \qquad p = 0.25 \ , \sqrt{s} = 540 \text{ GeV}$$

This shows that both p and ξ vary with energy. In particular the increase of chaoticity found in the simplified QS approach is confirmed, but one finds now a simultaneous dramatic increase of the rapidity correlation. (These results were obtained for $k = 1$ implying local charge conservation.) Since p plays in QS the role of an order parameter the change of p and ξ with energy is very suggestive for the neighbourhood of a phase transition. The natural candidate for this transition is that of hadronic matter - quark gluon plasma (QGP).

To investigate further this possibility one would have to understand the energy dependence of the chaoticity p. It is clear that the chances of producing a QGP increase with s or rather with $W = K\sqrt{s}$ (eq. (1)) and thus the experimentally observed P(n) might be considered as an average over events some of which reflect a pure QGP (large K) while others (corresponding to small K) not. There exists, however, another possibility, namely that from a certain value of the center of mass energy $\sqrt{s}$ on, there exist in each event two types of sources, one due to a QGP in thermal equilibrium and another due to a bremsstrahlung mechanism.

Such a two component model is suggested by the Pokorski-Van Hove-Carruthers-Minh picture in which in a hadronic collision the valence quarks practically do not interact and go through, emitting, through bremsstrahlung some gluons, while the bulk of the interaction cross section and multiplicity is due to the gluons left behind. If the gluon-gluon interaction is strong enough, this part of the system may equilibrate and form a QGP.

There are several experimental indications in favour of such a two component model - out of which we mention here the following two:

(a) The inelasticity distribution $\chi(K)$ and the leading particle spectrum calculated in such an approach seem to be consistent with available data (cf. G. Wilk, this volume).

(b) The multiplicity distributions measured in rapidity bins shifted one with respect to the other by an amount η_c are narrower at large η_c. This effect had been observed[35] for the first time in emulsion data at $\sqrt{s} \cong 20$ GeV and has been confirmed recently[36] by the UA-5 collaboration.

To see what the consequences of such a two component model are for the understanding of the possibility of formation of QGP, the following investigation has been carried out (ref. 34).

Assume that the multiplicity distribution P(n) is a convolution of a negative binomial $P_1(n_1, \langle n_1\rangle, k)$ and a Poisson distribution $P_2(n_2, \langle n_2\rangle)$.

$$P(n) = \sum_{n=n_1+n_2} P_1(n_1)P_2(n_2) \tag{23}$$

The factorial cumulant μ_j for $j \geqq 2$ are completely determined by P_1, since for a Poisson distribution μ_j ($j \geqq 2$) vanish. This means that the parameters of the problem are k and

$$p = \frac{\langle n_1\rangle}{\langle n\rangle}$$

where p is the effective chaoticity of the system. We remind that a negative binomial corresponds to a completely chaotic distribution and a Poisson to a completely coherent one. We also recall that in our QS interpretation of the negative binomial k is an integer and can take the values of 1 or 2 depending whether we have local charge conservation or not (we assume one single source or cluster emitting chaotically). By applying eq. (23) to the data we then determine p and hence $\langle n_1\rangle$ and $\langle n_2\rangle$ as a function of s and (pseudo) rapidity. The results are represented in figures 7 and 8. It is seen that

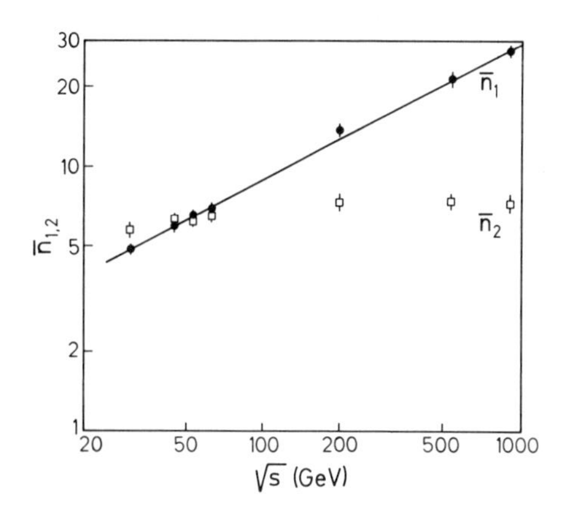

Fig. 7: Dependence of the estimated means $<n_1>$ and $<n_2>$ of the two components on $\sqrt{s}$. The chaotic component $<n_1>$ (circles) can be fitted by the power law $(0.72 \pm 0.08)(\sqrt{s})^{(0.52 \pm 0.01)}$.
The coherent component $<n_2>$ (squares) is shown here on the same log-log scale as $<n_1>$ for the sake of comparison. An excellent logs fit is obtained for $<n_2>$ as $\{1.43 \pm 0.62 + (1.46 \pm 0.16) \ln\sqrt{s}\}$. $\sqrt{s}$ is expressed in GeV.

$<n_1> \cong \sqrt{s}^{0.5}$ and that $<n_2> \cong 1.4 + 1.46 \ln\sqrt{s}$. This is exactly what one would expect from a thermal (QGP?) source and a bremsstrahlung source respectively. Furthermore the first source is concentrated at small y while the second one is more or less unformly distributed.

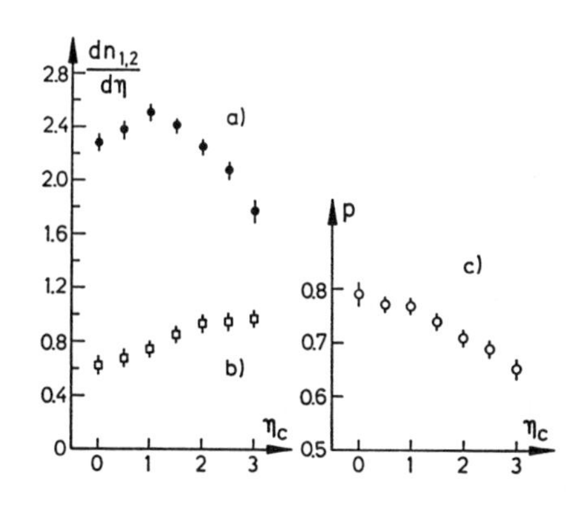

Fig. 8:

(a) and (b): Dependence of the estimated values of $<n_1>$ and $<n_2>$ on the center of η intervals (0.4 units wide) η_c.

(c): Dependence of the effective chaoticity p on η_c.

These results apparently confirm the conjecture that a QGP plasma may be formed in hadronic reactions and justify the two component model, although it is obviously an oversimplified scheme.

With this we also come to the conclusion of this review. It should be clear that multiparticle phenomenology must not limit itself to fitting data. This can be achieved by different models and is not a very rewarding exercise, unless the same model explains in a consistent manner all relevant observations. It should also be clear that a microscopic picture of strong interactions physics, although indispensable, is not a substitute for a global and economic understanding of multiparticle dynamics and that the cooperative behavior of hadronic matter is a subject of interest in itself. Those who often tend to see only the trees and not the wood may be reminded perhaps of Shakespeare's views on the subject quoted on page 3. That this opinion begins to prevail is due among others to the pioneering work of P. Carruthers and workshops of this kind, to the success of which Dan Strottman and Peter Carruthers have contributed so much.

This work was supported in part by the Bundesminister für Forschung und Technologie and the Deutsche Forschungsgemeinschaft.

I am grateful to my collaborators and especially G.N. Fowler, E.M. Friedlander and G. Wilk for the stimulating and fruitful collaboration on which most of the work reported here is based.

References

1) L.D. Landau, Izw.Akad.Nank., Ser.Fiz. 17, 51 (1953); for a review of the hydrodynamical model cf. e.g. E.J. Daibog et al., Fortschr. d.Phys. 27, 313 (1979).

2) For a recent review cf. R. Hagedorn in "Quark-Matter", editor Kajantie, Springer 1984.

3) K. Imaeda, Lett.Nuovo Cimento 1, 290 (1971); A. Mann and R. Weiner, Lett.Nuovo Cimento 2, 248 (1971).

4) G.N. Fowler, E.M. Friedlander, M. Plümer and R.M. Weiner, Phys. Lett. 145B, 407 (1984).

5) UA-1 collaboration, G. Ciapetti (1985) unpublished; G. Arnison et al., Phys.Lett. 118B, 176 (1982).

6) A. Breakstone et al., Phys.Lett. 132B, 463 (1983).

7) L. Van-Hove, Phys.Lett. 118B, 138 (1982).

8) Y. Hama and F.S. Navarra, Phys.Lett. 129B, 251 (1983).

9) P. Carruthers in proceedings of the New York Academy of Sciences 229, 91 (1974) and references quotes herein.

10) K. Wehrberger and R.M. Weiner, Phys.Rev. 31, 222 (1985) and K. Wehrberger in ref. 11.

11) "Local Equilibrium in Strong Interaction Physics", editor D.K. Scott and R.M. Weiner, World Scientific 1985.

12) E.J. Daibog et al., Fortschr.Phys. 27, 313 (1979); E.V. Shuryak, Phys.Rep. 61, 71 (1980).

13) N. Masuda and R.M. Weiner, Phys.Rev. D18, 1515 (1978).

14) Cf. ref. 11 and in particular the round table discussion of the Landau Model, p. 427. Cf. also R.M. Weiner, "Quark-Gluon Plasma and the Landau Hydrodynamical Model", to be published in "Problems of High Energy Density Physics", edited by the Academy of Sciences of the Ukrainian SSR.

15) P. Carruthers and D.V. Minh, Phys.Rev. D28, 130 (1983).

16) M. Basile et al., Nuovo Cimento A37, 329 (1983).

17) M. Basile et al., Nuovo Cimento A67, 244 (1982).

18) A. Mueller, Phys.Rev. D2, 224 (1970).

19) G. Goldhaber, this volume.

20) G.N. Fowler and R.M. Weiner, Phys.Rev. D17, 3118 (1978).

21) UA-5 collaboration; G.J. Alner et al., Phys.Lett. 138B, 304 (1984).

22) A. Capella, A. Staar and J. Tran Thanh Van, Phys.Rev. D32, 2933 (1985).

23) G. Pancheri, Y.N. Srivastava and M. Palotta, Phys.Lett. 151B, 453 (1985).

24) I.T. Chou and C.N. Yang, Phys.Lett. 116B, 301 (1982); S. Barshay, Phys.Lett. 127B, 129 (1983).

25) Meng Ta-chung in ref. 16. In this paper other references are contained; H. Banerjee, T. De and D. Syam, Nuovo Cimento 89A, 353 (1985).

26) For a review on quantum optical methods in particle physics cf. G.N. Fowler in ref. 11.

27) M. Biyajima, Phys.Lett. 137B, 225 (1984); C.S. Lam and M.A. Walton, Phys.Lett. 40B, 246 (1984); B. Durand and I. Sarevic, Univ. of Wisconsin preprint MAD/TH-85-7 (1985); A. Giovannini and L. Van Hove, CERN, Th report 4230/85.

27) M. Biyajima, Progr. Theor.Phys. 69, 966 (1983).

29) G. Fowler, E.M. Friedlander, R.M. Weiner and G. Wilk, Phys.Rev. Lett. 56, 14 (1986); G. Fowler, E.M. Friedlander, R.M. Weiner, J. Wheeler and G. Wilk in Proceedings of the XVI. Symposium on Multiparticle Dynamics, Kiriat-Anavim, June 1985.

30) UA-5 collaboration, G.J. Alner et al., Phys.Lett. 160B, 193, 199 (1985).

31) G.N. Fowler, E.M. Friedlander, R.M. Weiner, J. Wheeler, and G. Wilk, Quantum statistical connection between forward-backward correlations and multiplicity distributions and its striking consequences, Univ. of Marburg, preprint, February 1986.

32) K. Alpgard et al., Phys.Lett. 123B, 361 (1983).

33) S. Uhlig et al., Nucl.Phys. B132, 132 (1978).

34) G.N. Fowler, E.M. Friedlander, R.M. Weiner, and G. Wilk, Possible manifestation of quark-gluon plasma in multiplicity distributions from high energy reactions, to be published.

35) G.N. Fowler, E.M. Friedlander, and R.M. Weiner, Phys.Lett. 104B, 239 (1981).

36) UA-5 collaboration, G.J. Alner et al., CERN EP/85-61, unpublished.

INELASTICITY IN HIGH ENERGY REACTIONS

G. Wilk *)

Dept. of Physics, University of Illinois at
Chicago, Chicago, USA

ABSTRACT

The concept of inelasticity in high energy reactions is reviewed and discussed. The derivation of the inelasticity distributions in hadronic reactions is presented. Its importance in the analysis of multiparticle production processes is outlined and stressed.

1. INTRODUCTION

The high energy reactions result predominantly in the production of large number of particles, the mean multiplicity of which, $\bar{n}$, is large and increases with energy $\sqrt{s}$. That is then commonly understood as a manifestation of a strong coupling regime in the eventual quantum field theory of such processes. The succesfull candidate for such theory, which is quantum chromodynamics (QCD), has been mostly developed on the basis of the perturbative approach which is applicable only in the special, very limited, regions of the phase space.

Recently, however, the nonperturbative investigations of the QCD have revealed the possibility of existence of phase transition between the hadronic matter and the deconfined (on the large space-time scale) quark-gluon matter (QCD plasma) [1]. The experimental confirmation of such a phenomenon would provide decisive confirmation of QCD as a theory of strong interactions. That is the reason of the latest renew of interest in the multiparticle reactions between all possible types of projectiles.

---

*) On leave of absence from Institute for Nuclear Studies, Warsaw, Poland.

In addition to the large multiplicities the high multiparticle production processes present also certain other global features which impose strong constrains on their possible model descriptions. These are: the existence of leading particles and the corresponding inelasticity distribution, the very slow increase of the average transverse momentum $\bar{p}_T$ of secondaries as a function of $\sqrt{s}$, the violation of the KNO scaling in multiplicity distributions and the incompleteness of Bose-Einstein correlations.

2. THE STATISTICAL APPROACH AND INELASTICITY

Those global features are, in our opinion, strongly suggesting the necessity of a statistical approach in the description of multiparticle processes. One should stress, however, that this point of view has emerged only slowly[2], the possibility of the existence of QCD plasma (for which the statistical approach is the only one possible) has boosted the statistical approach considerably.

It should be stressed that one of the very first successfull models of the multiparticle production, the Landau Hydrodynamical Model (LHM)[3], correctly accomodated for all features of this processes (known at that time) and predicted some, which were confirmed by experiment only decades later. The remarkable increase of the plateau of the rapidity distribution, $[dN/dy]|_{y=0}$[4], as a function of $\sqrt{s}$ and the abovementioned very slow increase of the $\bar{p}_T$ were among them[5].

The LHM has been also applied succesfully to hadron-nucleus reactions[6] where again new effects were predicted before their experimental verification, e.g. the shape of $(dN/dy)_{pA}$ (c.f. Fig.1a). In fact, the other prediction of LHM presented at that time, namely the rapidity dependence of the energy fluxes with their peculiar behaviour (enhan-

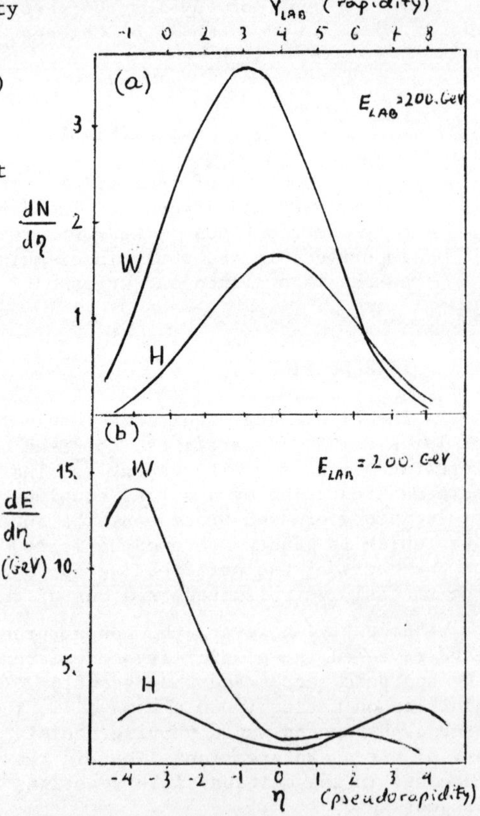

Fig.1. The rapidity distributions of secondary particles (a) and energy fluxes (b) in LHM for pp and pW scatterings (c.f. ref. 6 for details).

cement in the fragmentation regions, particularly in the nucleus fragmentation region) are still awaiting the experimental check (c.f. Fig. 1b). Notice, that in the hydrodynamical models the particle distributions measure the distribution of entropy which is different from the distribution of energy and bears different dynamical information. Because of this, the measurement of both quantities in the same experiment (eliminating thus the sources of biases) would be most usefull.

With LHM (as with every statistical model) is tighly connected the notion of inelasticity K. This quantity, introduced in connection with cosmic rays experiments[7], specifies the fraction of the initially available energy $\sqrt{s}$, which is next found in the produced secondaries:

$$W = K\sqrt{s}\ , \qquad K \cong 0.3 \div 0.6. \qquad (1)$$

Herefrom it then follows that in any statistical approach the relevant variable is W rather than $\sqrt{s}$ and, in fact, one should know the distribution of W, or of the inelasticity $K=W/\sqrt{s}$ in different events: $\chi(K)$.

The inelasticity K is closely connected with the existence of the leading particles which are supposed to take the rest of the available energy. The spectrum of leading particles, $f(x_L=p_L/P)$, is known up to ISR energies to be flat (c.f. ref. 8). This implies that the mean Feynman momentum $\bar{x}_L \sim 0.5$ and the fact that both leading particles are uncorrelated leads then to mean inelasticity $\bar{K}=\int_0^1 dK K\chi(K) \cong 0.5$ also. The only data on $\chi(K)$ present were obtained from buble chamber measurement performed by Brick at al.[9] at $\sqrt{s}$=16.5 GeV . Recently a remarkable new effect was observed at the SPS energies ($\sqrt{s}$=540 GeV) [10], namely an apparent decrease of $\bar{K}$ with $\sqrt{s}$ from $\bar{K}\cong 0.5$ at ISR to $\bar{K}\cong 0.25\div 0.3$ at SPS. (In fact an independent confirmation of this observation was obtained in quite different analysis of multiparticle production, c.f. ref. 11).

The inelasticity distribution $\chi(K)$ which emerges is a function with a maximum at $K\cong\bar{K}$, which shifts to smaller values of $\bar{K}$ at larger $\sqrt{s}$. First calculations [12,13] were merely relating $\chi(K)$ to the empirically known leading particle spectrum. The simple parametrization of $\chi(K)$ which resulted was then used to describe the $\sqrt{s}$ dependence of the $\bar{p}_T$ and the correlations between $\bar{p}_T$ and the multiplicity n [14].

Additional interest in the inelasticity was caused by the problem of the initial conditions in the LHM. To solve the equations of hydrodynamics one has to specify the initial enrgy density $\epsilon=W/V$. With the introduction of inelasticity it was obvious that $W=K\sqrt{s}$ with K distributed according to $\chi(K)$. But in what concerns the initial volume V the situation is far from clear. It is supposed to be the volume in which the initial stopping and equilibrating of the energy W takes place. (The unsatisfactory situation in this field and believe that finally we should have observed a clear plateau in the rapidity distribution has lead to a formulation of the, so called, scale-invariant hydrodynamical models [15]. In fact, this is the most commonly used and popular version of the hydrodynamical model although the assumption on the ultimate development of the central rapidity plateau is valid only approximately). In the original LHM the initial volume

$V=V_o/\gamma$ depends on the volume of the hadron at rest, V_o, and on the Lorentz contraction factor $\gamma=\sqrt{s}/(2m)$, where $\sqrt{s}$ is the total available energy and m is the mass of the hadron. The mean multiplicity of the reaction is then

$$\bar{n} \sim \mathcal{S} \sim V\,T^3 = \mathrm{const}\ K^{3/4} \cdot s^{1/4} \qquad (2)$$

where $\mathcal{S}$ is the total entropy, T - the temperature and $\varepsilon \sim T^4$.

Unfortunately, this relation was found [16] to be incompatible with data at fixed inelasticity K [8,17] (i.e. the $\sqrt{s}$ dependence of n for K=const was confirmed whereas the K dependence, for $\sqrt{s}$=const, not).The experimental data show rather that

$$\bar{n} = \mathrm{const} \cdot W^{1/2} = \mathrm{const}\ (\,K\sqrt{s}\,)^{1/2} \qquad (3).$$

This can immediately be interpreted as if the Lorentz factor γ would not be determined by the total available energy $\sqrt{s}$ but rather by the energy effectively used, $W=K\sqrt{s}$;

$$\gamma = \frac{W}{2m} = \frac{K\sqrt{s}}{2\,m} \quad . \qquad (4)$$

As the analysis in ref.16 was performed in the picture of the throughgoing valence quarks and stopping glue of the incoming hadrons, adapted to the LHM from the ref. 18, this would mean that only the gluonic part suffers Lorentz contraction which takes place after the leading particles (valence quarks) have left the system. The apparent contradiction with the parton model in which the valence partons, not the wee ones (identified with gluons),are Lorentz contracted is then obvious.

What is not so obvious is whether the Lorentz factor for a quark -gluon system has just its classical value. The analysis performed by Mueller [19] has shown that, in fact, γ depends on the dynamics of the system and can, in principle, take any value. Measurements performed at fixed K but at energies higher than those of the ISR would be most valuable here.

3. DERIVATION OF THE INELASTICITY DISTRIBUTION

The only derivation of $\chi(K)$ and its $\sqrt{s}$ dependence for central production [20] (and,at the same time,the leading particle spectrum $f(x_L)$) is based on the gluon-gluon interaction picture for hadronic reactions mentioned above [18]. Here we shall present the improved calculations which allow for more detailed study of multiparticle reactions [21].

Following the experimental indications mentioned before [8,13,17] let us assume that in every event at least two uncorrelated leading particles are produced and that they arise from the through-going valence quarks of the incoming hadrons. The glue components of the two hadrons are assumed to interact strongly and finally to produce the secondaries populating the central rapidity region. The dominance of the gluon interactions over constituent quarks interactions implicit here is just a cosequence of QCD.

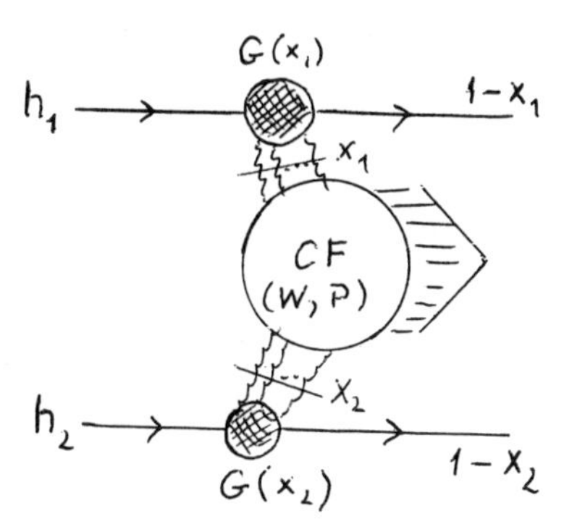

Fig.2. The schematic picture of the multiparticle process.

Effectively we have then the picture presented in Fig.2 with the central fireball (CF) formed by a (indefinite) number of "mini-fireballs" (MF) due to gluon-gluon interactions. The number of gluons is not controlable, their momentum distributions is, in the absence of a better guess, taken to be the same as in the deep inelastic estimations, namely $G(x) \approx \frac{1}{x}(1-x)^n$. The central assumption here is the demand of thermalization of the MF's forming the CF. The number of MF's is assumed to be given by Poisson distribution

$$P(n) = \frac{\bar{n}^n}{n!} e^{-\bar{n}}.$$

The energy W and momentum P of CF is then just equal to

$$W = \frac{\sqrt{s}}{2} (x_1+x_2),$$
$$P = \frac{\sqrt{s}}{2} (x_1-x_2), \qquad (5)$$

(where we are neglecting all rest masses) and accordingly the inelasticity K defined as the fraction of the invariant mass of the CF and the rapidity Δ of CF are respectively

$$K = \frac{M}{\sqrt{s}} = \sqrt{x_1 x_2}, \qquad (6)$$

$$\Delta = \frac{1}{2} \ln\frac{W+P}{W-P} = \frac{1}{2} \ln\frac{x_1}{x_2} ; \qquad (7)$$

$x_{1,2}$ are the fractions of energies taken off the incoming hadrons $h_{1,2}$ and deposited to the CF. The probability that this happens, $\chi(x_1,x_2)$, is

$$\chi(x_1,x_2) \sim \sum_{\{n_i\}} \delta(x_1 - \sum n_i x_{1i}) \delta(x_2 - \sum n_i x_{2i}) \prod_{\{n_i\}} P(n_i) \qquad (8)$$

where the sum extends over all MF's. Using the standard procedure of expressing δ functions as integrals one can perform all summations and one ends up with formula [22]

$$\chi(x_1,x_2) = \frac{1}{(2\pi)^2} \int_{-\infty}^{+\infty} dt_1 \int_{-\infty}^{+\infty} dt_2 \exp\left[i(x_1 t_1 + x_2 t_2)\right] \cdot$$
$$\cdot \exp\left[\int_{x_{10}}^{x_1} dx_1' \int_{x_{20}}^{x_2} dx_2' \omega(x_1',x_2') \left(e^{-i(x_1' t_1 + x_2' t_2)} - 1\right)\right]$$
$$\cdot \exp\left[- \int_{x_1}^{1} dx_1' \int_{x_2}^{1} dx_2' \omega(x_1',x_2')\right]. \qquad (9)$$

The last term represents the probability of there being no energy deposit from incoming hadrons beyond fractions x_1 and x_2. The spectrum of MF's

$$\omega(x_1,x_2) = \frac{dn}{dx_1 dx_2}$$

is assumed to have such form:

$$\omega(x_1,x_2) = C(x_1,x_2)G_{h_1}(x_1)G_{h_2}(x_2) \quad (10)$$

where $G_{h_{1,2}}(x_{1,2})$ are gluon distribution functions for $h_{1,2}$ and $C(x_1,x_2)$ represents the probability that MF's will form a CF. It can be then represented by the ratio of the total cross sections for gluon-gluon and hadron-hadron scatterings, which we shall parametrize as follows;

$$C(x_1,x_2) = \frac{\alpha}{x_1 x_2 s} + \delta \ln(x_1 x_2 \frac{s}{GeV^2}) \quad . \quad (11)$$

This parametrization is the simplest possible one which incorporates the most general trends of the cross sections as the function of the effective energy $W=\sqrt{x_1 x_2 s}$. As given MF must produce at least two particles (clusters) of mass μ each, the range of integration in eq. (9) is fixed by demanding that

$$x_{10}x_{20} \geq \frac{4\mu^2}{s} \quad . \quad (12)$$

Notice that in the limit $x_{1,2} \to x_{01,2}$ we have $C \sim \alpha/(4\mu^2)$, parameter α represents then roughly the dimension of the gluon cloud in the overlaping incoming hadrons. Because the number of gluons grows rapidly at small x's, the gluon distribution functions can be written as

$$G_{h_{1,2}}(x_{1,2}) \sim \frac{\beta}{x_{1,2}} = G(x_{1,2}), \quad (13)$$

other terms being unimportant. Parameter β controls here the amount of glue from each hadron involved in the interaction (the quantity uncontrolable by other means).

The same reasoning allows us to use the gaussian approximation for the integrals in the exponents in eq.(9) and to express the distribution $\chi(x_1,x_2)$ in terms of

$$I_{nm}(x_1,x_2) = \int_{x_{10}}^{x_1} dx_1' x_1'^n \int_{x_{20}}^{x_2} dx_2' x_2'^m \; G(x_1')G(x_2')C(x_1',x_2'); \quad n,m=0,1,2$$

$$I(x_1,x_2) = \int_{x_1}^{1} dx_1' \int_{x_2}^{1} dx_2' \; G(x_1')G(x_2')C(x_1',x_2'). \quad (14)$$

Defining

$$\hat{h}(x_1,x_2) = |I_{20}I_{02} - I_{11}^2|$$

we then have

$$\chi(x_1,x_2) = \frac{\exp\left[-I(x_1,x_2)\right]}{\hat{h}(x_1,x_2)} \times \exp\left\{-\frac{1}{\hat{h}(x_1,x_2)}\left[I_{20}(x_1-I_{10})^2+I_{02}(x_2-I_{01})^2 -2I_{11}(x_1-I_{10})(x_2-I_{01})\right]\right\} . \quad (15)$$

With $\chi(x_1,x_2)$ one can now calculate corresponding distributions in any other variables of interest, e.g.

$$\chi(x_1,x_2)dx_1dx_2=2K\chi(x_1=Ke^{\Delta},x_2=Ke^{-\Delta})dKd\Delta = \frac{2}{s}\chi(x_1=\frac{W+P}{s},x_2=\frac{W-P}{s})dWdP . \quad (16)$$

Any obserwable $\vartheta(\sqrt{s},y)$ (y being the rapidity) refering to the central region has to be then expressed as the convolution of $\vartheta(W=K\,s,y-\Delta)$ for the CF and $\chi(k,\Delta)$:

$$\vartheta(\sqrt{s},y) = \int_{K_{min}}^{1} dKK \int_{\ln K}^{-\ln K} d\Delta\,\chi(K,\Delta)\,\vartheta(w,y-\Delta) . \quad (17)$$

Notice that, in fact, in order to specify fully an event, one has to know the distribution of two variables (for example: inelasticity K and the position of the c.m. of the central fireball in rapidity, Δ , or $x_{1,2}$ etc.). The distribution of inelasticity only is the obtain by the averaging over the position of the fireball

$$\chi(K) = \int_{\ln K}^{-\ln K} d\Delta\,\chi(K,\Delta) . \quad (18)$$

In that way, however, some part of the dynamical information is averaged over.

The result of the fit to the data of Brick et al.[9] is shown in Fig.3.(The fit to the leading particle spectrum of ref. 8 ,not presented here, is equaly good). The values of parameters obtained are: μ=0.6 GeV, α=10, β=0.3, δ=0.2 GeV^{-2}. They are similar to those obtained already before in ref. 20 (notice the change in the parametrization of function $C(x_1,x_2)$, parameters α and β are then not directly comparable in both cases). As before, we observe decrease (or softening) of G(x) ($\beta<1$), much weaker, however, than before. The parameters μ, α and β were fixed by the shape of inelasticity and leading particle distributions, parameter δ is mainly responsible f the right energy dependence of the mean inelasticity $\bar{K}$.

In Fig.4 the results for different energies are presented. corresponding values of $\bar{K}$ are: 0.41, 0.36 and 0.26 for $\sqrt{s}$=16.5, 6 and 540 GeV. (For $\sqrt{s}$=40TeV $\bar{K}$ would be only 0.15).

In Fig.5 the $\chi(x_1-x_2)$ is plotted for two different energi

ce the big spread in momenta of the CF.(The effect is actually not so dramatic as the main variable in all physical applications is rather rapidity Δ of CF, not its momentum. The dispersion in Δ is, at most, of the order of 0.3÷0.4 for most of the values of $\vec{K}$).

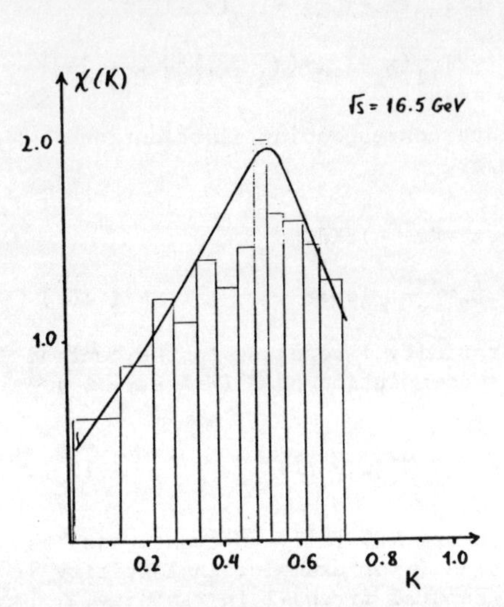

Fig. 3. The fit to (K) as measured by Brick et al.[9)]

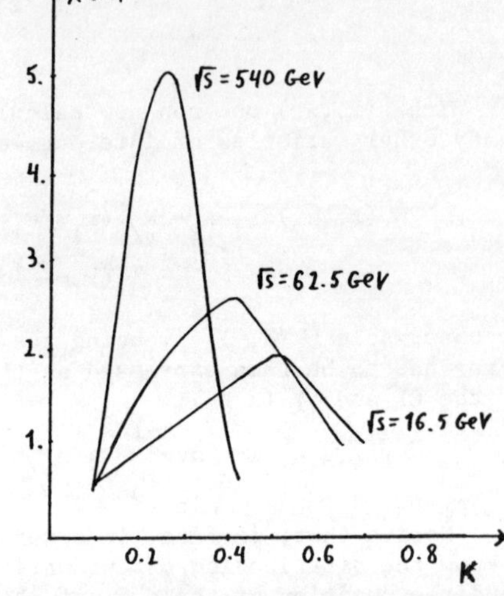

Fig.4. χ(K) for different energies.

Fig. 5. The distribution of the fireball (scaled) momenta. See text for values of parameters and discussion.

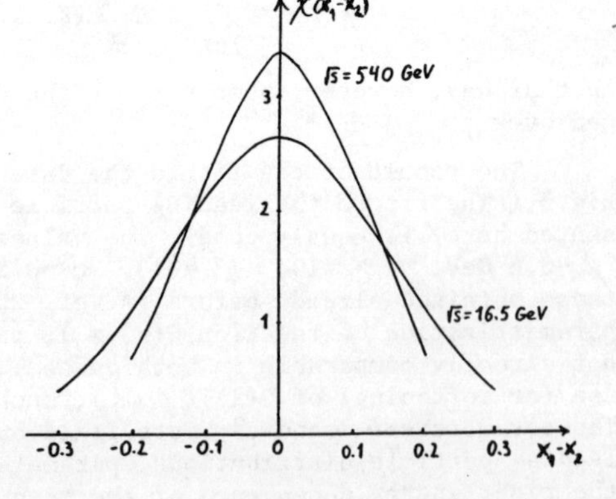

4. COMMENTS AND SUMMARY

The approach presented here can be regarded as an attempt to ·late that part of the dynamics which is responsible for the most of the production populating the central region. It is viewed here as the interaction of gluons present in every hadron, the eventual formation of the central fireball and its ultimate decay into observed particles. This is the usual subject of detail studies. But, and that was our crucial point, from many other investigations (mainly in the deep inelastic and hard scatterings) we have a fairly well established picture of hadron itself, its composition in terms of valence quarks and gluons and even some basic cross sections. This picture, we believe, should underlay every attempt aimed to describe the multiparticle processes. Our $\chi(K,\Delta)$ is supposed to do just this, to provide for every event in which energy $W=K\sqrt{s}$ is released at rapidity Δ, the probability of its occurence in terms of what is known about hadronic structure.

There are already other approaches using the concept of inelasticity, although none of them provides (or even utilizes) its distribution. It is, however, reassuring to find that the apparent decrease of mean inelasticity with energy, $\bar{K}\simeq 0.3$ at SPS energies, was also found in the independent analysis using the geometrical model and partition temperature concept[11]. The natural question which arises here is the interrelation between the impact parameter description of reactions (which is the basis of any geometrical model) and that with inelasticity. Both seem to be complementary, stressing different aspects of reactions and because of that there is apparently no one-to-one correspondence between them.

In what concerns the connection between the inelasticity and the leading particle, we would like to stress again the following important point: in drawing such connection in our case, both leading particles were supposed to be totally uncorrelated (modulo overall energy-momentum conservation) - in accordance with experimental findings (c.f. refs. 8,13). The opposit approach was presented recently, claiming that the mean inelasticity grows with energy rather than falls down[23]. This conclusion has been reached by analysing data on missing mass production: $pp\rightarrow M_X$ (missing mass)+p', p' being the recoiling proton. The missing mass M_X was then assumed to decay into another leading proton and the central fireball of mass M. Finally it was shown that $K_X=M_X/\sqrt{s}$ increases with $\sqrt{s}$ and, because from simple kinematics $K=M/\sqrt{s}\sim K_X^2$, so das also inelasticity K (from $K\sim 0.3$ at ISR to $K\sim 0.7$ at SPS). In such a construction, however, both leading particles (p' and the proton from fireball M) are maximally correlated. It is then just not our inelasticity which we were pesenting here. Most probably the inelasticity considered in ref.23 corresponds to selected sample of events only.(On the other hand such (growing with energy) inelasticity would result in mean multiplicity growing too fast with energy - at least in the statistical models).

One should stress here that the presence of one central fireball is but only an assumption (which is unavoidable if one wants to use the statistical picture). The other extreme would be to assume independence of all "mini-fireballs", one would arrive then at the structure characterictic to string models of multiparticle processes [24].

But in order to deal with QCD plasma one has to allow for some kind of thermalization and then we are again back with our concept of inelasticity. (One can, however, try to build a model for thermalization starting directly from the color string concept, c.f. ref. 25 and references therein).

As we have already indicated before, instead of leading particles one should rather talk about leading jets (composed with valence quarks in our case) fragmenting into a number of leading particles. Some properties of multiparticle distributions[26] also strongly suggest the presence of such component in the reaction. Altogether it leads to a two component picture of high energy multiparticle reactions in which we have central, thermalized (at least partially) fireball occuring in the phase space of (K,Δ) with probability $\chi(K,\Delta)$ and two valence jets producing leading particles plus some additional particles . Later would be most likely produced via some bresstrahlung-like mechanism with charachteristic Poisson-like distribution. Such a picture, gradually emerging as the most obvious and natural one from many analysis of data and models, is actually under consideration. (For other recent two-component approaches see ref. 27).

The fact that it is actually $\chi(K,\Delta)$ rather then $\chi(K)$ which seems to emerge naturally from the theory as a basic concept underlying statistical models of high energy reactions calls for additional checks. They could be performed by analysis of the observables crucially depending on the velocity (or rapidity) of the fireball as the forward-backward correlations in multiparticle distributions or the Bose-Einstein correlations. Those are actually also under consideration.

I would like to express my gratitudes to G.N.Fowler, A.Vourdas and R.M.Weiner on collaboration with whom this talk is based.

REFERENCES

1. Cleymans,J.,Gavai,R.V. and Suhonen,E., Phys.Rep. 130, 217 (1986); Shuryak,E.V., Phys.Rep.115, 151 (1984): "Local Equilibrium in Strong Interaction Physics", eds. Scott,D.K. and Weiner, R.M., World Scientific 1985; "Quark Matter'84", ed. Kajantie,K., Lecture Notes in Physics 221, Springer Verlag 1985; Mueller,B., Lecture Notes in Physics 225, Springer Verlag 1985; "Quark Matter'83", eds. Lundlam,T.W. and Wegner,H.E., North Holland 1985;

2. Feinberg,E.L., Sov.Phys.Usp. 26, 1 (1983);

Hagedorn,R., Rivista Nuovo Cim. 6, 1 (1983), c.f. also in "Quark Matter'84" and references therein;

3. Collected Papers of L.D.Landau, ed. Ter Haar, Gordon and Breach, New York 1965;
Carruthers,P., Ann.N.Y.Acad.Sci., 229, 91 (1974);
Cooper,F.,Frye,G. and Schonberg,E., Phys.Rev. D11, 192 (1975);
Daibog,E.I.,Rosental,I.L. and Tarasov,Yu., Fortsch. Phys. 27, 313 (1979); Shuryak,E.V., Phys.Rep. 61, 71 (1980);

4. Wehrberger,K. and Weiner,R.M., Phys.Rev. D31, 222 (1985);

5. Fowler,G.N.,Friedlander,E.M.,Plumer,M. and Weiner,R.M., Phys. Lett. 145B, 407 (1984);

6. Masuda,N. and Weiner,R.M., Phys.Rev. D18, 1515 (1978);

7. Cocconi,G., Phys.Rev. 111, 1699 (1958);

8. Basile,M. et al., Nuovo Cim. A73, 329 (1983), Lett.Nuovo Cim. 38, 359 (1983) and references therein;

9. Brick,D. et al., Phys.Lett. 103B, 242 (1981), Phys.Rev. D25, 2794 (1982);

10. Friedlander,E.M. and Weiner,R.M., Phys.Rev. D28, 2903 (1983);

11. Chou,T.T.,Yang,C.N. and Yen,E., Phys.Rev.Lett. 54, 510 (1985);
Chou,T.T. and Yang,C.N., Phys.Rev. D32, 1692 (1985);

12. Takagi,F., Z.Phys. C13, 301 (1982) and C19, 213 (1983);

13. Basile,M. et al., Lett.Nuovo Cim. 41, 298 (1984);

14. Plumer,M. in LESIP I, ref.1 and ref.5;

15. Chiu,C.B.,Sudarshan,E.C.G. and Wang,K.-H., Phys.Rev. D12, 902 (1975); Gorenshtein,M.I.,Zhdanov,V.I. and Sinyukov,Yu.M., Phys.Lett. 71B, 199 (1977) and Sov.Phys.JETP 47, 435 (1978); Bjorken,J.D., Phys.Rev. D27, 140 (1983); Kajantie,K.,Raitio,R, and Ruuskanen,P.V., Nucl.Phys. B222, 152 (1983);
von Gersdorff,H.,McLerran,L.,Kataja,M. and Ruuskanen,P.V., "Studies of the Hydrodynamic Evolution of Matter Produced in Fluctuations in $\bar{p}p$ Collider and in Ultra-relativistic Nuclear Collisions", Fermilab-Pub-86/18, c.f. also ref.1;

16. Carruthers,P. and Ming.D.V., Phys.Rev. D28, 130 (1983);

17. Basile,M. et al., Nuovo Cim. A67, 244 (1982);

18. Van Hove,L. and Pokorski,S., Nucl.Phys. B86, 243 (1975);

19. Mueller,A., Phys.Rev. D2, 224 (1970);

20. Fowler,G.N.,Weiner,R.M. and Wilk,G., Phys.Rev.Lett. 55, 173 (1985);

21. Fowler,G.N.,Vourdas,A.,Weiner,R.M. and Wilk,G. to be published;

22. This kind of approach dates back to the bremssthralung model of multiparticle production, Stodolsky,L. Phys.Rev.Lett. 28, 60 (1972);

23. Barshay,S. and Chiba,Y., Phys.Lett. 167B, 449 (1986);

24. Capella,A.,Pajares,C. and Ramallo,A.V., Nucl.Phys. B241, 75 (1984); Kaidalov,A.B. and Martirosyoan,K.A., Sov.J.Nucl.Phys. 39, 979 (1984) and 40, 135 (1984); Bopp,F.W.,Aurenche,P. and Ranft,J., Phys.Rev. D33, 1867 (1986);

25. Kerman,A.K.,Matsui,T. and Svetitsky,B., Phys.Rev.Lett. 56, 219 (1986);

26. Fowler,G.N.,Friedlander,E.M.,Weiner,R.M. and Wilk,G. to be published, see also R.M.Weiner, these proceedings;

27. Cai Xu,Chao Wei-qin,Meng Ta-chung andHuang Chao-shang, Phys. Rev. D33, 1287 (1986); Pancheri,G. and Srivastava,Y., Phys. Lett. 155B, 69 (1985); Banerjee,H., De,T. and Syam,D., Nuovo Cim. 89A, 353 (1985);

STOCHASTIC EVOLUTION OF MULTIPLICITY: IS THERE AN INITIAL STATE DISTRIBUTION?

Chia Chang Shih

Department of Physics, University of Tennessee

Knoxville, Tennessee 37996-1200

We have analyzed the stochastic evolution of the multiplicity distribution at finite $\bar{n}$ for a number of classical and quantum mechanic processes. Our solutions demonstrate the richness and flexibility of many possible pathways to hadronization. Phenomenological analyses based on the partially coherent distribution are summarized. Their implication on the necessity of the initial state distribution is also assessed.

1. INTRODUCTION

Recently there has been renewed interest on the stochastic nature of the multiplicity distribution in hadronic production at high energy.[1] As increasingly precise data at the Collider[2] and PEP[3] become available, we are headed toward a better understanding of the underlying dynamics and the statistical nature of the particle production processes[4].

In this talk, I shall discuss the nature of the stochastic evolution of multiplicity both as a classical and as a quantum mechanical process. I shall first explicitly analyze a number of simple models and use them to motivate phenomenological analyses. Comparing the resultant phenomenological analyses with the models enables us to gain further insight into the nature of the path to hadronization.

I have divided my talk into six sections. In Sect. 2, several simple examples of the classical production processes are given yielding to the Poisson distribution (PD) or the negative binomial distributions (NBD) used often in phenomenology. I shall use them to demonstrate various features of hadronization. Formulating the classical processes in terms of a Fokker Planck equation or a stochastic differential equation,[5] we readily observe the sensitivity of the long

long term solutions to the initial profile of the multiplicity. In Sect. 3, I shall examine multiplicity production as a quantum mechanical process.[6] The solutions are expressed through the use of the density matrix in a coherent state representation. The simplest case of coherent production is viewed as the solution of a harmonic oscillator driven by a deterministic force field.[5,7] With the addition of a stochastic force, chaos is then mixed naturally with coherence, leading to partially coherent laser-like distribution (PCLD).[1] Other generalization of the force field may also lead to distributions which are sub-Poissonian.[8] In Sect. 4, I shall present some highlights on the phenomenological fits with the surprisingly flexible PCLD, and demonstrate that neither the PD nor the NBD limit is the most likely solution. In Sect. 5, I shall discuss correlations on the consequence of conservation laws[9] and clustering on the multiplicity distributions. In Sect. 6, I briefly comment on the stochasticity contents of several dynamical models, such as Quantum chromodynamics (QCD) branching,[10] Lunds,[11] and phase space model.[12] I shall conclude with a discussion on the path to hadronization and reflect on the existence of the initial distribution in multiplicity.

2. SOME CLASSICAL PRODUCTION PROCESSES

It is useful to discuss briefly the analytical solutions of several very simple examples of multiplicity distributions. Possible features of hadronization can then be clearly demonstrated. These features sometimes become obscured in a more complicated model.

2.1 A Simple Markov Emission Process

It is well known in optics that the radiation by a coherent source leads to a PD, and radiation by a chaotic source leads to an NBD. Consider first the evolution of a Markov process in a dynamical variable t, which can either be time, rapidity, thickness of a jet, or some other appropriate variables. In an interval dt, the chance of producing a secondary is given by $dp = \lambda(t)\,dt$. With the secondaries not radiating, the change in P_n, the probability of observing n secondary in dt, is given by

$$dP_n(t) = P_{n-1}(t)\,\lambda(t)\,dt - P_n(t)\,\lambda(t)\,dt. \qquad (2.1)$$

For an initial condition $P_n = \delta_{n,0}$, we get the Poisson distribution

$$P_n\ (t) = PD\ (n,\bar{n}(t)),\ PD\ (n,\bar{n}) \equiv e^{-\bar{n}}(\bar{n})^n/n!,\ \bar{n}\ (t) = \int_0^t \lambda\ (t)\ dt. \quad (2.2)$$

With a more general initial condition at t = 0, $P_n = c_n$, we get

$$P_n\ (t) = \sum C_{n-m}\ PD\ (m,\bar{n}). \quad (2.3)$$

For a stationary λ, $\bar{n}$ (t) = ∞, when t = ∞. But even for a non-stationary source the form of the distribution remain Poissonian, whether $\bar{n}$ (t) becomes infinity or not. In the limit of a large $\bar{n}$, and fixed $z = n/\bar{n}$, we get the asymptotic Koba, Nielson, Olessen limit (KNO).[13]

$$\psi\ (z) = \lim \bar{n}\ P_n\ (t) = \delta\ (z - 1)\ , \quad (2\text{-}4)$$

which is singular and independent of C_n.

When λ (t) itself is subject to a random fluctuation, P_n is broader than a PD. An easy way to construct such a model is to assume a Gaussian distribution in λ,

$$p\ (\lambda) = e^{-\lambda/\lambda_o}/\lambda_o\ . \quad (2.5)$$

We then get P_n as the Bose-Einstein distribution

$$P_n\ (t) = BE\ (n,\bar{n}) \equiv (\bar{n})^n/(1 + \bar{n})^{n+1}, \quad (2.6)$$

with $\bar{n} = \lambda_0 t$, leading to the KNO limit

$$\psi\ (z) = e^{-z}. \quad (2.7)$$

2.2 Simple QCD Branching Process For Gluon Bremsstrahlung

Consider the secondary particles as the gluons with their production determined by the bremsstrahlung probability $dp(g \rightarrow gg) = A\ dt$, the branching equation for P_n is then given by

$$dP_n(t)/dt = A(n-1)\ P_{n-1} - AnP_n\ (t)\ . \quad (2.8)$$

For the initial condition $P_n\ (t=0) = \delta_{n,k}$, we get

$$P_n\ (t) = \frac{(n-1)!}{(n-k)!\ (k-1)!}\ (\frac{\bar{n}}{k})^{-n}\ (\frac{\bar{n}}{k} - 1)^{(n-k)}. \quad (2.9)$$

At t = ∞, Eq. 2.9 admits a solution similar to NBD,

$$P_n\ (t) \sim NB\ (n,\bar{n},k) = \frac{(n+k-1)!}{(n-1)!k!}\ (\frac{\bar{n}}{k})^n\ (1 + \frac{\bar{n}}{k})^{-(n+k)} \quad (2.10)$$

Since $\bar{n}(t) = ke^{At} = \infty$ at t = ∞, we may examine the KNO limit and get

$$\psi(z) = \psi_k(z) = \frac{k^k}{(k-1)!} z^{k-1} e^{-kz} \quad (2.11)$$

determined by the initial condition.

2.3 QCD Processes With Gluon Quark Bremsstrahlung And Pair Production

Including several other QCD processes ($dp\,(g \to q\bar{q}) = B\,dt$, and $dp(q \to qg) = C\,dt$), and with the n_q stationary, Eq. 2.8 is replaced by

$$dP_n(t)/dt = A(n-1)P_{n-1} - BnP_n + CP_{n-1} - (n \to n+1) . \quad (2.12)$$

I shall skip the detail, and simply write the solution in terms of a generating function $G(x,t) = \sum x^n P_n(t)$!:

$$G(x,t) = G_0(\tilde{x}(x))((B-A)/(B-Ax-A\zeta + Ax\zeta)^{C/A}, \quad (2.13)$$

$$\tilde{x}(x) = (B-Ax-B\zeta+Bx\zeta)/(B-Ax-A\zeta+Ax\zeta),\ \zeta = e^{-(B-A)t},$$

leading to $\bar{n}(t) = n_0\zeta + C(1-\zeta)$ and an KNO limit

$$\psi(z) = \psi_k(z),\ k = C/(B-A). \quad (2.14)$$

Notice that unlike Eq. 2.11, Eq. 2.14 is independent of the initial condition.

2.4 QCD Processes Including Recombinations

Here I shall also consider the possibility of the recombination processes including $dp(gg \to g) = D\,dt$, and $dp(q\bar{q} \to g) = E\,dt$. Eq. 2.12 is now replaced by

$$dP_n/dt = A(n-1)P_{n-1} - BnP_n + (C+E)P_{n-1} - Dn(n-1)P_n - (n \to n+1). \quad (2.15)$$

Since the complete solution is too complicated to present here, I shall only give the asympototic solution at $t = \infty$ as beta function,

$$P_n \sim \frac{1}{n!}\left(\frac{A}{D}\right)^v \beta(u,v),\ u = \frac{B}{D} - \frac{C+E}{A},\ v = \frac{C+E}{A} + n. \quad (2.16)$$

For very small values of D and E, the above solution is very similar to Eq. 2.13. However, the recombination process D always reduces the fluctuation in P_n. Remarkably, when the detail balancing condition is satisfied,

$$d = \frac{B}{D} - \frac{C+E}{A} = 0, \quad (2.17)$$

$P_n(t)$ actually recovers the Poisson distribution. Notice also when d is less than 0, Eq. 2.16 is actually sub-Poissonian.

2.5 Fokker-Planck and Stochastic Differential Equation

The above branching equation 2.15 is closely analogous to the master equation of the chemical reaction $A + X \underset{k4}{\overset{k2}{\rightleftarrows}} 2\,X$, and $B + X \underset{k3}{\overset{k1}{\rightleftarrows}} C$, and have been extensively analyzed.[5] With the Poisson representation for P_n,

$$P_n\,(t) = \int d\alpha\; f(\alpha,t) e^{-\alpha}\; \alpha^n/n!\;, \tag{2.18}$$

the dynamics of the master equation can be reformulated in terms of the Fokker-Planck equation of $f(\alpha,t)$:

$$\frac{\partial f}{\partial t} = [-\frac{\partial}{\partial\alpha}\,(k_3C+k_2A\alpha-k_1B\alpha\,-\,k_4\alpha^2)\,+\frac{\partial^2}{\partial\alpha^2}\,(k_2A\alpha-k_4\alpha^2)]f\;. \tag{2.19}$$

It is also equivalent to the stochastic differential equation,

$$d\alpha = [k_3C + k_2\;A\alpha\,-\,k_1B\alpha\,-\,k_4\alpha^2]dt + \sqrt{2(k_4A\alpha-k_4\alpha^2)}\;dW \tag{2.20}$$

where dW is a Wiener process. If the system is near equilibrium, satisfying detail balancing, the resultant distribution is generally a PD. However, near a phase transition, or when critical damping is observed, the nature of the multiplicity fluctuation can change very dramatically. Poisson distribution may not follow.[14]

3. STOCHASTIC PRODUCTION AS A QUANTUM MECHANICAL PROCESS

In this Section, I shall discuss the general features of secondary production as a quantum mechanical process. Formulation developed here shall be used later on in Sect. 4.

3.1 Coherent Production Of A Single Mode

In the early development of quantum optics, Glauber[6] considered the theory of photon detection by a many-atom system. In this system, the essential ingredient is the independence of the individual detection probability. Klauder and Sudarshan[5] preferred to consider the photon procedure as a semi-classical process similar to Eq. 2.1, where the Markov property is in the time variable t. I shall use the same rationale and divide a space-time region of the phase space into an large number of cells. Each cell may be in its ground state $|\,g\rangle$ or its hadronization state $|h\rangle$. The occupation of different cells is independent. Consider for the time being that all the hadronized states have the same energy. The collective behavior of all the cells can then be treated by an operator $a = \frac{1}{\sqrt{L}} \sum^{L} \tau_j$, where $\tau_j\, |h_j\rangle = |g_j\rangle$ is the lowering operator of the j-th cell. When the number of the cell, L, is much larger than the total number of the hadronized state, the dynamics of the system can be described as a forced harmonic oscillator:

$$H = \hbar\omega\ a^+ a + aV^*(t) + a^+V(t)\ . \tag{3.1}$$

Notice that in the right side of this expression I have explicitly assumed that all the cells in this region experience the same source. Thus the production by all the cells is not only independent but also coherent. The solution of the above problem is best treated in terms of the density matrix ρ. I shall further assume that ρ is diagonal with respect to the coherent states $|\alpha\rangle$. Here $|\alpha\rangle$ is an eigen state of the above operator a, so that $a|\alpha\rangle = \alpha|\alpha\rangle$, and is given by the well-known formula

$$|\alpha\rangle = \exp(-\tfrac{1}{2}\ |\ \alpha\ |^2)\ \sum\ (a^+)^n\ |o\rangle/n!\ . \tag{3.2}$$

If the density matrix is written in the form

$$\rho = \int d^2\alpha|\ P(\alpha)\ |\alpha\rangle\langle\alpha|\ . \tag{3.3}$$

then $P(\alpha)$ satisfies the von Neumann equation

$$i\hbar d\ \rho/dt = [H,\rho]\ . \tag{3.4}$$

For a linear and deterministic source term V(t), the S-matrix solution of Eq. 3.2 is given by

$$S = \exp(i(a^+V(w) + aV^*(w))/(\hbar w)^{1/2}),\ V\ (w)=\int V(t)e^{iwt}dt. \tag{3.5}$$

In terms of the number states $|n\rangle$, the transition probability between an initial state $|m\rangle$, and a final state $|n\rangle$ is given by, $x = |V(w)|^2/\hbar w$

$$Pm,n^* = Pn,m = e^{-x}x^{m-n}\ (L_m^{\ m-n}\ (x))^2\ n!/m!,\ m>n\ . \tag{3.6}$$

In particular, if m = 0, P_n is a Poisson distribution. The dynamics of the above forced oscillator possesses a somewhat stronger feature.[5,7] Starting from a coherent state, $P_n(t)$ is actually a PD throughout the evolution in analogous to Eq. 2.2. We get the equivalent expression expressed in terms of the density matrix,

$$P(\alpha) = \delta^2(\alpha - \alpha_v(t)), \tag{3.7}$$

where

$$\alpha_v(t) = e^{iwt} (\alpha_0 + i \int_0^t dt' \exp|-(iwt') V(t')/\hbar) . \qquad (3.8)$$

3.2 Chaotic Production Of A Single Mode

If the source is not deterministic, and m=0, we may also impose a gaussian distribution in the strength of the driven force. The distribution $P_n(t)$ is then modified as

$$P_n(t) = \text{const} \int e^{-\bar{n}/\lambda_0} d\bar{n} \; PD\,(n,\bar{n}) = BE(n,n_0) . \qquad (3.9)$$

Here, the situation is very analogous to those of Eq. 2.5 in the sense that a PD has been transformed into an NBD. However, the stochastic nature of the two processes are somewhat different. Notice that within a given value of $\bar{n}$, the coherence of $|\bar{n}\rangle$ is not completely destroyed. The randomization procedure just outlined is therefore somewhat restrictive.

A more general stochastic treatment can allow the coupling between a state $|\alpha\rangle$ and its environment, which may or may not be a heat bath. In this formulation, an effective Fokker-Planck equation can often be found.[5] The resultant equations are very similar to the branching equations discussed in Sect. 2.

3.3 Partially Coherent Processes

Consider again that the initial state is given by $|0\rangle$. But in this example, the driven force possesses a deterministic as well as a chaotic part,

$$P(\alpha) = \text{const.} \exp(-|\alpha-\nu|^2 / N) \qquad S = |\nu|^2 \qquad (3.10)$$

Eq. 3.7 should then generalize to

$$P_n = \exp(\frac{-S}{1+N/k}) \cdot \frac{(N/k)^n}{(1+N/k)^{n+k}} L_n^{k-1} (- \frac{kS/N}{1+N/k}) \qquad (3.11)$$

The above expression was first derived by Glauber[6] and used very successfully in quantum optics. Here, we shall refer to it a partially coherent laser-like distribution (PCLD). Notice that in the limit of noise N=0, the PCLD is reduced to a PD, and in the limit of total chaos S=0, the PCLD is reduced to an NBD. Equation 3.11 is therefore an interpolation formula between PD and NBD.

3.4 Sub-Poissonian Distribution Of A Single Mode

Consider again Eq. 3.1 Glauber noticed that as long as the source V is a deterministic function of a (and not a function of a^+), a coherent state remains a coherent state leading to a PD in P_n. A different direction of generalization is to include a linear function of a^+. This is equivalent to writing the Hamiltonian of the system as

$$H = \hbar w a^+ a + f a^2 + f^* a^{+2} + a V^*(t) + a^+ V(t) \tag{3.12}$$

so that

$$-i da/dt = wa + (V(t) + 2a^+ f)/\hbar . \tag{3.13}$$

Solution of the above equation is usually referred to in quantum optics as the two-photon coherent process. There, a new effective creation (destruction) operator b can be introduced through

$$b = \mu a + \nu a^+ , \quad |\mu|^2 - |\nu|^2 = 1 . \tag{3.14}$$

Many properties of the coherent state can be translated into b. For example, a coherent state of b, $b|\beta\rangle = \beta|\beta\rangle$, leads to

$$\bar{n} = |\nu|^2 + |\mu^*\beta - \nu\beta^*|^2 \tag{3.15}$$

$$\overline{n^2} - \bar{n}^2 = \bar{n} + 2|\beta|^2 \; |\nu|^2 \; (4|\nu|^2+3) - (\mu^*\nu^*\beta^2 + \mu\nu\;\beta^2) \\ (1 + 4|\nu|^2) + |\nu|^2 + 2|\nu|^4 . \tag{3.16}$$

allowing the possibility that the variance of n is less than $\bar{n}$, and implying a sub-Poissonian distribution in P_n.[15]

3.5 Production Of Many Modes

In general, the secondaries produced during a collision process have an energy spectrum. Different division of cells of the phase space should be used for individual modes of energy. If the secondaries within different modes are independently produced, we can construct a harmonic oscillator for each mode. The above formulation is equivalent to the formulation of a second quantized field coupled linearly to an external currently was introduced earlier from other dynamical consideration.[9]

With independent modes of the secondaries, the total multiplicity distributions should be convoluted. We therefore get

$$P_n = \sum P_{n1} P_{n2} \cdots P_{nk} \delta(n_1 + n_2 + \ldots + n_k - n) . \qquad (3.18)$$

In one extreme, when all the Pn's are PD, the resultant Pn is also a PD. In the other extreme, of an equally shared $\langle n \rangle$ among k modes of NBD, we again get an NBD. However, if the value of k is much larger than $\bar{n}$ ($\bar{n}$ shared by many modes), the resultant NBD would become very sharp, taking the PD as the limit distribution. For a phenomenological analysis, convolution of PCLD with different $\bar{n}$ should also be considered.

4. PHENOMENOLOGICAL ANALYSES

In this section I shall present very briefly phenomonological analyses for several situations. Qualitative features of these results remain the same when more possibilities are included.

4.1 Total Charge Multiplicity Distributions

I shall fit the P_n distribution in the form of the PCLD for finite $\langle n \rangle$, with the M (signal/noise) and the cell number k treated as free parameters. There remain several dynamical questions: (1) Shall we count the original hadrons, or use $n - n_\alpha$ (n_α being the effective number of leading hadrons)? (2) Shall we use $\langle n_{ch} \rangle$ or $n_+ = n_- = n_{ch}/2$ as the intrinsic parameter? (3) Shall we introduce a parameter corresponding to the clusters size C?

In Fig. 4.1.1, we present a χ^2 fit, using $\langle n_{ch} \rangle$-2 as the intrinsic parameter with k=4 and c=1. The only free parameter is the coherence parameter M. If we use $\langle n_{ch} \rangle$ as the intrinsic parameter with k=3 and c=1 we get Fig. 4.1.2. In Fig. 4.1.3, $\langle n_{ch} \rangle$2 as the intrinsic parameter with k=1 and c=2. Notice that all three fits are comparable and basicly acceptable. (At a high enough energy, when $\bar{n}$ is large, the effects on the finiteness of $\bar{n}$ is less important. However, at the current energy, the asymptotic KNO expression is not always adequate.)

4.2 Nonsingle-Diffractive Multiplicity Distributions

When appropriate selection criteria are applied (NSD), we may hope to exclude the bulk of single diffractive events.

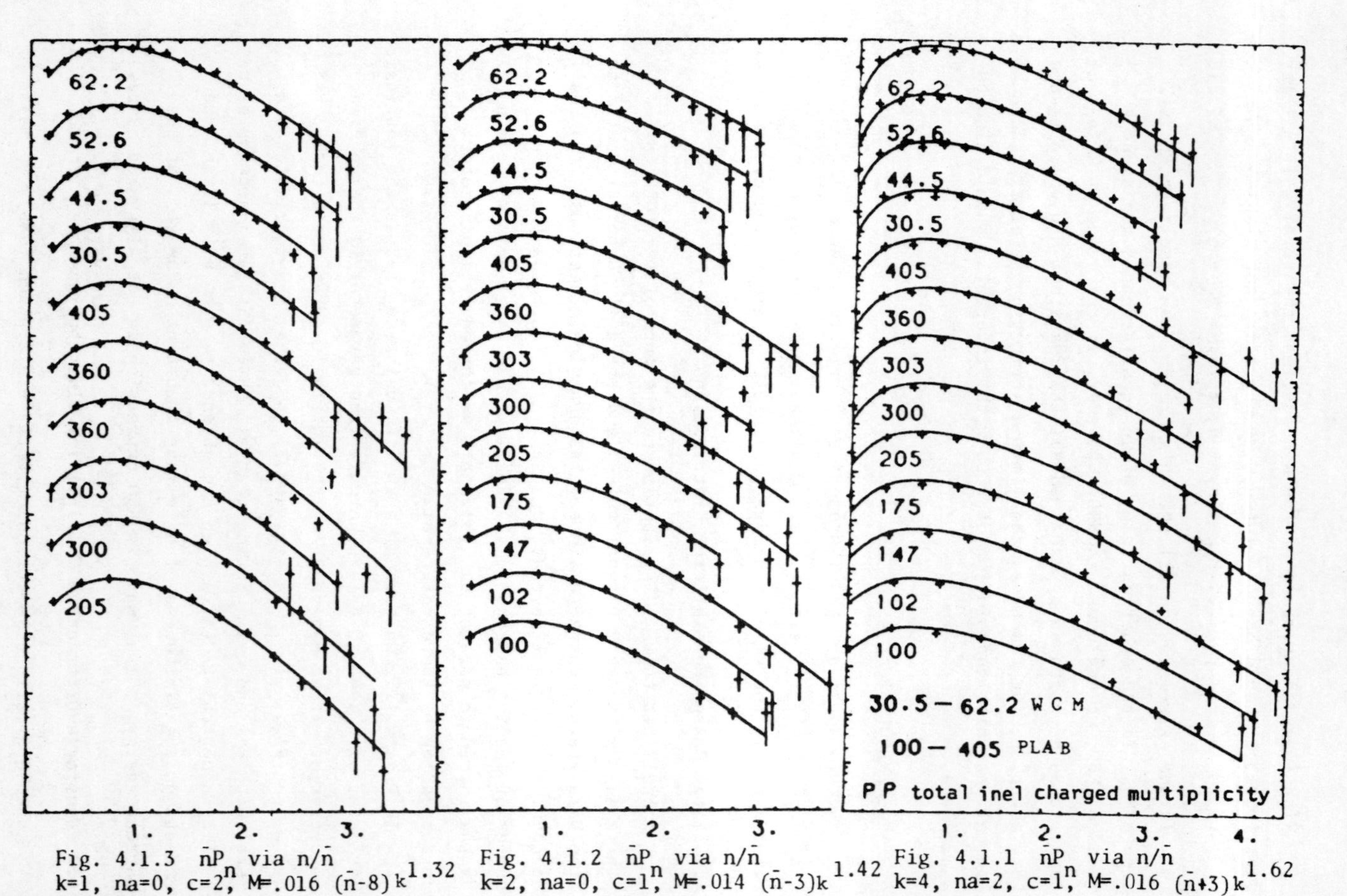

Fig. 4.1.3 $\bar{n}P_n$ via $n/\bar{n}$
$k=1$, $na=0$, $c=2$, $M=.016\ (\bar{n}-8)k^{1.32}$

Fig. 4.1.2 $\bar{n}P_n$ via $n/\bar{n}$
$k=2$, $na=0$, $c=1$, $M=.014\ (\bar{n}-3)k^{1.42}$

Fig. 4.1.1 $\bar{n}P_n$ via $n/\bar{n}$
$k=4$, $na=2$, $c=1$, $M=.016\ (\bar{n}+3)k^{1.62}$

The shape of the KNO curves are then different from those associated with the total multiplicity of 4.1. In a recent UA5 studies, negative binomials are used to fit data ranging from FNAL to the Collider.2]
The values of k thus obtained have a strong Energy dependence. However, the interpretation of such change in shape is by no means unique. Various PCLD fits can be found with the cumulant moment γ_j nearly identical to those obtained by UA5 using NBD. Detailed comparison with P_n also leads to the same conclusion. Clearly, we cannot easily distinguish the PCLD from the NBD fits. The dependence of M and/or k on energy is needed for a better understanding of the dynamics behind the production processes. Sensitivity of extrapolation of the multiplicity to even higher energy should be investigated.

4.3 Rapidity-Restricted Multiplicity Distributions

I shall only demonstrate briefly fits to an early set of data with the secondary hadrons restricted to a window in the rapidity $|y| < y_c$. In Fig. 4.3.1, k is fixed at 2 and M is fitted. In Fig. 4.3.2, $M = \infty$, (NBD), and k is fitted. Again, both fits are comparable and equally acceptable.

4.4 e^+e^- Two-jet Multiplicity Distributions

I would like to present here a sample of fits to the inclusive two-jet data which have been successfully fitted experimentally by negative binomials. Since the result of our fits are so close to each other, it is necessary to present them in the following table. Here, I chose the input value of k ranging from 1 to 100. A best-fitted M can then be given for each k value. There is indeed very little difference between these choices.

4.5 Freedom In The PCLD Distributions

So far most of the existing phenomenological analyses have used either Poisson or negative binomial distributions. Is the ambiguity of using PCLD I have presented here real? The answer is yes. To see this I shall first write down the cumulant moments associated with the PCLD

$$\gamma_2 = \frac{1}{\bar{n}} + \frac{a(1+b)}{k}, \quad \gamma_3 = \frac{1}{\bar{n}^2} + \frac{3a(1+b)}{\bar{n}k} + \frac{2a^2(1+2b)}{k^2} \, . \tag{4.1}$$

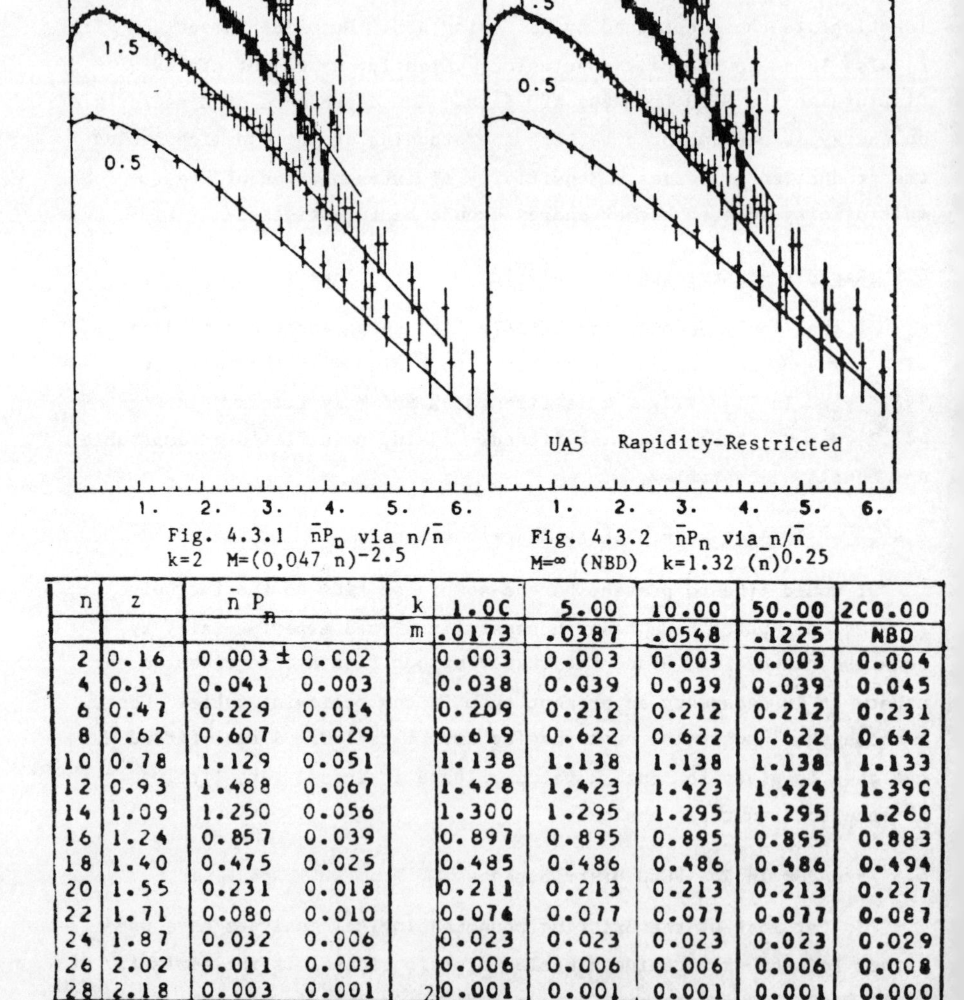

Fig. 4.3.1 $\bar{n}P_n$ via $n/\bar{n}$
k=2 $M=(0{,}047\,\bar{n})^{-2.5}$

Fig. 4.3.2 $\bar{n}P_n$ via $n/\bar{n}$
$M=\infty$ (NBD) $k=1.32\,(\bar{n})^{0.25}$

n	z	$\bar{n}P_n$		k	1.00	5.00	10.00	50.00	200.00
				m	.0173	.0387	.0548	.1225	NBD
2	0.16	0.003	± 0.002		0.003	0.003	0.003	0.003	0.004
4	0.31	0.041	0.003		0.038	0.039	0.039	0.039	0.045
6	0.47	0.229	0.014		0.209	0.212	0.212	0.212	0.231
8	0.62	0.607	0.029		0.619	0.622	0.622	0.622	0.642
10	0.78	1.129	0.051		1.138	1.138	1.138	1.138	1.133
12	0.93	1.488	0.067		1.428	1.423	1.423	1.424	1.390
14	1.09	1.250	0.056		1.300	1.295	1.295	1.295	1.260
16	1.24	0.857	0.039		0.897	0.895	0.895	0.895	0.883
18	1.40	0.475	0.025		0.485	0.486	0.486	0.486	0.494
20	1.55	0.231	0.018		0.211	0.213	0.213	0.213	0.227
22	1.71	0.080	0.010		0.076	0.077	0.077	0.077	0.087
24	1.87	0.032	0.006		0.023	0.023	0.023	0.023	0.029
26	2.02	0.011	0.003		0.006	0.006	0.006	0.006	0.008
28	2.18	0.003	0.001		0.001	0.001	0.001	0.001	0.000
				χ^2	12.94	11.13	11.13	11.16	11.51
				b	0.002	0.002	0.002	0.002	0.031

Table 4.4.1 PCLD fits to e^+e^- (inclusive whole events)
m = noise/signal (amp) = $0.0173\sqrt{k}$, $b=d\langle n_B\rangle_F/dn_F$

We then introduce a k_{eff}

$$k_{eff} = \frac{k}{a(1+b)} . \tag{4.2}$$

The cumulant moments can now be approximated to the first order using an NBD with $k = k_{eff}$; i.e.,

$$\gamma_j \ (PCLD(k,a,b)) = \gamma_j \ (NBD(k_{eff})) + O({k_{eff}}^{-2}) . \tag{4.3}$$

Since the multiplicity distributions allow a lot of compensation between different effects, it is natural that a large amount of freedom exists and ambiguities are left unconstrained by the experimental data. We are forced to look into more detailed features of a distribution such as the global or local correlations.

5. CORRELATIONS AND CONSERVATION LAWS

A number of very interesting measurements are available which put constraints on the P_n. For example, n may be made of two subpopulations, n_1 and n_2, which may stand for charge and neutral (or forward and backward) subpopulations, we can then discuss P_n in conjuction with the distribution of n_1 within n. More explicitly, we may write

$$P(n_1, n_2) = P(n) \ P(n_1 | n). \tag{5.1}$$

In the spirit of independency, the most natural $p(n_1 | n)$ is the binomial distribution

$$P(n_1 | n) = \frac{n!}{n_1 ! n_2 !} \ p^{n1} \ (1-p)^{n2}, \ n_2 = n-n_1 . \tag{5.2}$$

When Eq. 5.2 is combined with a NBD distribution in n, a fair amount of simple analytic results can be found. For example, in this special case the conditional distribution of n_2, with fixed n_1, is again an NBD.

$$P(n_2 | n_1) = NB(n_2, \bar{n}_2, k' = k+n_1) \tag{5.3}$$

so that

$$\bar{n}_2 = \langle n_2 \rangle \ \frac{k+n_1}{k+\langle n_1 \rangle} , \qquad \langle n_1 \rangle = p \ \bar{n} , \ \langle n_2 \rangle = (1-p) \ \bar{n} \tag{5.4}$$

Similarly,

$$\tilde{d}^2 \equiv \frac{4}{\bar{n}} \langle (n_1 - \bar{n}_1)^2 \rangle_n = 4p(1-p) < 1 . \tag{5.5}$$

For a binomial distribution, with p=1/2, the value of $\tilde{d}_n^2$ is always 1.

Experimental data indicate that a strong linear relationship indeed exists between $\bar{n}_2$ and n_1 for pp scattering. An NBD in n with a very large k is therefore not likely. On the other hand, the reduced variance of $\tilde{d}_n{}^2$, is definitely greater than 1, indicating the possibility of clustering.[1]

In the stochastic formulation, an exact global conservation law which imposes correlations is usually treated in a more primitive fashion. Consider what happens near the end of the stochastic evolution. The balance of a globally conserved physical quantity left by the produced secondary has to be shouldered by the rest of the population, making it non-Markovian. But, depending on the production mechanism, the deviation from Markov may not be very serious. Here I shall discuss only the conservation of two quantities, total energy and charge.

5.1 Conservation of Energy

Taking Eq. 2.1 as an example of the extreme case, we find that the energy constraint may be reflected in the t dependence of the production strength. This however does not change the shape of P_n, which remains to be a Poisson distribution at all t.

In a more realistic event, energy is shared by many partners. First of all, the leading particles may steal half of the total energy. The rest of the energy then has to be shared by the charged particles and the usually unobserved neutral particles. In the sector of the charged particles, we therefore do not expect the consideration of energy-momentum to be serious (except perhaps for the very high multiplicity fluctuations). The central rapidity region is even safer than the fragmentation region.

5.2 Conservation Of Charge

The conservation of charge is a very different matter and has potentially far-reaching consequences. There are two types of formulations leading to rather different predictions of the multiplicity distributions.

5.2.1 Local compensation via neutral clusters. As a typical example, we may consider the limiting case where pairs of charged secondaries are produced by neutral clusters. The clusters themselves are stochastically produced. In this situation, P_n may be characterized by the number of clusters $n_c = n_{ch}/2$. And the finiteness effect of the n_c is stronger than that of n_{ch}. For instance, a PD in n_c leads to a P_n,

$$P_n = \exp(-\bar{n}/2)\,(\bar{n}/2)^{n/2}/(n/2)!,\ n=n_{ch}, \tag{5.6}$$

which is rather different from a PD in n_{ch}, and can be easily distinguished experimentally.

5.2.2 Local compensation via generalized coherent states. A subspace of the coherent states discussed in Sect. 3 may contain a built-in local charge compensation.[9] Using two independent destruction operators, a(+) and a(−), we may construct

$$|\alpha_k> \sim \sum_n \frac{1}{(n!)^2}\, a(+)^n\, a(-)^n\ |0> \tag{5.7}$$

so that $a(+)\, a(-)|\alpha> = \alpha^2\,|\alpha>$. Notice that corresponding to this coherent state, the P_n distribution is given by

$$P_n \sim \bar{n}^{2n}/(n!)^2, \tag{5.8}$$

which is very different from a Poisson distribution.

5.2.3 Semi-global compensation processes. Consider a situation similar to those of Eq. 2.1. Assume that during this evolution, the source has a net charge at any given time. Its subsequent radiation tends to keep the source nearly neutral ($n_+ = n$, $n_- = m$). More pre-cisely, we shall include an additional term $d\, p(n{\to}m) = \mu(m-n)$. Since q (net) = n−m, the evolution of P_n can be written as

$$dP_{n,m} = \lambda(P_{n,m-1} + P_{n-1,m} - 2P_{n,m}) + \mu(n-m)\,(P_{n+1,m-1} - P_{n-1,m+1}), \tag{5.9}$$

which has the same structure as a truly stochastic process. We notice that the distribution in n_{ch} may still be close to a PD, but the deviation n−m is now small. Charge conservation is therefore almost satisfied with the local charge compensation. The individual distribution in n(+) and n(−), however may be sub-Poissonion. It would be interesting to examine separately the n(+) distributions in a restricted region.[15]

6. RELATION TO SOME DYNAMICAL MODES AND PATH TO HADRONIZATION

I shall briefly summarize the various stochastic properties of the production processes and on their relevance to a few dynamically oriented models.

6.1 Relevance to Dynamical Models

6.1.1 Internal and external secondaries. It would be desirable for a dynamical model to distinguish between the observed external hadron lines and the internal virtual lines, which may be hadrons, partons, or some other collective modes. For example, in a typical QCD branching model,[10] the final state gluons have to be linked to the observed hadrons. Including a parton fragmentation function could substantially increase the fluctuation of the observed P_n. Similarly, the phase space model[11] does not distinguish between the internal and external secondaries. The contribution to fluctuation either due to internal gluons, or to the external hadrons, could then be underestimated.

6.1.2 Recombination and final state in interaction. If hadronization occurs at a late stage of evolution, where many internal secondaries are participating, the tendency of recombination, or in other words, the final state interaction, cannot be totally ignored. This would reduce the fluctuation in P_n. In the limit of an equilibrium, and in detail balancing, the P_n may become Poissonian. Even a sub-Poissonian distribution cannot be ruled out completely.

6.1.3 Internal dynamics and KNO scaling. If the dynamical variable t is related to an observable such as rapidity, the dynamical evolution may be detectable. On the other hand, the internal evolution is most likely associated with the highly virtual state of the unstable gluon-quark plasma, and is not observable. We should therefore be cautious in identifying the behavior of the internally evolving $P_n(t)$ as a function of $\bar{n}(t)$ to the behavior of $P_n(t_f)$ as a function of the available energy of collision.

6.1.4 Pathway to hadronization. As we have demonstrated, there are many stochastically different pathways of hadronization. A limiting KNO behavior in one extreme may depend only on the fundamental structure of the dynamical equations, but in another extreme, it may reflect only a specific choice of the initial multiplicity profile. As we have demonstrated, using the multiparticle distribution P_n alone would not lead us to an unique answer.

6.1.5 Model dependence in simulations. Consider the example of a highly coherent dynamical model such as Lund,[11] where the breaking of a string linking a $q\bar{q}$ pair occurs independently, the mass shell condition and the quantum numbers associated with the final yo-yo states are externally imposed. At extremely high energy, we would anticipate that the production is coherent. Only if the background is small, a nearly Poissonian distribution of the P_n is natural. A Lund type of model used too heavily in a Monte Carlo model simulation to correct the raw data may suppress too much fluctuation and introduce bias in favor of a Poisson-like distribution.

6.2 Summary of Data

6.2.1 Coherence in data. Despite the non-uniqueness of the PCLD fits, it is clear that, under a variety of conditions, a substantial coherent component is present. Both the 2-jet, and the single-jet e^+e^-hadron processes are consistent with a highly coherent production process. This is further confirmed by a very weak forward/backward correlation. The recent observation of the very narrow distributions associated with the $\bar{p}p$ jet event may be interpreted in the same spirit. (Although at this point we cannot rule out the possibility of multi-mini-modes convolution leading to a very narrow multiplicity distribution.)[16]

6.2.2 Charge conservation. If the charge conservation or cluster properties need to be imposed for the total multiplicity data, and the n_{ch} distributions remain very narrow, the individual n(+) and n(-) must then be sub-Poissonian. We should be able to observe this by examining the fluctuation of the net charge in a limited domain of a rapidity window.

6.2.3 Chaos in data. The hadronic multiplicity distribution in $\bar{p}p$ is consistent with a PCLD, with k and/or M varying as a function of energy. Since the P_n distribution is not as sharp as a Poisson distribution, we may conclude that a fair amount of chaos is present in the processes.

6.2.4 Cluster consideration. Forward/backward distributions in $\bar{p}p$ hadrons strongly suggest that the production of clusters (or resonance) is needed at the intermediate stage of production. Should we see the clusters in the e^+e^- processes?

6.2.5 Initial profile. Do we need an initial profile of the multiplicity distribution? Consider that what we have observed are hadrons. An initial distribution is almost inevitable. Furthermore, the hadron-producing quark gluon matter may or may not produce a single coherent state of the hadronic operator. We may also interpret the widening of the P_n in $\bar{p}p$ is reflected in $f(\alpha,t)$ of the Poisson transform of such a distribution. I shall summarize the various pathways in a diagram for the possible relationship between the gluon and hadronic distribution.

QCD evolution --→ | ←-- hadronic evolution

		NBD	---------------- LOCAL --→
Collision	coherence	PD	NBD (chaos) observation
$g,Q,\bar{Q}$,			PCLD (correlation)
	branching	subPD	---------------- GLOBAL --→
hadron, A	Limiting P_n (QCD)?	clusterD	Limiting KNO (Exp)?

There are many exciting aspects of multiplicity distribution that I do not have time to include. To mention a few, the charge/neutral correlation, the charge/rapidity correlation, the charge/P_T correlation, the short range correlation, the multiplicity structures in neutrino induced interactions, and the structures in fragments of the leading particles may all lead to further understanding of the nature of the stochasticity of the strong interaction system. We look forward to real progress in the near future.

ACKNOWLEDGEMENTS

The research was supported in part by the National Science Foundation (NSF Phy-8417526). We are very grateful to Dr. P. Carruthers for many discussions and encouragement.

REFERENCES

1. Carruthers, P. and Shih, C. C., Phys. Lett. 127B, 242 (1983); ibid. Phys. Lett. 137B, 425 (1984); Phys. Lett. 165B, 209 (1985).
2. UA5 Collaboration, Alner, G. J. et al., Phys. Lett. 167B, 476 (1986).
3. Derrick M. et al. ANL.HEP-PR-85-105, preprint, and private communication.
4. Shih, C. C., Phys. Rev. to be published (1986); Biyajma, M. Phys. Lett. 139B 93 (1984); M. Biyajama and N. Suzuki, Phys. Lett. 143B, 463 (1983).
5. Klauder, J. R. and Sudarshan, E. C. G., "Fundamentals of Quantum Optics," (Benjamin, 1968); Saleh, B., "Photoelectron Statistics," (Springer-Verlag, 1982); Gardiner, C. W., "Handbook of Stochastic Methods," (Springer-Verlag, 1983).
6. Glauber, R. J. in "Physics of Quantum Electronics" ed., Kelley P. L., Lax, B. and Tannewald, P. F. (McGraw-Hill, New York, 1966).
7. Carruthers, P. and Nieto, M. M., Amer. J. Physics 33, 537 (1966); Glauber, R. J., Phys. Lett. 21, 650 (1966).
8. Yuan, H. P., Phys. Rev. A13, 2226 (1976).
9. Horn, D. and Silver, R., Ann of Phys. 66, 509 (1971).
10. Durand, B. and Sarcevic, I., University of Wisconsin preprint, 1985.
11. Anderson, B., Gustafson, G., Ingelman, G. and Sjostrand, T., Phys. Reports, 97, 31 (1983).
12. Hwa, R. C. and Lam C. S., McGill University preprint 1985; Lam, C. S. and Walson, M. A., Phys. Lett. 140B, 246 (1984).
13. Koba, Z., Nielson, H. B. and Olesen, P., Nucl. B40 317 (1972).
14. Schenzle, A. and Brand, H., Phys. Rev. A20, 1628 (1979).
15. Shih, C. C., to be submitted for publication.
16. For Example, "Coherent States" ed., Klauder, J. R. and Skagerstam, Bo-Sture, (World Scientific, 1985).

MULTIPLICITY DISTRIBUTIONS AND KNO SCALING IN THE PARTON BRANCHING MODEL

Ina Sarcevic[+]
School of Physics and Astronomy, University of Minnesota
116 Church Street S.E.
Minneapolis, Minnesota 55455

ABSTRACT

We present a parton branching model for the multiparticle production in $p\bar{p}$ collisions. We obtain a new non-scaling law for the probability distribution.[1] In the high energy limit, scaling is approached from below in agreement with experimental data. We contrast our branching model with the stochastic picture.[2]

1. INTRODUCTION

In 1972 Koba, Nielsen and Olesen[3] predicted , that at sufficiently high energies

$$\bar{n} \frac{\sigma_n}{\Sigma\sigma_n} = \psi \left(z = \frac{n}{\bar{n}}\right) \qquad (1)$$

where σ_n is the partial cross section for producing a state of multiplicity $\bar{n}$ and $\psi(z)$ is energy independent function. This scaling seemed to hold approximately for energies up to ISR[4], but at CERN collider[5] energies scaling violations have been observed. On Fig. 1 we see that KNO scaling violations are manifested in rising and

---

[+]Address after September 1, 1986.: Los Alamos National Laboratory, Theoretical Division T-8, MS B285, Los Alamos, NM 87545

broadening of the multiplicity distributions suggesting that if there is any scaling, it is approached from below as energy increases. The parton branching model that we derive in the next section systematicaly deviate from KNO scaling by approaching the scaling from below in the high z tail in agreement with the experimental data.

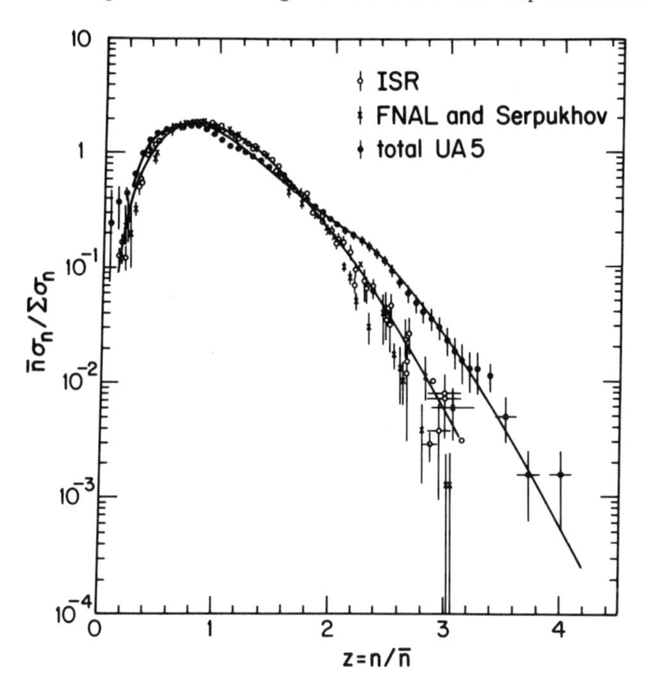

Fig. 1 The multiplicity distributions $\bar{n}\sigma_n/\Sigma\sigma_n$ plotted as a function of $z = \frac{n}{\bar{n}}$, for energy ranges from FNAL (s∼10 GeV) through ISR (s∼63 GeV) to CERN Sp$\bar{p}$S collider (s∼540 GeV), Ref. 3.

2. MODEL BASED ON QCD BRANCHING PROCESSES[1)]

a) <u>The coupled quark-gluon equations</u>. We assume that hadrons collide producing n_o initial gluons and m_o initial quarks with the corresponding probability distribution $P_{mn}(n_o,m_o,t)$. These gluons and quarks branch loosing their energy. When they reach the hadronization energy, branching stops and quarks and gluons hadronize. Scaling violations can come from any of these stages. In this paper we assume that number of partons is proportional to the number of hadrons and we concentrate on the branching stage. We consider the following branching processes: quark bremsstralung, 3-gluon branching, $q\bar{q}$ pair production and 4-gluon branching with probabilities $\tilde{A}$, A, B and C respectively.[6] In the leading logarithm approximation these probabilities are obtained from integrating the splitting functions $P_{g\to g}$, $P_{q\to g}$ and $P_{g\to q}$ where

$$P_{g\to g} = \frac{2N_c}{x'} \, , \; P_{q\to g} = \frac{(N_c^2-1)}{2N_c}\frac{2}{x'} \, , \; P_{g\to q} = \frac{N_f}{2}(x'^2 + (1-x')^2). \quad (2)$$

over the fractional momentum x'. We note that in the large N_c limit $A = 2\tilde{A}$.

The probability distribution for getting m quarks and n gluons satisfies the following evolution equation

$$\frac{\partial P_{mn}}{\partial t} = -AnP_{mn} - BnP_{mn} - \tilde{A}mP_{mn} - CnP_{mn} + A(n-1)P_{m\,n-1}$$
$$+\tilde{A}mP_{m\,n-1} + B(n+1)P_{m-2\,n+1} + C(n-2)P_{m\,n-2}. \quad (3)$$

where t is the evolution parameter related to the parton energy, $t \sim \ln\ln Q^2$. This equation can be solved exactly only in some limiting cases. If we assume that P_{mn} is a smooth function of m and n and nP_{mn} (mP_{mn}) varies slowly between n and n+1 (m and m+1) Eq. (3) becomes a differential equation for the probability distribution P(m,n,t):

$$\frac{\partial P(m,n)}{\partial t} = -[A+2C-B]P(m,n) + [-(A+2C-B)n - \tilde{A}m]\,\frac{\partial P(m,n)}{\partial n} -$$
$$- 2Bn\,\frac{\partial P(m,n)}{\partial m} + \ldots \quad (4)$$

where we have neglected terms higher than second order in the Taylor expansions. Eq. (4) can be solved with the assumption of n_o initial gluons and m_o initial quarks. In that case we obtain a new non-scaling law for the probability distribution P(m,n):

$$(\bar{m} - \frac{2B}{\lambda^+}\bar{n})\,(\bar{n} + \frac{\tilde{A}}{\lambda^+}\bar{m})\,P(m,n) = \Psi\left\{\frac{m - \frac{2B}{\lambda^+}n}{\bar{m} - \frac{2B}{\lambda^+}\bar{n}},\ \frac{n + \frac{\tilde{A}}{\lambda^+}m}{\bar{n} + \frac{\tilde{A}}{\lambda^+}\bar{m}}\right\} \tag{5}$$

where

$$a_o = A - B + 2C,\ \lambda^{\pm} = \frac{a_o}{2}\,(1 \pm \sqrt{1+8\tilde{A}B/a_o^{\,2}}). \tag{6}$$

and ψ is an arbitrary function. In the limit when $\tilde{A} = B = 0$, $\bar{n}P_n$ scales. When we nelect 4-gluon branching and consider the large N_c limit ($A = 2\tilde{A}$) we can solve Eq. (3) exactly for the generating function

$$G(x,y,t) = \sum_{m,n} x^m y^n P_{mn}(t). \tag{7}$$

and the corresponding cumulant moments

$$\gamma_m = \frac{\overline{(n - \bar{n})^m}}{\bar{n}^m} \tag{8}$$

with two different initial conditions: n_o gluons or m_o quarks .

b) <u>The decoupled gluon equation</u>. At high energies gluons dominate ($\bar{m}/\bar{n} \sim \frac{2B}{A}$, $B<<A$). Therefore we can neglect quark evolution ($m = m_o$ = const). The evolution equation for the probability distribution P(n,t) is

$$\frac{\partial P_n}{\partial n} = -AnP_n - BnP_n - \tilde{A}mP_n - CnP_n + A(n-1)P_{n-1}$$
$$+B(n+1)P_{n+1} + \tilde{A}mP_{n-1} + C(n-2)P_{n-2}, \tag{9}$$

This equation can not be solved exactly. We can make the same approximation as we did for the coupled quark-gluon equation and obtain the differential equation for the probability distribution P(n,t). We find the following solution

$$P(n,t) = \int_0^\infty d\lambda w(\lambda) e^{-\lambda t} \Psi(a,c,y). \tag{10}$$

where $\Psi(a,c,y)$ is the confluent hypergeometrical functions regular for large y, $w(\lambda)$ is an unspecified weight function,

$$a = 1 - \lambda/a_o, \quad c = \frac{(a_1 - \tilde{A}m)}{(a_1/2)}, \quad y = \frac{a_o}{(a_1/2)} n, \quad a_1 = A + B + 4C. \tag{11}$$

For large y, P(n,t) has a scaling and a non-scaling piece. In the tail of the distribution non-scaling piece has a negative sign and scaling is approached from below as observed experimentally.
The only case when we can solve Eq. (9) exactly is when we neglect 4-gluon branching and assume n_o initial gluons. Then we get the following probability distribution

$$P_n(n_o,t) = [1 + \frac{1}{A-B}(e^{(A-B)t} - 1)]^{-n-n_o-a'} [e^{(A-B)t} - 1]^{n_o+n}$$

$$x \left(\frac{B}{A-B}\right)^{n_o} \left(\frac{A}{A-B}\right)^n \frac{(n+n_o+a'-1)!}{n!(n_o+a'-1)!} \, {}_2F_1(-n, -n_o; -n-n_o-a'+1; u). \tag{12}$$

where

$$a' = \frac{\tilde{A}m}{A} \quad \text{and} \quad u = 1 - \frac{(A-B)^2 e^{(A-B)t}}{AB(e^{(A-B)t} - 1)^2} \tag{13}$$

The corresponding multiplicities are given by:

$$\bar{n} = \frac{\tilde{A}m}{A-B}(e^{(A-B)t} - 1) + n_o e^{(A-B)t} \tag{14}$$

In the limit when we consider only gluons ($m = m_o = o$) and neglect $g \to ggg$, the probability distribution is the 3-gluon branching distribution[6)]

$$P_n^{n_o}(\bar{n}) = \frac{(n-n_o)!}{n!(n_o-1)!} \left(\frac{n_o}{\bar{n}}\right)^{n_o} \left(1-\frac{n_o}{\bar{n}}\right)^{n-n_o} \tag{15}$$

where n_o is the initial number of gluons. For large n and $\bar{n}$, it approaches the KNO scaling function:

$$\bar{n}P_n \underset{n,\bar{n} \to \text{large}}{\longrightarrow} \psi\left(z = \frac{n}{\bar{n}}\right) = \frac{n_o^{n_o}}{(n_o-1)!} e^{-n_o z} z^{n_o-1} \tag{16}$$

All correction terms have negative signs indicating the approach from below to this scaling form, in agreement with data.
Finally, we consider the case when there are m_o initial quarks and no gluons. Neglecting 4-gluon branching and $q\bar{q}$ pair production we can solve Eq. (9) exactly. The solution is the negative binomial distribution

$$P_n^k(\bar{n}) = \frac{(n+k-1)!}{n!(k-1)!} \left(\frac{k}{\bar{n}}\right)^k \left(1 + \frac{k}{\bar{n}}\right)^{-n-k} \tag{17}$$

where

$$k = \frac{\tilde{A}m_o}{A} \quad \text{and} \quad \bar{n} = k(e^{At} - 1) \tag{18}$$

Since in the leading logarithm approximation $\tilde{A}/A$ is a constant, the only possible energy dependence of the parameter k can come from the energy dependence of the initial number of quarks m_o.

2. COMPARISON WITH THE STOCHASTIC PICTURE

The negative binomial distribution has been derived in many statistical models.[8,9)] The parameter k has a different meaning in the various approaches. In the stochastic model proposed by Carruthers and Shih [8], k is the number of sources (clusters) emitting fields behaving as Gaussian random variables. The UA5 group used this distribution to fit experimental data remarkably well from 10 GeV up to 900 GeV with k decreasing from 20 to 3 [10]. This behavior for k does not fit into any physical picture, but it is the only way to get the observed broadening in the high z tail. In the large n and $\bar{n}$ limit, the negative binomial distribution approaches <u>the same</u> KNO scaling function as the 3-gluon branching distribution. On Fig. 2 we see that if we identify k with n_o, for fixed k the negative binomial distribution approaches scaling from <u>above</u>, while the three-gluon branching distribution approaches the same limit from <u>below</u>.

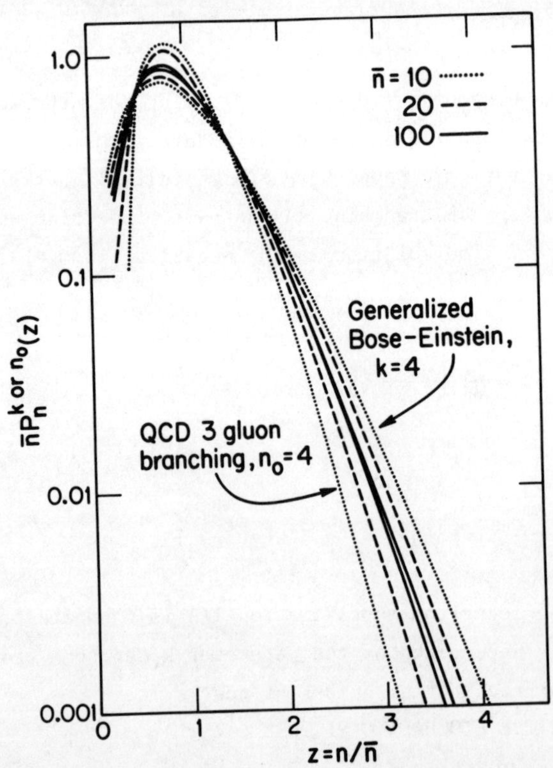

Fig. 2 The multiplicity distributions $\bar{n}P_n(z)$ plotted as a function of $z = \frac{n}{\bar{n}}$ for $k = n_o = 4$, Ref. 6. The values for $\bar{n}$ correspond roughly to

FNAL, early $Sp\bar{p}S$, and SSC energies.

We note that the negative binomial distribution is the stationary solution of the following equation

$$\frac{\partial P_n(t)}{\partial t'} = r_{n+1}P_{n+1}(t') - r_nP_n(t') - g_nP_n(t') + g_{n-1}P_{n-1}(t') \qquad (19)$$

where t' is the evolution parameter (time)
and

$$r_n = \beta n$$
$$g_n = \alpha n + \delta \qquad (20)$$

This equation is recognized as an evolution equation for the maser amplification[11)]where α, β and δ are probabilities for stimulated emission, spontaneous emission and stimulated absorption respectively. The parameters k and $\bar{n}$ which appear in the negative binomial distribution are related to these probabilities by

$$k = \frac{\delta}{\alpha} \qquad \text{and} \qquad \bar{n} = \frac{\delta}{\beta-\alpha} \qquad (21)$$

We note that Eq. (19) is the same as our branching Eq. (9), but with different physical meaning. Namely, in the branching model the evolution parameter is energy while in the stochastic picture it is the time. In the stochastic picture the energy dependence of the multiplicities has to be put in by hand while in the parton branching model it is a prediction.

3. SUMMARY AND CONCLUSION

In the parton branching model with coupled quarks and gluons there is no KNO scaling in the lowest approximation. When four-gluon branching is neglected, exact solution for the generating function and the corresponding moments can be obtained. At high energies, gluons dominate and we can neglect quark evolution. The probability distribution in this case approaches scaling from below in the high z tail in agreement with data. If we consider only three-gluon branching the probability distribution approaches the KNO scaling function given by Eq. (16) from below. To compare this simple branching model with the stochastic picture we can identify initial number of quarks n_o with the number of sources k in the negative binomial distribution. In that case the negative binomial distribution approaches the same scaling function as a three-gluon branching distribution, but from above. The only way that negative binomial distribution can describe the experimental data is if parameter k decreases drastically with energy which does not fit any physical picture.

Finaly we summarize the disparities between our branching model and the stochastic picture. Although the branching model and stochastic model can be described with the same type of evolution equation they have a different physical meaning. Namely, in the parton branching model the evolution is in energy, while in the stochastic picture it is time. Multiplicities are different: in the branching model they come naturally from the evolution equation as energy dependent, while in the stochastic model $\bar{n}$ is a parameter. The stochastic picture is concerned with the stationary (equilibrium) solution of the evolution equation. Multiplicity distributions in the stochastic picture can also be derived from the coherent states and density matrix.

The negative binomial distribution can also be obtained in the branching model in the case of a quark jet. However, one has to keep in mind that the parameter k has a different physical meaning than in the stochastic picture, which is a consequance of a different evolution parameter.

One way of distinguishing between different models is by looking at the large z tail. Unfortunately, the uncertainities of the data are largest in that region. We also note that cumulant moments of Eq. (8) are more sensitive to the shape of the distribution than the moments

$$C_m = \frac{\overline{n^m}}{\bar{n}^m} \tag{22}$$

used by the UA5 group to present their data.

The recent discoveries of the UA1 group[12] that "jet" events have a much narrower distribution and different mean multiplicity ($\bar{n}_{jet} \sim 2\bar{n}_{non-jet}$) than "no jet" events are offering a real challenge to many models. The shape of the distribution is also very sensitive to different cuts in the fractional momenta, different rapidity regions, distinction between the diffractive and nondiffractive events and the separation of the leading particles.[13] All this shows the importance of investigating the initial conditions in the parton branching model. Furthermore one has to keep in mind that our model is applicable for non-diffractive events in the region where $p_\perp > p_{\perp\,min}$ and x not too near 1 and it does not include soft gluons.

AKNOWLEDGEMENTS

The work presented here was done in collaboration with B. Durand. It is also a pleasure to thank P. Carruthers, L. Durand, S. Gasiorowicz, S. Rudaz and C. C. Shih for many useful discussions. Research was supported in part by the Department of Energy Grant no. DE-AC02-83ER40105 and in part by a University of Minnesota Graduate School Doctoral Disertation Fellowship.

REFERENCES

1) B. Durand and I. Sarcevic, MAD/TH-85-7.
2) B. Durand and I. Sarcevic, MAD/T8-85-19, to be published in Phys. Lett. B.
3) Z. Koba, H. B. Nielsen and P. Olesen, Nucl. Phys. B40, 317 (1972).
4) W. Thome et al. Nucl. Phys B129, 365 (1972); J. Firestone et al., Phys. Rev. D14, 2902 (1976)
5) UA5 Collaboration, G. J. Alner et al., Phys. Lewt. 138B, 304 (1984), ibid. Phys. Lett. 160B, 193 (1985).
6) B. Durand and S. D. Ellis, Proc. of the 1984 Summer Study on the Design and Utilization of the S.S.C., Snowmass, CO, 1984; S. Ellis, Proc. of 11th SLAC Summer Institute, SLAC Report #267 (1983).
7) A. Giovannini, Nucl. Phys. B161, 429 (1979).
8) P. Carruthers and C. C. Shih, Phys. Lett. 127B, 242 (1983), 137B, 425 (1984);
9) M. Biyajima, Prog. Theor. Phys. 69, 966 (1983), A. Giovannini and L. Van Hove, CERN-TH4230/85.
10) UA5 Collaboration, G. J. Alner et al., CERN-EP/85-197.
11) N. G. Van Kampen, "Stochastic Processes in Physics and Chemistry" (North-Holland Personal Library, Amsterdam, 1982).
12) UA1 Collaboration, G. Ciapetti, Proc. of 5th Topical Workshop on $p\bar{p}$ collider physics, editor M. Greco, World Scientific, 488 (1985).
13) M. Derrick, ANL-HEP-CP-85-17 (1985); K. Goulianos, Phys. Reports 101 No. 3, 169 (1983); G. Pancheri, CERN-EP/84-96; S. Rudaz and P. Valin, UMN-TU-521/85; T. K. Gaisser, F. Halzen and A.D. Martin, MAD/PH/258(1985); A. Capella, Orsay Preprint, LPTHE-85/42

HADRON-NUCLEAR COLLISIONS

NUCLEAR STOPPING POWER AT RELATIVISTIC AND ULTRA-RELATIVISTIC ENERGIES

Madeleine Soyeur

CEN de Saclay, Service de Physique Théorique
F-91191 Gif-sur-Yvette Cedex, France

ABSTRACT

The concepts of stopping power and momentum degradation length for protons incident on nuclear matter are defined and discussed. The interpretation of proton-nucleus data at relativistic ($E_p^{Lab} \simeq$ a few GeV) and ultra-relativistic energies ($E_p^{Lab} \gtrsim 100$ GeV) in terms of these concepts is reviewed.

Contents

1. STOPPING POWER AND MOMENTUM DEGRADATION LENGTH : DEFINITIONS

The concept of stopping power of matter to fast particles is frame dependent and defined for the ideal scattering experiment represented in Fig.1. Let $\bar{p}$ be the momentum of the incident particle along the z axis. We want to study the evolution of the energy E and longitudinal momentum p(z) of this particle as it propagates through a semi-infinite slab of matter ($z \geqslant 0$). This evolution is a characteristic property of the matter contained in the semi-infinite slab.

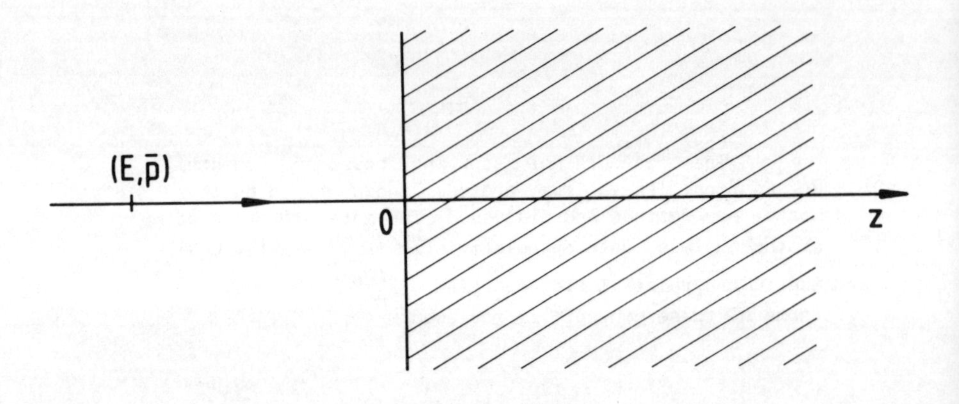

Figure 1 : *Fast particle of momentum p propagating freely for $z < 0$ and through a semi-infinite slab of matter for $z > 0$.*

The stopping power S is the average energy loss of the incident particle along the z axis per unit length,

$$S = - \frac{dE}{dz} . \tag{1}$$

The degradation length Λ_p of the incident particle momentum is defined by the equation

$$\frac{dp(z)}{dz} = - \frac{p(z)}{\Lambda_p} , \tag{2}$$

i.e.

$$p(z) = p(0)\, e^{-\frac{z}{\Lambda_p}} . \tag{3}$$

The stopping power and degradation length discussed in this paper refer to the propagation of a high energy proton incident on a semi-infinite slab of symmetric nuclear matter (with equal number of protons and neutrons).

2. NUCLEAR STOPPING POWER AT HIGH ENERGY ($E_p \gtrsim 1$ GeV)

2.1. Stopping dynamics : general features

The definitions given in the previous section have a very simple interpretation if the fast incident particle propagates through matter by successive elastic collisions (billiard ball model). In this case, the stopping power is a measure of the transformation of longitudinal energy into transverse energy.

For protons incident on nuclear matter at high energy ($E_p \gtrsim 1$ GeV), the interpretation of the stopping power is more complicated.

We consider first the relativistic regime, i.e. proton energies of the order of 1 GeV. At these energies, the elastic and the inelastic proton-proton cross-sections become comparable as illustrated in Fig.2.[1] The main inelastic channel is the reaction $pp \to pn\pi^+$. The data[2] suggest that this reaction proceeds dominantly (~ 80 %) through the intermediate $[n\Delta^{++}(1232)]$ state for $E_p^{Lab} > 650$ MeV ; other possible mechanisms are the $[p\Delta^+]$ intermediate state or direct pion production. Therefore, a 1 GeV proton incident on nuclear matter will deposit energy not only by transforming longitudinal energy into transverse energy but also by exciting nucleons into Δ's or, less frequently, by direct pion production. In proton-neutron collisions, charge exchange reactions are also possible.

In the ultra-relativistic energy regime ($E_p \gtrsim 100$ GeV), the dynamics of stopping is expected to be even more different from the billiard ball model. Because of Lorentz time dilatation, the formation time of a secondary particle increases linearly with its energy and therefore the fast particles, and in particular the leading baryon, are produced far from the collision point, at distances of the order of $\left(\frac{E}{m}\right)$fm (E and m are the energy and the mass of the particle)[3]. The incident proton does not retain its identity after the first collision and proton fragments rather than the proton itself rescatter in the subsequent collisions. Therefore, the detailed understanding of stopping power and energy deposition at ultra-relativistic energies requires models involving the dynamics of the nucleon constituents.

2.2. Stopping power and baryon densities achieved in central heavy ion collisions

A major motivation for studying central heavy ion collisions at high energies is the possibility of creating temporarily a quasi-macroscopic system of nuclear matter at high density. One hopes to be able to determine the equation of state of matter in this regime from the reaction products.

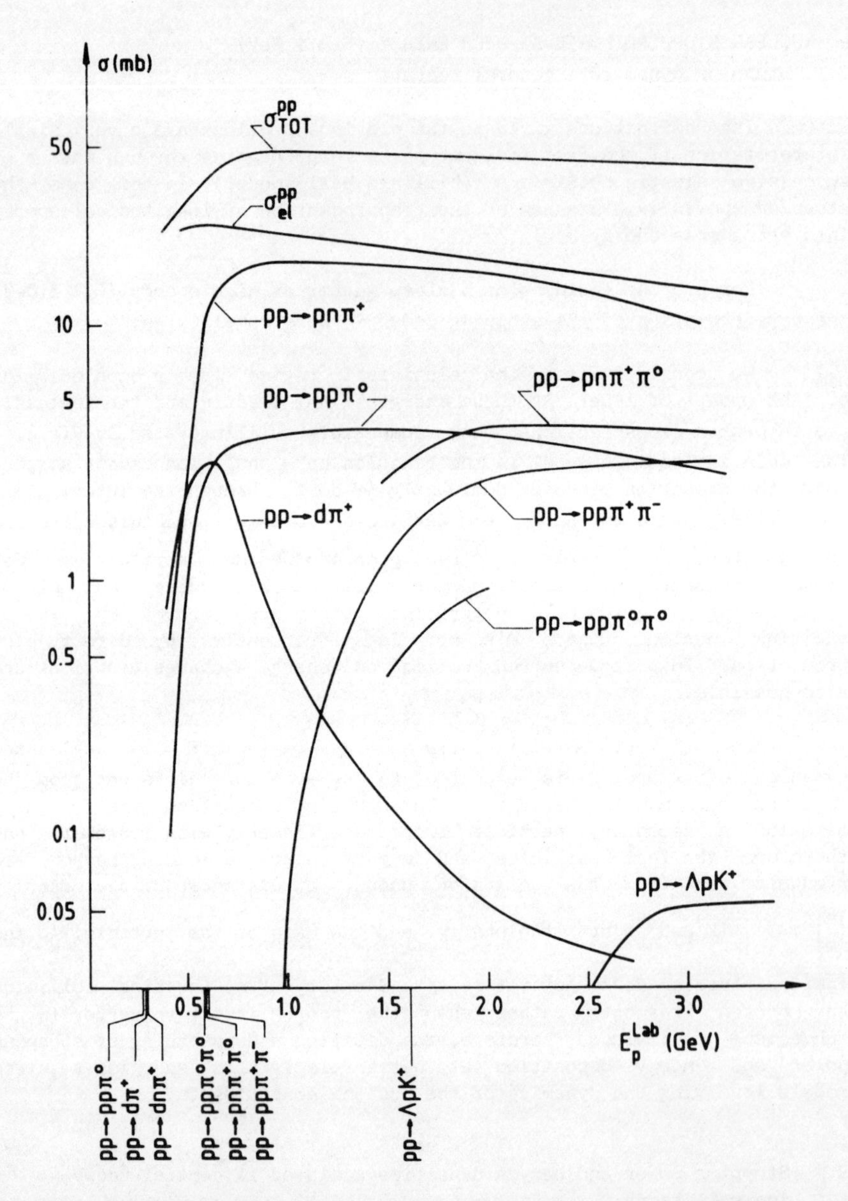

Figure 2 : *Proton-proton scattering cross-sections for* $0.5 < E_p^{Lab} < 3$ *GeV. The contributions from the inelastic channels are shown together with the total and elastic cross-sections. The thresholds for the inelastic channels are indicated.*

The stopping power of nuclear matter to fast nucleons is closely related to the baryon densities attainable in central heavy ion collisions. In order to built up high baryon densities, nucleons have to be stopped in the interaction region.

At incident energies of the order of 1 GeV, stopping of heavy ions in central collisions has been observed at the Bevalac, in event by event analyses using 4π detectors.[4,5] The experimental signature of the complete stopping of the ions is the isotropy of the momentum flux of the event.

The reaction $^{40}Ar + ^{208}Pb$ at E/A = 0.772 GeV has been studied at the Streamer Chamber.[4] It was triggered by requiring that only high multiplicity events be accepted. The energy and momentum carried by the protons only were measured. All neutrons, free or bound in clusters, were disregarded in this analysis. The isotropy of the momentum distribution of each event is characterized by the ratio of the mean longitudinal momentum over the mean transverse momentum,

$$R \equiv \frac{2}{\pi} \frac{\langle p_t \rangle}{\langle p_{\|} \rangle}$$

$$= \frac{2}{\pi} \sum_{\nu} |p_{t\nu}| \Big/ \sum_{\nu} |p_{\|\nu}| \ , \qquad (4)$$

in which $p_{t\nu}$ and $p_{\|\nu}$ are the transverse and longitudinal momenta of the detected protons; for an isotropic distribution, R is unity. The distribution of the mean longitudinal and transverse momentum of the central collision events is shown in Fig.3. The contour lines of event frequency appear rather symmetric around the diagonal which corresponds to R=1, i.e. to events with spherical momentum flux. Similar results were observed at the Plastic Ball for high multiplicity events created in the reaction $^{93}Nb + ^{93}Nb$ at E/A = 0.4 GeV[5].

Estimates of the baryon density reached in central collisions of heavy ions at Bevalac energies range from 2 to 4 times nuclear matter density (i.e. from 0.3 to 0.7 baryon/fm^3)[7].

At energies higher than those available at the Bevalac, there are at present no experimental data on central heavy ion collisions. Therefore, it is important to understand the stopping power of nuclear matter at these energies to predict the conditions that could be created in the interaction region.

As the incident energy per nucleon increases, one expects that the complete stopping of the ions will become less and less probable and that the transparency regime will set in. In the transparency regime, the

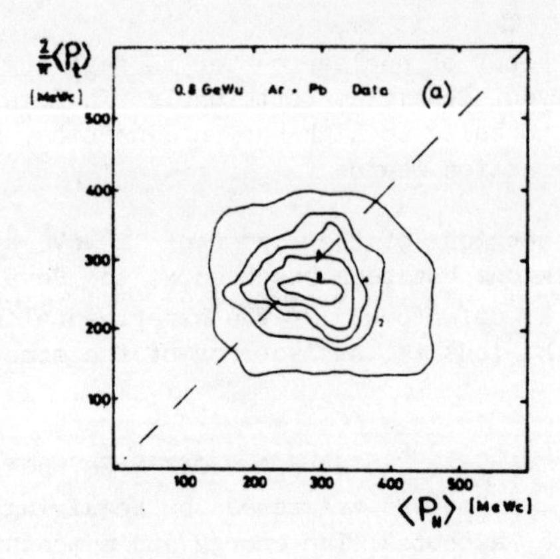

Figure 3 : *Contour lines of event frequency for central collisions in the reaction Ar+Pb at E/A = 0.772 GeV*[6]. $\langle p_t \rangle$ *and* $\langle p_\parallel \rangle$ *are the mean transverse and longitudinal momenta of the event.*

ions "pass through" each other and emerge after the collision as two excited fragments receding from each other at the speed of light; in between the ions, a central region of very high energy density but little baryon density develops [8]. It is important to know the energy domain in which this transition from stopping to transparency takes place in order to evaluate the possibility of creating a high baryon density quark-gluon plasma in heavy ion collisions. This issue depends on both the stopping power and the energy deposition in the energy range 10-100 GeV per nucleon. Indeed, a large fraction of the longitudinal energy lost by the incident nucleons is not deposited in the target: the fast fragments of the projectile are emitted outside the nucleus. In order to produce a high density quark-gluon plasma, one needs complete stopping of the ions and sufficient energy deposition to provide the latent heat necessary to induce the deconfinement phase transition. This latent heat is expected to be of the order of a few GeV fm^{-3}.[9]

Information on nuclear stopping power comes primarily from proton-nucleus experiments. We devote the next two sections to a discussion of semi-exclusive data taken for incident protons of a few GeV[10-14] and to various analyses of inclusive data at ultra-relativistic energies (E_p^{Lab} = 100-400 GeV)[15-17].

3. STOPPING POWER AT RELATIVISTIC ENERGIES : SEMI-EXCLUSIVE STUDIES OF PARTICLE PRODUCTION IN A FEW GeV PROTON-NUCLEUS REACTIONS

We discuss in this section the semi-exclusive proton-nucleus data obtained at the 12 GeV KEK proton synchrotron[10-14].

The aim of these experiments is to measure the stopping cross-section for several GeV protons incident on various nuclear targets. A stopping event is defined as a highly inelastic process with no forward leading particle; it is characterized by multiparticle emission in the target rapidity region. Stopping events are identified by a two component spectrometer [12,13] consisting of a cylindrical drift chamber and a forward spectrometer. The cylindrical drift chamber covers the angular range $\theta = 30°$ to $120°$ for all azimuthal angles; it detects the particles emitted in the target rapidity region. The forward spectrometer covers the angular range $\theta = 0°$ to $6°$; it allows detection of the forward leading particles. The trigger condition for inelastic events is that at least one particle be detected in the cylindrical drift chamber. The forward spectrometer allows then to separate inelastic events with and without leading particles. The latter are the stopping events.

The semi-exclusive reactions studied are therefore of the type,

$$p + A \rightarrow N_c^l + N_c^{tf} + X \ , \tag{5}$$

in which N_c^l and N_c^{tf} are the charged particles observed respectively in the forward spectrometer (leading particles) and in the cylindrical drift chamber (target fragmentation). The incident proton momentum varies from 1 to 4 GeV/c (i.e. $0.43 < E_p^{Lab} < 3.17$ GeV). The targets are ^{27}Al and ^{208}Pb.

The observation of nuclear stopping is illustrated in Fig.4 for 4 GeV/c protons incident on Al and Pb targets[13]. This figure shows the cross-sections for inelastic events as a function of the charged particle multiplicity. The squares and circles refer to different trigger conditions (at least one and at least two particles observed in the cylindrical chamber respectively). Together with the cross-sections are shown the fractions of events (in percent) with leading protons. One observes that the fraction of leading proton coincidences decreases very fast with increasing multiplicity and goes to zero for events with multiplicity larger than 5.

Stopping cross-sections can be derived from these data by subtracting from the inelastic cross-sections the contribution from events with leading baryons. Only leading protons are measured. Therefore, to deduce stopping cross-sections, one needs to estimate the forward leading neutron component. Neutrons come from charge exchange reactions or from the decay of Δ's and N^*'s produced in the projectile fragmentation region. A very rough estimate[13] of the leading neutron to proton ratio gives

$$\frac{n}{p} = 0.75 \pm 0.25 \ . \tag{6}$$

Using this ratio, En'yo derived the following values of the stopping cross

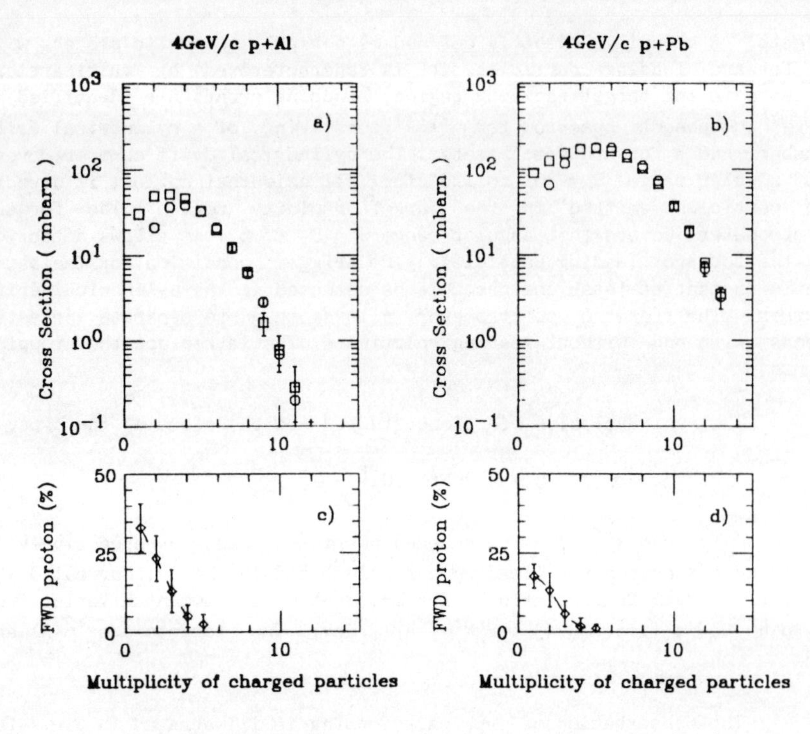

Figure 4 : *Charged multiplicity distributions for the 4 GeV/c a) p+Al and b) p+Pb reactions. The fraction of events with leading protons are shown in c) for the p+Al reaction and in d) for the p+Pb reaction*[13].

sections for 4 GeV/c protons[13]:

$$p + {}^{27}\mathrm{Al} : \sigma_{stop} = (200 \pm 17)\ \mathrm{mb}\ , \tag{7}$$

$$p + {}^{208}\mathrm{Pb} : \sigma_{stop} = (1160 \pm 110)\ \mathrm{mb}\ . \tag{8}$$

These stopping cross-sections represent respectively 50 % (for Al) and 80 % (for Pb) of the geometrical cross-section. Therefore, to a large extent, heavy nuclei at these energies (~ 3 GeV) are not transparent.

To estimate the energy deposition in the target nucleus, one has to evaluate the contribution from the neutral particles. For a target with N neutrons and Z protons, the number of emitted neutrons is taken to be (N/Z) times the observed number of protons and the number of π°'s is assumed to be the same as the number of π^-'s. The energy deposition in Al and Pb calculated with these assumptions is shown as a function of the mean multiplicity in Fig.5[13]. The incident proton momentum is 4 GeV/c. In high multiplicity events, a very large fraction (more than 80 %) of the incident energy (3.17 GeV) is deposited in the nucleus.

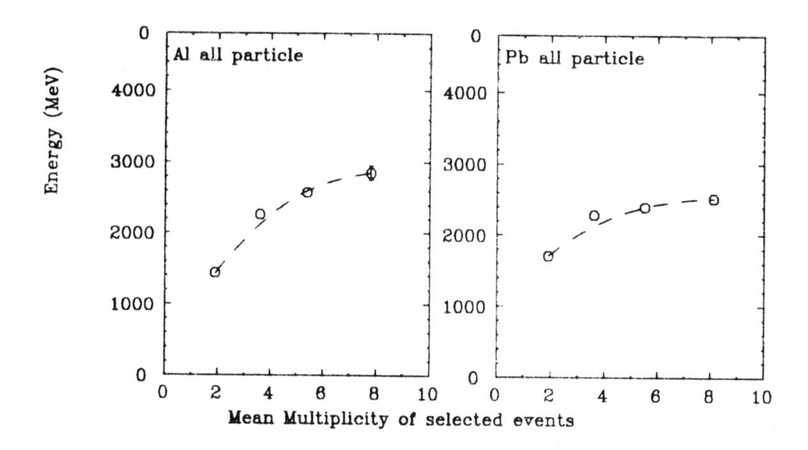

Figure 5 : *Multiplicity dependence of the energy deposition in the 4 GeV/c p + Al and p + Pb reactions*[13].

4. STOPPING POWER AT ULTRA-RELATIVISTIC ENERGIES

4.1. Inclusive proton-nucleus cross-sections at 100 GeV

Estimates of the nuclear stopping power at ultra-relativistic energies ($E_p^{Lab} \gtrsim 100$ GeV) are based on proton-nucleus inclusive data, i.e. on differential cross-sections for the reaction

$$p + A \rightarrow p + X \, , \tag{9}$$

with various nuclear targets.

Fig.6 shows the data of Barton et al[15] obtained with 100 GeV protons incident on p, ^{12}C, ^{27}Al, ^{63}Cu, ^{108}Ag and ^{208}Pb. The invariant cross-sections are plotted as functions of Feynman x which is approximately the ratio of the outgoing proton momentum over the incident proton momentum. These results cover the kinematic range $30 < p < 88$ GeV/c and $p_t = 0.3$ GeV/c.

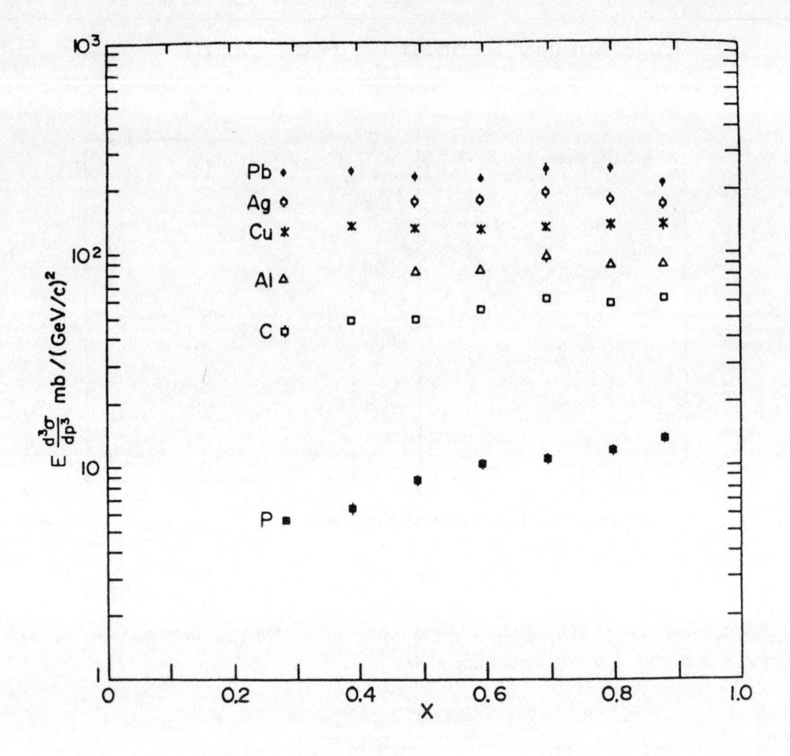

Figure 6 : *Invariant differential cross-sections for* $p+A \to p+X$ *as a function of Feynman* $x \simeq \frac{p_{out}}{p_{in}}$. [15]

Recently, data on the reaction

$$p + A \to n + X \tag{10}$$

at 400 GeV have become available[16]. Fig.7 shows the x dependence of the $p+p \to n+X$ cross-section; the 400 GeV $p+p \to p+X$ cross-section is also plotted for comparison. It is about twice larger than the neutron inclusive cross-section. Fig.8 shows the x dependence of the $p+Pb \to n+X$ cross-section at 400 GeV compared to the $p+Pb \to p+X$ cross-section at 100 GeV.

Proton-nucleus inclusive data at 120 GeV for targets ranging from Be to U have also been measured at CERN[17]. They cover the kinematic range $15 < p < 70$ GeV/c and are averaged over p_t.

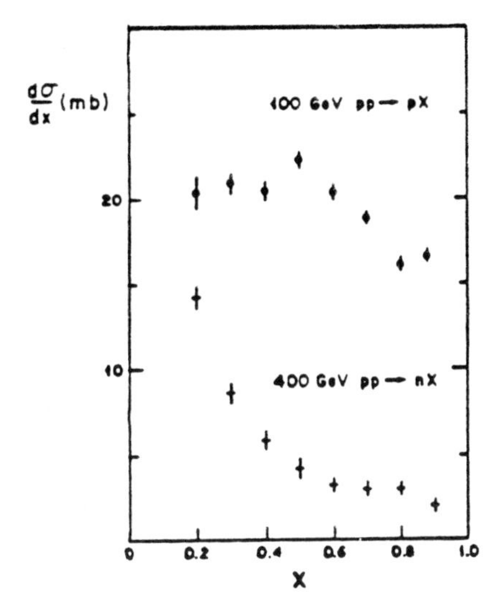

Figure 7 : *x dependence of the inclusive proton and neutron cross-sections for proton-proton collisions at 400 GeV*[16].

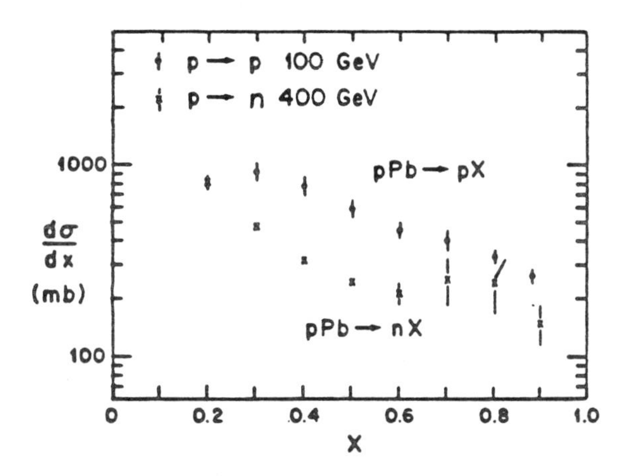

Figure 8 : *x dependence of the p+Pb → n+X cross-section at 400 GeV compared to the x-dependence of the p+Pb → p+X cross-section at 100 GeV* [16].

4.2. Empirical determination of the mean proton rapidity loss in the center of a heavy nucleus

The Barton et al data[15] described in the previous paragraph have been used by Buzsa and Goldhaber[18] to estimate the stopping power of nuclear matter. They transform the invariant inclusive cross-section of the p+Pb → p+X reaction shown in Fig.6 into a probability that the incident proton loses rapidity ($-\Delta y$) after traversing the inner part of the Pb nucleus. This transformation rests on a few simplifying assumptions and on a simple model for the nuclear surface. The assumptions are that the inclusive p_t distributions are independent of x and A and that the ratios of production of the various baryons are independent of A. The outer half of the lead nucleus is described as n nuclei of small mass[18]. The result of this work is shown in Fig.9. The average rapidity loss in the inner part of the Pb nucleus for a 100 GeV proton is $\Delta y = -2.5 \pm 0.2.$[18]

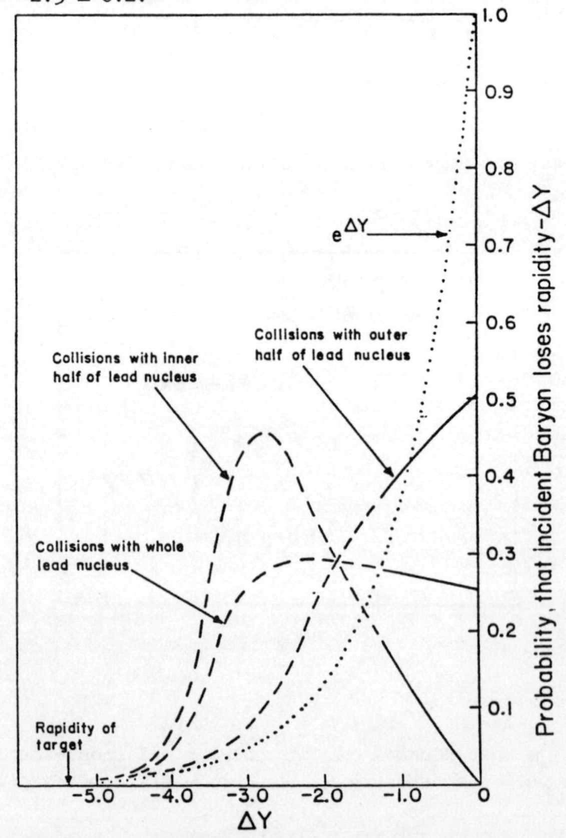

Figure 9 : *Probability distributions for rapidity loss of protons in Pb nucleus at 100 GeV. The contributions from the inner and outer parts of the nucleus are indicated*[18].

4.3. Multiple scattering expansions

4.3.1. Sequential nucleon-nucleon scattering along straight line trajectories

The simplest approach to describe the stopping of a high energy proton by a heavy nucleus is to assume that the incident proton undergoes successive collisions with the target nucleons on a straight line trajectory. This approach neglects the property that the leading baryon is produced outside the nucleus and cannot rescatter as a nucleon (see the discussion of section 2). The sequential nucleon-nucleon scattering picture should therefore be viewed as an effective analysis of the data whose main interest is to provide order of magnitude estimates of the momentum degradation length of a high energy proton in nuclear matter.

The sequential scattering evolution model was initially proposed by Hwa[19] and later on refined by a number of authors[20-22].

We recall first briefly the geometry of the model. The incident proton collides on average with a number of nucleons $N_A(b)$ given by

$$N_A(b) = \sigma_{NN} \int dz\ \rho_A(z,b) , \tag{11}$$

where σ_{NN} is the nucleon-nucleon cross-section and ρ_A the nuclear density distribution. The cross-section for collision with N target nucleons in a row is then[23]

$$\sigma_A(N) = \int d^2b \frac{1}{N!} [N_A(b)]^N \exp[-N_A(b)]. \tag{12}$$

The dynamics of the model enters through the evolution equation for the momentum degradation of the incident proton in successive collisions. If $H(x,N)$ is the probability that the nucleon has a fraction x of the initial momentum after N collisions, the evolution equation[19] reads

$$H(x,N+1) = \int_x^1 \frac{dx'}{x'} H(x',N)\ Q\left(\frac{x}{x'}\right) , \tag{13}$$

in which $Q(x)$ is the probability that the incident nucleon has momentum fraction x after a collision with one more target nucleon.

The various analyses using the sequential evolution model[19-22] differ by the choice of $Q(x)$, i.e. by the relative probability of hard and soft collisions. Fig.10 (from Csernai and Kapusta[22]) shows the different functions used for $Q(x)$.

Fitting the Barton et al data[15] with the sequential evolution model, one can determine the momentum degradation length Λ_p [eqs.(2)-(3)] in the nuclear interior. The first calculation of Λ_p by

Hwa[19] used an approximate solution of the evolution equation (13) and gave Λ_p= 17 fm. Solving eq.(13) exactly, Csernai and Kapusta[22] found, with the same function Q(x), Λ_p= 4.9 fm. Wong[20] obtained Λ_p= 5.3 fm and Hüfner and Klar[21] got $\Lambda_p \sim$ 7.3-8.8 fm (for a more detailed discussion see ref.22).

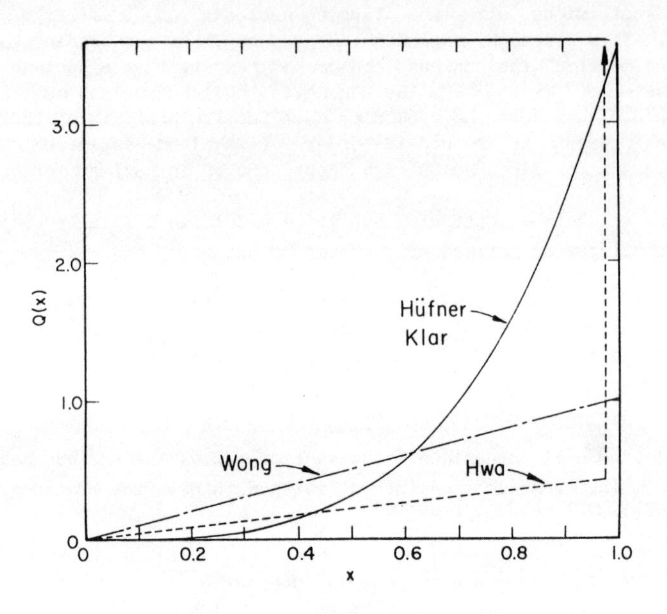

Figure 10 : *Probability that the incident nucleon has momentum fraction x after a collision with one more target nucleon. The various curves refer to the values used by the authors of refs. 19-21. This figure is taken from ref.22.*

Taking into account the difference between the proton and neutron inclusive data, Csernai and Kapusta [22] fitted both the Barton et al[15] and the Forrest [16] data and obtained $5.7 < \Lambda_p < 6.9$ fm. This range of values is consistent with the Busza-Goldhaber analysis[18].

4.3.2. Successive interactions of nucleon constituents

As emphasized in section 2, the ultra-relativistic proton-nucleus dynamics should be described at the level of the nucleon constituents to take into account the nucleon fragmentation.

We review briefly in this section the results obtained in the framework of the dual parton model which provides a unified and quite successful description of particle production in nucleon-nucleon,

nucleon-nucleus and nucleus-nucleus collisions[24,25].

In this model, the incident proton splits during the first collision into 2n colored fragments (n is the number of rescatterings). Each of these fragments together with a complementary colored fragment of another nucleon generates a color flux tube which gives rise asymptotically to a chain of hadrons[24]. This is illustrated in Fig.11.

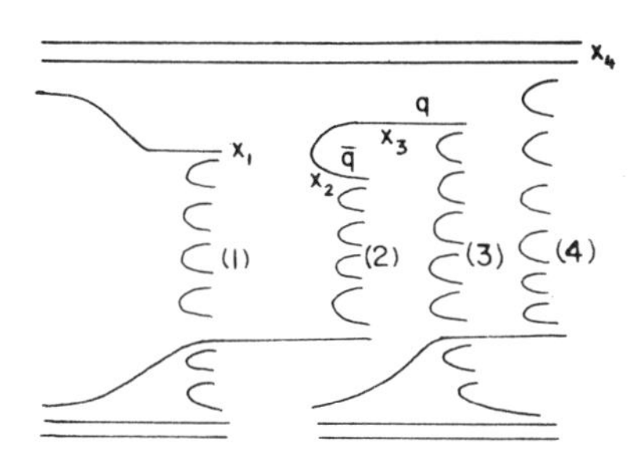

Figure 11 : *Proton-nucleus interaction in the dual parton model*[24].

The momentum distribution functions of the valence and sea quarks are given in terms of the dominant Regge singularities and the hadronization of the chains is described by quarks fragmentation functions[24].

The leading baryon is this picture is produced in the chain formed by the diquark of the incident proton and a valence or sea quark of a nucleon from the target[25].

It is interesting to relate the results of such model to the stopping power parameters defined previously. This is shown in Fig.12. The average rapidity loss of a 200 GeV/c proton is plotted as a function of the number of successive collisions. $\langle \Delta Y \rangle_m$ is defined by

$$\langle \Delta Y \rangle_m = Y_{max} - \langle Y \rangle_m , \qquad (14)$$

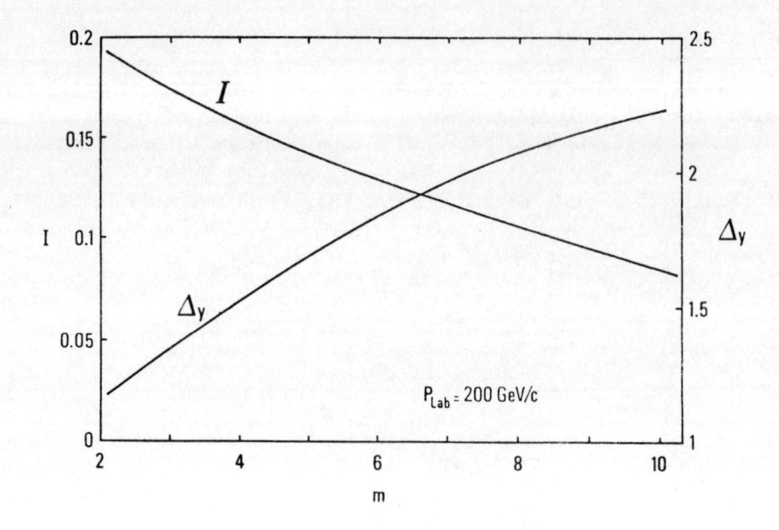

Figure 12 : *Average rapidity loss of a 200 GeV/c proton and average inelasticity coefficient as a function of the number of collisions m*[25].

where $\langle Y\rangle_m$ is the average baryon rapidity after m collisions. $\langle \Delta Y\rangle_1$ is approximately 1. If $\langle E\rangle_m$ is the mean energy of the leading baryon after m collisions, the inelasticity coefficient I_m is defined by

$$\langle E\rangle_m = (1-I_m)\langle E\rangle_{m-1} \quad . \tag{15}$$

I_1 is ~ 0.5[25]. The value of the momentum degradation length Λ_p is 7.5 fm, a value quite comparable to the estimates of the previous section.

Similar values have also been obtained using the multi-chain and additive quark models[26].

5. CONCLUDING REMARKS

We have briefly reviewed in this talk our present understanding of nuclear stopping power at relativistic and ultra-relativistic energies.

At relativistic energies (a few GeV), it seems important to be able to reproduce the semi-exclusive data[10-14] on particle production in proton-nucleus collisions. These data provide strong constraints on the conditions created in relativistic heavy ion collisions and need to be understood in order to relate particle production in central nucleus-nucleus collisions to the equation of state of nuclear matter.

At ultra-relativistic energies (~ 100 GeV), a detailed description of the energy deposition in proton-nucleus collisions has to involve the nucleon constituents. Progress in this approach is important to evaluate the possibility of creating a high baryon density quark-gluon plasma in ultra-relativistic heavy ion collisions.

The author thanks K.Nakai for sending her the most recent proton-nucleus data obtained at KEK and G.Baym for many useful discussions on stopping power at ultra-relativistic energies.

REFERENCES

1. The data are taken from V.Flaminio, W.G.Moorhead, D.R.O.Morrison and N.Rivoire, Compilation of cross-sections, p and $\bar{p}$ induced reactions, CERN-HERA 84-01.

2. F.Shimizu et al, Nucl.Phys. A389, 445 (1982).

3. See for example K.Gottfried, Proc. 5th Int.Conf.on High Energy Physics and Nuclear Structure, June 18-22, 1973, Uppsala, Sweden.

4. R.E.Renfordt et al, Phys.Rev.Lett. 53, 763 (1984).

5. H.A.Gustafsson et al, Phys.Lett. 142B, 141 (1984).

6. This figure is taken from R.Stock, Lectures given at the Erice Nuclear Physics School, April 20-22, 1985.

7. R.Stock, GSI preprint 85-60, November 85.

8. J.D.Bjorken, Phys.Rev. D27 (1983) 140.

9. J.Engels and H.Satz, Phys.Lett. 159B (1985) 151.

10. T.A.Shibata, Doctor Thesis, University of Tokyo, April 1982.

11. T.A.Shibata et al, Nucl.Phys. A408 (1983) 525.

12. K.Nakai, Nucl.Phys. A418 (1984) 163_c.

13. H. En'yo, Doctor Thesis, University of Tokyo, January 1985.

14. K.Nakai, paper presented at the Second Asia Pacific Physics Conference, Bangladore, India, January 13-17, 1986.

15. D.S.Barton et al, Phys.Rev. D27 (1983) 2580.

16. P.Forrest et al, unpublished.

17. R.Bailey et al, Z.Phys. C29 (1985) 1.

18. W.Busza and A.S.Goldhaber, Phys.Lett. 139B (1984) 235.

19. R.C.Hwa, Phys.Rev.Lett. 52 (1984) 492.

20. C.Y.Wong, Phys.Rev.Lett. 52 (1984) 1393.

21. J.Hüfner and A.Klar, Phys.Lett. 145B (1984) 167.

22. L.P.Csernai and J.I.Kapusta, Phys.Rev. D29 (1984) 2664 ; Phys.Rev. D31 (1984) 2795.

23. R.J.Glauber and C.Matthiae, Nucl.Phys. B21 (1970) 135.

24. A.Capella and J.Tran Thanh Van, Z.Phys. C10 (1981) 249.

25. A.Capella et al, preprint LPTHE Orsay 86/10.

26. S.Date, M.Gyulassy and H.Sumiyoshi, Univ.of Tokyo preprint, INS-Rep-535 (1985).

HADRON-NUCLEAR INTERACTIONS AT 100 GEV

W.D. WALKER, P.C. BHAT, W. KOWALD, W.J. ROBERTSON
Duke University, Durham, N.C. 27706

P.A. ELCOMBE, M.J. GOODRICK, J.C. HILL, W.W. NEALE
University of Cambridge, Cavendish Laboratory,
Cambridge CB3 OHE, England

P.W. LUCAS, L. VOYVODIC, R. WALKER
Fermilab, P.O.Box 500, Batavia, IL 60510

R. AMMAR, D. COPPAGE, R. DAVIS, D, DAY, J. GRESS,
S. KANEKAL, N. KWAK, L. HERDER
University of Kansas, Lawrence, KS 66045

J.M. BISHOP, N.N. BISWAS, N.M. CASON, V.P. KENNEY,
M.C.K. MATTINGLY, R.D. RUCHTI, W.D. SHEPHARD, S.J.Y. TING
University of Notre Dame, Notre Dame, IN 46550

J.WHITMORE, R.A. LEWIS, R. MILLER, B.Y. OH,
G.A. SMITH, W. TOOTHACKER
Penn State University, University Park, PA 16802

We are continuing our studies of the interactions, p, $\bar{p}$, π^+, π^-, with targets of Mg, Ag and Au. We present arguments that indicate that we can categorize the events in Au and Ag according to impact parameter. This is done by dividing the events according to the number of grey protons that are observed. On the basis of this division we study the characteristics of the interactions. We estimate $\frac{dE}{dx}$ for the incident particle. By measuring the amount of energy deposited we estimate $\frac{dE}{dx}$ to be 3 GeV/Fermi over a considerable range of nuclear matter (1-10 Fermi). We also discuss the possibility of a formation length. Estimations are given of the energy deposited in nucleonic knock-ons.

I. Introduction

A. History

The study of hadronic interactions in nuclei dates back to the time of the first work using nuclear emulsions as the target material and particle detector.[1] The nuclear targets were used as a substitute for hydrogen targets. The last ten years have seen a renaissance of interest in hadronic interactions with nuclei.[2,3,4] The nucleus itself furnishes the experimenter a laboratory with which to probe interactions with closely packed targets.

B. Outline of Paper

This note has as its focus two main points. First the space-time development of what is likely a quark-gluon shower. Second, we estimate the rate of energy loss of the incident particle as it progresses through the target. The latter point has relevance to the question of the production of a quark-gluon plasma in a nucleus-nucleus collision.

An important aspect of the analysis is the division of collisions on a given element by their impact parameter. For the remaining inelastic events we have attempted to divide the events into low impact parameter and high impact parameter collisions. This division is accomplished on the basis of the number of visible protons. One would expect that the number of visible protons, Np, would be greater for the low impact parameter collisions. In this paper we often exclude the diffractive events in which the number of visible protons (Np) is zero.

C. Description of the Experiment

This experiment was done using the Fermilab 30" hydrogen bubble chamber. For this experiment the bubble chamber was fitted with three pairs of small foils of silver, gold and magnesium. The thickness were such as to be a small fraction of a radiation length in order to minimize the effect of photon conversion. The bubble chamber was run to allow the determination of ionization and good π-proton discrimination up to momenta of about 1 GeV/c. The chamber was exposed to π^+, p and π^--p beams at 100 GeV/c. A short portion of the run was devoted to 320 GeV/c π^-. The incident particles were tagged by Cerenkov counter and wire chambers upstream of the bubble chamber. The experimental set-up is shown in Figure 1. In this note, the downstream chambers were used only for shapening momentum determinations.

The division of events by their impact parameters is of crucial importance for the interpretation of the experiment. The key to the division has to do with the number of identified protons. Experiments using Ne[2,5,6] as a target have shown that the distribution in the number of protons, Np, is independent of bombarding energy[4,5] and of projectile. An experiment done at the CERN SPS[6] has also shown this to be the case for collisions with elements for elements heavier than neon. (The Np is thus a nearly universal function of the target nucleus.) We believe that the Np is indicative of the volume of nuclear matter in which the reaction takes place. The quantity Np is on the average closely related to the chord length of the axis of the shower in the nuclear matter. In this experiment we have divided the heavy element collisions into two groups, Np $\leq$ 4 and Np $\geq$ 5.

This division placed comparable numbers of events in the peripheral and central collision categories. Such a division is necessarily imprecise in that the incident projectile can strike the target with small impact parameter yet travel several mean free paths before interacting. Such a collision will produce only a small number of recoil protons and be classified as a peripheral collision. The physics we wish to address has to do with the path length of the secondary particles in the nuclear matter.

We show in Table I the effect of the selection on the basis of Np. The tracks counted had momenta less than 10 GeV./c. The conclusion that we draw from the table is that peripheral collisions on Ag and Au nuclei are nearly equivalent to central collisions on Mg. This is to be expected if one calculates the average chord lengths of the particle paths in nuclear matter.

We have constructed a simple model to check on the validity of our division of the heavy nucleus events by impact parameter. We show in Figure 2 the frequency spectrum of the number of identified protons. The curves are the result of estimating the number of protons in the volume swept out by the production of the first collision in the nucleus. In particular, we assume the volume to be a truncated cone of radius 1 fermi and half angle .3 radians and then calculate the distribution in number of protons to be found in such a cone. The result of these calculations is compared with our data in Figure 2. The agreement between the model and the data can be improved if we correct for the protons among the secondary particles which are not identified by ionization. The presence of these protons is indicated by the excess of positive particles which appear above momenta of 1

GeV/c. We have tabulated the number of what are presumed to be unidentified protons below.

We have elaborated this consideration by looking at KNO multiplicity distributions for negative particle production using these cuts on Np. We have included data from a 200 GeV π^--neon run[5] as well. These results are shown in Figure 3. The distribution from peripheral Ag-Au collision and the π^--Ne collisions agree rather well with the hydrogen KNO distribution (π^--p at 200 GeV). The multiplicity distribution for the central Au-Ag collisions seems to give a slightly different distribution which is a bit higher in the region of $n_-/\langle n \rangle = 1$ and lower on the edges of the distribution. The central collisions have nearly twice the shower particle multiplicity as the peripheral collisions. Considering that the average pion multiplicity is nearly twice that for the peripheral collisions and four times that for hydrogen collisions, the agreement is good. The deficit of events for small n_s is likely due to the lack of diffractive events in the high Np sample.

General Characteristics

Table II and III contain a great deal of the information available from this experiment. We have divided the tabulations into energy bins. The number of + and - particles (excluding identified protons have been tabulated. The sum of the longitudinal momenta carried by these particles has also been given. If one looks at the particles which have momenta greater than 20 GeV/c then one sees that characteristics in terms of number of particles and the momenta carried are very similar to interactions with proton targets.[7] We conclude that for interactions with Mg (or peripheral collisions with Ag or Au) the results are nearly identical with nucleon interactions. This result can also be drawn from the experiment of Barton et al.[8] Their experiment goes only as low as 30 GeV/c for the reaction products but they have much more comprehensive data. For the central collisions with Ag and Au, there is considerable attenuation of the fast products of the reaction. We must conclude that the interaction of the fast particle "turn on" beyond about 4 Fermis in nuclear matter.

If one looks at the products of momenta less than 10 GeV/c we see a more gradual rise in the number of secondary particles with the increasing A of the target. We show in Figure 4,5 the rapidity distributions for the data with the various projectiles and cuts on Np. The central collisions on Ag or Au show a large enhancement at low rapidities for produced pions starting at about 3.5 units of rapidity. The objects producing these secondaries must have momenta somewhat above the 5 GeV or so at which the enhancement begins to show. We note that the low rapidity enhancement begins to show for flight paths greater than 4 or 5 Fermis.

We believe that there is indeed evidence for a formation length as proposed by Nicolaev.[4] The formation length might be parameterized in the form $a_f = b\gamma\frac{\hbar}{mc}$ (with b = .1). It is possible that the effects seen have nothing to do with the Landau-Pomerancuk effects but rather with some sort of color shielding because of the angular collimation of the products.

Energy Transfer To Struck Nucleons

In order to understand the mechanisms of pion production

these complex central collisions we have investigated the momentum spectrum of the protons emerging from the collisions. In the work of Yeager et al and Band on π-Ne collisions, the spectrum of struck nucleons was determined. The differential for 200 GeV π^--Ne collision momentum spectrum is given approximately by $A/(P+.3)^2$. The momentum spectrum from the central Ag and Au collisions seems to be slightly steeper.

A positive excess is found in the momentum spectrum of minimum ionizing particles with momenta between 1 and 4 GeV/c. The amount of the excess varies from about 1/2 a particle per event in the peripheral collisions to a bit over 1 particle per event for central collisions in Ag and Au. The onset of the effect occurs where visual identification of the protons begins to fail. We presume that the excess is produced by the unidentified protons. We also tabulate the excess momentum carried by the positive tracks. The estimated momentum per event carried varies from 1.0 to 4.0 GeV/c as one goes from peripheral to central collisions. We add this "positive excess" momentum to the longitudinal momentum carried by the identified protons. The average momentum carried by identified and unidentified protons varies between 2.2 and 5-6 GeV/c in going from peripheral to central collisions. From the above numbers we find on average that 5 GeV/c is transferred to nucleons and 14 GeV/c for peripheral and central collisions on Ag and Au. Assuming an average of .8 GeV/c transferred to nucleons in an inelastic collisions we estimate averages of 6 and 17 collisions in peripheral and central interactions respectively.

In the past it has been assumed that most of the nucleons are the result of a nucleonic cascade in which one nucleon strikes another with the result that many nucleons are knocked out. Although this may be the source of the very low energy end of the nucleon spectrum, the main source of nucleons is in the particle production process. It may be that there are processes that are peculiar to hadron nuclear processes in which much more than the usual 800 or 900 MeV/c is transferred to a nucleon. However, 15 to 20 is the most reasonable estimate of the number of interactions for the central collisions.

We give a summary of the data in Table IV. The number of protons are estimated from the positive excess of particles which is so characteristic of nuclear collisions.

Estimation of $\frac{dE}{dx}$

We have attempted to calculate the deposition of energy in the target nucleus.[9,10] The motivation for doing this is two-fold. First, one would like to know how the hadronic cascade develops as the hadrons propogate through the nuclear matter. Second, one would like to know the energy deposited in the nuclear target in order to make estimates of energy deposition in nucleus-nucleus collisions. This information allows one to look at the possibility of producing a quark-gluon plasma.

We have used our data to measure directly the amount of energy deposited in pions of momentum less than 10 GeV/c. In order to do this we use the data presented in Tables II and III. With our division of events by impact parameter we can calculate the average distances that the hadronic products will propagate through nuclear matter. The first interaction is assumed to occur approximately one mean free path into the nuclear matter.

The results of these considerations are shown in Tables IV,

V, VI and VII. The part of the table dealing with the decrease in the amount of energy carried by projectiles which have more than 30 GeV/c momentum are taken from the tables of Barton et al.[8] In constructing this table, we have used elements that have similar A's to those of our experiment. By taking differences between the entries in the energy columns we have estimated the loss of energy from the interval above 30 GeV. By using the data from A we have estimated the energy loss in central collisions on Pb. For the energy interval less than 10 GeV we have calculated the energy flow <u>into</u> that region as the projectile traverses more nuclear matter. We have done the same for the energy interval 10–20 GeV by finding the flow into that energy interval. We have found nearly the same value of $\frac{dE}{dx}$ for both pions and nucleons incident. We have data in effect for two length intervals, between 0–4 Fermis and between 0–9 Fermis. The energy deposited in charged pions in the 0–10 GeV interval seems to be close to 2 GeV/F for both sets of intervals. The energy interval 10–20 GeV shows a smaller acquisition and it seems to decrease as the projectile traverses the 4–10 Fermis distance. One can correct for the presence of π^0's approximately in the lower energy intervals by multiplying $\frac{dE}{dx}$ values by 1.5 which raises the value of $\frac{dE}{dx}$ to close to 3 GeV/F. With more data we would be able to deduce more about the character of the cascade process by making smaller energy intervals. From the data presented we see a steady attrition of the incident energy. The energy interval 10–20 GeV seems to be a "flow through" region with particles flowing in from the top of the interval and out at the bottom. One can note that in the 4–9 F region there is essentially no change in the energy content of the energy interval. This is intrinsically a different method of computing $\frac{dE}{dx}$ from those employed previously.[9,10] The energy loss of protons has been estimated by looking at emerging protons that have been degraded. If one does this then a correction must be made for fast neutrons. We avoid this by looking at the energy that is deposited at low energy. We feel that the correction for missing π^0's can be made with modest precision.

Conclusions

1. Leading particles are nearly identical for light nuclei and hadron-nucleon collisions.
2. Leading particles are strongly absorbed for central collisions.
3. Large momentum and energy transfer to the nucleons.
4. KNO scaling for the peripheral collisions and nearly so for the central collisions.
5. $\frac{dE}{dx}$ of about 3 GeV/F transferred from the leading particles to the low energy particles.

We believe that the above results are quite evident from the data. Whether or not there is a formation length or not is more difficult to ascertain. From these data and those of Barton et al[8] it appears that there is an enhanced absorption of the leading particles in the distance range of 4-10 Fermis. This effect might be associated with a Landau-Pomerancuk type interference.

The absorption might simply be the effect of the geometry of the production process. The particles peel off the axis of the shower as it progresses. Eventually a large effective cross section is developed by the whole ensemble of secondaries. At small distances from the shower origin it is likely that there will be some degree of color shielding, this results in a smaller than asymptotic cross section for the secondaries. Such an effect would probably account for the decrease in leading particles in the 4-10 Fermis range.

REFERENCES

1. U. Camerini, Lock, D. Perkins - Prog. In. C.R. Physics Vol. 1 North Holland Publishing (1952).

2. W.M. Yeager et al, Phys. Rev. D **16**, 1294 (1977).

3. J. Elias et al, Phys. Rev D **22**, 13 (1980).

4. There have been several reviews of hadron-nuclear collisions. I. Otterlund, Nuclear Phys. A418, 87 (1984). N.N. Nikoloev Zh ETF **81**, 814 (1981).

5. H. Band, PhD. Thesis, Duke University (1980) unpublished. Also, S.A. Azimov et al, Phys. Rev. D **23**, 2512 (1981).

6. K. Braune et al, Zeit fur Phys. **13**, 191 (1982).

7. A.E. Brenner et al, Phys. Rev. D **26**, 1497 (1982).

8. D. Barton et al, Phys. Rev. D **27**, 2580 (1983).

9. A.S. Goldhaber, W. Busza, Phys Lett. **139B**, 235 (1984).

10. S.Date', M.Gyulassy, H.Sumiyoshi, Phys.Rev. D **32**, 619 (1985).

TABLE I

Comparison of collisions with Mg and peripheral collisions with heavy elements

% of Interactions	Material	Np	$\langle n_\pm \rangle$	$\langle y_\pm \rangle$	$\langle Np \rangle$
40%	Au-Ag	1,2,3,4	10.4±.5	2.1±.1	2.3±.1
68%	Mg	≥ 1	9.6± .8	2.1±.1	2.0±.2
40%	Au-Ag	≥ 5	17.7±.8	1.7±.1	9.3±.4

TABLE II

$P(\bar{P})$ Collisions

No. of Protons		Np ≥ 1	Np ≤ 4		Np ≥ 5	
	$P(\bar{P})$-P	$P(\bar{P})$-Mg	$P(\bar{P})$-Ag	$P(\bar{P})$-Au	$P(\bar{P})$-Ag	$P(\bar{P})$-Au
P ≥ 20 GeV/c						
No.of +(−)	.62 (.56)	.48 (.58)	.59 (.53)	.76 (.57)	.44 (.36)	.26(.27)
No.of −(+)	.10 (.12)	.22 (.17)	.11 (.22)	.12 (.13)	.10 (.07)	.05(.14)
$\Sigma P_{+(-)}$	36.4(28.2)	16.3(24.4)	24.8(23.8)	32.2(27.1)	15.0(13.5)	9.7(8.6)
$\Sigma P_{-(+)}$	6.2 (3.7)	7.3 (6.2)	4.0 (6.8)	4.4 (5.1)	4.7 (3.2)	2.4(5.0)
10 GeV/c ≤ P ≤ 20 GeV/c						
No.of +(−)	.39(.37)	.44(.62)	.58(.47)	.49(.46)	.60(.53)	.47(.49)
No.of −(+)	.19(.20)	.41(.35)	.32(.33)	.23(.38)	.35(.40)	.34(.39)
$\Sigma P_{+(-)}$	4.9(6.3)	7.1(9.2)	8.8(7.4)	7.8(6.8)	8.7(7.8)	6.8(7.8)
$\Sigma P_{-(+)}$	3.5(2.9)	6.7(5.2)	5.1(5.1)	4.2(5.8)	5.2(6.1)	5.3(6.4)
P ≤ 10 GeV/c						
No.of +(−)	2.89(2.8)	5.4 (4.2)	4.5 (4.5)	3.8 (3.7)	8.8 (8.1)	10.0 (9.0)
No.of −(+)	1.9 (3.0)	4.3 (4.6)	3.7 (4.9)	3.4 (3.8)	7.3 (9.3)	7.9 (10.1)
$\Sigma P_{+(-)}$	7.4 (6.1)	14.2(10.8)	10.7(11.0)	10.1(9.1)	17.4(14.4)	19.8(15.3)
$\Sigma P_{-(+)}$	3.9 (6.1)	10.5(10.0)	8.2 (11.2)	8.0 (9.6)	12.9(17.5)	11.5(18.0)
No. of Events		27(60)	202(116)	86(112)	86(70)	58(69)

TABLE III

$\pi^+(\pi^-)$ Collisions

	$N_p \geq 1$		$N_p < 4$		$N_p > 5$	
	$\pi^+(\pi^-)$-P	$\pi^+(\pi^-)$-Mg	$\pi^+(\pi^-)$-Ag	$\pi^+(\pi^-)$-Au	$\pi^+(\pi^-)$-Ag	$\pi^+(\pi^-)$-Au
$P \geq 20$ GeV/c						
$N_{+(-)}$	.59 (.49)	.59 (.44)	.59 (.53)	.57 (.50)	.31 (.25)	.24 (.30)
$N_{-(+)}$	.20 (.21)	.18 (.29)	.23 (.25)	.27 (.25)	.19 (.22)	.16 (.21)
$\Sigma P_{+(-)}$	26.9(23.3)	22.7(17.3)	23.4(21.2)	24.0(19.5)	10.6(12.3)	8.9(12.1)
$\Sigma P_{-(+)}$	7.3 (7.1)	6.8 (10.2)	9.0 (9.6)	10.1.(9.5)	7.5 (7.6)	7.2 (6.5)
10 GeV/c $\leq P_L \leq$ 20 GeV/c						
$N_{+(-)}$	.66(.33)	.56(.53)	.56(.55)	.51(.64)	.50(.40)	.54(.74)
$N_{-(+)}$	.27(.24)	.38(.48)	.30(.44)	.31(.37)	.38(.38)	.36(.53)
$\Sigma P_{+(-)}$	6.7(5.8)	8.8(8.1)	8.6(8.3)	7.8(9.6)	7.6(6.3)	8.5(10.7)
$\Sigma P_{-(+)}$	3.5(3.5)	6.0(7.6)	5.0(7.1)	5.2(5.8)	6.1(5.9)	5.6(8.3)
$P_L \leq$ 10 GeV/c						
$N_{+(-)}$	3.0(2.5)	4.3 (4.2)	4.3 (4.0)	4.2 (3.7)	8.2 (7.6)	9.2 (7.1)
$N_{-(+)}$	1.9(2.6)	3.7 (4.0)	3.4 (4.1)	3.2 (4.0)	6.2 (8.6)	7.3 (8.4)
$\Sigma P_{+(-)}$	5.6(5.6)	11.4(10.6)	10.5(10.0)	11.3(9.3)	16.2(14.6)	17.7(12.6)
$\Sigma P_{-(+)}$	3.8(5.3)	8.6 (9.1)	8.6 (9.6)	7.8(10.0)	10.4(16.5)	12.3(15.8)
No. of Events		92(62)	230(174)	173(128)	108(85)	94(86)

In π^--Ag and π^--Au ($N_p \leq 4$) analysis the cuts are (i) $n_s \geq 4$ and (ii) for $n_s = 4$ and 5 momenta of all the tracks should be > GeV/c.

Table IV

Number of Minimum Ionizing Protons*

Incident Particle	Number between $1 \leq P \leq 2$ GeV/c $n_+ - n_-$		Number between $2 \leq P \leq 4$ GeV/c $n_+ - n_-$		Number between $4 \leq P \leq 8$ GeV/c $n_+ - n_-$	
Number of protons						
π^- Ag,Au $N_p \leq 4$	.10		.03		−.0022	
(Number of protons)		(.19)		(.11)		(.12)
π^+ Ag,Au $N_p \leq 4$	.27		.19		.22	
π^- Ag,Au $N_p \geq 5$	.61		.17		.21	
(Number of protons)		(.71)		(.30)		(.28)
π^+ Ag,Au $N_p \geq 5$	.81		.42		.35	
π^- Ag,Au $N_p \geq 9$	.83		.40		.15	
(Number of protons)		(.97)		(.52)		(.26)
π^+ Ag,Au $N_p \geq 9$	1.10		.63		.38	
P–Ag,Au $N_p \leq 4$	.16		.27		.01	
(Number of protons)		(.29)		(.24)		(.09)
P–Ag,Au $N_p \leq 4$	.42		.21		.16	
P–Ag,Au $N_p \geq 5$	.54		.48		.07	
(Number of protons)		(.66)		(.60)		(.21)
P–Ag,Au $N_p \geq 5$	.78		.72		.35	
P–Ag,Au $N_p \geq 9$	.54		.37		.044	
(Number of protons)		(.83)		(.50)		(.32)
P–Ag,Au Np.9	1.12		.63		.59	

* The estimated number of protons comes from averaging the positive excess as we change the sign of the projectile. The effect of the charge of the leading particle is cancelled by this procedure. The errors can be estimated by using the number of events in Tables II and III.

TABLE V

Energy Flow Down P > 30 GeV/c

	energy in particles of P > 30 GeV/c	$\frac{dE}{dx}$ in GeV/F
P + P	32.9	GeV/F
P + Al	27.5	2.1
P + Ag	21.1	2.2
P + Pb	17.5	2.2
$P + Pb)_{cent}$	10.9	2.2
$\pi^+ + P$	26.2	GeV/F
$\pi^+ + Al$	25.2	.35
$\pi^+ + Ag$	20.6	1.2
$\pi^+ + Pb$	17.6	1.4
$\pi^+ + Pb)_{cent}$	12.5	1.9

TABLE VI

Energy Flow Into P < 10 GeV/c

	energy in pions P < 10 GeV/c	$\frac{dE}{dx}$ in GeV/F
P − P	10.65	
P − Mg	22.9	2.9 (2.0)*
P − Ag ($Np \leq 4$)	17.7	1.7 (2.4)
P − An ($Np \leq 4$)	16.4	1.4 (1.9)
P − Ag ($Np \geq 5$)	28.9	2.4
P − An ($Np \geq 5$)	30.3	2.0

* deduced from P incident

$\pi^+ + P$	9.4	
$\pi^+ + Mg$	18.5	2.2
$\pi^+ + Ag$ ($Np \leq 4$)	18.1	1.8
$\pi^+ + An$ ($Np \leq 4$)	17.3	1.8
$\pi^+ + Ag$ ($Np \geq 5$)	23.3	2.0
$\pi^+ + An$ ($Np \geq 5$)	26.5	1.9

* Results from P collisions

TABLE VII

Energy Flow Into $10 \leq P \leq 20$ GeV/c

	Energy in interval in GeV	$\frac{dE}{dx}$ in GeV/F
P – P	8.4	
P – Mg	13.8	1.3
P – Ag ($Np \leq 4$)	13.9	1.4
P – Au ($Np \leq 4$)	12.0	.90
P – Ag ($Np \geq 5$)	13.9	.72
P – Ag ($Np \geq 5$)	12.1	.38
π^{+}–P	10.2	
π^{+}–Mg	14.8	1.11
π^{+}–Ag ($Np \leq 4$)	13.6	.72
π^{+}–Au ($Np \leq 4$)	13.0	.64
π^{+}–Ag ($Np \geq 5$)	13.7	.50
π^{+}–Au ($Np \geq 5$)	14.1	.44

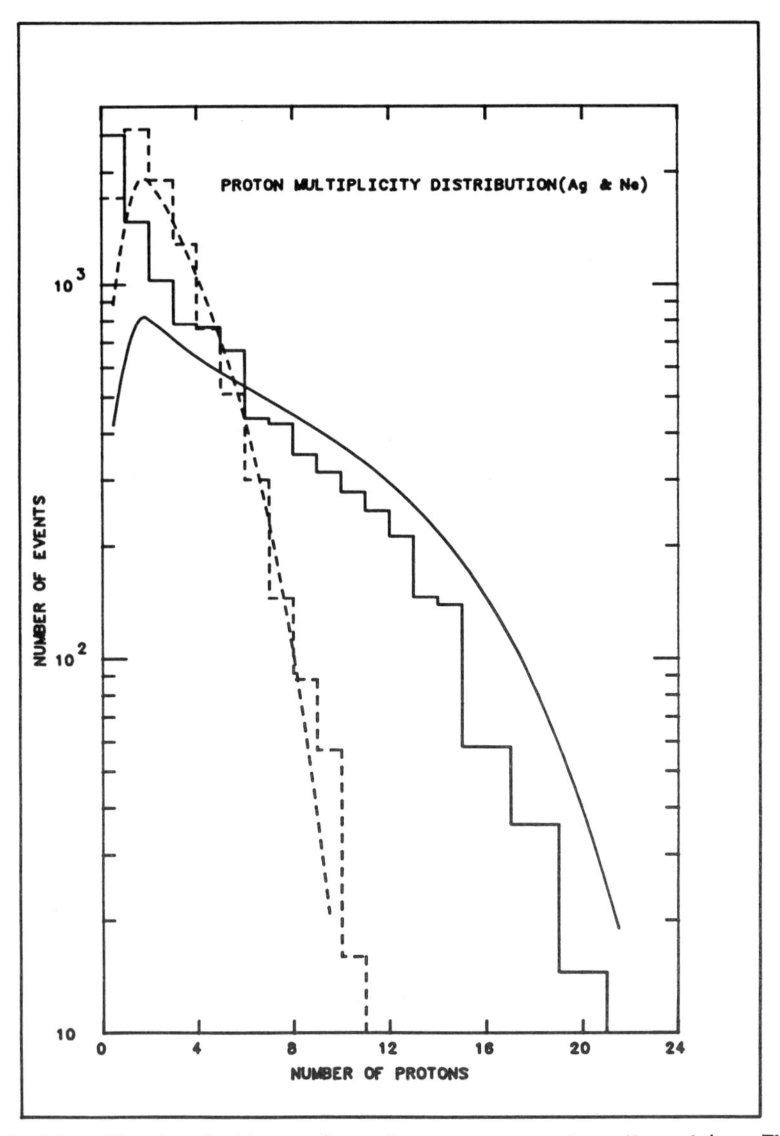

Figure 1: Distribution in the number of grey protons from Ne and Ag. The curves were calculated by a method described in the text.

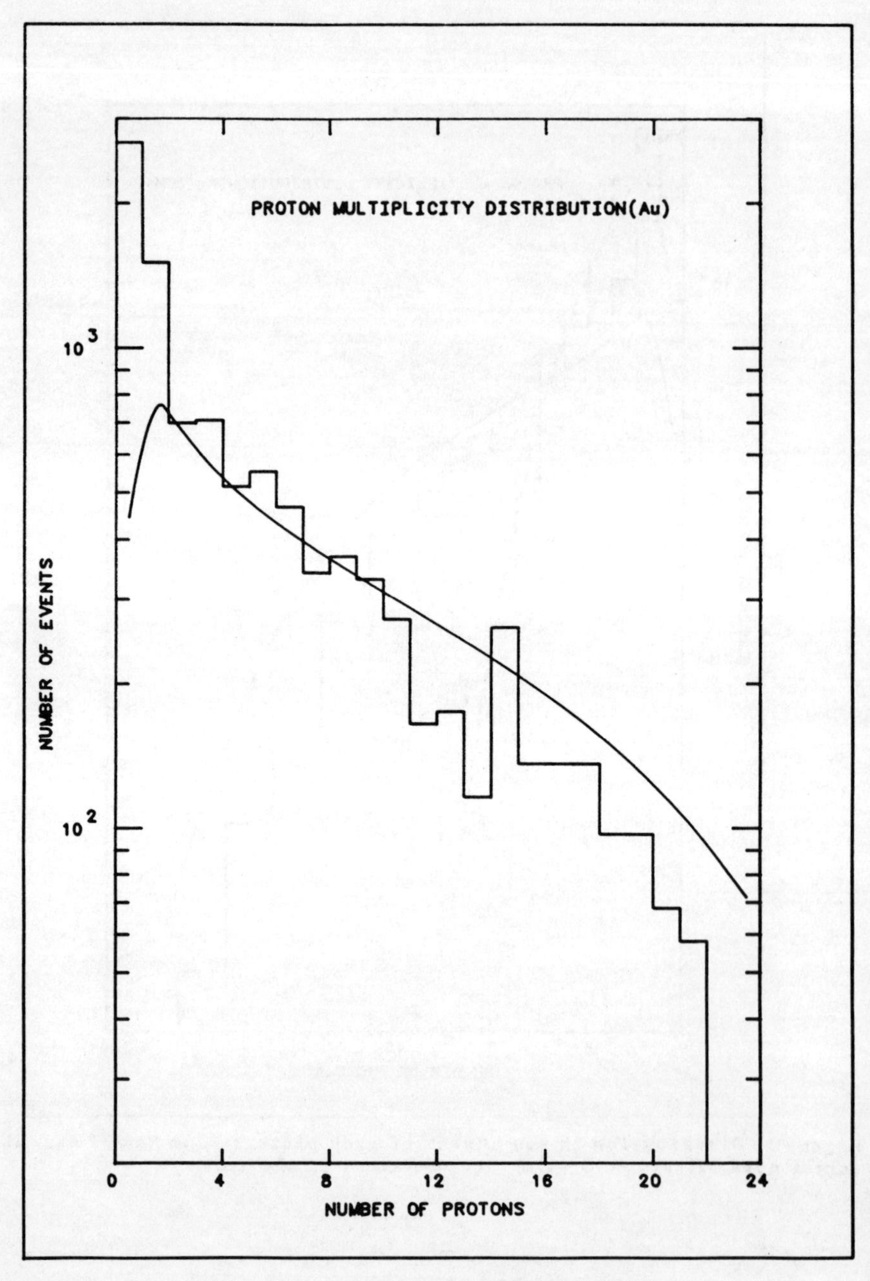

Figure 2: Distribution in the number of grey protons from Au.

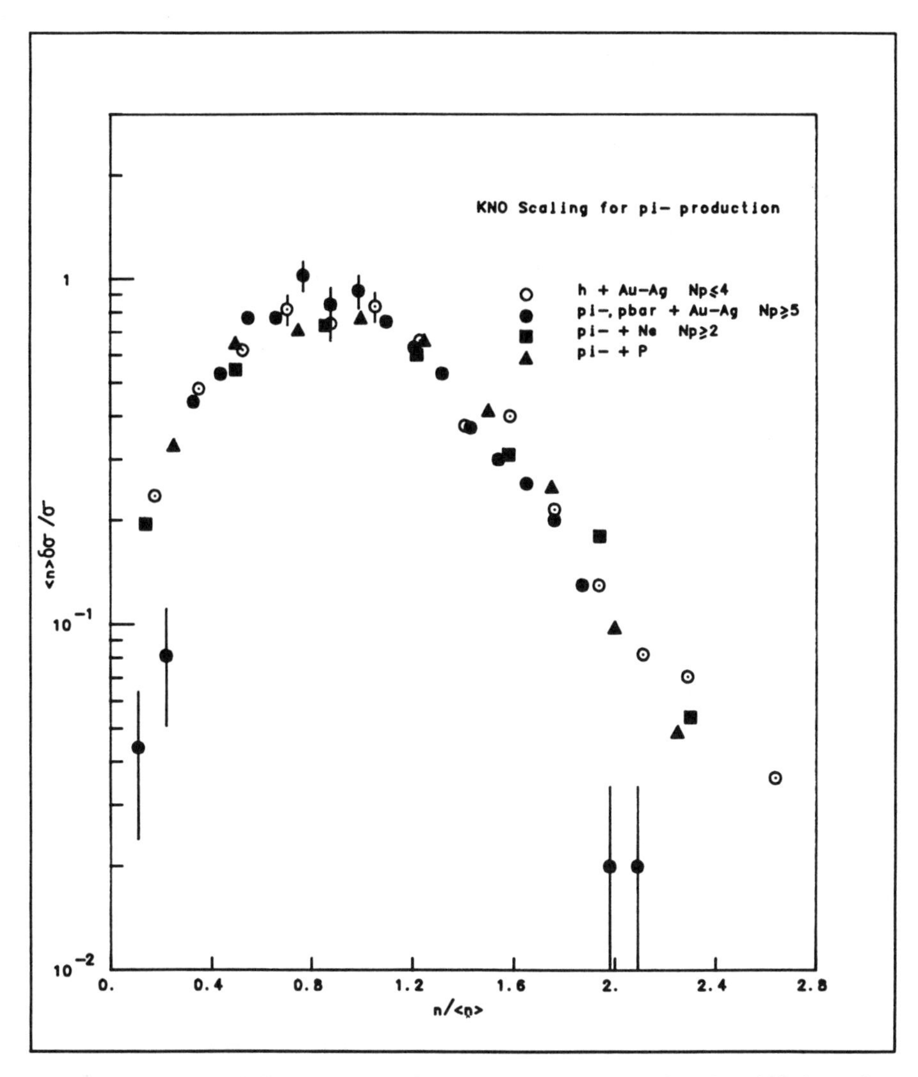

Figure 3: KNO multiplicity scaling curves from peripheral collisions in Au, Ag, 200 π^{-}-Ne collisions, central collisions in Ag, Au, π^{-}-p at 200 GeV.

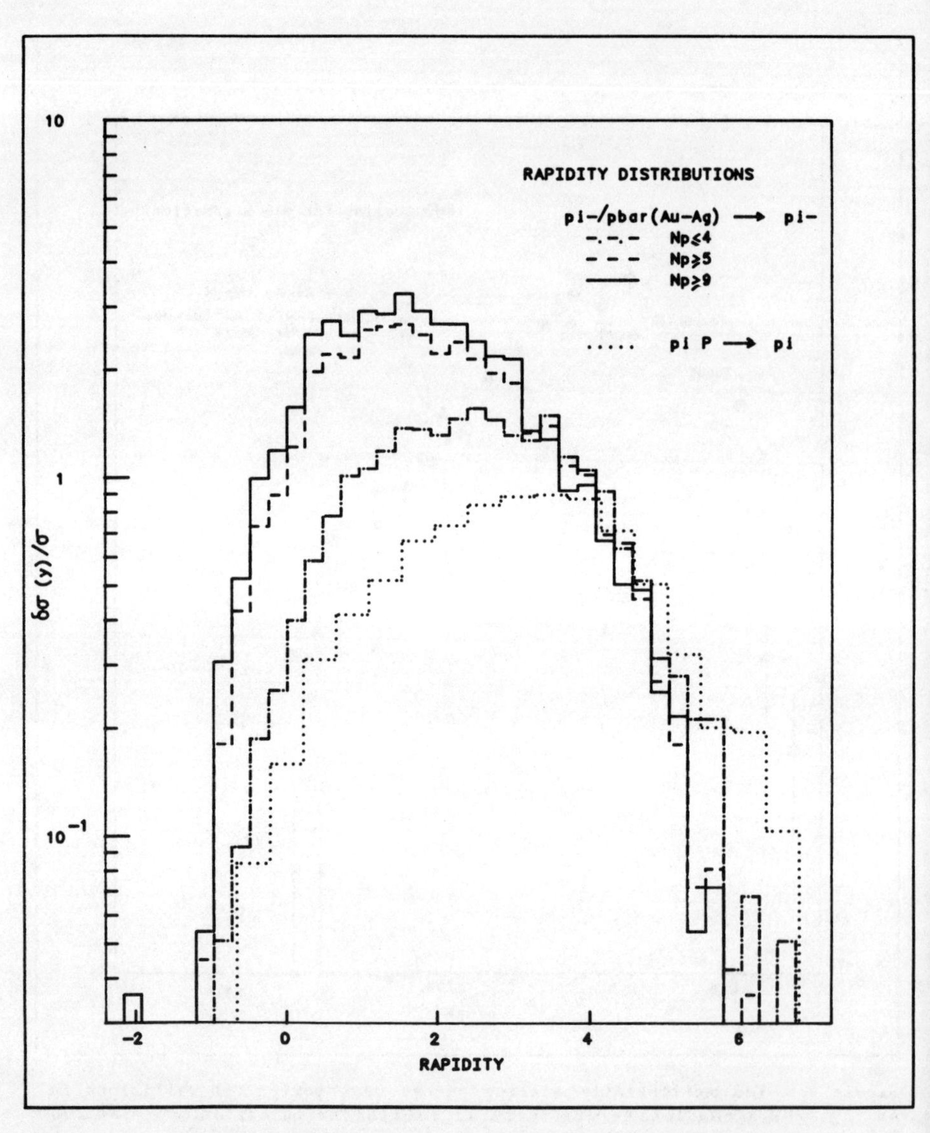

Figure 4: Rapidity distributions of π^- for π^--p, and peripheral and central collisions with Ag and Au. We have lumped together π^-, $\bar{p}$ primary projectiles.

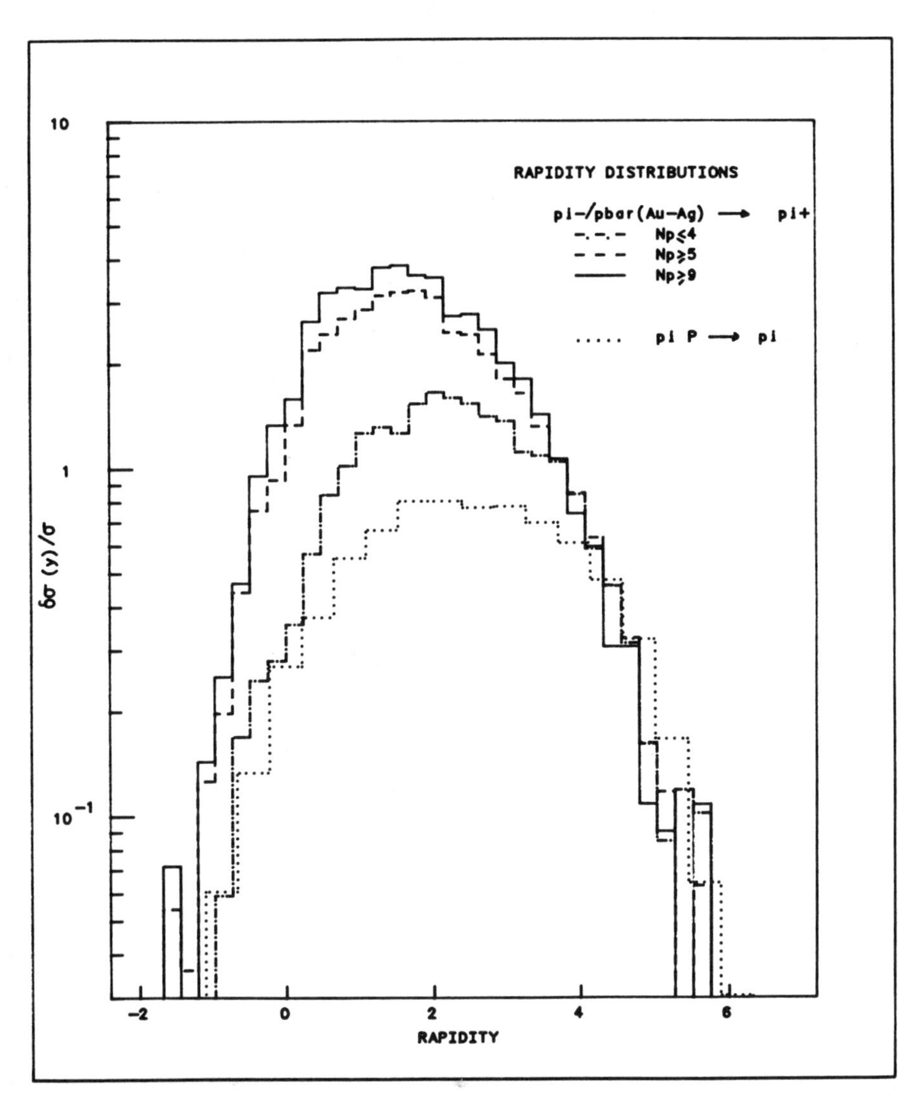

Figure 5: Rapidity distributions of π^+ for π^--p and peripheral and central collisions with Ag and Au. We have lumped together cases with π^- and $\bar{p}$ primary projectiles.

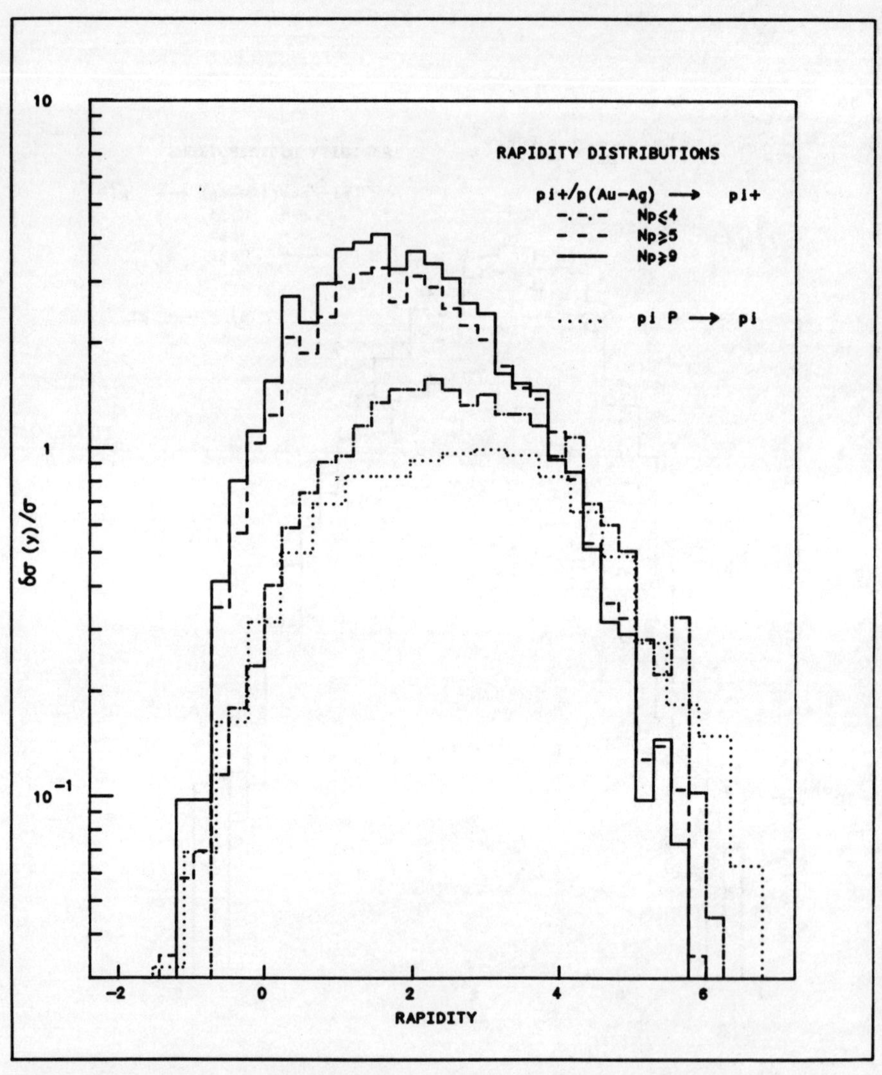

Figure 6: Rapidity distribution of π^+ for π^+-p and central and peripheral collisions of π^+ with Ag and Au.

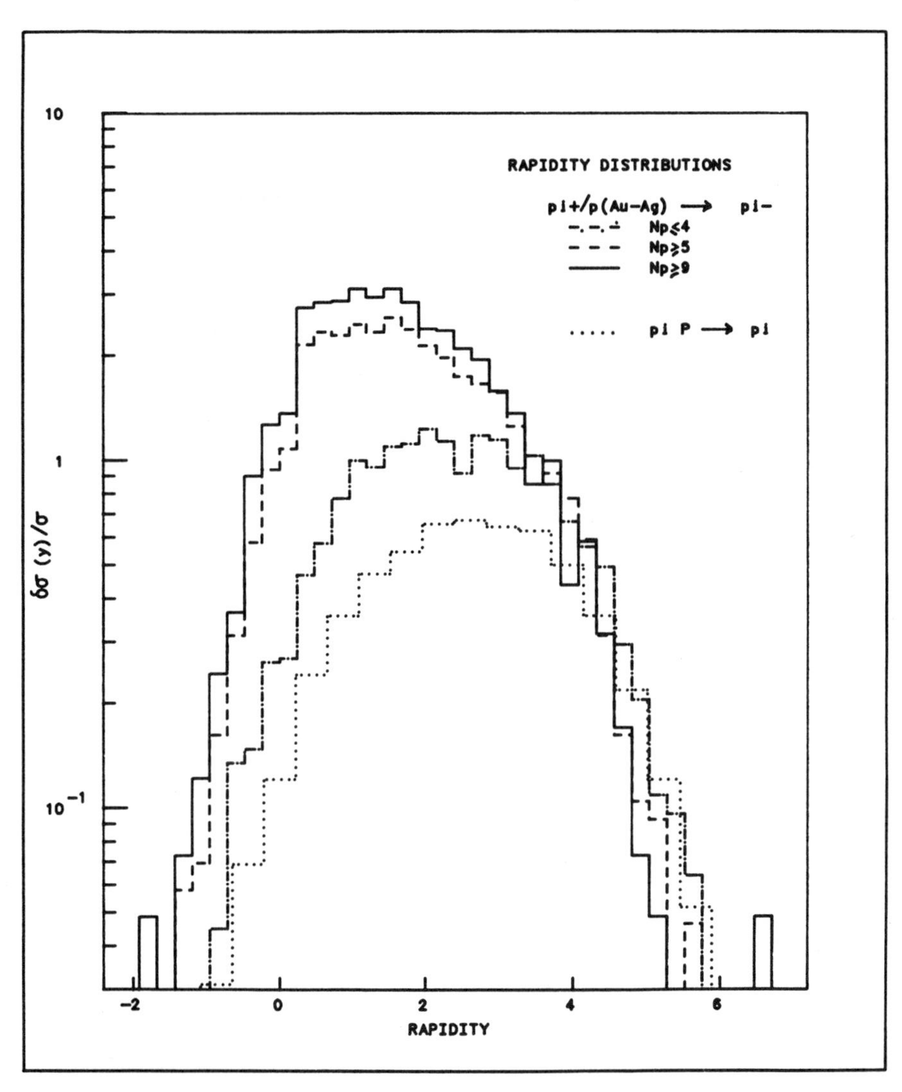

Figure 7: Rapidity distribution of π^- for π^+-p and central and peripheral collisions of π^+ with Ag and Au.

STOPPING POWER AND ENERGY DEPOSITION IN p+A COLLISIONS

R. J. Ledoux, M. Bloomer, H. Z. Huang
Physics Department
Massachusetts Institute of Technology
Cambridge, MA 02139

1. Introduction

There has been renewed interest in p+A collisions in the last few years.[1-4] This has resulted from theoretical speculation that a new state of matter, a Quark Gluon Plasma (QGP), may be the high temperature limit of QCD.[5] Could a QGP be formed for short periods of time in nucleus-nucleus collisions? This question can not be answered until such collisions are actually studied. However, by studying p+A collisions, it may be possible to estimate the conditions that exist in the initial stages of A+A collisions. In particular it has been suggested that by studying the energy loss of the incident proton in p+A collisions an estimate can be made of the maximum baryon and energy densities in central A+A collisions. In this talk I will concentrate on presenting the available data on inclusive leading baryon distributions ($pA \rightarrow pX$) as well as data which is relevant to the question of where the energy lost from the leading particle goes. There have been a number of recent experiments which makes such a review relevant at this time. This talk will be divided into two sections. I will begin by discussing the state of affairs in inclusive leading baryon spectra and then proceed to a discussion of energy deposition.

2. Leading Particle Spectra

How can the maximum baryon density of A-A collisions be estimated? I can not do justice to this question in a short period of time, but it has been discussed extensively by many authors,[1-4,16,17] and a variety of talks in this conference will address this question (see talk by M. Soyeur). Schematically the argument is the following: as viewed in the c.m. frame a central A-A collision consists of two interacting Lorentz contracted nuclei. The valence quarks as they interact lose energy (or rapidity) and hence slow down. The simplest estimate of the maximum baryon density can be obtained by considering a central Pb+Pb collision at a beam energy corresponding to the median rapidity loss measured in p+Pb. In such a collision in the so called "stopping regime," the quarks in the two nuclei will on the <u>average</u> stop in the c.m. frame. The Lorentz contracted densities would then be at least partially preserved and the average nuclear matter density would increase by $\approx 2\gamma_{beam}$. Thus a measurement of the average energy loss of a proton in a proton-nucleus collision allows an estimate of the maximum baryon densitites which may exist in the initial stages of such collisions. What we would like to measure, therefore, is the energy loss distribution of protons as a function of the amount of nuclear matter they traverse. Usually, rather than energy loss, the rapidity loss of the leading particle is extracted since the shape of this distribution is Lorentz invariant. Such a measurement is referred to as a measurement of the "nuclear stopping power."

It should be emphasized that there are two important assumptions in going from a definition of the "nuclear stopping power" to the experimental observable. First, in pA → pX measurements it is assumed that the observed proton is in fact the projectile fragment. This is a reasonable assumption for protons which have a momentum greater than 1/2 that of the beam. For lower momentum protons it is possible that the leading projectile fragment may be another particle. In particular, the limited available data suggests that a neutron is the leading particle a significant fraction of the time (see below).

The second assumption in our definition of nuclear stopping power is that there exists an observable which is a measure of the impact parameter in p+A collisions. By choosing different mass targets it is possible to study the A dependence of the inclusive leading particle spectra. This A dependence is usually expressed in terms of the variable $\bar{\nu} = A(\sigma_{pp}^{inel}/\sigma_{pA}^{inel})$. $\bar{\nu}$ is referred to as the "mean number of collisions" since its definition is consistent with taking the mean path length through the target nucleus divided by the p-p mean free path. Experimentally, $\bar{\nu}$ is found to be proportional to $A^{1/3}$ and that the number of produced particles (π's or "shower particles" in emulsion experiments) is linearly proportional to $\bar{\nu}$.[6,7] $\bar{\nu}$ will be used as our scaling variable in A but no fundamental importance will be attributed to it. $\bar{\nu}$ is found to vary from 1 to 3.8 in p+A collisions for A from 1 to 208 (p to Pb).

It would be desirable to obtain an event-by-event measure of the impact parameter since even interactions with heavy targets are dominated by peripheral collisions making it difficult to extract the energy loss for central collisions only. Such an observable has been suggested by many authors and is extracted from "target" protons (n_p) by the relationship $\bar{\nu}(n_p) = C \times \sqrt{n_p}$ where n_p is the number of protons with momenta less than approximately 1 GeV/c (grey tracks).[8] C is chosen such that the mean $\bar{\nu}(n_p)$ is equal to $\bar{\nu}$ for a given target. Fig. 1 is an example of the $\bar{\nu}$ dependence of the R value (ratio of multiplicity for a given $\bar{\nu}$ to that in p-p collisions). As can be seen both definitions of $\bar{\nu}$ are consistent and that $\bar{\nu}(n_p)$ varies from 1 to 7. $\bar{\nu}(n_p)$ is a statistical variable with a width estimated to be approximately ±1.

The body of data on leading baryon inclusive distributions is rather limited, spanning an energy range from 19.2 GeV to 400 GeV.[9-13] Table 1 lists the energy, targets, baryon measured and

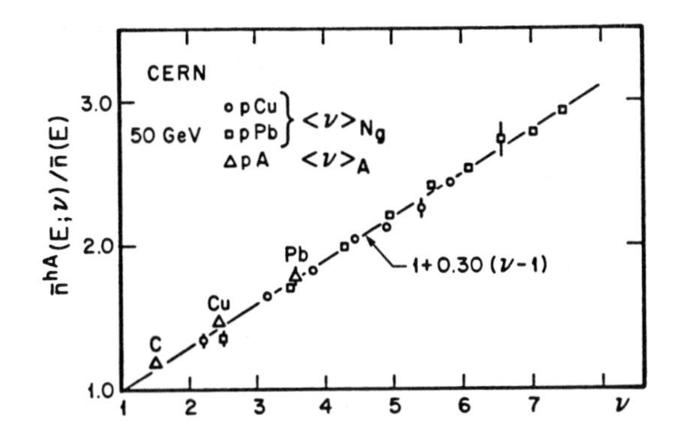

Fig. 1: Ratio of the multiplicity of produced particles in pA collisions divided by the multiplicity in pp collisions as a function of $\bar{\nu}$ and $\bar{\nu}(n_p)$. [8]

TABLE 1: Inclusive Leading Baryon Data for pA → Baryon + X

E_{beam}(GeV)	Ref.	Targets	Baryon Measured	x^F_{range}	$P_\perp^{Max}$(GeV/c)	Comments
19.2	Allaby, et al.[9]	Be,C,Al,Cu, Pb	p	.23-.80	0.5-1.1	$P_\perp$ range covered depends on x, and target
24	Eichten, et al.[10]	Be,C,Al,Cu, Pb	p	.04-.70	0.5-1.5	" "
100	Barton, et al.[11]	P,C,Al,Cu, Ag,Pb	p	.25-.9	0.3,0.5	Invariant X-sec measured at 2-values of $P_\perp$
120	Bailey, et al.[12]	Be,Cu,Ag, W,U	p	.07-.60	Full(?)	dσ/dx Tabulated
400	Forrest, et al.[13]	P,Be,Cu,Pb	n	.2-.9	Full	dσ/dx plotted by authors
200	IHSC[15]	p,Mg,Ag,Au	no PID	Full	Full	Hybrid Spectrometer 4π-Visual, no PID

the kinematic range covered by electronic measurements which have particle identification (PID). As can be seen, only the 19.2, 24, and 120 GeV data span a significant range of Feynman x and $P_\perp$. There is overlap for different energy data for $0.2 < x < 0.6$. There is only one preliminary pA → nX measurement. Included in Table 1 is a preliminary set of hybrid spectrometer data taken by the International Hybrid Spectrometer Collaboration (IHSC-FNAL E565/570).[15] It has been included in Table 1 even though there is no PID since it is a set of data which includes a global measurement of pA interactions. (There are other sets of data of a similar nature, namely, streamer chamber data and other hybrid spectrometer data. However, the streamer chamber data[18] has not yet been analyzed to obtain leading particle data. W. Walker will discuss a similar set of hybrid spectrometer data later in this conference.)

In order to compare data of different energies, it is desirable to use a scaling variable. A variable which is often used to discuss inclusive leading baryon data is the Feynman x variable ($x = P_L^{c.m.}/P_{L,max}^{c.m.}$). The value of Feynman x is related to the longitudinal momentum fraction of the leading particle in the c.m. frame and, therefore, a measurement of $d\sigma/dx$ is related to the probability of finding a baryon with a certain fraction of the initial beam momentum. If Feynman x is truly a scaling variable for inclusive leading particle distributions in pA collisions then data obtained at different beam energies within the "scaling" regime should agree if plotted versus x. However, the definition of Feynman x is dependent on a unique definition of the c.m. system--usually the N-N frame. If multiple gluon exchange or multiple scattering contribute, the use of Feynman x as a scaling variable may not be appropriate.

Why is it so difficult to obtain $d\sigma/dx$ distributions? The answer can be seen by considering plots of the invariant cross

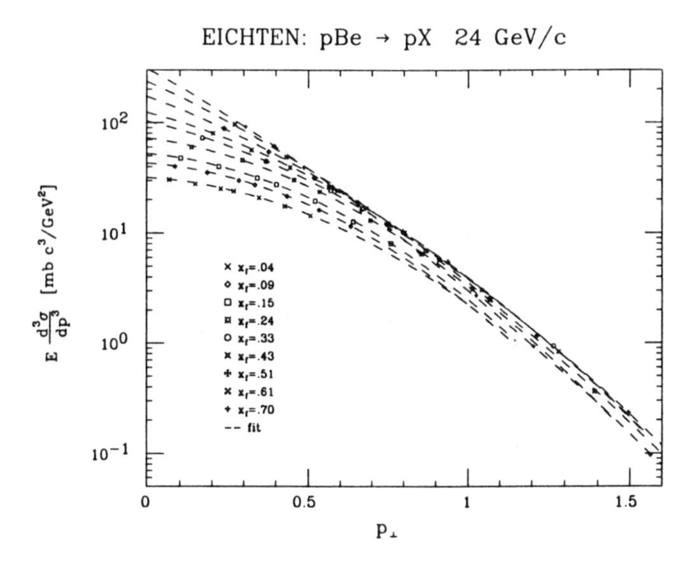

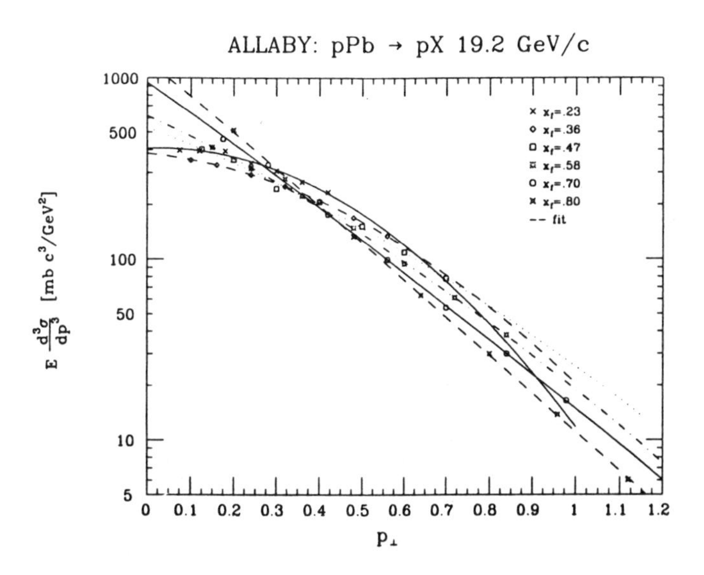

Fig. 2: Invariant cross section at different Feynman x for a) 24 GeV/c pBe [10] and b) 19.2 GeV/c pPb [9] collisions. The dashed curves were used in extrapolating the cross section.

sections for different x values versus $P_\perp$. Fig. 2 displays distributions for pBe at 24 GeV and pPb at 19.2 GeV. The cross sections vary dramatically with x and the $P_\perp$ distributions have a rather pronounced dependence on x. Thus, to obtain accurate $d\sigma/dx$ distributions it is necessary to measure and identify protons over a large range of momenta and at large $P_\perp$. We have integrated the invariant cross sections at 19.2 GeV (Allaby, et al.[9]) and 24 GeV (Eichten, et al.[10]) by making a least-square fit and extropolating where necessary to $P_\perp \geq 1.0$ GeV/c. The data of Barton, et al.[11] has been converted to $d\sigma/dx$ using the measured pp transverse momentum distribution. The recent 120 GeV/c[12] data of Bailey, et al. contains tabulated $d\sigma/dx$ distributions.

Fig. 3 displays $d\sigma/dx$ for pp → pX at 100 GeV and pp → nX at 400 GeV.[14] The inclusive proton distribution is relatively flat whereas the inclusive neutron distribution is much softer. The $\langle x \rangle \approx 0.4 \pm 0.05$ which indicates that even in pp collisions there is a 60% energy loss of the leading baryon. Figures 4a-d show the various $d\sigma/dx$ measurements for Be, Al, Cu and Pb targets. The shape of the $d\sigma/dx$ distributions for the Be targets differ dramatically. The differences in shape may be attributed to a breakdown of scaling. It should be noted, though, that σ_{Be}^{inel} as measured by Bailey, et al., is ≈ 167 mb, whereas, other have found it to be ≈ 220 mb.[10] This may contribute to the difference in absolute normalization for x > 0.4. The data for the Al, Cu and Pb targets agree fairly well for x > 0.3. $d\sigma/dx$ for Bailey, et al., and Barton, et al., for the Cu and Pb targets increase approximately exponentially with decreasing x, qualitatively indicating that the proton is losing more energy on the average as compared to pp, pBe and pAl.

The arithmetic mean ($\langle x \rangle$) and median x value ($x_{1/2}$) obtained from the data of Bailey, et al., are contained in the first

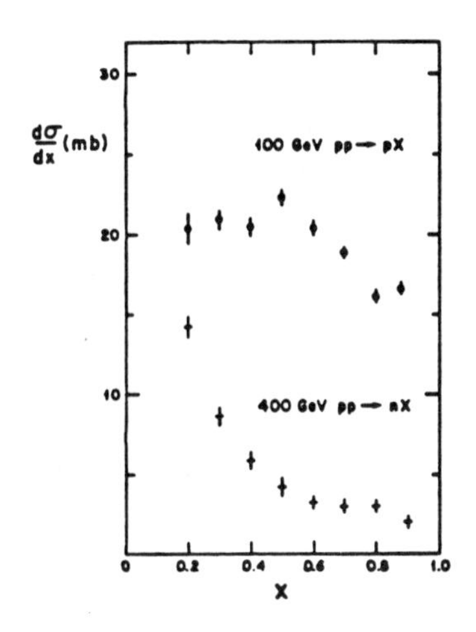

Fig. 3: $d\sigma/dx$ for pp → pX at 100 GeV/c and pp → nX at 400 GeV/c (preliminary data of Forrest et. al.[13]). From W. Busza, ref. 14.

columns of Table 2 for the Be, Cu and U targets. The data was smoothly extropolated for $x < 0.06$ and $x > 0.6$ (see Figure 8 below). $\bar{\nu}$ was obtained using the measured σ_{pA}^{inel} and $\sigma_{pp}^{inel} = 30$ mb. Error bars were obtained from estimates of the error caused in extropolating the data. As can be seen the <x> for Be is, within errors, equal to that for pp. Both the <x> and

Table 2: Average (<x>) and Median ($x_{1/2}$) of Leading Particle in pA Collisons

Bailey, et al., 120 GeV [12] pA → pX				Preliminary IHSC, 200 GeV [15] Leading Positive Part.				Preliminary IHSC, 200 GeV[15] Leading Postive Part.		
Targ.	$\bar{\nu}$	<x>	$x_{1/2}$	Targ.	$\bar{\nu}$	<x>	$x_{1/2}$	$\bar{\nu}(n_p)$	<x>	$x_{1/2}$
Be	1.4	.39±.05	.35±.05	Mg	2.0	.32±.05	.23±.05	1.6±1	.36±.03	.28±.03
Cu	2.5	.34±.05	.27±.05	Ag	3.1	.34±.04	.29±.04	3.8±1	.30±.05	.21±.04
U	3.8	.30±.05	.22±.05	Au	3.65	.28±.04	.23±.04	6.3±1	.20±.03	.13±.04

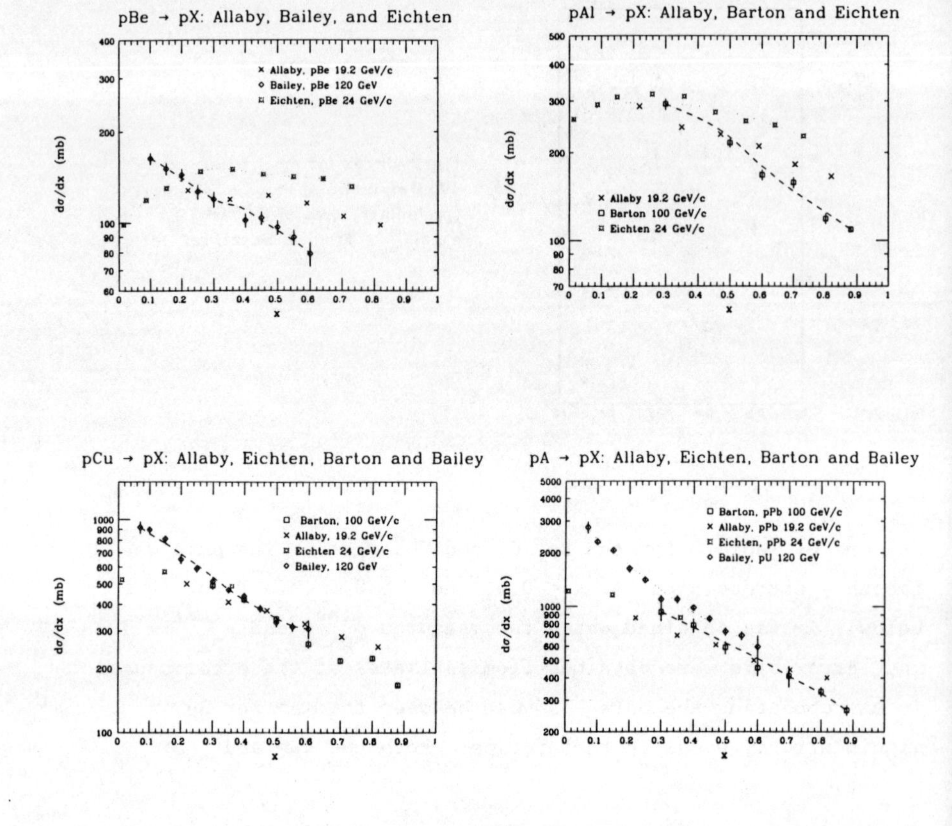

Fig. 4: dσ/dx for pA → pX at various energies for a) Be, b) Al, c) Cu and d) Pb and U targets. The invariant cross sections have in some cases been extrapolated (see text).

$x_{1/2}$ decrease with increasing mass of the target with the median always lower. The $x_{1/2}$ for pU $\approx$ 0.22±0.05 which represents a 78% energy loss. It should be noted that $x_{1/2}$ can be significantly affected by the exact nature of $d\sigma/dx$ for $x<0.1$. The extropolations of Busza and Goldhalber discussed below indicate that there is substantial cross section in this region.

Included in Table 2 are preliminary results from FNAL E565/570 taken by the IHSC.[15] There is no PID in these measurements for momenta greater than 1.5 GeV/c. Events were selected in which the highest momentum poarticle was positive and with the assumption that it is in fact a proton. Only events with multiplicity ≥ 3 were scanned. This may lead to a bias at large x. The reason for including these events is to compare them to the data of Bailey, et al., and since these events were measured in a bubble chamber, they cover the full kinematic range and make it possible to gate on $\bar{\nu}(n_p)$. The $\langle x \rangle$ and $x_{1/2}$ are tabulated by target (Mg, Ag and Au) and also by $\bar{\nu}(n_p)$. The $\langle x \rangle$ and $x_{1/2}$ for the Ag and Au targets are comparable to the values obtained for Cu and U. The data for the Mg target appear to be low. The reason for this discrepancy for the Mg data is not known but it should be noted that the error bars are large. The $\langle x \rangle$ and $x_{1/2}$ for $\bar{\nu}(n_p)$ = 1.6 and 3.8 agree well with those obtained using $\bar{\nu}$. $\bar{\nu}$ = 6.3 corresponds to a collision with the center of a heavy nucleus and the value of $x_{1/2}$ = 0.13±0.04 represents a 87% energy loss. These results are, admittedly, based on low statistics ($\approx$ 75 events per target) and no PID; however, the good agreement between this set of data and that of Bailey, et al., allows a reasonable extropolation from an average over all impact parameters for a heavy target to $x_{1/2}$ = 0.13 for a central collision.

There have been a number of theoretical attempts[16] to fit the inclusive proton data of Barton, et al. The first of these was that of Busza and Goldhaber[17] (B+G) which used the measured A-dependence of the inclusive proton data of Barton, et al., to predict the rapidity loss distribution for <u>central</u> pPb collisions. As already mentioned the data of Barton, et al., was taken at fixed $P_{\perp}$, therefore, extrapolation to regions of phase space not measured must be made. This was done by assuming that the inclusive $P_{\perp}$ distributions are independent of x and A, and that the ratio of the production of various baryons is independent of A. The nucleus was divided into a peripheral region composed of 3.3 nuclei with A=15 containing 1/2 of the total inelastic cross section and a central part which has no contribution at x=1. Their extrapolated fits for a peripheral ($\bar{\nu}\approx1.5$), mean ($\bar{\nu}\approx3.8$) and central ($\bar{\nu}\approx5.8$) pPb collision are shown in Figure 5. These calculations yield a median rapidity loss ($\Delta y_{1/2}$) for the whole Pb nucleus of -1.9±0.3 and $\Delta y_{1/2}$ = -2.5±0.5 for a central collision.

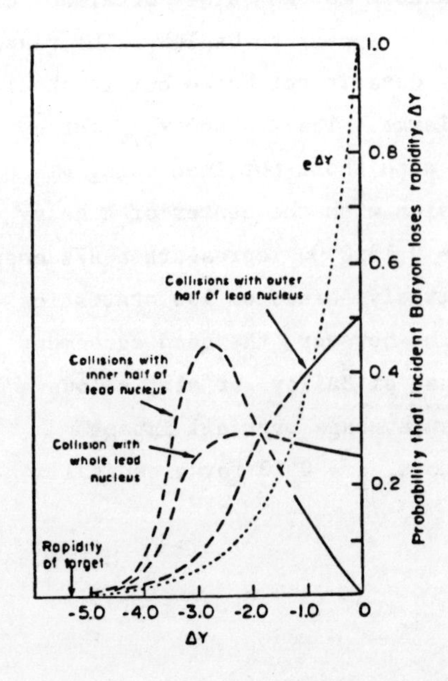

Fig. 5: The extrapolation of P(Δy) for pPb collisions as obtained by Busza and Goldhaber [17] from the data of Barton, et. al. [11]

I would like to compare the extrapolation of Busza and Goldhaber with some of the data plotted in Figures 4a-d. The invariant cross sections for the 19.2 and 24 GeV data can be converted into $d\sigma/dy$ distributions using the same extrapolations of the data as discussed above in connection with $d\sigma/dx$ and σ^{inel} from measurements performed by other authors at similar energies. The 120 GeV data cannot be directly converted into $d\sigma/dy$ since only $d\sigma/dx$ has been published. However, if a constant $<p_{\perp}>$ is assumed as a function of x, an approximate $d\sigma/dy$ can be obtained from the expression $d\sigma/dy \approx dx/dy*d\sigma/dx$. The probability of losing Δy ($P(\Delta y)$) curves of Busza and Goldhaber have been normalized to fit the data of Bailey, et al., for the whole Pb nucleus. This is necessary since $P(\Delta y) \neq (1/\sigma^{inel})*d\sigma/dy$ due to the fact that a proton is not always the leading baryon.

The comparison of $(1/\sigma\ )*d\sigma/dy$ for a Be target ($\bar{\nu}\approx1.4$) and a Pb target (U in the case of the 120 GeV data; $\bar{\nu}\approx3.8$) and the calculations of Busza and Goldhaber are shown in Figures 6a and b. The data for the Be target drop monotonically, and roughly agree with the predictions of B+G for peripheral pPb collision. I believe it is fair to say that the data from light targets is similar to that for pp collisions and the extrapolations of B+G. The comparison of pPb and pU collisions is shown in Figure 6b. All the data agree well in shape were they overlap. The data are in good agreement with the extrapolation of B+G and, therefore, reinforces their estimate of $\Delta y_{1/2}\approx-1.9$ for a mean pPb collision. It should be cautioned, however, that the $d\sigma/dy$ obtained from $d\sigma/dx$ of Bailey, et al., is only approximate.

There presently does not exist data that allows a direct measurement of $P(\Delta y)$ for central pPb collisions. Hybrid spectrometer data does exist in which it is possible to select the leading positive particle as a function of $\bar{\nu}(n_p)$ (see above). An estimate of $\Delta y_{1/2}$ can be obtained by assuming the leading

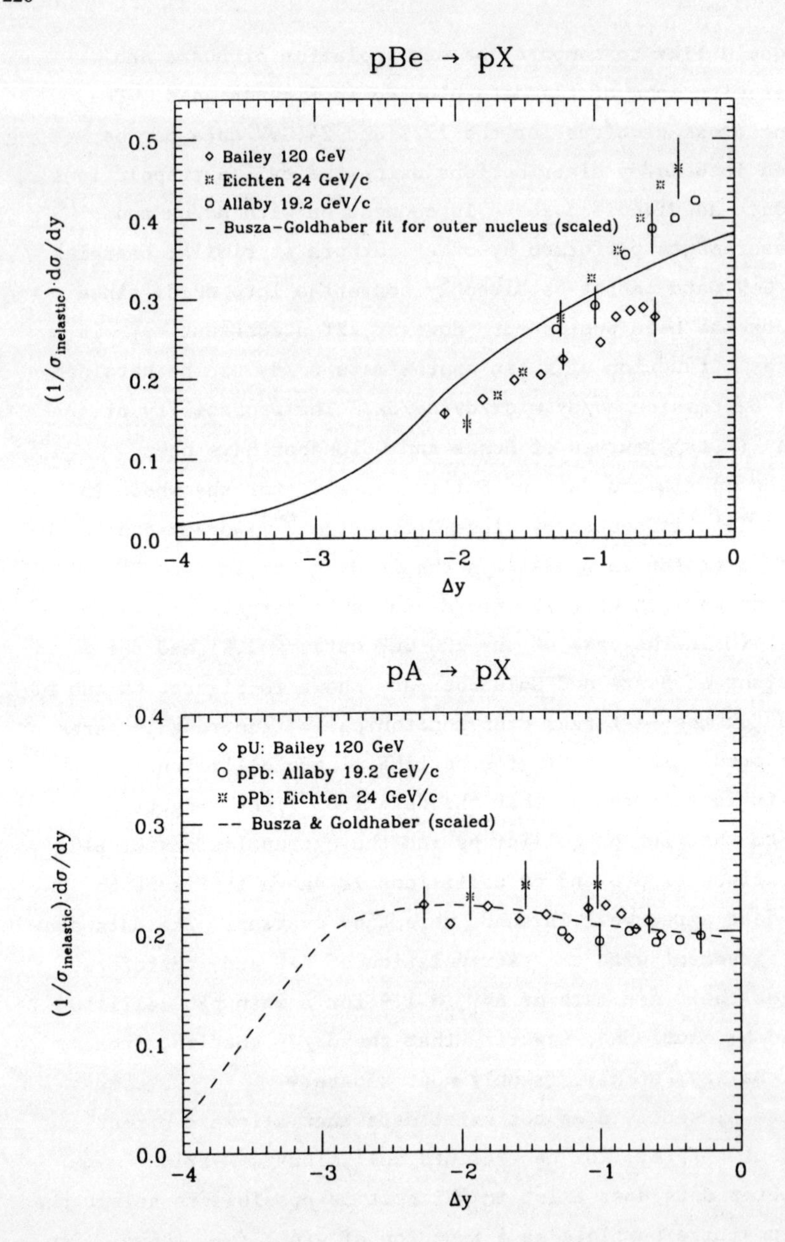

Fig. 6: Comparison of $(1/\sigma)\cdot(d\sigma/dy)$ for a) Be and b) Pb and U targets extracted from the data in Figure 4 a) – d) (see text) to the extrapolation of Busza and Goldhaber.

positive particle is a proton and gating on events with $\bar{\nu}(n_p)\approx5.8$. The preliminary results of such a calculation using the data from FNAL E565/570 are shown in Figure 7 along with the extrapolated $P(\Delta y)$ for central pPb collisions from B+G. These measurements suggest that there is an increased rapidity loss of $\Delta y_{1/2}\approx-2.1$ as more central collisions are selected. This is to be compared with $\Delta y_{1/2}$ of -2.5 obtained by B+G. It would be expected that the hybrid spectrometer estimate of $\Delta y_{1/2}$ be smaller since it may include events in which the leading positive particle is a pion. These results are provocative but can only be confirmed by measurements that include PID and central triggers. Such measurements will be performed in the next year.

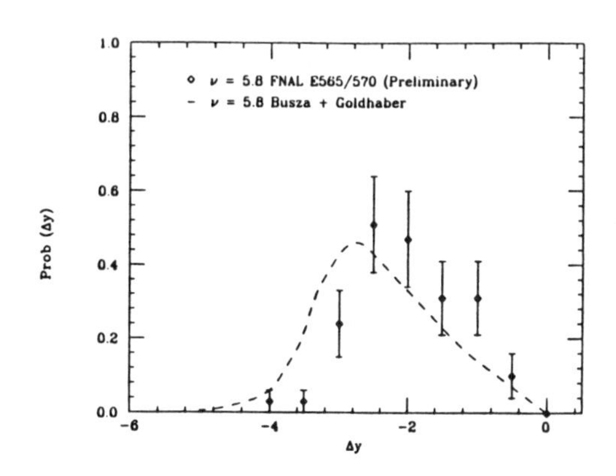

Fig. 7: $P(\Delta y)$ extracted from preliminary hybrid spectrometer data [15] for $\bar{\nu}(n_p)\cong5.8$. Only events in which the leading particle was positive were selected, and it was assigned the mass of a proton.

A mean $\Delta y_{1/2}$ can be obtained from $x_{1/2}$ if a constant $\langle p_\perp \rangle$ is assumed. Table 3 summarizes the results of the calculation performed on the data of Bailey, et al. Also included in Table 3 are the extrapolations of B+G and the preliminary results from FNAL E565/570. The $\Delta y_{1/2}$ obtained from $x_{1/2}$ of Bailey, et al., are lower than the estimates of B+G, but as Figure 6b illustrates there may be a significant amount of cross section for $x < .07$ (i.e., $|\Delta y| > 2$).

Table 3: Comparison of Busza Goldhaber Estimates of Rapidity Loss with those Extracted from $x_{1/2}$

Data	$\bar{\nu}$	$x_{1/2}$	$\langle\Delta y_{1/2}\rangle^*$	$\bar{\nu}$	extrapolation Busza and Goldhaber
Bailey, et al.[12]	1.4	.35±.05	-1.1±.15	1	-0.95±.15
" "	3.8	.22±.05	-1.5±.35	3.8	-1.9±0.3
IHSC†[15]	5.8	.13±.04	-2.1±0.6	5.8	-2.5±.0.5

*assuming $\langle p_\perp\rangle$=0.4 GeV/c independent of x.

†data does not include PID.

So far, we have concentrated on measurements of the inclusive proton spectra. It is to be expected, however, that the leading baryon can also be a neutron. A preliminary set of 400 GeV $pA \rightarrow nX$ data has been obtained by Forrest, et al., [13] and they are compared to the inclusive proton distributions of Bailey, et al., in Figure 8. The data of Forrest, et al., have been multiplied by a factor of 2 to aid in comparison. The inclusive neutron cross section is on the order of a factor of two smaller than the proton cross section. An extrapolation of the neutron cross section to $x < 0.2$ would indicate a somewhat softer spectrum. Since $x \approx 0.2$ represents only a $\Delta y = -1$, it is difficult to estimate precisely how the baryon rapidity loss distribution would be affected by a soft neutron distribution. However, if more complete measurements do yield a softer neutron distribution with a non-negligible cross section then the rapidity loss for pA collisions may be larger than estimates made from inclusive proton

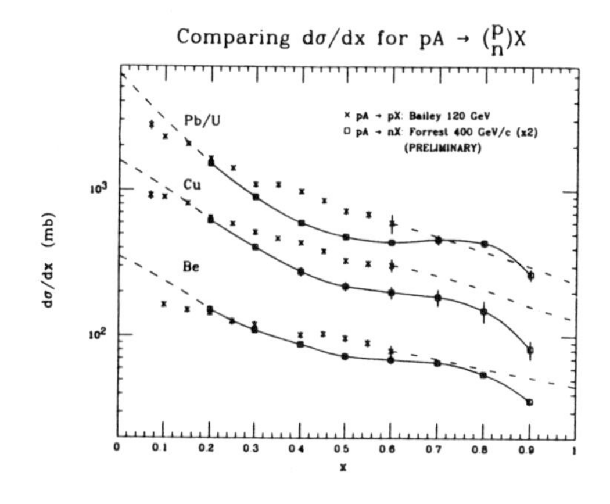

Fig. 8: $d\sigma/dx$ for $pA \to pX$ at 120 GeV and $pA \to nX$ at 400 GeV/c [13] (multiplied by 2.0).

distributions alone. The relative ratios of proton and neutron cross sections in different rapidity regions may be important in ascertaining what reaction mechanism contributes to large rapidity losses. Hopefully, measurements of the neutron channel will also be made in the near future.

It is difficult to summarize all of the data presented above, mainly due to the fact that the data are often incomplete. However, I will attempt an admittedly subjective summary of the existing data.

1) Scaling: The $d\sigma/dx$ distributions from light targets differ appreciably at small values of x for data at different energies. The agreement is better for heavier targets. The rapidity loss distributions also show variations with energy, however, the general characteristics of the distributions are similar at all energies as a function of target mass, i.e., $1/\sigma * d\sigma/dy$ for light targets are monotonically decreasing whereas those for Pb are flat. Thus, the present data suggest that rapidity loss is a more proper scaling variable than Feynman x. Of course, more complete measurements at lower x (i.e., greater Δy) should be made.

2) Rapidity Loss in Central Collisions: The rapidity loss for central collisions has so far only been estimated from extrapolations of the A dependence of inclusive proton distributions. The rapidity loss from Pb targets appears to be consistent with the extrapolations of B+G. The extrapolation of B+G for the rapidity loss in central pPb collisions from the inclusive proton data as well as the hybrid spectrometer data (with no PID) gated on central collisions suggests that $\Delta y_{1/2} \approx -(2.0 - 2.5)$ for such collisions. Thus, increases of 5-10 times normal nuclear matter densities may be achieved in central pPb collisions at collider energies of ≈ 4 GeV or fixed target energies of ≈ 30 GeV.

3) Comparison of Proton and Neutron Distributions: A comparison of $d\sigma/dx$ for 120 GeV $pA \to pX$ and 400 GeV $pA \to nX$ indicates that the inclusive neutron cross sections for $x>0.2$ are on the order of 1/2 of the proton cross section. Extrapolation to $x<0.2$ indicates that the neutron distribution may be softer. Consideration of these preliminary data illustrates the importance of the neutron channel in any estimate or model of the nuclear stopping power.

3. Energy Deposition

The incident proton loses a significant fraction of its initial energy in a central pA collision. It is found that on the average only 15% of the incident beam energy remains in a leading proton. Where does this energy go? What is known is that pions are the most abundantly produced particles and that in pp collisions they are produced with a relatively flat rapidity distribution centered on the N-N c.m. frame. In pA collisions, pseudo-rapidity and rapidity distributions suggest that: 1) the total number of produced particles (i.e., neglecting recoil protons) increases linearly with the parameter $\bar{\nu}$ (see Figure 1);

2) the most probable rapidity shifts to a value less than $y_{NN}^{c.m.}$ with increasing values of $\bar{\nu}$ (or target mass) and $R(y) \equiv (d\sigma/dy)^{pA}/(d\sigma/dy)^{pp}$ decreases with increasing y; and 3) the region near the beam rapidity shows a depletion of particles and energy. This is presumably the projectile fragmentation region and extends for approximately -2 units of rapidity from the beam. R(y) obtained in 200 GeV pA collisions measured in a steamer chamber by DeMarzo, et al.,[18] is shown in Figure 9.

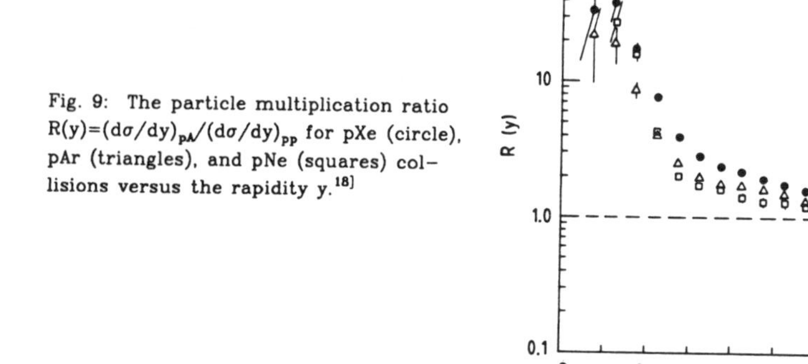

Fig. 9: The particle multiplication ratio $R(y)=(d\sigma/dy)_{pA}/(d\sigma/dy)_{pp}$ for pXe (circle), pAr (triangles), and pNe (squares) collisions versus the rapidity y.[18]

Existing data, which is summarized above, gives many of the general features of the produced particle distributions. However, if an estimate of the maximum energy density is to be made it is important to know in greater detail how the available energy is distributed in central collisions. The discussion which follows will be based on a purely phenomenological analysis of the hybrid spectrometer data, FNAL E565/570. As mentioned above this preliminary data is a global measurement of each event with PID only for p<1.5 GeV/c. In principle, a similar analysis could be performed on existing bubble and streamer chamber data. W. Walker will present a complimentary analysis of 100 GeV hybrid spectrometer data taken at FNAL.

In Section 2 we discussed inclusive proton distributions. The <x> decreases with an increase in $\bar{\nu}$. An important question is how does the distribution of leading pions change relative to the leading baryon. To answer this question, we have taken the hybrid spectrometer data and, on an event-by-event basis, determined $\bar{\nu}(n_p)$ and have also ordered the particles by their total momentum. It is then possible to extract the mean momentum fraction ($x \equiv p^{total}/p_{beam}$, approximately Feynman x for leading particles) for the n'th leading particle gated on $\bar{\nu}(n_p)$. Figure 10 displays such a distribution for three gating regions of $\bar{\nu}$ corresponding to peripheral ($\bar{\nu}\approx1.6$), mean ($\bar{\nu}\approx3.8$) and central ($\bar{\nu}\approx6.3$) pPb collisions. The decrease in the 1'st leading particle with increasing $\bar{\nu}$ is consistent with the decrease in the <x> of the leading proton obtained from inclusive cross sections. It should be noted that in Figure 10 no reference has been made to the particles charge in ordering the particles, i.e., there are negative particles in the leading particle distributions.

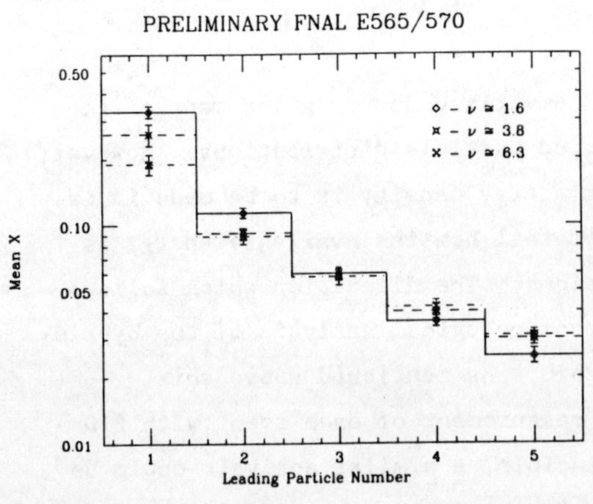

Fig. 10: The $\langle x\rangle \equiv \langle p_{out}/p_{beam}\rangle$ of the n'th highest momentum particle for different cuts on $\bar{\nu}(n_p)$. Preliminary data from FNAL E565/E570. [15]

There is less $\bar{\nu}$ dependence for the second leading particle as compared to the leading particle. The third leading particle has no $\bar{\nu}$ dependence within the present statistics. The $\langle x \rangle$ of the third particle is approximately 0.05, independent of $\bar{\nu}$. The $\langle x \rangle$ for the fourth and fifth particles appear to be increasing with $\bar{\nu}$. These results indicate that for 200 GeV collisions, energy is decreased from particles with $x > 0.05$ and added to lower momentum particles. There are approximately three particles with $x > 0.05$ independent of $\bar{\nu}$. The number of particles with $x < 0.05$ must be proportional to $\bar{\nu}$ so as to conserve energy.

These results are very interesting because they suggest that although the $\langle x \rangle$ of the leading particle has a significant dependence on $\bar{\nu}$ there is a relatively constant energy component contained in the second and third leading particle which has a weak dependence on $\bar{\nu}$. We will define for the purposes of discussion $x > 0.05$ ($x=0.05$ corresponds to $p=10$ GeV for $p_{beam}=200$ GeV) as the projectile fragmentation region. This definition is motivated by the observation that energy decreases from this region for all values of $\bar{\nu}$. Table 4 contains the total momentum fraction contained in charged particles with $x > 0.05$. Approximately 50% of the total momentum in peripheral collisions is contained in charged particles with $x > 0.05$, while 35% is contained in the leading particle. Thus, approximately 15% of the total momentum is found in the second and third leading particles. The same is true for collisions with $\bar{\nu} \approx 3.8$ and 6.3. Also contained in Table 4 are the numbers of charged particles in the projectile fragmentation region. There are approximately three charged particles on the average, one of which is negative. This demonstrates that the charge of the beam is accounted for in our definition of the projectile fragmentation region.

The above numbers will, of course, be affected if neutral particles are included. As discussed above, the inclusive neutron

Table 4. Comparison of Total Momentum and Particle Number Contained in Particles by Cut on 10 GeV/c Momentum. Preliminary Data from FNAL E565/570.[15]

	P > 10 GeV/c			P < 10 GeV/c		
$\bar{\nu}(n_p)$	$\Sigma P^{\pm}/P_{beam}$	$n^{\pm}$	n^{-}	$\Sigma P^{\pm}/P_{beam}$	$n^{\pm}$	n^{-}
1.6	0.51±.02	2.8±.1	1.04±.07	.12±.01	9.4±0.07	3.9±.3
3.8	0.43±.03	2.9±.2	1.13±.12	.18±.02	20.0±2.4	7.0±.9
6.3	0.35±.03	2.8±.2	1.09±.12	.25±.03	36.0±5.0	11.±1.5

cross section is on the order of 1/2 that of the proton. There is on the average one negative particle with $x > 0.05$ which is presumably a π^-. This suggests that there may also be a π° in this region. Both these effects indicate that the total momentum fraction should be multiplied by a correction factor on the order of 3/2. For the present discussion we will focus on the relative momentum in charged particles, keeping in mind that the total observed momentum will not sum to the beam momentum.

The total momentum and number of negative particles contained in particles with $x < 0.05$ are shown in Figure 11. Both the number and total momentum in this region increase linearly with $\bar{\nu}$. These results are tabulated in Table 4 where the contributions of both negative and positive particles are included. As mentioned above, the linear increase in the number of produced particles is consistent with many previous measurements. The total momentum fraction contained in charged particles with $x < 0.05$ varies from 0.12 to 0.25 for $\bar{\nu}$=1.6 and 6.3, respectively. Once again these numbers should probably be

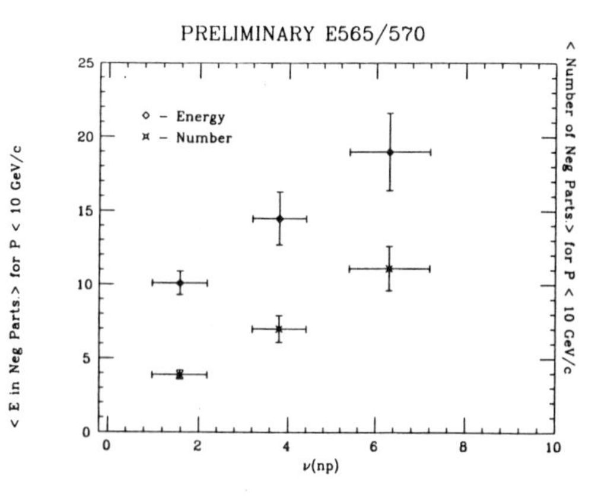

Fig. 11: The total energy and the number of negative particles with p < 10 GeV/c as a function of $\bar{\nu}(n_p)$. Preliminary data from FNAL E565/570. [15]

multiplied by 3/2 to take into account the contribution from π°'s. Even with the addition of π°'s, the total momentum fraction in particles with $x < 0.05$ for central collisions is on the order of 35%. Since this region has the highest dn/dy it is probably the most relevant for estimating energy densities in A+A collisions. (For a beam energy of 200 GeV, $x < 0.05$ spans approximately 4 units of rapidity.)

The results from the above analysis of energy deposition are based on preliminary data and as such its conclusions must be verified. At a minimum, these results suggest that caution should be used in estimating energy densities from the leading particle energy loss distributions, since there may be a component of energy lost to "fast" pions which is independent of impact parameter. If these results are confirmed by future experiments, then they may be an important clue in determining the mechanism of energy loss in pA collisions. What is needed are new experiments with full PID which can correlate the leading particles with global event descriptors (egs. total multiplicity, total

$P_\perp$, n_p). Hopefully, these measurements will be performed in the next couple of years at BNL and CERN as part of their heavy-ion programs.

4. Summary

Examination of existing inclusive pA → pX distributions indicates that there is a significant increase in energy loss from pp to pPb collisions with no significant change in the rapidity loss for measurements taken at different beam energies. The extrapolations of Busza and Goldhaber agree favorably with existing data for up to 2 units of rapidity loss. Analysis of preliminary hybrid spectrometer data gated on central collisions is also qualitatively consistent with the extrapolations of B+G. Preliminary analysis of the energy distribution of particles in pA collisions suggests that there is a component of energy in leading particles which has a weak impact parameter dependence. This cautions against the use of the total energy loss from the projectile in estimates of the energy densities which may be achieved in A+A collisions.

Acknowledgements

The authors would like to thank W. Busza, L. Grodzins, M. Guylassy, A. Kerman, T. Matsui, B. Svetitsky, S. Steadman, and P. Stamer for their many helpful discussions.

References

1. Quark Matter Formation and Heavy Ion Collisions: Proceedings of the Bielefeld Workshop, May 1982, ed. (Jacob, M. and Satz, H.) (Singapore: World Scientific Publishing Company, 1982).

2. Proceedings of the 3rd International Conference on Ultra-Relativistic Nucleus-Nucleus Collisions (Quark Matter 83), Brookhaven National Laboratory, September 26-30, 1983, Nuclear Physics A418, 1-678 (1984).

3. Proceedings of the 4th International Conference on Ultra-Relativistic Nucleus-Nucleus Collsions (Quark Matter 84), Helsinki, Finland, June 17-21, 1984, ed. Kajantie, K. (Berlin: Springer-Verlag, 1985).

4. Proceedings of the 2nd International Conference on Nucleus-Nucleus Collisions, Visby, Swededn, June 10-14, 1985, Nuclear Physics A447, 1-724 (1986).

5. See e.g., theory section on latice gauge results in Ref. 2, p. 447.

6. Elias, J. E., et al., Phys. Rev. D22, 13 (1980).

7. Otterlund, I., et al., Physica Scripta 31, 328 (1985).

8. Chao, W. Q., et al., Nucl. Phys. A395, 482 (1983).

9. Allaby, J. V., et al., CERN Report 70-12.

10. Eichten, T., et al., Nucl. Phys. B44, 333 (1972)

11. Barton, D. S., et al., Phys. Rev. D27, 2580 (1983).

12. Bailey, R., et al., Z. Phys. C 29, 1 (1985).

13. Forrest, P., et al., private communication from K. Heller (unpublished).

14. See Ref. 3, "Nuclear Stopping Power," Busza, W., p. 77.

15. IHSC, FNAL Proposal E656/570.; Brick, D. H., et al., Proceedings XIII International Symposium on Multiparticle Dynamics, Volendam, 1982; "Stopping Power and Energy Deposition in 200 GeV/c P + A Interactions," Ledoux, R. J., et al., Proceedings of the Gross Properties of Nuclei and Nuclear Excitations International Workshop XIV, Hirschegg, Austria, January, 1986 (to be published).

16. The following articles attempt a theoretical understanding of nuclear stopping power:

    a. Hwa, R. C., Phys. Rev. Lett. 52, 492 (1984); Hwa, R. C., and Zahir, M. S., Phys Rev D 31, 499 (1985).

    b. Hüfner, J., and Klar, A., Phys. Lett. 145B, 167 (1984); Klar, A., and Hüfner, J., Phys. Rev. D 31, 491 (1985).

c. Wong, C. Y., Phys. Rev. Lett. 52, 1393 (1984).

d. Csernai, L. P., and Kapusta, J., Phys. Rev. D 29, 2664 (1984); Csernai, L. P., and Kapusta, J., Phys. Rev. D 31, 2795 (1985).

e. Date, S., Gyulassy, M., and Sumiyoshi, H., INS-Rep.-535, 1985.

17. Busza, W., and Goldhaber, A. S., Phys. Lett 139B, 235 (1984); see also Ref. 2, "Fragmentation Data in Hadron-Nucleus Collisions: Implications about Nuclear Stopping Power," Busza, W., p. 635.

18. De Marzo, C., et al., Phys. Rev. D29, 2476 (1984).

HADRON-NUCLEUS INTERACTIONS AT HIGH ENERGY

Clive Halliwell

Physics Department
University of Illinois at Chicago
829 Taylor St
P.O. Box 4348
Chicago, Illinois 60680
U.S.A.

Properties of energetic secondaries produced at large angles using 800 GeV incident protons are presented. H_2, Be, C, Al, Cu and Pb targets were used for the study. The yields for producing such secondaries vary as A^{α} where A is the atomic mass number of the target and α attains values as large as 1.6. There is evidence that jet-like events have α values approaching unity, indicating a hard scattering mechanism may be occuring. Events with large values of target-fragmentation energy have, on average, large values of energy in the central region and small values of forward-going energy. Energy flows and number of secondaries are independent of the target when events with similar amounts of energy in the central region are studied.

INTRODUCTION

Interactions of hadrons with nuclei (hA interactions) have been studied in the past primarily to investigate the production and subsequent space-time evolution of the produced secondaries[1)]. This has been possible because the nucleus interferes with the secondaries immediately after the primary interaction, an occurence that cannot take place in hadron-nucleon (hN) interactions.

From these hA studies it has been found that low energy secondaries, produced mostly in the target-fragmentation region, are formed inside the nucleus and therefore cascade within the nucleus causing a multiplicity enhancement when compared to data from hN interactions. In contrast to this, the average number of secondaries formed in the central region is approximately proportional to the average number of collisions expected within a nucleus[2)]. This result has lead to the idea that energetic secondaries are produced outside of the nucleus and therefore are unable to take part in a hadronic cascade. Finally, leading particles from hA interactions have less energy compared to those produced in hN interactions because of momentum degradation caused by collisions within the nucleus[3)]

Recently, interest has focussed on the production of leading hadrons from hA interactions. By measuring the energy lost by the incident hadron, it is hoped that an estimate can be made of the energy deposited within the nucleus. This in turn may be useful in estimating the incident hadron energies required to study the formation of the quark-gluon plasma[4)]. An analysis[5)] using this approach has been carried out recently using data from an experiment that triggered on leading particles[6)].

An alternative method for studying energy deposited in the nucleus is to directly measure the energy associated with secondaries produced in the central region. This method was used in Fermilab experiment E557 which was designed to trigger on large amounts of energy at wide angles (in the vicinity of 90° as measured in the proton-proton centre-of-mass frame) to the incident beam direction ("transverse energy", E_t). One of the aims of E557 was to measure the momentum distribution of the secondaries as a function of polar angle[7)]. A large enhancement of positive charged particles (as compared to negative ones) was found at small angles, corresponding to energetic leading protons. This occurred in events with small amounts of E_t. The leading proton enhancement dispersed over a larger angular region as bigger E_t values were required, signifying sizeable energy lost by the protons.

The E557 data were obtained using 400 GeV/c incident protons. The experiment concentrated on hN interactions and obtained little hA data. A more recent data collection run of E557* was carried out in 1984 using 800 GeV/c protons and several nuclear targets. One of the aims of this recent run was to study the properties of rare events that occur when the incident proton loses a large fraction of its momentum. This paper reports on the analysis of these new data.

EXPERIMENTAL METHOD

The layout of the E557 apparatus is shown in Fig. 1. The experiment was performed using the Fermilab MT beam line. The apparatus consisted of several highly segmented calorimeters that detected photons and hadronic secondaries.

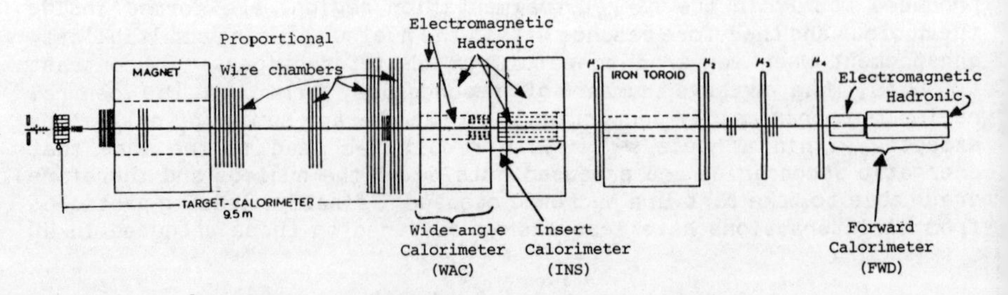

Figure 1. E557/672 apparatus at Fermilab.

The properties of these calorimeters (the wide-angle, WAC, the insert (INS), the forward, FWD, and the beam, BM) are listed in Table 1.

Table 1.

Calorimeter Properties

Calorimeter	material	acceptance in pp cm frame $\Delta\theta^*$	Δy^*	Number of modules
WAC, electromagnetic	lead-scintillator	60->145	-1.1->.5	126
WAC, hadronic	iron-scintillator	60->145	-1.1->.5	126
INS, electromagnetic	lead-glass	22->55	.7->1.6	84
INS, hadronic	iron-scintillator	20->50	.8->1.7	24
FWD, electromagnetic	lead-glass	5->30	1.3->3.1	114
FWD, hadronic	iron-scintillator	5->35	1.2->3.1	60
BM, electromagnetic	lead-scintillator	0->5	>3.1	1
BM, hadronic	iron-scintillator	0->5	>3.1	1

Their segmentation and approximate acceptances, as measured in the proton-proton centre-of-mass frame, are shown in Fig. 2. The proportional wire chambers shown in Fig. 1 were not used in this analysis except for reconstructing the position of the interaction vertex upstream of the spectrometer magnet. No particle identification of the produced secondaries was attempted.

The calorimeters served as triggering devices as well as detectors of the produced secondaries. In this paper the results from two groups of triggers will be reported. Firstly, an inelastic collision was detected by demanding that a large pulse height had occurred in a counter placed immediately downstream of the target. This trigger, termed the "interacting beam" trigger, was sensitive to approximately 90% of the total inelastic proton-proton cross section. Secondly, a group of triggers consisting of the "interacting beam" trigger with an additional requirement that at least a certain amount of E_t was present in a preset region of the WAC and INS calorimeters, was also used. The modules used in the WAC and INS calorimeters to form two of these triggers are shown in Fig 3. To form the E_t triggers the output from each calorimeter module was weighted by the sine of the polar angle that the module subtended at the target. E_t sums for several different configurations of calorimeter modules were formed simultaneously. Data from three of these configurations are presented in this paper: two full azimuthal acceptance ("global", $45^o < \theta^* < 135^o$, and "limited global", $60^o < \theta^* < 120^o$) triggers and a limited azimuthal acceptance ("small aperture") trigger.

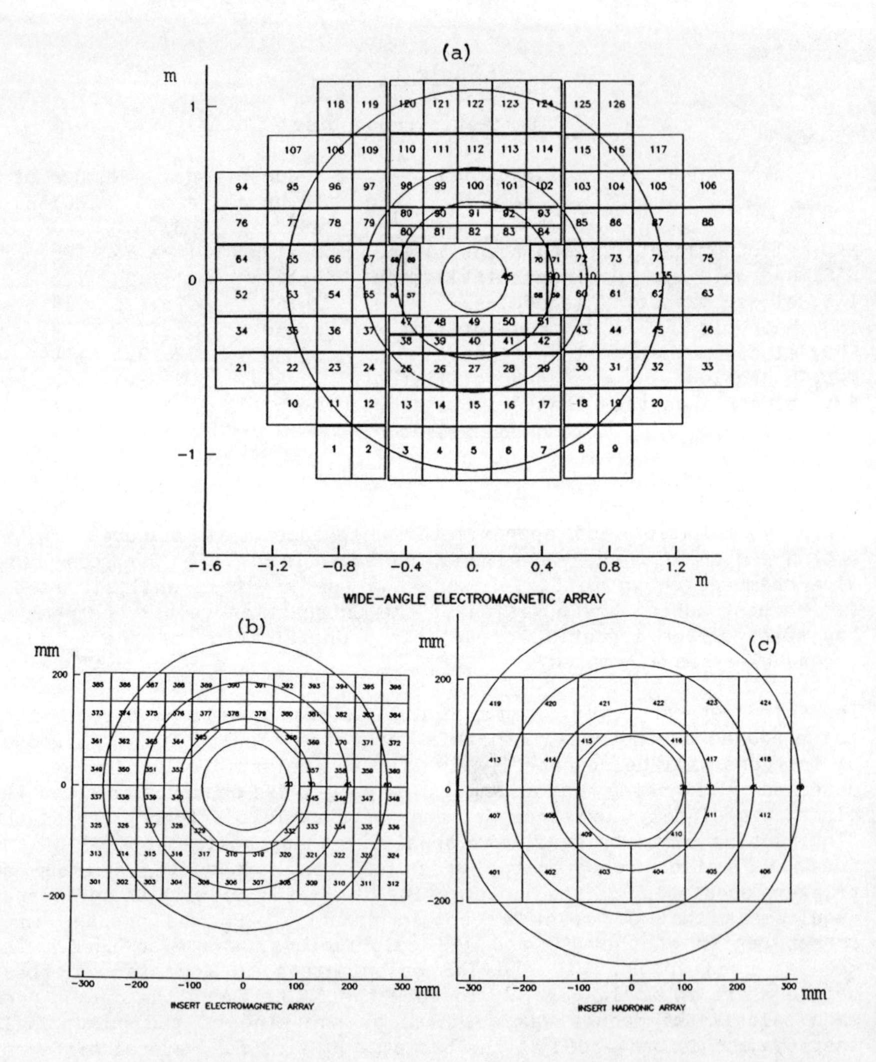

Figure 2. Segmentation and polar acceptances for massless secondaries for the (a) WAC hadronic and electromagnetic, (b) INS electromagnetic, (c) INS hadronic calorimeters. The circles are $\theta^* =$ (a) 45, 90, 110, 135, (b) 20, 30, 45, 60, (c) 20, 30, 45, 60 degrees in the proton-proton centre-of-mass frame.

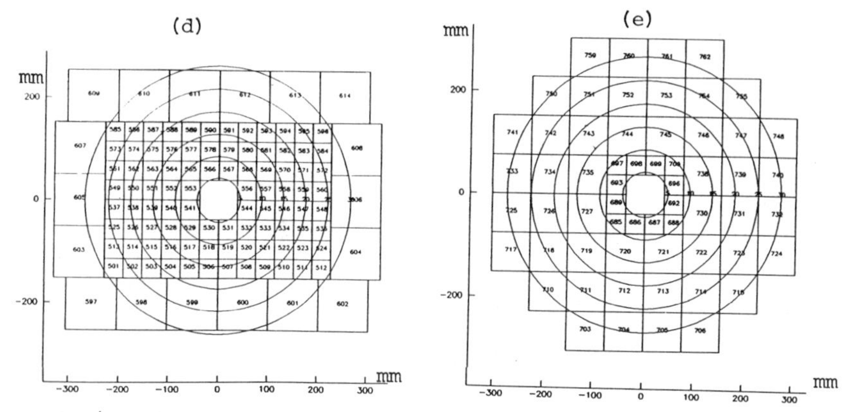

Figure 2. (cont)

Segmentation and polar acceptances for massless secondaries for the (d) FWD electromagnetic and (e) FWD hadronic calorimeters. The circles are $\theta^* =$ (d) 5, 10, 15, 20, 25, 30 and (e) 5, 10, 15, 20, 25, 30 degrees in the pp centre-of-mass frame.

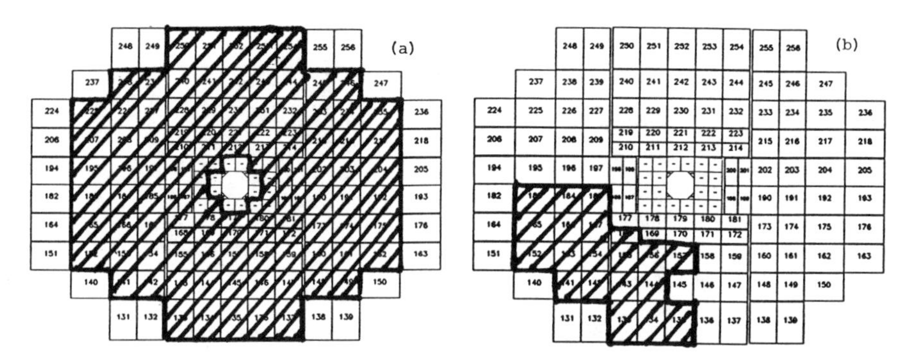

Figure 3. Modules in the WAC and INS calorimeters used to form (a) "global" and (b) a "small aperture" trigger.

ANALYSIS

The analysis presented in this paper was performed using the calorimeter module outputs directly. No correction was made for the response of the electromagnetic section of the calorimeters being approximately 20% larger for incident electrons than hadrons. It is estimated that the E_t determination has to be decreased by 17% due to this source, the p_t kick from the spectrometer magnet, the leakage from the downstream face and sides of the calorimeters and the energy resolution of the calorimeter modules. No attempt has been made to form "energy clusters" from the calorimeter data.

YIELDS

Inelastic cross sections were obtained using the "interacting beam" trigger. These cross sections were typically 10% lower than those measured in previous experiments[8]. This was due mainly to the requirement that at least two charged particles had traversed the trigger counter. Yields of events with E_t values up to approximately 16 GeV as detected in the "global" trigger region (see Fig. 3) of the WAC and INS calorimeters were measured using this trigger.

Yields of events with higher E_t values were obtained by imposing an ever increasing E_t requirement from the WAC and INS calorimeters at the trigger level. By doing this it was possible to obtain events with E_t values in the "global" trigger region up to approximately 36 GeV. The results of this analysis are shown in Fig 4.

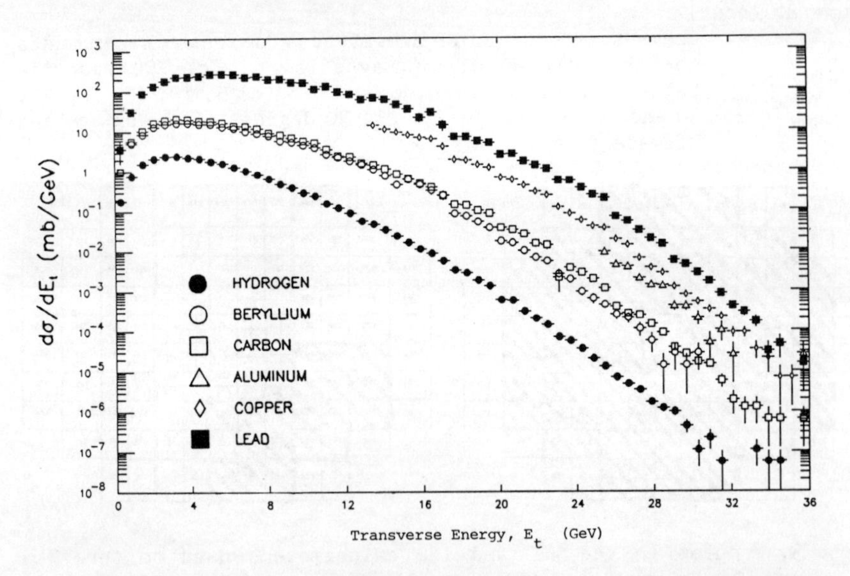

Figure 4. Yields as a function of E_t detected in the "global" triggering region. Data for hydrogen and various nuclear targets are shown. The 17% correction mentioned in the text has not been applied.

Events that contributed to this plot were required to have a reconstructed interaction vertex in the target region. The reconstruction was accurate enough to easily distinguish between the three nuclear targets (see Fig. 5). The efficiency for reconstructing vertices for events with small E_t values was somewhat lower than for events with large E_t values because the number of secondaries was typically lower for the former class of events. This inefficiency contributes to the dip in the yield curve (Fig 4) at low values of E_t (the total inelastic cross section obtained by integrating the hydrogen data is only 16 mb).

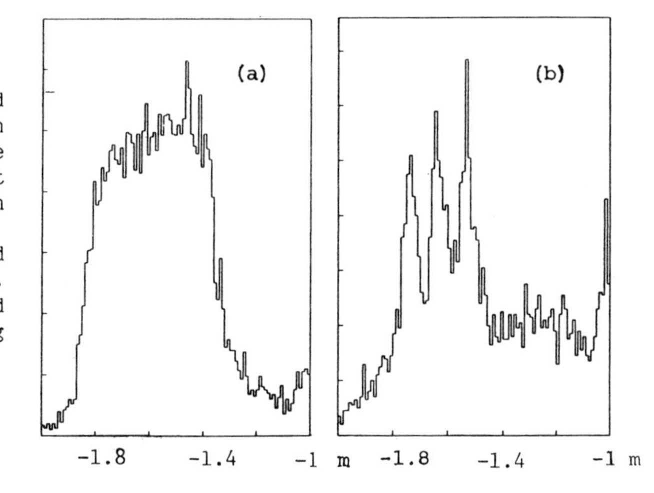

Figure 5.

Reconstructed interaction vertex coordinate along the incident beam direction for:
(a) hydrogen and
(b) 3 lead targets.
The trigger used was "interacting beam".

The probability of producing events with E_t values greater than, say, 32 GeV, is approximately 10^4 times greater from a lead nucleus than from hydrogen. This means that events where the incident proton loses at least 80% of its momentum (as measured in the proton-proton centre-of-mass frame) are readily accessible if nuclear targets are used.

To quantify the enhancement that nuclear targets produce compared to hydrogen, values of $d\sigma/dE_t$ for several E_t ranges for the various nuclear targets were fitted to the form $d\sigma/dE_t \propto A^{\alpha(E_t)}$. The fits did not include the hydrogen data as they typically lay well below the straight line extrapolation (see Fig. 6).

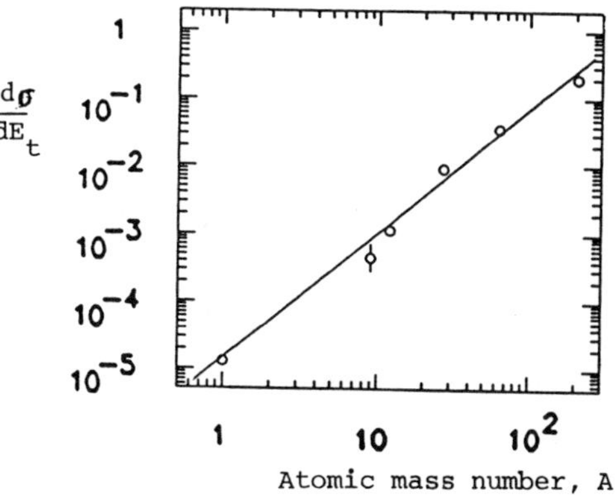

Figure 6.

Fit of $d\sigma/dE_t$ to $A^{\alpha(E_t)}$.
The E_t range was $25.2 < E_t < 25.8$ GeV.

The variation of α as a function of E_t is shown in Fig 7. Even if the α values for low E_t values are ignored, it is still apparent that α increases significantly above unity. This confirms similar results from other calorimeter based experiments[9] and single-particle experiments[10]. The same trend was apparent in the "reduced global" and "small aperture" data.

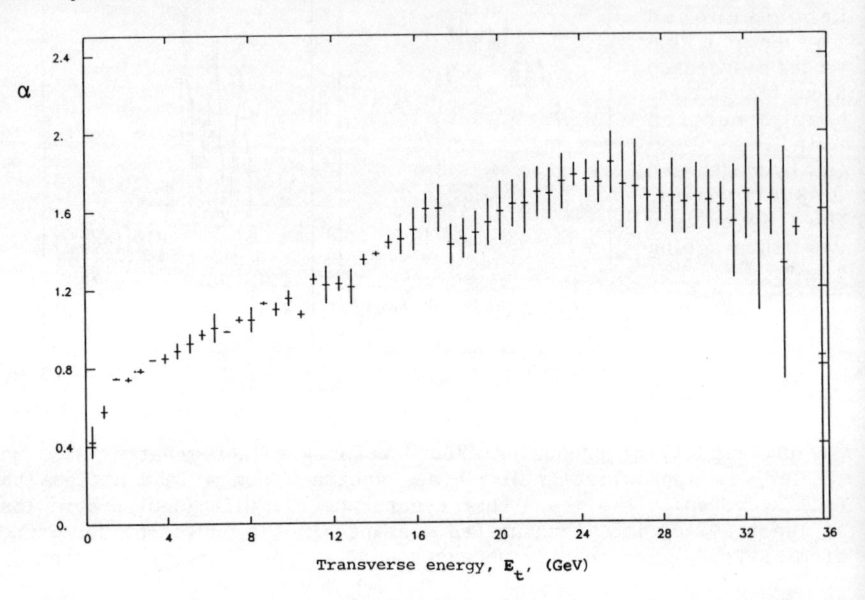

Figure 7. Variation of α with E_t detected in the "global" triggering region.

EVENT STRUCTURE

In past calorimeter experiments [11], event structure was studied by employing the planarity variable, P. In the plane transverse to the beam direction, the principle axis of an event was found and the p_t vector for each module was decomposed into components parallel and transverse to this axis. With the sum of the squared components along and transverse to the principle axis denoted as A and B, the planarity is defined as $P = (A-B)/(A+B)$. For pencil-like back-to-back jets, P approaches 1, while for large multiplicity isotropic events it approaches 0. Fig. 8 shows the observed planarity distributions for the high E_t "global" data. Only modules that were included in the "global" triggering region contributed to the planarity calculation. It is evident from Fig. 8 that the majority of the events obtained using this trigger are nonplanar.

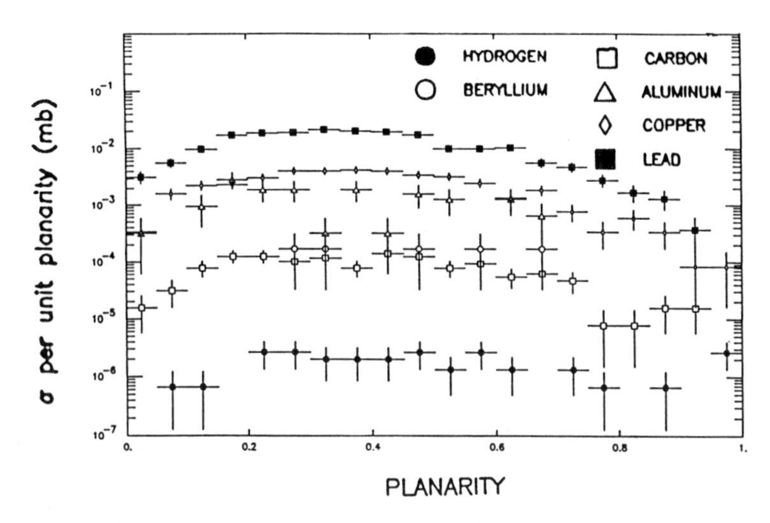

Figure 8. Planarity distributions for data collected using the "global" trigger. The E_t threshold was 28.8 GeV. Statistical errors only are shown.

The data shown in Fig. 8 can be paramatrized as $A^{\alpha(P,E_t)}$. Integrating over an E_t range, one can study how α varies with planarity. Fig. 9(a) shows such a variation for the "global" data. There is a slight tendency for α to decrease to unity from 1.6 for events with high values of planarity. This could be evidence that "jetty" events are produced by a hard scattering mechanism. This view is somewhat supported by the variation of α with planarity for the "reduced global" data (this trigger was designed to diminish the effect of forward-going secondaries being included in the scattered jets detected at wide angles) and for the "small aperture" data (see Figs. 9(b) and 9(c)). Several sets of data from various "small aperture" triggers were studied and all showed a significant decrease of α with planarity.

To see if events produced using the "small aperture" trigger are created by a different mechanism from data collected using the "global" or "interacting beam" triggers, data from a detector surrounding the target was analysed. The energy deposited in it was, hopefully, strongly correlated with the number of collisions that had occurred within the nucleus. The acceptance of the detector, as measured in the proton-proton centre-of-mass frame was $159^{\circ} < \theta^{*} < 179^{\circ}$. The layout of it is shown in Fig 10.

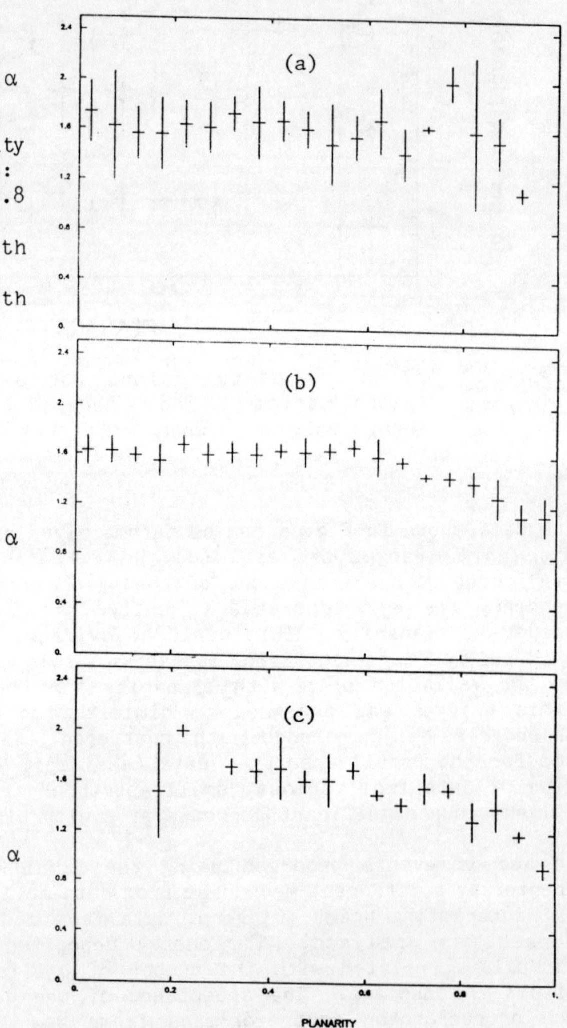

Figure 9.

Variation of α with planarity for 3 different triggers: (a) "global" with a 28.8 GeV threshold, (b) "reduced global" with a 19.1 GeV threshold, (c) "small aperture" with a 8.4 GeV threshold.

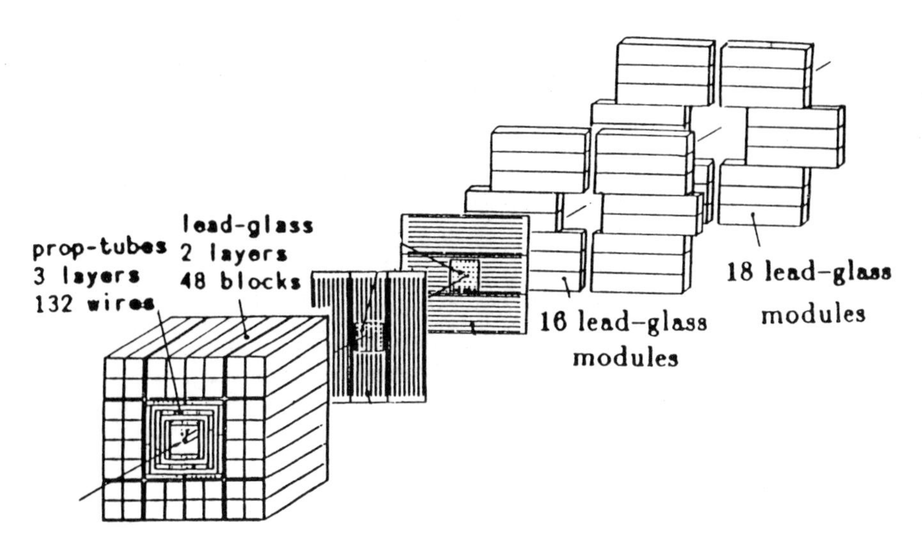

Figure 10. Exploded view of the target-fragmentation region detector.

Only data from the lead-glass blocks immediately surrounding the target (the "barrel") will be presented here.

The ratio of total energy detected in the "barrel" from hA interactions as compared to that from hN interactions is shown in Fig 11.

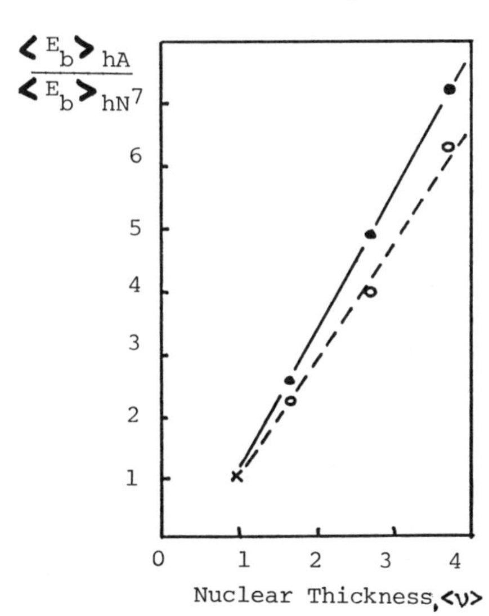

Figure 11.

Variation of the total energy in the "barrel" for hA interactions normalized to the total energy from hN interactions as a function of nuclear thickness, $\langle \nu \rangle$. $\langle \nu \rangle$ is defined as $A\sigma_{pN}/\sigma_{pA}$ where σ_{pN} and σ_{pA} are the total inelastic proton-nucleon and proton-nucleus cross sections and A is the number of nucleons in the nucleus. Data obtained using the "interacting beam", o, and "small aperture", ●, triggers are shown.

The enhancement of the energy in the "barrel" produced in hA interactions compared to hN ones increases faster than $\langle\nu\rangle$. This trend also occurs for the charged multiplicity detected in the proportional tubes surrounding the target. However, the multiplicity of secondaries is expected to increase in nuclei faster than the nuclear thickness due to the hadronic cascade within the nucleus; an increase in energy is not easily explained in naive multiple-interaction models, such as, for example, that described in Ref 12.

It was also found (see Fig. 12) that the E_t in the central region (as detected as the WAC and INS calorimeters) was linearly correlated with the average neutral target-fragmentation region energy. Also, the amount of energy detected in the "barrel" is far greater for hA interactions than for hN ones, if events with similar E_t are examined. These two observations indicate that events with large E_t values are produced by multiple interactions within the nucleus, assuming that the energy detected in the "barrel" is directly correlated with the number of collisions within the nucleus. This would add creadance to the view that values of α exceeding unity are related to multiple scattering within the nucleus (see Fig. 4).

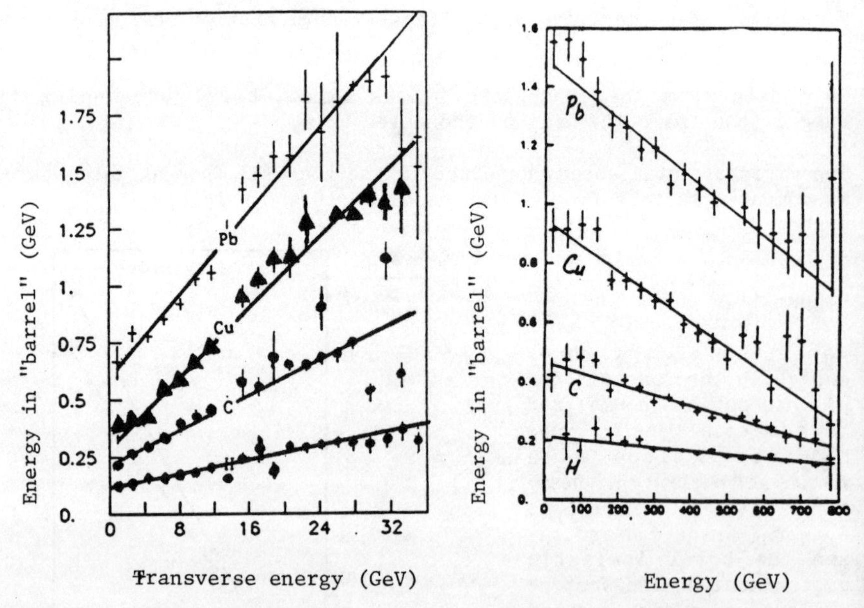

Figure 12. Correlation between energy detected in the "barrel" and (a) E_t detected in the WAC and INS calorimeters and (b) forward going energy detected in the FWD and BM calorimeters, using "interacting beam" and "global" triggers.

To see if the detailed structure of events produced in hA and hN interactions differ significantly, the energy flows (that is, average energy per event in a particular polar angular region) in the WAC, INS, FWD and BM calorimeters were combined over the angular range $\theta^* < 145^o$ as measured in the proton-proton centre-of-mass frame. To diminish systematic errors it has been customary in the past to divide data obtained from hA interactions by the corresponding data from hN ones. This approach could not be applied directly to the data obtained using the "interacting beam" trigger. As mentioned above, a large fraction of low E_t hN events were lost because the interaction vertex was not reconstructed efficiently. Therefore, if the hA data obtained using this trigger was divided by the corresponding hN data, a serious bias would have resulted. To overcome this problem, data from the same E_t ranges had to be compared.

In Fig. 13 comparisons of the polar angle dependence of the energy flows from hA and hN interactions are shown for events obtained using the "interacting beam " trigger. Some effects of the granularity of the calorimeters are apparent in the plots. However, it can be seen that the energy flows from hA and hN interactions are similar (to within 10%) when events with similar values of E_t are compared.

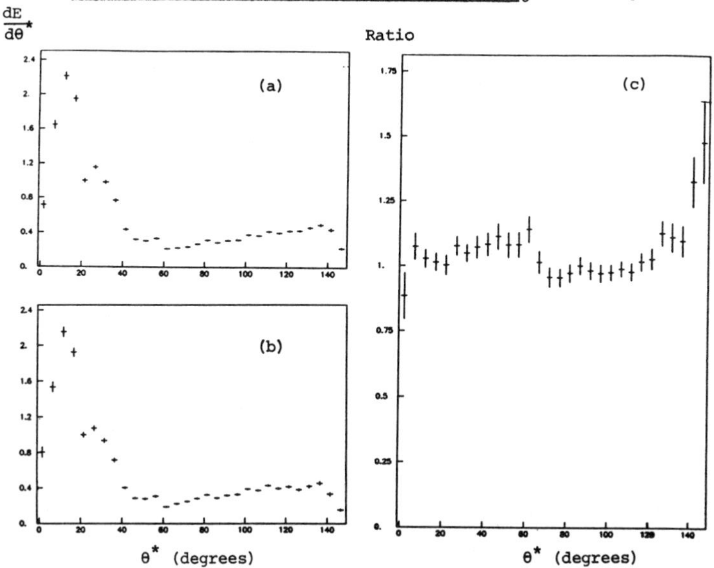

Figure 13. Polar angle dependence of energy flows (in arbitrary units) for (a) lead and (b) hydrogen data. θ^* is measured in the proton-proton centre-of-mass frame. The E_t range was $6 < E_t < 4$ GeV. E_t was measured in the "global" triggering region. (c) shows the ratio (lead/hydrogen) of the energy flows. Data from the WAC, INS and FWD calorimeters contributed to these plots.

In Fig. 14, similar data are shown for events with large values of E_t. The trigger condition produces a large energy enhancement at wide angles when this data is compared to that in Fig. 13. Correlated to this is the decreasing of the energy flow in the forward direction. Even with these dramatic changes in event structure, the energy flows from hA and hN collisions appear to be very similar (again to within 10%).

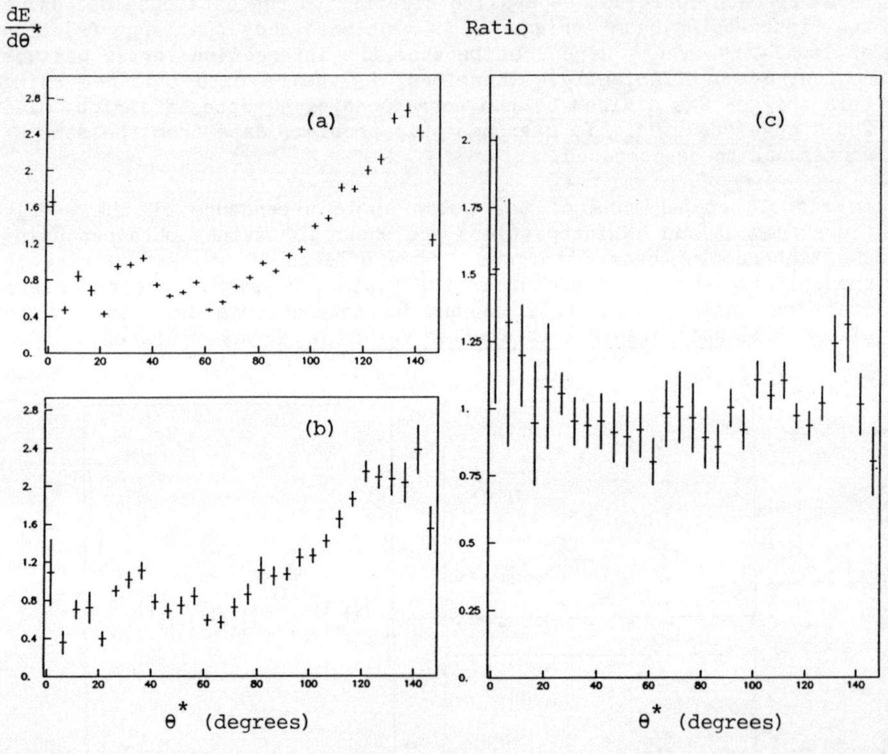

Figure 14. Polar angle dependence of energy flows (in arbitrary units) for (a) lead and (b) hydrogen data. θ^* is measured in the proton-proton centre-of-mass frame. The E_t range was $22 < E_t < 20$ GeV. E_t was measured in the "global" triggering region. (c) shows the ratio (lead/hydrogen) of the energy flows.

In addition to this energy measurement, a preliminary attempt was made to study the multiplicity of secondaries. At this time, no attempt has been made to form "energy clusters" from the calorimeter data as was done in the 400 GeV/c run of E557[11]. Therefore, the average number of <u>modules</u> in the WAC calorimeter per event containing a certain amount of module E_t was studied as a function of total E_t in the event.

In Fig. 15 the distributions and their ratios are shown. It can be seen that the events from hA and hN interactions have a similar number of modules firing for events with similar E_t. The data shown in Fig. 15 was obtained using an "interacting beam" trigger. Fig. 16 shows similar data using the "global" trigger with a high E_t threshold. Although the distributions for higher E_t events are broader, signifying an increase in the tranverse momentum per secondary, the structure of the lead and hydrogen events remains similar.

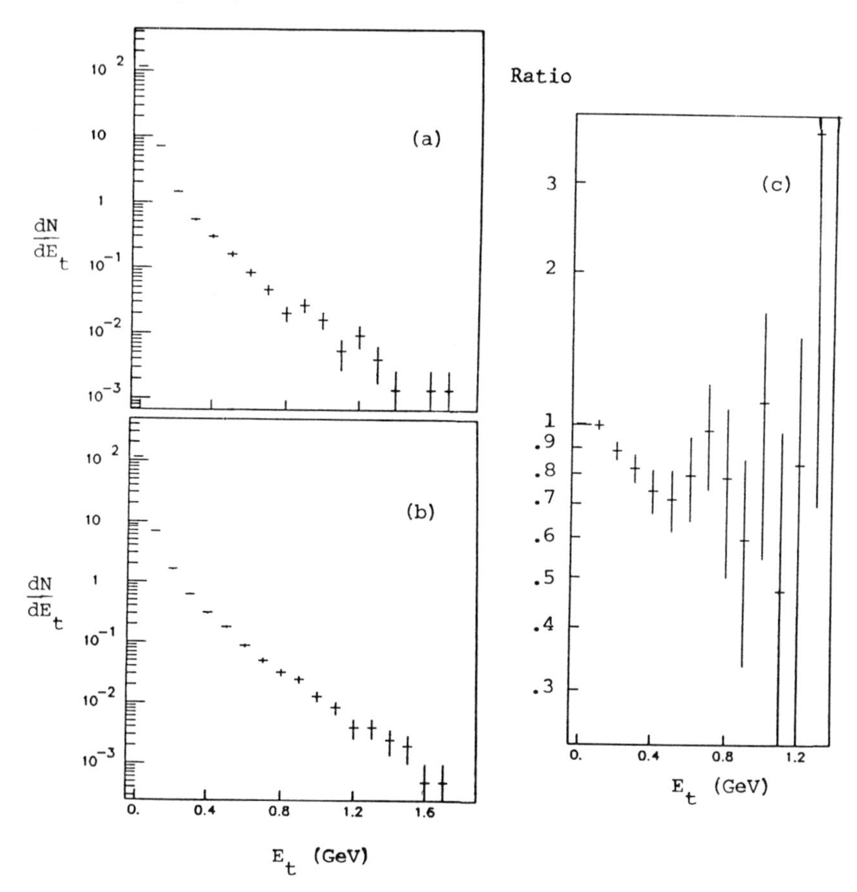

Figure 15. Average number of modules in the hadronic portion of the WAC calorimeter having E_t deposited in them for (a) lead and (b) hydrogen targets. The integral under the separate histograms is the total number of modules in the WAC, that is, 126. (c) shows the ratio of the lead to hydrogen data. The total E_t in the "global" triggering region was in the range $3 < E_t < 4$ GeV.

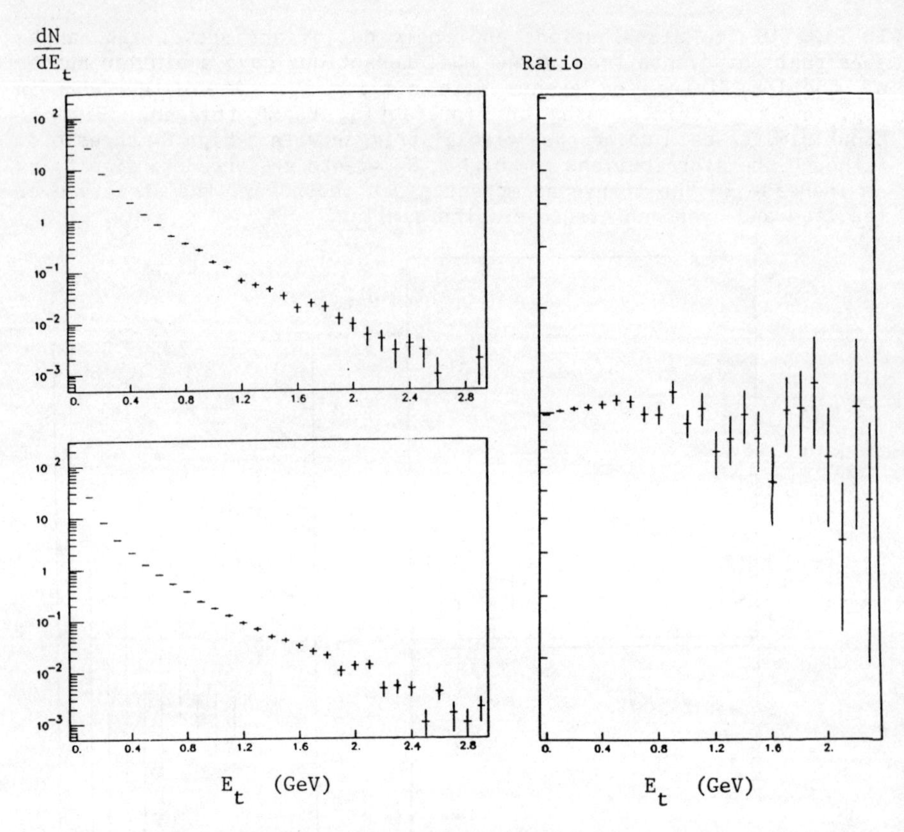

Figure 16. Average number of modules in the hadronic portion of the WAC calorimeter having E_t deposited in them for (a) lead and (b) hydrogen targets. (c) shows the ratio of the lead to hydrogen data. The total E_t in the "global" triggering region of the WAC and INS calorimeters was in the range $16 < E_t < 18$ GeV

Several previous experiments have measured the multiplicity of secondaries in the central region from hA and hN interactions. The average multiplicity from proton-lead collisions is 2 to 2.5 times that from proton-proton. The results mentioned above are not in contradiction to these previous measurements. The new results are quoted for events with similar E_t values. The average E_t value for a proton-lead interaction is approximately twice that from a proton-proton one (see the yield curves in Fig. 4). Therefore, because the multiplicity of modules with substantial E_t increases linearly with E_t as measured in the "global" triggering region (see Fig. 17), the average multiplicity from lead is approximately twice that from hydrogen.

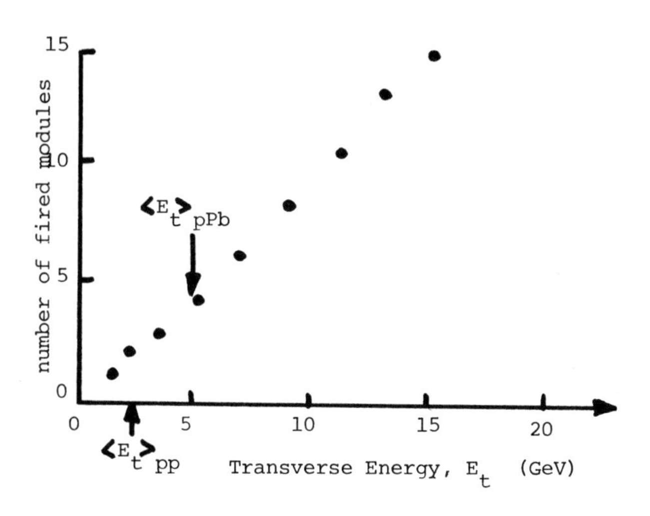

Figure 17. The multiplicity of modules in the hadronic porion of the WAC calorimeter having E_t values greater than 50 MeV in them. The data is for a hydrogen target. Similar results were obtained for proton-lead events. The value of the slope of the straight line was not sensitive to the 50 MeV E_t requirement; cuts of upto 250 MeV on module E_t produced no discernible change.

Similar results were obtained using the electromagnetic portion of the WAC calorimeter. A similar study was carried out using the modules in the FWD calorimeter. Again, as a function of E_t in the "global" triggering region, the energy flows and multiplicity of modules firing appear to be similar for hN and hA interactions.

CONCLUSIONS

The production of events with large amounts of E_t using nuclear targets is greatly enhanced over hydrogen. The majority of these events not jet-like. The yield enhancement increases faster than the number of nucleons within a nucleus. There is evidence that when "jet-like" events are produced, the enhancement varies as the number of nucleons in the nucleus, indicating a hard scattering mechanism.

To see if the enhancement is due to multiple scattering within the nucleus, correlations between the amount of energy in the central region and the energy detected in the target fragmentation region were studied. From these studies it appears that high E_t hA events are produced by multiple collisions within the nucleus.

The events from hydrogen and nuclear targets have similar event structure (multiplicity, energy flow) when events with <u>similar E_t values</u> are compared. It appears that the method by which E_t is produced (single vs. multiple collisions) is of secondary importance. This suggests that E_t may be a relevant parameter needed to describe the production of events.

* Members of the E557/672 collaboration are:

CALTECH: R. Gomez.
FERMILAB: L. Dauwe, H. Haggerty, E. Malamud, N. Nikolic .
FLORIDA STATE UNIVERSITY: S. Hagopian, J. Lanutti.
GEORGE MASON UNIVERSITY: R. Ellsworth.
INDIANA UNIVERSITY: S. Blessing, R. Crittenden, P. Draper, A. Dzierba, R. Heinz, T. Marshall, J. Martin, A. Samamurti, P. Smith, A. Snyder, C. Stewart, T. Sulanke, A. Zieminski.
RUTGERS: T. Watts.
SERPUKHOV: V. Abramov, Y. Antipov, B. Baldin, S. Denisov, V. Glebov, V. Kryshkin, O. Michailov, S. Polovnikov, R. Sulyaev.
UNIVERSITY OF ARIZONA: B. Pifer.
UNIVERSITY OF ILLINOIS AT CHICAGO: R. Abrams, J. Ares, H. Goldberg, C. Halliwell, S. Margulies, D. McLeod, A. Salminen, J. Solomon, G. Wu.
UNIVERSITY OF MARYLAND: R. Glasser, J. Goodman, S. Gupta, G. Yodh.

REFERENCES

1) Otterlund, I., Nucl. Phys. A418, 87c (1984);
Busza, W., Acta. Phys. Pol. B8, 333 (1977)
2) Elias, J., et al., Phys. Rev. D22, 13 (1980)
3) De Marzo, C., et al., Phys. Rev. D26, 1019 (1982)
De Marzo, C., et al., Phys. Rev. D29, 2476 (1984)
4) For a review of quark-gluon plasma see topics "Quark Matter '84", ed. Kajantie, K., Lecture Notes in Physics 221, Springer Verlag, 1985
5) Busza, W. and Goldhaber, A.S., Phys. Lett. 139B, 235 (1984)
6) Barton, D. et al., Phys. Rev. D27, 2580 (1983).
7) "Valence quark effects in beam remnants in high E_t pp collisions at $\sqrt{s}$ = 27.4 GeV", submitted to Phys. Rev. D.
8) Carrol, A. S., et al.,Phys. Lett. 80B, 319 (1979)
9) Brown, B., et al., Phys. Rev. Lett. 50, 11 (1982);
Bromberg, C., et al., Phys. Rev. Lett. 42, 1202 (1979);
Bromberg, C., et al., Nucl. Phys. B171, 38 (1980);
10) Cronin, J., et al., Phys. Rev. D11, 3105 (1975);
Kluberg, L., et al., Phys. Rev. Lett. 38, 670 (1977);
Antreasyan, D., et al., Phys. Rev. D19, 764 (1979)
11) De Marzo, C., et al., Phys. Lett. 112B, 173 (1982);
De Marzo, C., et al., Nucl. Phys. B211, 375 (1983);
Brown, B., et al., Phys. Rev. Lett. 49, 711 (1982);
Brown, B. C., et al., Phys. Rev. D29, 1895 (1984)
12) Halliwell, C., Acta. Phys. Pol. B12, 141 (1981)

SOME PROSPECTIVE 800 GEV PP AND PA PARTICLE YIELDS AND ASSOCIATED MULTIPLICITIES

L. Voyvodic
Fermi National Accelerator Laboratory[1]
Batavia, IL 60510

ABSTRACT

Recent Tevatron spectrometer experiments on heavy flavour production by 800 GeV protons on visible targets also include data on more conventional particle production processes. Features considered here for additional data analysis which may be undertaken are the central region yields of produced particles (negative pions), spectra of leading particles (protons and neutrons), and observable associated multiplicities.

1. INTRODUCTION

Two recent Fermilab experiments on heavy flavour production with 800 GeV protons have featured high resolution target chambers, wide aperture spectrometers and relatively loose on-line event triggers. Here we discuss how these experiments may also provide useful and complementary data on production of conventional particles. In particular, our focus is on semi-inclusive yields of both centrally produced pions and leading particle nucleons in the Tevatron energy region for pp and pA collisions. Such data, combined with results from earlier Fermilab and ISR experiments, should help towards critical tests of multiparticle hadroproduction mechanisms and models.

Experiment #743, a Europe-India-U.S. collaboration, employed the small hydrogen bubble chamber LEBC and a multiparticle detector system which included wire chamber magnetic spectrometer, plus multi-celled Cerenkov counters and transistion radiation detector for particle identification.

Experiment #653, a Japan-Korea-U.S. collaboration, employed nuclear emulsion track chambers, silicon strip vertex localizing detectors, wire chamber magnetic spectrometer, hadronic and electromagnetic calorimeters and muon detector.

In both experiments the downstream systems covered acceptance apertures considerably in excess of $\theta=2/\sqrt{s}=51$ mrad, the 90^0 cms angle for 800 GeV nucleon-nucleon collisions. This should allow both individual particle analysis and charged multiplicities over the forward hemisphere to be obtained directly from magnetic tape records of electronic tracking data. More complete studies of individual event structure would also make use of multiplicities visible in the track chambers. Both experiments also took some runs with metal foil targets. However, schedule changes prevented the carrying out of a more complete series of nuclear target measurements to be made with the same apparatus.

1- Operated by Universities Research Association for the U. S. Department of Energy

For the conventional particle production studies considered here, intitial samples of ~10^4 events appear to be sufficient, representing less than 1 percent of the recorded events in each experiment. For establishing trigger biases, comparison studies with untriggered events need to be made. For the pp experiment this is currently in progress, using LEBC pictures taken without interaction trigger. For the p-emulsion experiment, which used both interaction and muon trigger, comparison could also be made with data on untriggered emulsion events, such as the high statistics study of bare emulsion expsoures to 800 GeV protons currently underway in E-508, a Poland-U.S.-U.S.S.R. collaboration.

2. INCLUSIVE PARTICLE YIELDS

2.1 Inclusive Cross-Sections

For the comparisons to be made with expected 800 GeV proton results, typical particle yields from pp collisions are shown in Fig. 1, from a lower energy Fermilab spectrometer experiment[1], integrated over P_T[2].

The two main features of interest here are the steeply falling yields of negative produced particles (predominantly pions), and the nearly flat spectrum for leading particle protons. Intermediate behaviour of positive pion and kaon yields (associated in part with leading particle neutrons and hyperons) would require more complex analyses.

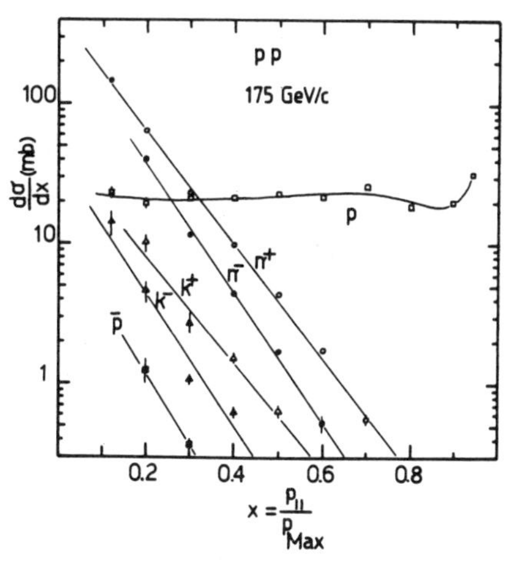

Fig. 1. Single particle inclusive cross sections for charged particles.[1,2]

2.2 Inclusive Leading Particles

From comparison with the total pp inelastic cross-section of 33 mb, the leading protons in Fig. 1 contribute ~22 mb, or a leading proton should appear in about two-thirds of the events, with a leading neutron in most of the remaining events. As in analysis of ISR single hemisphere data[3], about one-half of the protons can be recognized directly (when $0.4 \lesssim x \lesssim 0.85$). In E-743 the transition radiation detectors should help to identify the lower energy protons, down to $x \sim 0.2$. In E-653 good detection is expected for both the high energy protons ($x \gtrsim 0.4$) and particularly for forward neutrons of all energies ($0.1 \lesssim x \lesssim 0.9$). Further discussion of leading particle spectra is made in the last section on semi-inclusive yields and inelasticity.

2.3 Inclusive Pion Production

Spectrometer tracking data in E-653 and E-743 can be expected to provide direct studies of central region inclusive rapidity distributions for produced pions (negative particles), ideally in the form shown in Fig. 2 for ISR spectrometer results. As a minimum, the 800 GeV pp data would provide more details of the scaling breakdown in the central 'plateau' region.

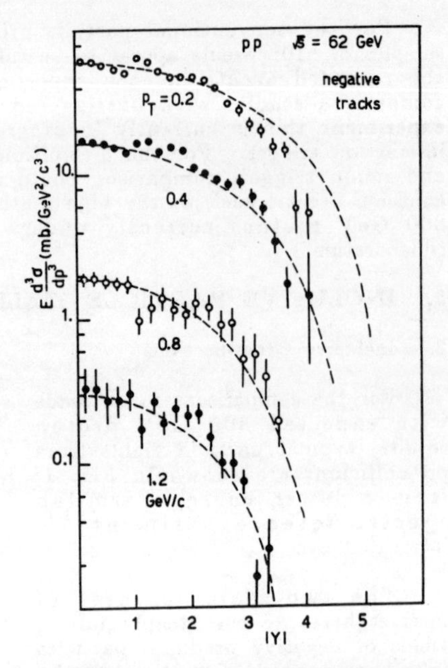

Fig. 2. Inclusive cross-sections for negative particles in ISR spectrometer experiments[2].

The ratios $R(y,P_T)$ of inclusive particle densities, and resultant $A^{\alpha(y,P_t)}$ behaviour of cross-sections, will be of considerable interest when 800 GeV p-emulsion and pp data are compared. Existing data from proton energies up to 400 GeV appear consistent with a simple linear fit $\alpha(y) \simeq 1\text{-}0.4y/\ln s$, whereas a flattening of R and α values at central rapidities ($y \approx \ln \sqrt{s}$) can be expected to set in at some sufficiently high energy.

3. SEMI-INCLUSIVE PARTICLE YIELDS

3.1 Observable Associated Multiplicities

For pp events in E-743 a sample of $\sim 10^4$ triggered events of known multiplicity, $n = n_{ch}$, in the bubble chamber are being used for comparison with untriggered events for efficiency calibrations. This same sample of events provides natural candidates for associating with the spectrometer data. In addition, the spectrometer tracking chambers should independently provide the forward hemisphere multiplicities, n_F, for the same events.

For p-emulsion events in E-653 the spectrometer tracking chambers should also provide forward hemisphere multiplicities, n_F. When event vertices are examined in emulsions the two parameters of interest are $n = n_s$, the multiplicity of fast produced particles, and N_h, the nuclear response multiplicity of visible proton tracks. The latter are particularly useful for estimating values of $<\nu>$, the number of intranuclear projectible collisions on an event by event basis[4].

From differences, $n_B = n - n_F$, in both experiments one can study forward-backward correlations, reported to be particularly strong in nuclear targets in earlier experiments[5] at proton energies up to 400 GeV.

3.2 Semi-inclusive Pion Production

Combining event multiplicities with spectrometer data for negative particles should allow detailed studies of pion production by 800 GeV protons, extending lower energy data[6][7] such as that shown in Fig. 3 from Fermilab 30 inch bubble chamber experiments.

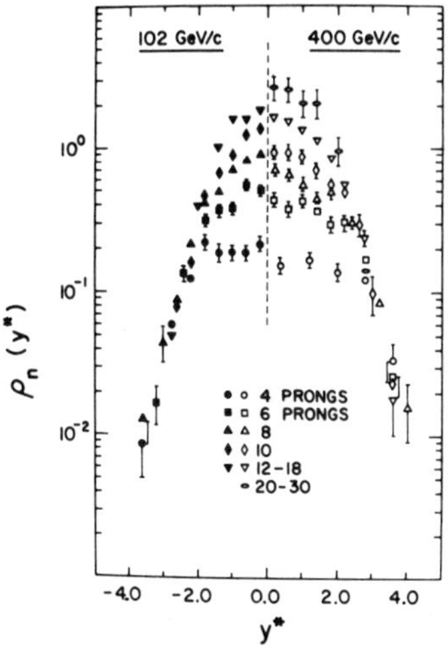

From E-743 pp data one can explore further detailes of KNO semi-inclusive scaling and its violations, using as dependent variables both $z=n_{ch}/<n_{ch}>$ and $z=n_F/<n_F>$. More detailed comparisons can also be made of the approximately Gaussian rapidity distributions with predictions of hydrodynamic models.

From E-653 p-emulsion data, analogous studies would provide novel data for comparison of particle production in pp and pA collisions, especially if the emulsion events are characterized in bins of $<\nu>$ from the visible proton N_h data.

Fig. 3. Semi-inclusive single-particle densities for negative particles in pp collisions[6].

3.3 Semi-inclusive Leading particles and Inelasticities

There are increasing indications that in hadronic multiparticle production it is the leading particle component which carries critical information on the underlying processes.
From this point of view it may be $d\sigma_n/dx_{LP}$, the semi-inclusive leading particle spectrum, which relates for a given on-going energy fraction x_{LP} the inelasticity fraction $K=1-x_{LP}$ available for producing n particles, thus characterizing the multiplicity distribution $P_n=\sigma_n/\sum\sigma_n$ and semi-inclusive production yields $(d\sigma_n/dy)$. At the constituent level the leading particle effect is assumed to originate with fragmentation of valence quarks, whereas the 'consumed' energy fraction K results in central region multiparticle production from gluon and/or sea quark interactions.

Experimental observations of pp leading particle effects at Fermilab and ISR have shown universality for average net multiplicities $<n>$ when studying protons in one hemisphere[3][8], in both hemispheres[9][10], as well as the absence of correlations in proton energies between the two hemispheres[11].

More direct information on $d\sigma_n/dx_{LP}$ may be found in invariant inclusive cross-sections for pp->px at 100 GeV incident energy.[1] After dividing these by x_{LP}, it would appear that the integrated flat $d\sigma/dx$ spectrum for protons in Fig. 1 may be fortuitous, deriving from the addition of rising spectra at low multiplicities and falling proton spectra at high multiplicities. (The flat proton inclusive spectrum at Fermilab and ISR energies also implies an average $<x_{LP}>=0.5$ and inelasticity $<K>=0.5$).

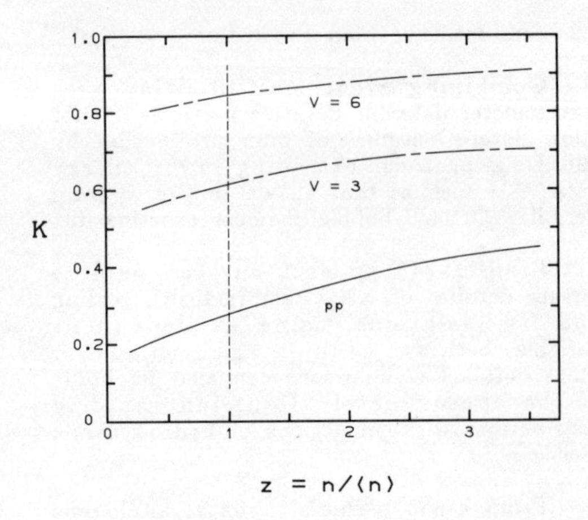

Fig. 4. Calculated inelasticies versus the KNO variable z. The pp curve is from[12]. The pA curves assume a leading particle intranuclear cascade of ν collisions.

The recent model of[12] suggests KNO-type semi-inclusive scaling of event inelasticities in pp interactions, shown in Fig. 4. The curves shown for nuclear target inelasticities are based on the assumption that K values for the first collision are as in the pp case, and use of their average value $<K>=0.28$ for $(\nu$-1) successive intranuclear collisions. Other inelasticity predictions have also been made, with fits to available experimental data, and with additional features of predicting either decreasing[13] or increasing[14] values of inelasticity parameters at increasing particle energies.

It is evident from the above comments that special emphasis is placed on obtaining leading proton and neutron data from the two Tevatron spectrometer experiments, in association with event multiplicities.

For example, from 800 GeV pp collisions in E-743, scatter plots of x_{LP} from leading proton versus charged multiplicity n_{ch} from LEBC, and versus forward hemisphere multiplicity n_F, would provide input for inelasticity behaviour studies and comparison with models.

Similarly, from 800 GeV p-emulsion collisions in E-653 both the high energy protons and wide energy range of neutrons would be examined semi-inclusively with the associated event multiplicities n_s, n_F and N_h. The resultant inelasticities, stopping powers and energy deposition estimates could provide new tests of high energy hadronic interaction mechanisms in nuclei. Comparison of proton to neutron ratios of inclusive and semi-inclusive yields in this experiment would additionally provide tests of specific quark fragmentation models[15].

ACKNOWLEDGEMENTS

It is a pleasure to acknowledge helpful discussions with my colleagues on experiment #743 and with members of experiment #653 collaboration.

REFERENCES

[1] Brenner, A.E., et al, Phys. Rev. D26, 1497 (1982).
[2] Palmonari, F., Erice Lecture, CERN-EP/82-176.
[3] Basile, M. et al, Phys. Lett. 99B, 247 (1981).
[4] Stenlund, E. and Otterlund, I., Nuc. Phys. B198, 407 (1982).
[5] Azimov, S.A. et al, Zeit. Phys. C 10, 1 (1981).
[6] Bromberg, C. et al, Nucl. Phys. B107, 82 (1976).
[7] Oh, B.Y. et al, Nuc. Phys. B116, 13 (1976).
[8] Barish, S.J. et al, Phys Rev. Lett. 31, 1080 (1973).
[9] Basile, M. et al, Nuovo Cimento A67, 244 (1982).
[10] Brick D. et al, Phys. Lett. 103B, 242 (1981).
[11] Basile, M. et al, Nuovo Cimento A73, 329 (1983).
[12] Chou, T.T. and Yang, C.N., Phys. Rev. D32, 1692 (1985).
[13] Fowler, G.N. et al, Phys. Rev. Lett. 55, 173 (1985).
[14] Barshay, S. and Chiba, Y. Kyoto Preprint RIFP-620 (1985).
[15] Hwa, R. Private communication (1986).

NUCLEAR FRAGMENTATION

MULTIFRAGMENTATION IN INTERMEDIATE ENERGY HEAVY ION COLLISIONS

B.V. Jacak,[*] H.C. Britt,[*] G. Claesson,[**] K.G.R. Doss,[**] R. Ferguson,[‡‡]
A.I. Gavron,[*] H.-A. Gustafsson,[**] H. Gutbrod,[‡] J.W. Harris,[**]
K.-H. Kampert,[‡] B. Kolb,[‡] A.M. Poskanzer,[**] H.-G. Ritter,[**] H.R. Schmidt,[‡]
L. Teitelbaum,[**] M. Tincknell,[**] S. Weiss,[**] H. Weiman,[**] J. Wilhelmy[*]

[*]Los Alamos National Laboratory,
Physics Division
Los Alamos, NM 87544

[**]Lawrence Berkeley Laboratory,
1 Cyclotron Road
Berkeley, CA 94720

[‡]Gesellschaft fur Schwerionenforschung,
Postach 11 05 41
D-6100 Darmstadt 11, West Germany

[‡‡]Oak Ridge National Laboratory
P. O. Box X, Bldg. 6003
Oak Ridge, TN 37831

There has been considerable recent interest in the production of intermediate mass fragments (A>4) in intermediate and high energy nucleus-nucleus collisions. The mechanism for production of these fragments is not well understood and has been described by models employing a variety of assumptions. Some examples are: disassembly of a system in thermal equilibrium into nucleons and nuclear fragments,[1,2] liquid-vapor phase transitions in nuclear matter,[3] final state coalescence of nucleons[4] and dynamical correlations between nucleons at breakup[5,6].

Previous studies of fragment production, with one exception,[7] have been single particle inclusive measurements; the observed fragment mass (or charge) distributions can be described by all of the models above. To gain insight into the fragment production mechanism, we used the GSI/LBL Plastic Ball/Wall detector system[8] to get full azimuthal coverage for intermediate mass fragments ($Z<10$) in the forward hemisphere in the center of mass system while measuring all the light particles in each event.

The complete measurement of light particles (p, d, t, ^3He, ^4He) allows us to determine the charge multiplicity of participant baryons, which increases as the impact parameter between the projectile and target nuclei gets smaller. In addition, the complete measurement of light particles allows us to perform global analyses of the events and look for collective effects in fragment emission by comparing to flow effects seen in the light particles. Lastly, the large acceptance for intermediate mass fragments in the Plastic Ball allows us to measure their multiplicities event-by-event. This experiment should help to differentiate the various models by allowing a better characterization of the system breaking up, as well as yielding information on the impact parameter of the collision. We measured fragmentation for 200 MeV/nucleon Au + Au and Au + Fe, using the inverse kinematics in the latter case to focus fragments from the large reaction partner into our detectors. Presented here are first results from the experiment. These will not yet allow us to determine the exact fragmentation mechanism, but will provide many constraints on the results of the various models.

The observed participant charge multiplicity distribution for Au + Au is shown in figure 1. We have used this quantity to sort the events into groups according to impact parameter, as indicated by the lines in the figure. The events with the fewest observed charges (labeled "mul1") correspond to the most peripheral collisions, while the events with the highest charge multiplicities ("mul5") arise from central collisions. The drop in the number of events with very low

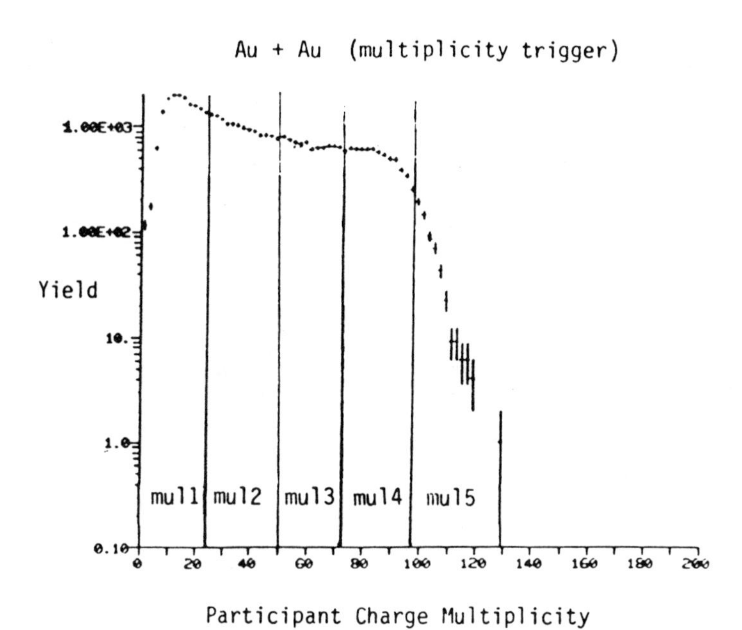

Fig. 1. Participant charge multiplicity distribution for Au + Au.

multiplicities is a result of the trigger used in the experiment, which was designed to discriminate against the most peripheral collisions.

Figure 2 shows a density plot of the invariant cross section for lithium fragments from Au + Au, as a function of the rapidity and the perpendicular momentum per nucleon. The five parts of the figure correspond to the five cuts on the participant charge multiplicity indicated in figure 1. No corrections for the angular and energy cutoffs in the detector have been applied to the data. Thus two distinct sections are visible in each plot, corresponding to the two subsections of the detector system which were sensitive to intermediate mass fragments.

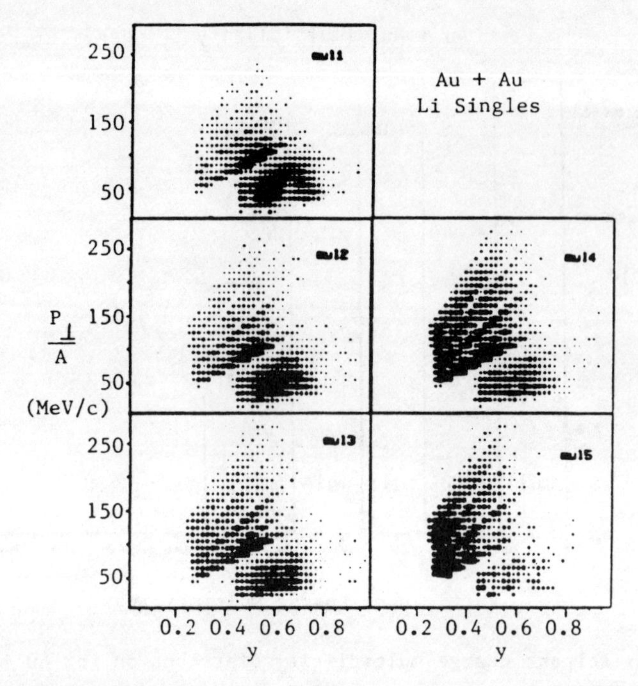

Fig. 2. Multiplicity selected rapidity plots for lithium fragments for Au + Au.

It is evident from the figure that peripheral collisions give rise to fragments with rapidities very close to the beam rapidity, consistent with expectations for fragmentation of a slightly excited projectile. In the "mul1" plot we see a hole in the yield at exactly the beam rapidity, corresponding to coulomb repulsion between the emitted lithium fragment and a heavy remnant of the Au projectile. In events where a projectile remnant is observed, the azimuthal angles of the remnant and fragments are correlated, supporting the picture of fragments evaporated from a large projectile residue.

In the more central collisions in the highest multiplicity bins, we see many lithium fragments emitted with smaller rapidities, intermediate between those of the target and the projectile. In contrast to the forward peaked angular distributions arising from peripheral collisions, these fragments are emitted relatively isotropically in the center of mass system. Such behavior is what one might expect when the projectile nucleons impart more energy to the target and create an excited region which moves at a velocity approximately halfway between that of the projectile and the target.

The transition between the peripheral and central collisions is very smooth, with a gradual shift in the rapidities of the observed fragments away from the projectile rapidity. In order to check if both projectile-like and midrapidity fragments are formed in the same event, we chose events with multiple fragments and required that at least one fragment fall into a midrapidity window. We made a rapidity plot similar to figure 2 of the other fragments in the event, and found that the mixing occurs event-by-event; even in the central collisions where midrapidity fragments are formed, we observe some associated projectile rapidity fragments.

In order to check our detection efficiency for multiple fragments, we summed the charges observed in the forward c.m. hemisphere for the Au + Au reaction and compared this with the charge of one Au nucleus. The result is shown in figure 3, plotted as a function of the participant charge multiplicity. For peripheral collisions, we observe a large projectile remnant in many events, these correspond to the small lobe on the left (low multiplicity) side of the figure. As the charge of the projectile remnant is not well measured, it is assigned one half the charge of the projectile. It is clear that the remnant charge is underestimated for the most peripheral collisions and overestimated as the impact parameter and projectile remnant become

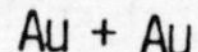

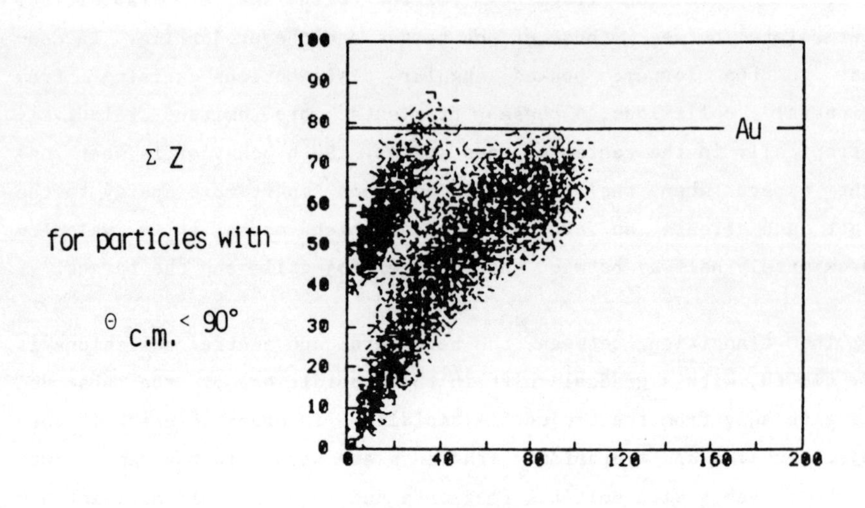

Fig. 3 Sum of charges observed in forward c.m. hemisphere, as a function of the participant charge multiplicity.

smaller. For central collisions (large multiplicities), we observe almost all of the Au charge in the form of light and intermediate mass fragments. These results indicate that in central collisions the system breaks up into rather small pieces with no large remnant. Also, we gain confidence that we may study fragment multiplicities event by event due to the high efficiency of the detector system.

The total intermediate mass fragment multiplicity is found to increase with increasing participant baryon charge multiplicity. In peripheral collisions, the few fragments which are formed are preferentially emitted at angles less than ten degrees in the lab. More fragments are produced in central collisions, and these fragments are found at larger angles.

Figure 4 shows the multiplicity distributions for fragments observed at angles between 10 and 30 degrees in the lab. It should be noted that for the Au + Au system, 30 degrees in the lab corresponds to 90 degrees in the center of mass frame. The five curves correspond to the same five cuts on the participant charge multiplicity that were described in figure 1 and used in figure 2. In central collisions, the most probable fragment multiplicity is 3-4, with a tail out to as many as 9 fragments. For peripheral collisions the distribution peaks at zero to one fragment, with a smaller tail to high multiplicities. The multiplicities of fragments emitted to small angles are very low, regardless of impact parameter.

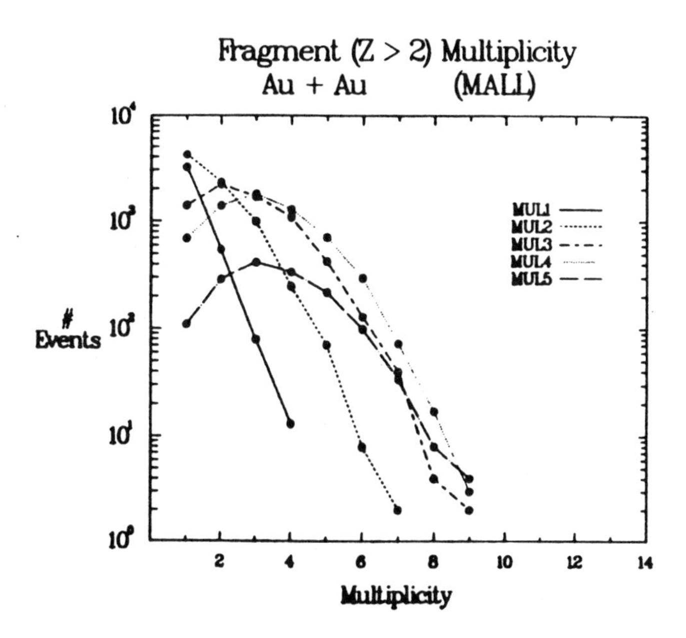

Fig. 4. Multiplicity distributions of intermediate mass fragment detected at 10-30 degrees in the laboratory.

Collective flow of nuclear matter upon re-expansion has been proposed as an important signature of compression effects. Two collective effects, the side-splash of the participants and the bounce-off of the spectator nucleons have been observed at the Plastic Ball[9]. A sensitive test for collective flow can be found in the transverse momentum analysis of Danielewicz and Odyniec[10]. This method determines the reaction plane from the transverse momentum transfer between the forward and backward hemispheres in the center of mass, and then examines the mean transverse momentum per nucleon <px/A> in this reaction plane as a function of the center of mass rapidity. By removing auto-correlations (i.e. calculating the reaction plane for each individual particle from the transverse momentum sum of all the other particles), this method is sensitive to the real dynamic correlations.

Figure 5 shows the mean transverse momentum per nucleon projected

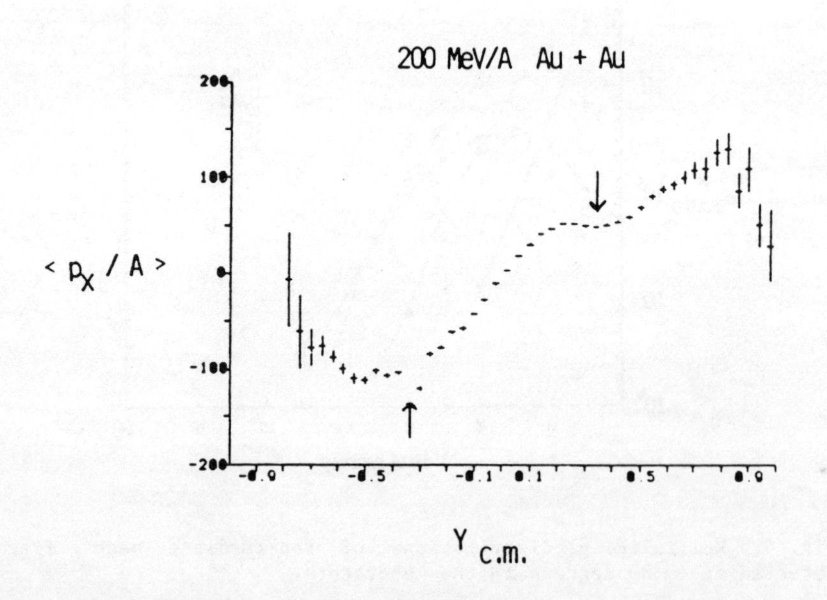

Fig. 5. Mean transverse momentum per nucleon in the reaction plane, as a function of c.m. rapidity for Au + Au.

into the flow plane for the Au + Au reaction. The calculation includes only Z=1 and Z=2 particles, and may be compared to analyses of earlier Plastic Ball experiments[11]. All observed light particles are included in the figure; those originating from the projectile or target spectators were not removed. The curve shows the s-shape typical for the collective transverse momentum transfer between the forward and backward hemispheres. Close to the target and projectile rapidities the curve bends towards the axis because of contamination from spectator matter.

We have examined the intermediate mass fragments for flow effects by looking at the azimuthal correlation between the reaction plane and the emission angle of the fragments. We find that fragments are preferentially emitted in the same direction as the flow direction, and that this correlation gets stronger for more massive fragments.

As a first step in studying the origins of multiple fragments, we construct the "effective source" of the fragments by summing the 4-momenta of the observed fragments. This source moves with a velocity similar to the beam velocity in the case of peripheral collisions, and slows down as the impact parameter decreases. We again calculated the reaction plane from the light particles, and looked at the projection of the "effective source" transverse momentum per nucleon in and out of the reaction plane. As with the fragment singles, the source azimuthal angle is correlated with the reaction plane. The transverse momentum per nucleon in the reaction plane increases for more central collisions, reaching a maximum of about 150 MeV/c. Perpendicular to the reaction plane, the source transverse momentum shows no preferred direction.

We are currently studying whether the intermediate mass fragment energy distributions and multiplicities can be reproduced with models incorporating the various assumptions mentioned earlier. These models must also reproduce the observed flow effects, thus rather stringent criteria have already been established by these experiments.

REFERENCES

1. J. Randrup and S. E. Koonin, Nucl. Phys. A356, 223 (1981). G. Fai and J. Randrup, Nucl. Phys. A404, 551 (1983).

2. J. P. Bondorf, R. Donangelo, I. N. Mishustin, C. J. Pethick and K. Sneppen, Phys. Lett. 150B, 57 (1985).

3. G. Bertsch and P. J. Siemens, Phys. Lett. 126B, 9 (1983).

4. H. H. Gutbrod, A. Sandoval, P. Johansen, A. M. Poskanzer, J. Gosset, W. G. Meyer, G. D. Westfall, and R. Stock, Phys. Rev. Lett. 37, 667 (1976).

5. B. Strack and J. Knoll, Z. Phys. A315, 249 (1984).

6. G. E. Beauvais, and D. H. Boal, Univ. of Ill. preprint P/86/2/26, 1986.

7. A. I. Warwick, H. Wieman, H. H. Gutbrod, M. R. Maier, J. Peter, H. G. Ritter, H. Stelzer, F. Weik, M. Freedman, D. J. Henderson, S. B. Kaufman, E. P. Steinberg, and B. D. Wilkins, Phys. Rev. C27, 1083 (1983).

8. A. Baden, H. H. Gutbrod, H. Lohner, M. R. Maier, A. M. Poskanzer, T. Renner, H. Riedesel, H. G. Ritter, A. Warwick, F. Weik, and H. Wieman, Nucl. Inst. Meth. 203, 189 (1982).

9. H. A. Gustafsson, H. H. Gutbrod, B. Kolb, H. Lohner, B. Ludewigt, A. M. Poskanzer, T. Renner, H. Riedesel, H. G. Ritter, A. Warwick, F Weik, and H. Wieman, Phys. Rev. Lett. 52, 1590 (1984).

10. P. Danielewicz and G. Odyniec, Phys. Lett. 157B, 146 (1985).

11. H. A. Gustafsson, H. H. Gutbrod, B. Kolb, H. Lohner, B. Ludewigt, A. M. Poskanzer, T. Renner, H. Riedesel, H. G. Ritter, T. Semiarczuk, J. Stepaniak, A. Warwick and H. Wieman, Z. Phys. A321, 389 (1985).

Fragmentation and the Liquid-Gas Phase Transition

C. J. Pethick
Department of Physics, University of Illinois at Urbana-Champaign
1110 West Green Street, Urbana, IL 61801
and
NORDITA, Blegdamsvej 17
DK-2100 Copenhagen Ø, DENMARK

and

D. G. Ravenhall
Department of Physics, University of Illinois at Urbana-Champaign
1110 West Green Street, Urbana, IL 61801

ABSTRACT

The development of hydrodynamic instabilities in hot nuclear matter is considered, and its relevance to fragmentation in heavy-ion collisions is discussed.

In recent years much effort has been devoted to trying to learn about the properties of nuclear matter from experiments on heavy ion interactions. One aspect of this is the question of what, if any, is the relationship between the liquid-gas phase transition and nuclear fragmentation. Previous discussion has been confined to long wavelength instabilities associated with bulk nuclear matter, and in this paper we extend these considerations to allow for the finite size of the system, and for the finite duration of the collision.

We wish to apply the ideas of fluid mechanics to heavy ion collisions. Information about hydrodynamic variables, such as the average density and temperature, cannot be obtained directly from experiment, and we shall therefore be guided by some recent simulations.[1-6] The particular ones we have found useful are those of Pandharipande and collaborators[1-3], which are especially well suited for our analysis since detailed results for thermodynamic variables are given. What we shall do is to take the simulations and use them as a basis for an attempt to understand the physics of the fragmentation process. By doing this we hope to uncover the properties of matter important for fragmentation, which will enable one to make more realistic models of fragmentation in nuclear matter.

In these simulations one studies, by molecular dynamics, the evolution of compressed matter consisting of argon atoms. The particles are taken to be classical, and the interaction is a truncated Lennard-Jones potential,

$$V(r) = 4\varepsilon_o\left(\left(\frac{a}{r}\right)^{12} - \left(\frac{a}{r}\right)^{6}\right) - 4\varepsilon_o\left(\frac{1}{3^{12}} - \frac{1}{3^{6}}\right) \qquad \text{for } r < 3a,$$

$$= 0 \qquad \text{for } r > 3a, \tag{1}$$

For our purposes, the most important features of this system are that its particle-particle interaction and its equation of state resemble those of nuclear matter. There is a liquid-gas phase transition, and the coexistence curve is shown in Fig. 1. The temperature is in units of ε_o, and the density in units of a^{-3}. For comparison we show, in Fig. 2, the coexistence curve for nuclear matter with equal numbers of neutrons and protons calculated by one of us,[7] who fitted a Skyrme-type interaction to the microscopic equation of state determined by Friedman and Pandharipande.[8] Notice that the coexistence curves for the liquid-gas transition have rather similar shapes in the two cases. For argon, there is a solid phase which has no analogue for nuclear matter. The absence of a solid nuclear-matter phase is due partly to the short-range repulsion of the nucleon-nucleon interaction being less strong than for a Lennard-Jones potential, and partly to the quantum statistics, which gives rise to a large Fermi kinetic energy that tends to counteract the

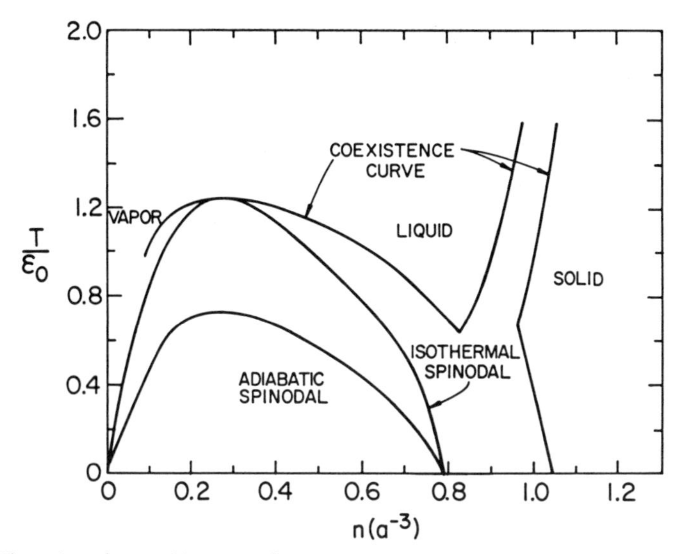

Fig. 1 Phase diagram for argon with the truncated 6,12 interaction, taken from Ref. 1.

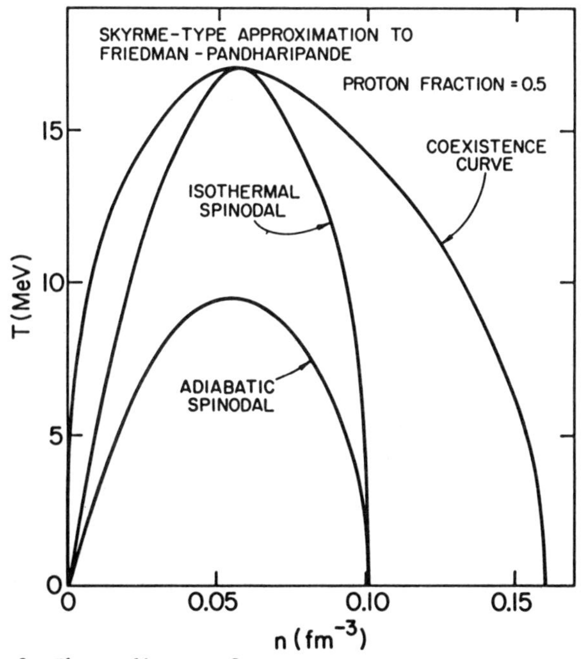

Fig. 2 Phase diagram for symmetric nuclear matter, taken from Ref. 1.

localizing effect of the potential. The existence of a solid phase does not appear to have a dominant influence on the dynamics of interest in the study of fragmentation.

One may ask how well nuclear matter is represented by argon. The critical temperature of argon is 151K, while in the calculations one finds $T_c \simeq 1.3\ \varepsilon_o$. Consequently the choice $\varepsilon_o \simeq 120K$ enables one to fit the critical temperature. For the Friedman-Pandharipande equation of state for nuclear matter, one finds $T_c \simeq 17.5$ MeV and therefore to fit this by a suitably scaled Lennard-Jones potential, one must have $\varepsilon_o \simeq 14$ MeV. Measured in these units the binding energy of argon is ~ 8 ε_o per particle, while that for nuclear matter is ~ 1.1 ε_o. This difference reflects the fact that for nuclear matter the binding energy (~ 16 MeV/nucleon) is very much less than the potential energy per particle (~ 50 MeV binding per nucleon) as a consequence of the large Fermi kinetic energy (~ 35 MeV per nucleon). In the case of argon, the particles are classical, and consequently the binding energy is much larger relatively speaking. Another quantity one may compare for the two cases is the compressibility parameter at zero temperature, $K = 9\ \partial P/\partial n$, where n is the particle number density and P the pressure. For argon, K is large, ~ 300 ε_o, again due to the very repulsive core of the potential, while for nuclear matter it is only ~ 14 ε_o (200 MeV). Cross sections are also rather different for the two systems, being a few times larger for argon than for nuclear matter.

The calculations reported in Ref. 1 describe the disassembly of hot spheres of argon. The initial sphere is constructed by performing molecular dynamics calculations for particles in a cube for a sufficiently long time that thermal equilibrium is established. At time t = 0, all particles outside a sphere whose diameter is equal to the cube edge are removed, and the subsequent evolution of the particles is followed. In Ref. 2 the effects of Coulomb interactions were included in simulations of the disassembly of a drop, and in Ref. 3 the collision of two cold spheres was considered, but we shall not discuss these in detail, since in their gross features, these calculations resemble those for disassembly of an uncharged drop.

To analyze the results in terms of average quantities we shall, following Ref. 1, focus our attention on particles near the center, where it is most likely that hydrodynamic concepts are applicable. One defines a density, n, of the central matter as the average density of particles in a sphere centered on the center of mass which contains 50% of the total number of particles. Particle velocities are resolved into components parallel and perpendicular to the line joining a particle to the center of mass, and an average central temperature, T, is defined as the average transverse kinetic energy per particle in the central matter. This definition reduces to the usual one for matter in thermal equilibrium, with no net flows, and eliminates the effect of bulk radial motion in the case of expanding drops.

The results of the simulations may be divided into four classes, three of which are represented by the trajectories in the n, T plane shown in Fig. 3. In simulations that start at high temperatures, the drop vaporizes completely, leaving mainly single atoms. For low initial temperatures, and initial densities up to a few times the saturation density, one expects the drop to expand, to lose some atoms by evaporation, to cool, and subsequently to settle down at the saturation density. No simulations have been carried out for such cases since they require lengthy computations. When initial conditions are intermediate between these two, there are two possible types of events. In the first, violent evaporation, the drop expands and becomes unstable. Density fluctuations develop and give rise to viscous dissipation, which heats the drop. Some particles evaporate, and the drop becomes stable again and eventually settles down at the saturation density. The final sort of event is fragmentation, in which the average density of the central matter decreases continuously, and the drop breaks up into a number of fragments.

In looking at the trajectories of the central matter it is important to bear in mind that the starting point does not uniquely specify the trajectory. This is because states with the same initial

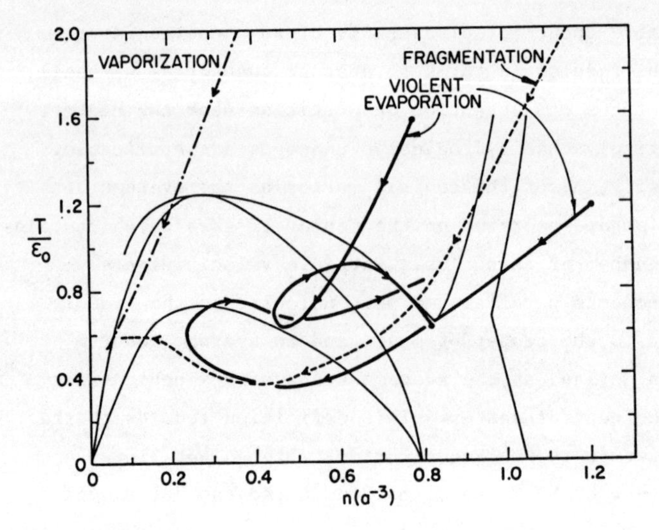

Fig. 3 Schematic trajectories of the central matter in the disassembly of hot drops of argon (Ref. 1).

density and temperature will generally differ on a microscopic scale. How a drop evolves depends on the initial fluctuations, as well as on the average properties, and can therefore be different even though the initial average properties are the same.

To develop a framework for discussing the conditions under which instabilities can develop, we first consider long wavelength instabilities of bulk, electrically neutral, systems. While uniform matter at densities and temperatures that lie within the coexistence curve is thermodynamically unstable to breakup into two phases, it is not always unstable to small density perturbations. If one considers isothermal density fluctuations, matter is unstable to small long-wavelength perturbations within the curves in Figs. 1 and 2 marked "isothermal spinodal," which is named after the analogous process of spinodal decomposition in binary alloys. On this line the isothermal sound velocity, c_s^{iso}, given by

$$c_s^{iso^2} = \frac{1}{m} \left(\frac{\partial P}{\partial n}\right)_T , \tag{2}$$

vanishes. Here m is the particle mass. As far as one can see from Fig. 3, the isothermal spinodal line does not seem to play an important role in the physics of fragmentation, since the trajectories show no special features there.

In the initial stages of evolution of a hot drop, entropy is conserved to a good approximation,[9-11] and this has led to the suggestion that perturbations may grow adiabatically rather than isothermally. The adiabatic sound velocity is given by

$$c_s^{ad^2} = \frac{1}{m}\left(\frac{\partial P}{\partial n}\right)_s, \tag{3}$$

where the subscript on the derivative denotes that it is to be evaluated at constant entropy per particle, s. The condition for adiabatic instability is that $c_s^{ad^2}$ is negative, and the lines marked "adiabatic spinodal" in Figs. 1 and 2 show where matter first becomes unstable to small adiabatic density fluctuations. It is interesting that in the simulations instabilities in the drops set in only if the matter is within the region bounded roughly by the adiabatic spinodal line in Fig. 3. Does this mean that the fluctuations develop adiabatically?

For fluctuations to develop adiabatically, the characteristic time for the disassembly process must be small compared with the time for heat to be conducted over a distance comparable to the length scale of the instability. The latter may be estimated from the thermal diffusion equation

$$C_p \frac{\partial T}{\partial t} = \underset{\sim}{\nabla}\cdot(K\underset{\sim}{\nabla}T) , \tag{4}$$

where K is the thermal conductivity, and C_p is the specific heat at constant pressure. The characteristic time for heat diffusion is therefore given by

$$\tau_T = \frac{1}{D_T q^2} , \tag{5}$$

where $D_T = K/C_p$ is the thermal diffusivity, given in order of magnitude by

$$D_T \sim \frac{1}{3} v \ell, \tag{6}$$

where v is a typical particle velocity and ℓ a typical mean free path. The wavenumber of the perturbation, q, is larger than the wavenumber $\sim \pi/R$ for the lowest mode in a sphere of radius R, where R is the radius of the drop. Temperature fluctuations therefore die out on time scales

$$\tau_T \lesssim \frac{1}{\pi^2} R^2/D_T . \tag{7}$$

The mean free path is given by

$$\ell = \frac{1}{n\sigma} , \tag{8}$$

where σ is the particle-particle cross section and n is the density. Dynamical time scales are usually long compared with the time, τ_s, for sound to travel a distance equal to the drop radius,

$$\tau_s \sim \frac{R}{c_s} . \tag{9}$$

We use the sound velocity near saturation density, rather than that at a lower density, since the time the drop spends in the unstable region is determined by earlier phases of the event. (In the simulations in Refs. 1-3, it is found that characteristic dynamical time scales are at least R/c_s.) For argon we use in Eq.(6) the thermal velocity $v = (3T/m)^{1/2}$, with $T \sim \varepsilon_o/2$, and for σ in Eq.(7) the geometric cross section πa^2. We then find that

$$\frac{\tau_T}{\tau_s} \approx 4 \frac{R}{a} , \qquad \text{(argon)} \tag{10}$$

so that even for disturbances whose length scale is of order the inter-particle spacing the process is adiabatic.

For nuclear matter, v is about c/3. For a density $n \simeq 0.17\ \mathrm{fm}^{-3}$ and $\sigma \sim 30$ mb, one finds $\ell \sim 2$ fm. At low energies, scattering is inhibited by Pauli blocking effects, which increase ℓ by a factor of order $(T_F/T)^2$, where T_F is the Fermi temperature. Also, if one takes into account the non-locality of the optical potential,[12,13] one finds mean free paths that are typically as much as twice our crude estimate. It is therefore clear that the estimate of 2 fm for a mean free path is on the low side, and will tend to give unrealistically long times for thermal conduction. Inserting these estimates, one finds $D_T \sim 0.2$ fm c, and

$$\frac{\tau_T}{\tau_s} \lesssim \frac{R}{6\mathrm{fm}} \, . \qquad \text{(nuclei)} \tag{11}$$

Thus in nuclear disassembly, one would expect the matter to be isothermal to a good approximation on all length scales less than the size of the largest nuclei. The fact that the total entropy is conserved to a good approximation in the early stages of expansion does not imply that small perturbations will develop with the entropy per particle conserved locally. They will be isothermal.

So far we have considered long-wavelength instabilities, and in finite drops of nuclear matter the conditions for instability will be modified by the finite size, and by the Coulomb interaction. In the spirit of a differential Thomas-Fermi or Landau-Ginzburg model we may write the free energy density of a small density fluctuation as

$$F = F_o + \frac{1}{2}\frac{1}{n}\frac{\partial P}{\partial n}\,\delta n^2 + \frac{1}{2}\,\beta(\nabla n)^2 + F_{\mathrm{Coulomb}}, \tag{12}$$

where n is the particle number density, which varies in space, and β is a coefficient which determines, among other things, the surface energy. (For simplicity we assume that neutrons and protons are locked together, so that the proton fraction remains fixed. It is straightforward to relax this condition at the expense of slightly longer equations.) The dispersion relation for an infinite medium is[†]

$$\omega^2 = \omega_p^2 + c_s^2 q^2 + \frac{\beta n q^4}{m} , \tag{13}$$

where $\omega_p = (4\pi n x^2 e^2/m)^{1/2}$ is the plasma frequency, with x the proton fraction. c_s is the sound velocity, which depends on the degree to which the fluctuations are adiabatic or isothermal. In general it is given by

$$c_s^2 = c_s^{iso^2} + \frac{(c_s^{ad^2} - c_s^{iso^2})}{1 + \frac{i}{\omega\tau_T}\frac{C_p}{C_V}} , \tag{14}$$

where τ_T is given by Eq.(5). At frequencies high compared with $1/\tau_T$, c_s reduces to the adiabatic sound velocity, while at low frequencies it tends to the isothermal one.

Now let us estimate the magnitude of the terms in (13). For $n = 0.17\ \text{fm}^{-3}$ and symmetrical nuclear matter,

$$\omega_p \approx 3 \times 10^{-2}\ \text{fm c}^{-1} . \tag{15}$$

The gradient term is of order

$$\omega_g = \sqrt{\frac{\beta n}{m}}\, q^2$$

$$\approx 3 \times 10^{-2} \left(\frac{q}{0.5\ \text{fm}^{-1}}\right)^2 \text{fm}^{-1} c, \tag{16}$$

since $\beta \simeq 60$ MeV fm^5 for the Skyrme interaction fitted to the nuclear matter results of Ref. 8. For the lowest mode in a sphere of radius R, q is of order $\pi/R \sim 0.5\ \text{fm}^{-1}$ for $R \sim 6$ fm. This shows that on all

† Readers familiar with the discussion of the Jeans instability in astrophysics will realize that our treatment of the Coulomb term is somewhat cavalier, since the configuration about which we are making perturbations is not an equilibrium state. However, one knows from the gravitational problem that the simple treatment yields the right order of magnitude for the characteristic frequencies.

length scales less than the drop size, Coulomb effects will be less important than the β term. In Fig. 4, we show the line along which

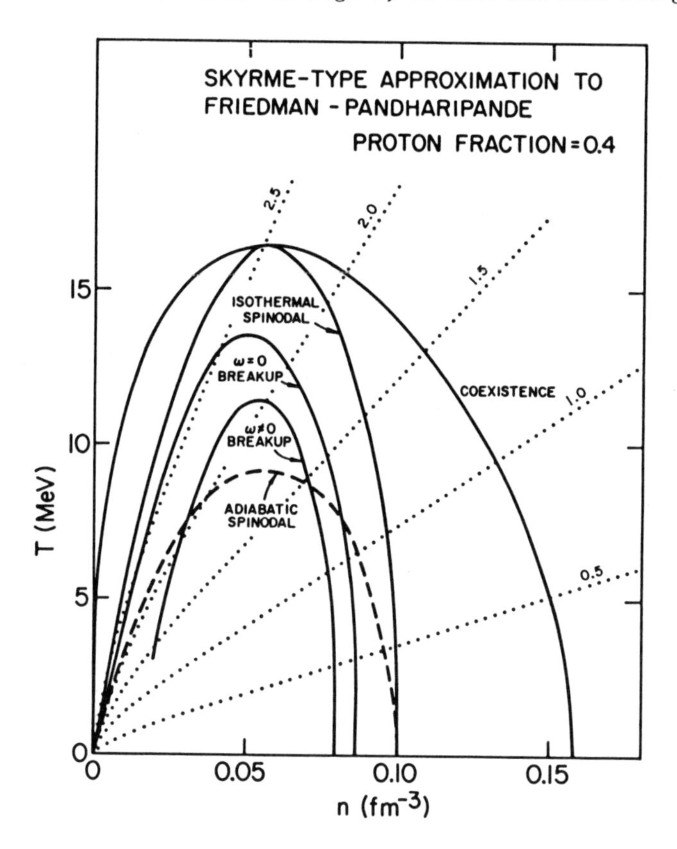

Fig. 4 Phase diagram and instability boundaries for nuclear matter with a proton fraction 0.4. On the line marked "$\omega = 0$ breakup" matter is unstable to a perturbation with $q = 5.76/R$, and $R = 6$fm, while on the line "$\omega \neq 0$ breakup", the growth rate of the mode is equal to $\tau_u^{-1} = c/(3R)$. The dotted lines are adiabats, labelled with the entropy per baryon.

$\omega^2 = 0$ for a q equal to $5.76/R$, with $R = 6$fm, when Coulomb effects are neglected. This wavenumber corresponds roughly to the value expected for a quadrupolar density fluctuation, which would lead to formation of two fragments. For fragmentation into more pieces, q is larger and the instability region in Fig.4 is smaller.

In our discussion we have not considered modes that change the shape of the drop. These surface modes will generally have components that couple to density fluctuations within the drop, and consequently they too will be affected by the liquid-gas phase transition. We remark in passing that, in the simulations, one sees the effects of the Coulomb interaction on surface modes: sausage-shaped drops return to a spherical shape in the absence of Coulomb forces, but may undergo fission when the Coulomb interaction is included.

Another effect we have not included is viscous dissipation, which has the effect of reducing growth rates of instabilities. For longitudinal waves this may be taken into account by replacing c_s^2 in Eq. (13) by $c_s^2 - i\omega(\zeta + 4/3\ \eta)/(mn)$, where ζ is the bulk viscosity and η is the shear viscosity. The characteristic frequency associated with viscous dissipation is

$$\omega_v = \tau_v^{-1} = \frac{(\zeta + \frac{4}{3}\eta)}{mn} q^2, \tag{17}$$

which is of order the frequency for temperature inhomogeneities to be smoothed out, $\omega_T = \tau_T^{-1}$ (Eq. 5).

In our discussion above we have assumed that ordinary hydrodynamics may be applied. This is questionable when the mean free path becomes long compared with the length scales of a perturbation, and in such cases numerical estimates of growth rates based on the use of transport coefficients cannot be relied upon quantitatively. When, for a particular length scale L, estimates of τ_T and $\tau_v = \omega_v^{-1}$ based on transport coefficients become smaller than the time, τ_f, for a free particle to travel a distance L, these characteristic times are in fact of order $\tau_f = L/v$.

There are two further effects that limit the region of the phase diagram over which expanding droplets will fragment. The first is that, for fragmentation to occur, instabilities must have a growth time less than the time, τ_u, the system spends in the unstable region of the phase diagram, i.e.

$$\omega_q^2 < -\frac{1}{\tau_u^2} \, . \tag{18}$$

In Fig. 4 we show the curve along which $\omega_q^2 \tau_u^2 = -1$ for $q = 5.76/R$. τ_u is taken to be a typical sound travel time R/c_s, with $c_s = c/3$ and $R = 6F$. The second effect which may inhibit growth of instabilities is that the level of fluctuations in the initial drop may be so small that the fluctuations do not develop. While this may be a significant effect in simulations of droplet disassembly, it is much less important for collisions of droplets, where one would expect higher levels of fluctuations.

Our considerations suggest a number of directions for future work. The first is to investigate transport properties of matter used in simulations, and to make detailed hydrodynamic analyses of the results. The second is to compare these transport coefficients with those of nuclear matter. As we have seen, even though two sorts of matter may have similar critical temperatures, their transport coefficients may be very different. A third direction is to study the extent to which hydrodynamic ideas may be used to investigate processes under conditions when the mean free path is comparable to or larger than characteristic length scales, a very real possibility for nuclear matter if mean free paths are as large as 5 fm. One knows from studies of quantum liquids, such as the helium liquids, that hydrodynamic concepts are still valuable under conditions when mean free paths are long, because coherence between particles can be enforced by the molecular fields produced by the interactions between particles.

To summarize, our calculations strongly suggest that instabilities in nuclear matter will develop isothermally under the conditions for which one expects fragmentation to occur. Instabilities are suppressed significantly by the finite size of nuclei, and they will be unimportant if their growth times are less than the time matter spends in the unstable region of the phase diagram. These effects lead to the conclusion that in heavy ion reactions one will not see significant effects associated with the critical point.

We should like to express our thanks to Gordon Baym for stimulating us to think about the problems discussed here, and for many valuable discussions. In addition we are grateful to David Boal and Vijay Pandharipande for useful comments, and to Vijay Pandharipande for generously making available to us the results of the simulations he and his collaborators carried out. This work was supported by the National Science Foundation under grants DMR82-15128, PHY84-15064 and PHY86-00377.

REFERENCES

1. Vincentini, A., Jacucci, G., and Pandharipande, V. R., Phys. Rev. C 31, 1783 (1985).
2. Lenk, R. and Pandharipande, V. R., Phys. Rev. C (in press).
3. Schlagel, T. J. and Pandharipande, V. R., to be published.
4. Boal, D. and Beauvais, G. E., Illinois preprint P86/2/26.
5. Aichelin, J. and Bertsch, G., Phys. Rev. C31, 1730 (1985).
6. Knoll, J. and Strack, B., Phys. Lett. 149B, 45 (1984).
7. Ravenhall, D. G., unpublished.
8. Friedman, B. and Pandharipande, V. R., Nucl. Phys. A361, 502 (1981).
9. Bertsch, G. and Siemens, P. J., Phys. Lett. 126B, 9 (1983).
10. Bertsch, G., Nucl. Phys. A400, 221 (1985).
11. Lopez, J. A. and Siemens, P. J., Nucl. Phys. A431, 728 (1984).
12. Negele, J. W. and Yazaki, K., Phys. Rev. Lett. 47, 71 (1981).
13. Fantoni, S., Friman, B. L., and Pandharipande, V. R., Phys. Lett. 104B, 89 (1981).

DO INTERMEDIATE MASS FRAGMENTS SHOW EVIDENCE FOR A LIQUID-GAS PHASE TRANSITION?

R. Trockel, K.D. Hildenbrand, U. Lynen, W.F.J. Müller, H.J. Rabe,
H. Sann, H. Stelzer, W. Trautmann, R. Wada
GSI, 6100 Darmstadt, Germany
N. Brummund, R. Glasow, K.H. Kampert, R. Santo
University of Münster, 4400 Münster, Germany
D. Pelte, J. Pochodzalla
University of Heidelberg and MPI Heidelberg, 6900 Heidelberg, Germany
E. Eckert
University of Frankfurt, 6000 Frankfurt, Germany

ABSTRACT

Inclusive cross sections for the production of intermediate mass fragments have been measured for several heavy-ion reactions in the energy range E/A=30 MeV to 84 MeV. The systematics of the temperature and velocity parameters, derived from a moving source parameterization of the differential cross sections, and the total yield ratios for neighbouring isotopes indicate that preequilibrium light particles and intermediate mass fragments originate from different mechanisms. The results do not support a thermodynamic interpretation on the basis of a liquid-gas phase transition of nuclear matter.

1. INTRODUCTION

The mechanism responsible for the production of intermediate mass fragments (IMF, $Z \geq 3$) in heavy-ion reactions at energies $E/A > 30$ MeV represents a topic of high current interest. As a fascinating possibility it has been suggested that IMF may result from the passage of excited nuclear matter into the region of liquid-gas coexistence where droplet formation due to dynamical instabilities or breakup near the critical temperature may lead to large probabilities for complex particle production. The conditions as well as the consequences of a liquid-gas phase transition have been widely discussed in the literature [1] and have been covered by several speakers at this conference [2]. The question to be discussed in this talk is whether the characteristics of IMF production in intermediate energy heavy-ion reactions support such an interpretation in thermodynamical terms. The discussion is based on inclusive data measured by the GSI-Frankfurt-Heidelberg-Münster collaboration in experiments with heavy-ion beams from the CERN SC and the SARA facility in Grenoble.

2. EXPERIMENTAL

The majority of the experiments was performed with heavy-ion beams from the CERN synchrocyclotron. Targets of ^{58,64}Ni, ^{nat}Ag and ^{197}Au were bombarded with ^{12}C and ^{20}Ne projectiles of $E/A = 48$ MeV and ^{12}C and ^{18}O projectiles of $E/A = 84$ MeV. More recently, additional data from the bombardment of various targets with a $E/A = 30$ MeV ^{40}Ar beam from the SARA facility at Grenoble were obtained.

Light particles and intermediate mass fragments were detected and identified by five-element telescopes which consisted of an axial-field ionization chamber followed by three silicon detectors of thickness 50 μm, 300 μm, and 1000 μm, respectively, and backed by a 1 cm diameter x 1 cm BGO detector. The dynamic range of the telescopes extended from $E/A \sim 1$ MeV (energy necessary to reach the 50 μm detector) to $E/A \leq 70$ MeV (energy of protons and α particles stopped in the BGO detector). Full isotope separation for ions up to carbon is achieved at $E/A > 3.5$ MeV at which energy particles pass through the 50 μm detector (Fig. 1).

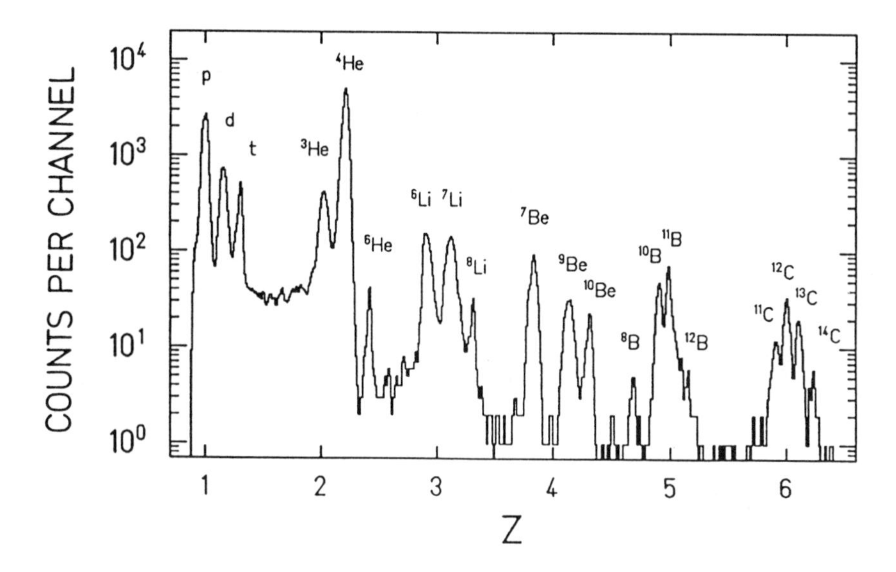

Fig. 1: Linearized ΔE spectrum measured with the 50 μm detector in the reaction $^{18}O + {}^{nat}Ag$ at E/A = 84 MeV at $\vartheta_{lab} = 41°$.

3. THERMAL-SOURCE ANALYSIS

Examples of spectra of beryllium ions measured at various angles between $\vartheta_{lab} = 40°$ and 120° are given in Fig. 2. The range of angles is outside the forward region where projectile fragments from quasielastic interactions dominate the spectra. The increase of the differential cross section towards smaller angles and the approximately exponential decrease with energy are characteristic for reactions at intermediate energies. At small energies the intensities decrease as a result of the Coulomb repulsion from the emitting nuclei.

The fits to the spectra shown in Fig. 2 were obtained with a parameterization that assumes emission from an equilibrated source at temperature T moving in beam direction with velocity v. The emission spectra in the moving frame are assumed to be proportional to $\sqrt{E}\ \exp(-E/T)$. The yields are then transformed into the laboratory frame where a Coulomb energy randomly distributed in the range of 0.5 to 1.5 times V_C is added. V_C is a fit parameter and turns out to be about one half of the Coulomb

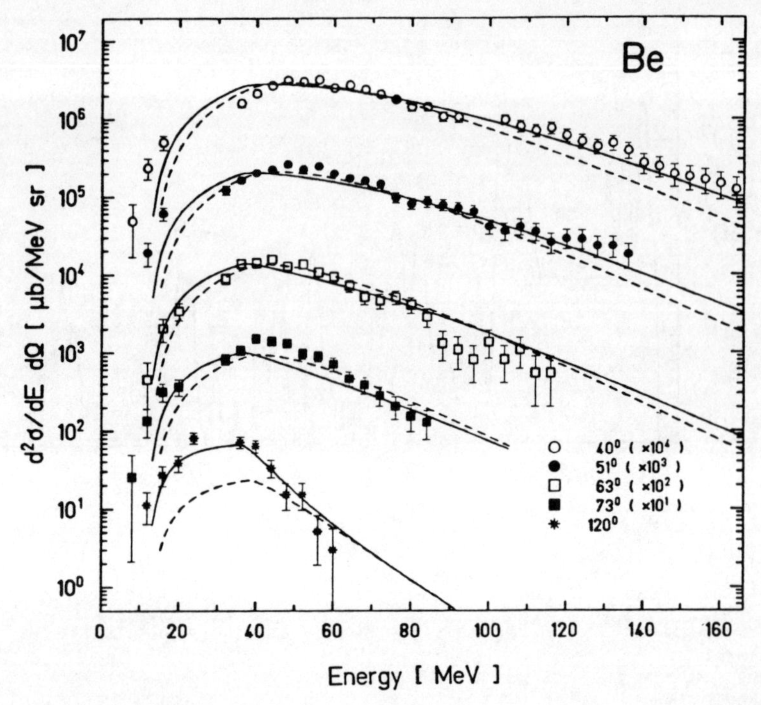

Fig. 2: Energy spectra of beryllium ions from the reaction ^{18}O + ^{197}Au at E/A = 84 MeV , measured at angles ϑ_{lab} = 40°, 51°, 73°, and 120°. The dashed and full lines give the results of thermal-source fits to the data, assuming one and two independent sources, respectively. Note the changes in the ordinate scale by factors of 10.

repulsion between the fragment and the target nucleus if taken as two touching spheres.

To obtain a satisfactory description in the whole range of θ_{lab} = 40° to 120° it was found to be necessary to introduce a second source that moves with rather slow velocity and has a lower temperature (T = 6 MeV, v = 0.015c). This source mainly contributes at backward angles. Since its intensity is only about 10 % of that of the intermediate source the addition or omission of this slow source has a very small influence on the parameters derived for the intermediate velocity source; the extracted temperatures change by at most 1 MeV.

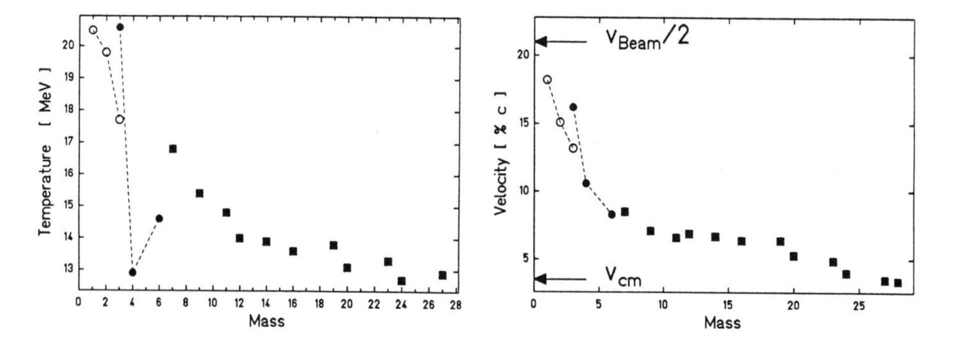

Fig. 3: Temperature and velocity parameters of the moving-source parameterization for the reaction ^{18}O on ^{197}Au at E/A = 84 MeV. Hydrogen and helium isotopes and IMF are distinguished by different symbols.

4. SYSTEMATICS OF SOURCE PARAMETERS

The fit parameters temperature T and velocity v that were derived from the thermal source analysis depend considerably on the mass or atomic number of the detected ions. This is demonstrated in Fig. 3 in which the results for the reaction $^{18}O + ^{197}Au$ at E/A = 84 MeV are shown. The temperatures for light particles of mass $A \leq 3$ are of the order T = 20 MeV. They drop to values near T = 14 MeV for mass A = 5 and decrease continuously with increasing mass. The discontinuity at A = 4 is most likely due to contributions from sequential alpha decay of excited particle unstable fragments of larger mass. In the case of small decay Q values the apparent temperature of the decay products is considerably smaller than that of the primary fragment.

The source velocities are largest for protons. They are equal to about one half of the beam velocity, i.e. equal to the velocity of the nucleon-nucleon center-of-mass system. The source velocities decrease with increasing mass and, for large fragments, level off at approximately 5 % of the velocity of light (Fig. 3).

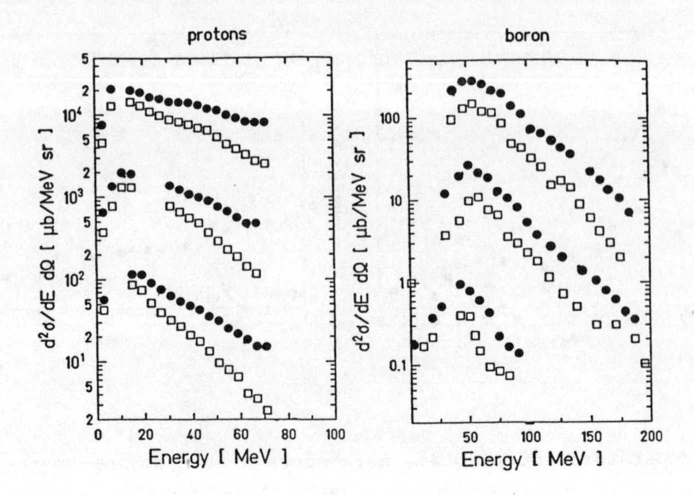

Fig. 4: Energy spectra of protons and boron ions, measured at three angles, from the bombardment of a ^{197}Au target with beams of ^{12}C with E/A = 48 MeV (open squares) and of ^{18}O with E/A = 84 MeV (full points).

The parameter variation shown in Fig. 3 is typical for the range of energies and reactions studied in the CERN SC experiments. It is clear from the figure that light particles and IMF cannot come from the breakup of the same equilibrated nuclear subsystem. At this point, the results could still be considered as consistent with a picture in which larger fragments are emitted at later times when the source has cooled and slowed down, e.g. due to accretion of nuclei from the target. We note, however, that a reduction of the velocity by a factor of four, corresponding to an increase in mass by a factor of four, should lead to a decrease in temperature of at least a factor of two which is not observed.

The results obtained for two reactions with different projectile mass and velocity are compared in Figs. 4 and 5. The main difference is already apparent in the spectra (Fig. 4). Whereas the exponential slopes of the proton spectra in the angular range dominated by preequilibrium emission change as expected in a thermal model this is not the case for

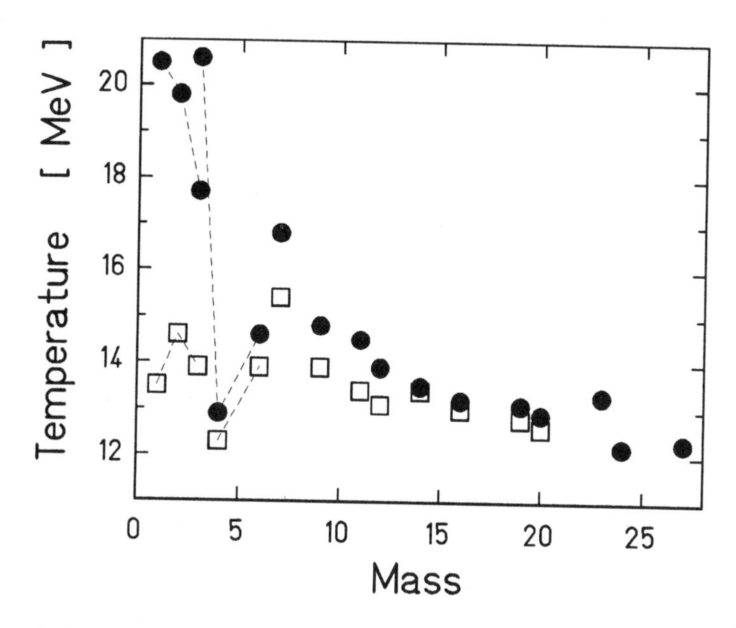

Fig. 5: Temperature parameters derived for the two reactions ^{12}C (E/A = 48 MeV, open squares) and ^{18}O (E/A = 84 MeV, full points) on Au.

the boron ions. The apparent temperature is not influenced by the change in projectile energy. The yields, however, are considerably larger at the higher energy.

The apparent temperatures as a function of ejectile mass are shown in Fig. 5. A thermal interpretation is limited to masses $A \leq 3$ for which the temperatures vary with bombarding energies in an expected fashion. For IMF the differences are negligible. The smooth decrease of the temperature for mass $A \geq 5$ has been interpreted as result of a recoil effect. In this case the apparent temperature T of a fragment with mass A is equal to $T = T_o(1 - A/A_s)$ where T_o and A_s are the intrinsic temperature and the mass number of the emitting source. T_o can be obtained from an extrapolation of T to mass A = 0. It turns out that T_o is about the same not only for the two cases shown in Fig. 5 but for the whole range of reactions studied. All values fall in the range $T_o \sim 15$ MeV to 17 MeV, including the value obtained for the reaction ^{40}Ar + Au at E/A = 30 MeV (Table 1). The apparent temperatures of this latter reaction fully

TABLE 1

Temperature parameter T_o, exponent τ of power law fit to the yield curves, and production cross section σ_{prod} (summed over $Z \geq 3$) for the investigated reactions.

System	E_{lab} [GeV]	T_0 [MeV]	τ	σ_{prod} [mb]
^{18}O + Au	1.51	15.4 ± 1.5	2.19 ± 0.15	580 ± 120
^{18}O + Ag	1.51	17.4 ± 1.5	2.29 ± 0.15	520 ± 100
^{12}C + Au	1.08	15.1 ± 2.0	2.71 ± 0.20	260 ± 50
^{12}C + Ag	1.08	14.8 ± 2.0	2.46 ± 0.20	275 ± 60
^{20}Ne + Au	0.96	14.5 ± 1.5	2.41 ± 0.15	632 ± 130
^{20}Ne + Ag	0.96	16.0 ± 1.5	2.20 ± 0.15	651 ± 130
^{12}C + Au	0.58	15.0 ± 2.0	3.30 ± 0.20	190 ± 40
^{12}C + Ag	0.58	14.2 ± 2.0	2.90 ± 0.20	207 ± 40
^{40}Ar + Th	1.01	15.2 ± 2.0	1.72 ± 0.15	1770 ± 80
^{40}Ar + Au	0.98	15.8 ± 2.0	1.72 ± 0.15	1730 ± 80

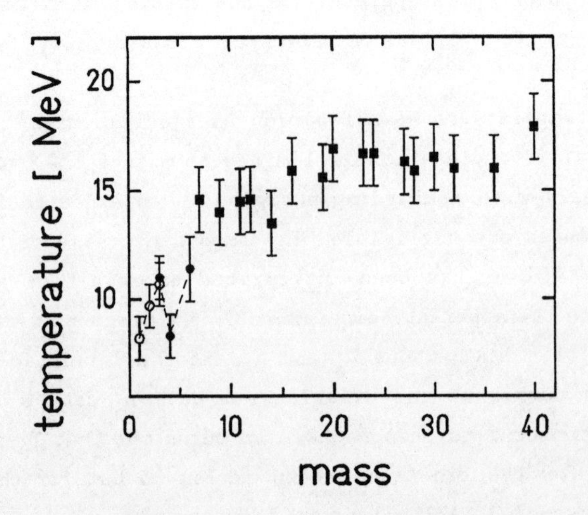

Fig. 6: Temperature parameters as a function of fragment mass for the reaction $^{40}Ar + ^{197}Au$ at E/A = 30 MeV.

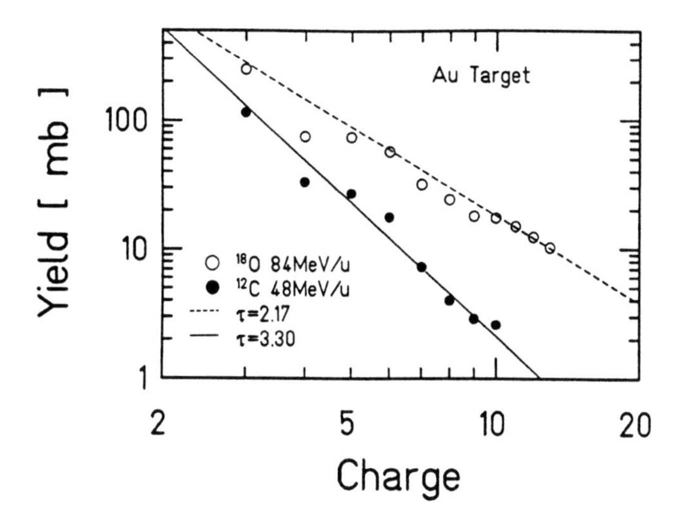

Fig. 7: Total element yields for the two reactions ^{12}C (E/A = 48 MeV) and ^{18}O (E/A = 84 MeV) on ^{197}Au. The full and dashed lines are the results of power law fits to the yields which gave τ = 3.30 and 2.17 for the two reactions.

rule out the possibility of a thermodynamic origin of the temperature parameters. As shown in Fig. 6, T is larger for IMF than for light particles and has values that, just on the basis of energy and momentum conservation, cannot be reached in a reaction at E/A = 30 MeV.

To conclude this section, we note that the invariance of the apparent temperature supports models in which this parameter is related to e.g. the Fermi momentum k_f as in the model of Aichelin, Hüfner and Ibarra [3] in which $T_o = k_f^2/5\,m$ = 12.5 MeV (m is the nucleon mass). Other aspects of the data are also consistent with this model of cold spallation which, however, will not be explicitly demonstrated in this talk.

5. YIELD SYSTEMATICS

Total cross sections for the production of specific elements or isotopes were obtained by integrating over 4π the intensity of the thermal source whose parameters were determined by the fit. Fig. 7 gives the total element yields for the two reactions ^{18}O (E/A = 84 MeV) and ^{12}C (E/A = 48 MeV) on ^{1}[illegible]Au that were compared in the preceeding section (Figs. 4 and 5). The measured yields decrease continuously with increasing atomic number Z and can be fitted with a power law $\sigma(Z) \propto Z^{-\tau}$. The parameter τ is found to be smaller at the higher energy where the yields are larger. Inspection of Table 1, which lists τ and the summed production cross sections for elements with $Z \geq 3$, shows that this correlation holds for all of the investigated cases.

A pure power law dependence has been predicted for disintegration of the expanding equilibrated nuclear matter at the critical temperature [1]. According to some approaches the parameter τ should reach a value of 2.3 under these conditions [4]. We find that τ actually assumes values close to that in some of the investigated cases. It seems to be doubtful, however, that this behaviour should be interpreted as a sign of a liquid-gas phase transition. The main reason is that a power law dependence is predicted by several other models as well, such as the cold spallation [3] and percolation [5] models and therefore has limited significance in itself as long as the predicted correlation with a source temperature is not established.

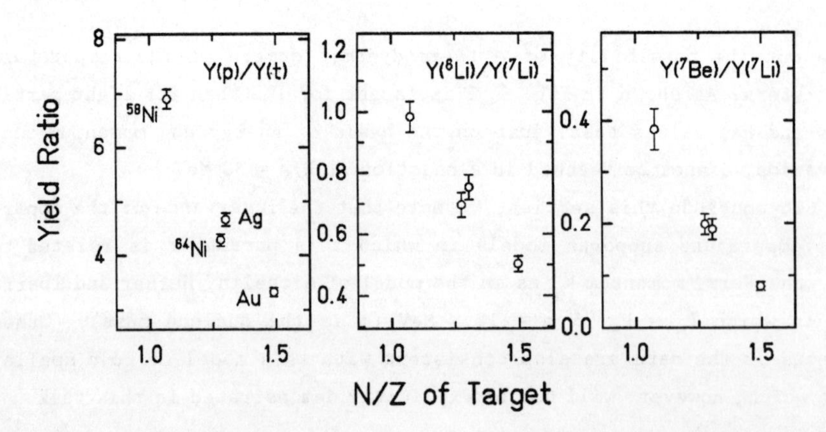

Fig. 8: Ratios of total yields of protons versus tritons (left), ^{6}Li versus ^{7}Li (middle), and ^{7}Be versus ^{7}Li (right) for ^{18}O (84 MeV/A) induced reactions on various targets (see left panel) as a function of the neutron-to-proton ratio of the target nucleus.

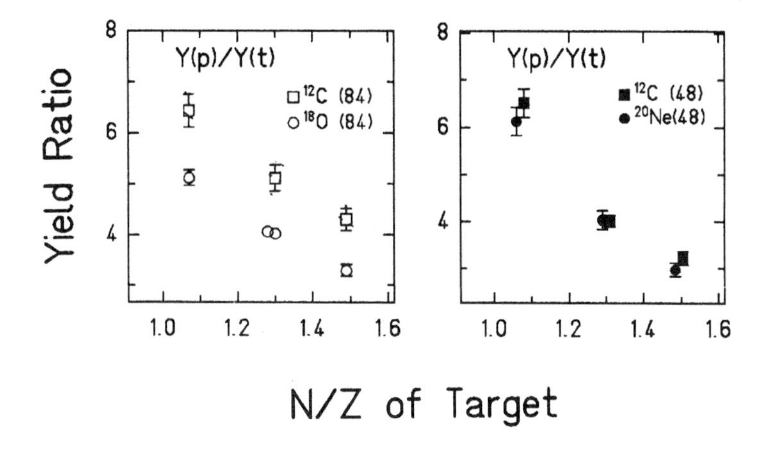

Fig. 9: Proton-to-triton yield ratios for the bombardment of various targets (see Fig. 8) with ^{12}C (open squares) and ^{18}O (open circles) of E/A = 84 MeV (left panel) and with ^{12}C (full squares) and ^{20}Ne (full circles) of E/A = 48 MeV (right panel).

6. ISOTOPE RATIOS

The cross section ratios for the production of different isotopes of the same element or of neighbouring isobars were found to depend strongly on the reaction. Fig. 8 shows yield ratios for hydrogen and lithium isotopes and for mass A = 7 isobars from ^{18}O (E/A = 84 MeV) induced reactions on several targets, plotted as a function of the neutron-to-proton ratio N/Z of the target. The ratios which indicate the probability for producing neutron poor over neutron rich isotopes decrease dramatically with increasing N/Z of the target. The significance of N/Z as the relevant parameter is emphasized by the fact that the ratios for the ^{64}Ni and ^{nat}Ag targets which have the same N/Z turn out to be identical within the errors.

The yield ratios of hydrogen isotopes also depend on the neutron-to-proton ratio of the projectile as is demonstrated in Fig. 9. The proton over triton ratios are larger for the ^{12}C induced than for the ^{18}O induced reaction whereas they are the same for ^{12}C and ^{20}Ne induced

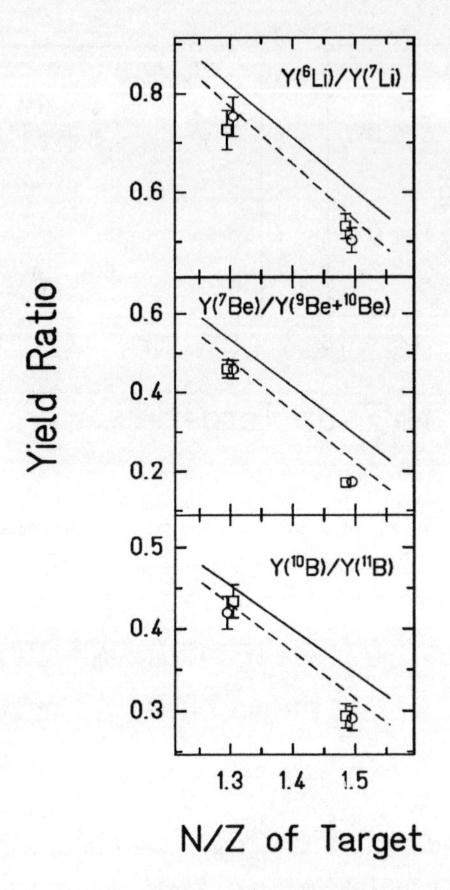

Fig.10: Isotope ratios from the bombardment of Ag (N/Z ~ 1.3) and Au (N/Z ~ 1.5) targets with ^{12}C (open squares) and ^{18}O (open circles) beams of E/A = 84 MeV. The full and dashed lines indicate the displacement of the ratios obtained with the two projectiles that would be expected if the N/Z ratio of a source consisting of four times as many nucleons from the target as from the projectile were the relevant order parameter.

reactions on a given target. The values for the ^{18}O and ^{12}C induced reactions fall on the same line, however, if they are plotted as a function of the neutron-to-proton ratio of a source that consists of equal numbers of projectile and target nucleons chosen according to the N/Z ratios of the two nuclei. This result indicates that equal parts from the projectile and target nuclei are involved in the emission of preequilibrium light particles which is consistent with the temperature and velocity parameters obtained from the moving source analysis. It does not necessarily require the formation of a thermally equilibrated subsystem of nuclear matter, however, since fast nucleons which are emitted after their first nucleon-nucleon collision would exhibit similar characteristics.

The yield ratios of IMF do not depend on the N/Z ratio of the projectile. Fig. 10 shows that, in contrast to the observation for light particles, the lithium, beryllium, and boron isotope ratios are identical for ^{18}O and ^{12}C induced reactions and only depend on the N/Z ratios of the target or, which here is indistinguishable, of the compound nuclei. The full and dashed lines given in Fig. 10 mark the expected variation of the isotope ratios if the N/Z ratio of a source consisting of four times as many nucleons from the target as from the projectile were the relevant parameter (the factor of 4 being suggested by the derived source velocity, see Fig. 3). This variation is clearly outside the range consistent with the experiment. The observed behaviour rather suggests that the whole target nucleus takes part in the IMF production. This would be the case if, e.g., emission from a hot compound nucleus [6], after a deep-inelastic collision involving N/Z equilibration of the dinuclear system [7], or as the result of cold breakup of the target spectator matter [3] were the dominating process.

7. SUMMARY

In this talk results from a systematic investigation of intermediate mass fragment production in intermediate energy heavy-ion collisions were presented. The discussion included the temperature and velocity parameters derived from a thermal-source analysis of inclusive differential cross sections, of total production yields by the mechanism that is identified by the thermal-source parameterization and of ratios for the production of neighbouring isotopes.

It was found that preequilibrium light particles and IMF result from different mechanisms. Whereas the characteristics of light particle production were found consistent with emission from a localized and equilibrated excited source of nuclear matter, originating in equal parts from the projectile and the target, this was not the case for the IMF production. The heavier fragments rather come from a mechanism that involves the major part of the target, such as a compound, a deep-inelastic or a cold spallation mechanism. In particular, it was demonstrated that the temperature parameters derived from the moving-source analysis of the IMF cross sections do not exhibit the features expected for thermodynamic temperatures of an equilibrated subsystem of nuclear matter. It also was pointed out that the characteristics of light-particle production are equally well explained by assuming that they result from the hard nucleon-nucleon collisions during the initial stage of the reaction.

REFERENCES

1. Siemens, P.J., Nature 305, 410 (1983);
   Curtin, M.W., et al., Phys.Lett. 123B, 289 (1983);
   Panagiotou, A.D., et al., Phys.Rev.Lett. 52, 496 (1984);
   Boal, D.H., Nucl.Phys. A447, 479c (1986).
2. see contributions to this workshop by Scott, D., Pethick, C., Grant, C., Strack, B., Gelbke, C.K., Koonin, S., Aichelin, J., and Boal, D.
3. Aichelin, J., et al., Phys.Rev. C30, 107 (1984).
4. Goodman, A., et al., Phys.Rev. C30, 851 (1984).
5. Campi, X., et al., Nucl.Phys. A428, 327c (1984);
   Bauer, W., et al., Nucl.Phys. A, in press.
6. Charity, R.J., et al., Phys.Rev.Lett. 56, 1354 (1986).
7. see, e.g., Gobbi, A., and Nörenberg, W., in Heavy Ion Collisions, ed. by R. Bock (North-Holland, Amsterdam, 1980), Vol. 2, p. 127.

DYNAMICS OF DROPLET FORMATION IN NUCLEAR MATTER*

Cheryl Grant

School of Physics and Astronomy
University of Minnesota
116 Church Street S.E.
Minneapolis, MN 55455

ABSTRACT

A system of coupled rate equations which predict the time evolution of the droplet distribution in dilute nuclear matter as it approaches equilibrium is proposed. The coupled equations are solved numerically giving information on the time scale involved at different temperatures.

---

*Invited paper presented at the "Second International Workshop on Local Equilibrium in Strong Interaction Physics", Sante Fe, New Mexico, April 1986.

1. INTRODUCTION

In the study of medium energy heavy-ion reactions it is important to consider the possible effects of the nuclear liquid-gas phase transition. Up to the present time, however, very few nuclear physicists have been concerned with the dynamics involved in going from one phase to the other. The problem of estimating how fast a nuclear system approaches equilibrium seems a relevant one, and one that I will address in this paper.

2. MOTIVATION

If a nuclear system, by means of a central collision of heavy nuclei, has attained an intermediate state of high temperature and density, by necessity it cools and expands. It will eventually cross over the phase transition boundary separating stable uniform nuclear matter from the metastable regions. See Fig. 1.

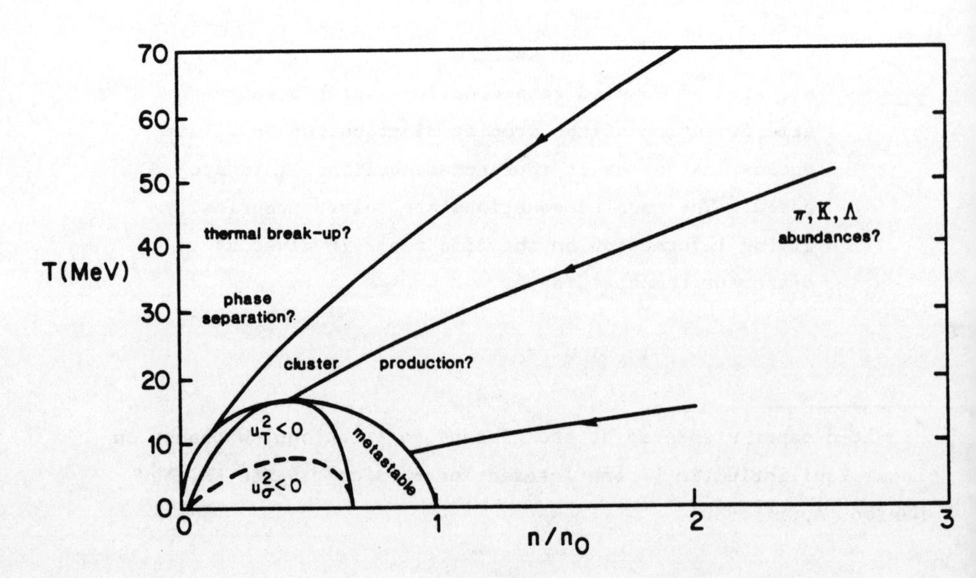

Figure 1.

The phase transition boundary is given approximately by a semi-circular parametrization:

$$\left(\frac{T}{T_c}\right)^2 + \left(\frac{n}{n_c}-1\right)^2 = 1, \tag{1}$$

where T_c is the critical temperature (taken to be 16 MeV), and n_c is the critical density (taken to be one-half normal nuclear density n_o). If the boundary is approached from the high density region, $n>n_c$, gas bubbles could form in the liquid phase, and if the boundary is approached from the low density region, $n<n_c$, then liquid droplets could form in the gas phase. Only one phase of nuclear matter can exist above the Maxwell curve (gas for $n<n_c$, and liquid for $n>n_c$), but inside the metastable regions, the two phases can coexist.

The problem involves creating a metastable system, initially formed as a supersaturated vapor with $n_G<n<n_c$ and with fixed temperature $T<T_c$, and watching its evolution towards equilibrium. (If one chooses to consider the evolution of the system at a fixed temperature, energy will not be conserved, which is unphysical for heavy-ion collisions, but the calculations will be simplified.) As the system evolves droplets of the liquid phase will form and, given infinite reaction time, an equilibrium distribution for the various size droplets should be attained. To determine this equilibrium distribution we employ the following droplet model scenario.[1)]

3. DROPLET MODEL

Consider a fixed volume of gas containing A+B particles (nucleons). If, by a chance fluctuation, an A-particle droplet (a droplet consisting of A nucleons) is formed within the volume, the system will suffer a change in free energy. It is known that the probability for a fluctuation to occur is given by $P \propto \exp[-\Delta \text{ free energy}/T]$, so that the yield, or density, of droplets of size A can be written as:

$$Y(A) \propto \exp\,[-\Delta G(A)/T]. \tag{2}$$

Here the Gibbs free energy, G, is taken to be the relevant free energy involved in droplet formation. This is because the droplet and the gas are assumed to be in kinetic equilibrium; that is, they share the same temperature and pressure. Therefore, given that $\Delta G = G_{\text{with drop}} - G_{\text{no drop}}$, where

$$G_{\text{no drop}} = \mu_G\,(A+B) \tag{3}$$

$$G_{\text{with drop}} = \mu_G B + \mu_L A + 4\pi R_A^2 \sigma + T\tau \ell n A, \tag{4}$$

the yield of A-particle droplets can be written as:

$$Y(A) \propto A^{-\tau} \exp[(\frac{\mu_G - \mu_L}{T})\,A - \frac{4\pi r^2 \sigma}{T}\,A^{2/3}]. \tag{5}$$

Here μ_G and μ_L are the chemical potentials of the gas and liquid phases, $R_A = rA^{1/3}$ is the radius of the droplet, and $(n_L)^{-1} = \frac{4}{3}\pi r^3$ is the density of the liquid phase. The $\sigma(T)$ is the surface tension, and τ is Fisher's constant, taken to be 7/3 in the mean field approximation. The last term in eq. (4) is added because the droplet closes on itself reducing the total entropy associated with surface fluctuations.

Now, to obtain the equilibrium density for A-size droplets, we make the approximateion that the proportionality constant in eq. (5) is the equilibrium gas density, and assume that the reaction occurs along the phase boundary so that $\mu_G = \mu_L$. Adjusting the exponential to insure continuity at A=1, one can write an expression for the equilibrium concentration for droplets of size A.

$$Y^{EQ}(A) = Y^{EQ}(1)\;A^{-\tau}\;\exp\,[\frac{-4\pi r^2 \sigma}{T}\,(A-1)^{2/3}] \tag{6}$$

Eq. (6) gives the distribution that a metastable vapor phase should converge towards given sufficient reaction time. Now one must derive an expression which estimates the rate at which this evolution proceeds.[2-4]

4. RATE EQUATION

4.1 Gain and Loss Terms

The time rate of change of A-particle droplet concentrations can simply be approximated by

$$\frac{dY(A)}{dt} = G_A - L_A. \tag{7}$$

Here G_A defines the rate of gain of A-particle droplets, and L_A the rate of loss of A-particle droplets. To clarify, illustrations will be made for the G_A terms.

To gain an A-size droplet consider the following simple reaction processes. There could be "I" processes in which two pre-existing species X and Z react to produce a species A: (X+Z→A+W). Likewise there could be "J" processes where three species X', Z', and K' interact to form a species A: (X'+Z'+K'→A+W'). One must also consider possible reactions where a species "X" spontaneously decays to produce a species A: (X"→A+W"). There could be "L" such decays. G_A would then be the sum of the rates for all such processes.

$$G_A = \sum_{i=1}^{I} [\text{Rate } (X+Z \to A+W)]_i + \sum_{j=1}^{J} [\text{Rate } (X'+Z'+K' \to A+W']_j$$

$$+ \sum_{\ell=1}^{L} [\text{Rate } (X'' \to A+W'')]_\ell \tag{8}$$

To simplify Eq. (8), suppose that for each process the rate is a product of a rate coefficient and the concentration of the species involved in the reaction.

$$G_A = \sum_{i=1}^{I} [\text{Rate Coefficient } (X+Z\rightarrow A+W)]_i [Y(X)Y(Z)]_i$$

$$+ \sum_{j=1}^{J} [\text{Rate Coefficient } (X'+Z'+K'\rightarrow A+W')]_j \; [Y(X')Y(Z')Y(K')]_j$$

$$+ \sum_{\ell=1}^{L} [\text{Rate Coefficient } (X''\rightarrow A+W'')]_\ell \; [Y(X'')]_\ell \qquad (9)$$

One must then determine what these rate coefficients are.

If one assumption is made, that there exists no cluster-cluster reactions, only gas-gas reactions or gas-cluster reactions, then constraints are imposed on the products and the reacting species in eq. (9). That is to say, droplets may grow or shrink only by reactions which involve the addition to or the subtraction of single nucleons. Given this assumption there are four possible processes by which an A-size droplet can be gained:

Condensation	$1+ (A-1) \rightarrow A$	$C_{A-1,1}$
Decay	$(1+ A) \rightarrow A+1$	$D_{A+1,1}$
Formation	$1+ 1+ (A-1) \rightarrow A+1$	$F_{A-1,1,1}$
Break-up	$1+ (A+1) \rightarrow A+1+1$	$B_{A+1,1}$.

A complete expression for G_A can now be written:

$$G_A = D_{A+1,1}\, Y(A+1) + C_{A-1,1}\, Y(A-1)\, Y(1)$$

$$+ B_{A+1,1}\, Y(A+1)\, Y(1) + F_{A-1,1,1}\, Y^2(1)\, Y(A-1). \qquad (10)$$

Likewise similar arguments can be used to obtain an expression for the loss term:

$$L_A = D_{A,1}\, Y(A) + C_{A,1}\, Y(A)Y(1) + B_{A,1}\, Y(A)\, Y(1)$$

$$+ F_{A,1,1}\, Y^2(1)Y(A). \tag{11}$$

The rate coefficients C, D, B, and F are not independent, however. Consider three physical relationships which can be used to define them: detailed balance, collision cross sections, and particle conservation.

4.2 Rate Coefficients

The principle of detailed balance requires that a steady state must exist at equilibrium: $\dot{Y}(A) = 0$ or $G_A^{EQ} = L_A^{EQ}$. For illustration consider the following reaction process:

$$1 + (A\text{-}1) \overset{EQ}{\rightleftarrows} A$$

or

$$C_{A-1,1}\, Y^{EQ}(A\text{-}1)\, Y^{EQ}(1) = D_{A,1}\, Y^{EQ}(A). \tag{12}$$

Eq. (12) imposes constraints on C and D, and equations resulting from the three other reaction processes likewise constrain F and B.

The second limitation on the rate coefficients is that of the collision cross section. Consider a typical break-up reaction $1+A \rightarrow 1+1+(A\text{-}1)$ where a fast gas particle collides with an A-size droplet breaking off a piece. If one equates the break-up coefficient $B_{A,1}$ with a thermal-averaged break-up cross section $\overline{\sigma_{A,1\ \text{Break up}}}$ where $\overline{\sigma_{A,1\ \text{Break-up}}} = \langle\sigma_{A,1\ \text{B-up}} \cdot \upsilon\rangle$ and

$$\sigma_{A,1\ \text{B-up}} = \pi R_A^2\,(1 - e^{-m\upsilon^2/2T}) \tag{13}$$

one obtains a nice form for $B_{A,1}$.

Eq. (13) is used as the relevant parametrization for $\sigma_{A,1\ \text{B-up}}$ since it illustrates the expected dependence of the cross section on a geometric term πR_A^2, and on an energy-dependent term $(1-e^{-m\upsilon^2/2T})$ which increases the likelihood of break-up as the relative velocity increases. If the particles are given a Maxwell-Boltzmann energy distribution one obtains an expression for $B_{A,1}$:

$$B_{A,1} = 4\pi \left(\frac{m}{2\pi T}\right)^{3/2} \int \upsilon^3 e^{-m\upsilon^2/2T} \sigma(\upsilon)\, d\upsilon$$

which can be integrated to give

$$B_{A,1} = 3 \left(\frac{\pi T}{2m}\right)^{1/2} R_A^2. \tag{14}$$

Likewise, if one equates a thermal averaged condensation cross section, $\overline{\sigma_{A,1}}_{\text{condensation}}$, with the condensation rate coefficient for a reaction $1+A \rightarrow (A+1)$ where a low-energy gas particle collides and fuses with an A-size droplet, a nice form for $C_{A,1}$ can be obtained. Here we express $\sigma_{A,1\ \text{condensation}}$ as:

$$\sigma_{A,1\ \text{condensation}} = \pi R_A^2\, e^{-m\upsilon^2/2T} \tag{15}$$

so that the sum of the break-up and condensation cross sections gives the standard geometric cross section πR_A^2. $C_{A,1}$ integrates in a similar manner to $B_{A,1}$ to obtain:

$$C_{A,1} = \left(\frac{\pi T}{2m}\right)^{1/2} R_A^2. \tag{16}$$

The two other rate coefficients can now be related back to $C_{A,1}$ and $B_{A,1}$ through detailed balancing.

The third constraint on C, B, D, and F is that of particle conservation. This requiares that the baryon density must remain constant--no particles may be added or taken away at any step in the evolution of the system.

$$\sum_{A=1}^{N} A\,\dot{Y}(A) = 0 \tag{17}$$

Here N is the total number of gas particles in the system initially. This constraint essentially restricts the incremental expansion or reduction of the rate coefficients as droplet size is increased or decreased, yielding expressions for the rate coefficients for droplets of size other than A.

4.3 Full Rate Equation

Finally, employing all the tactics required to solve the above three relationships simultaneously, a full rate equation can be written for the change of A-particle droplet concentrations in time.

$$\begin{aligned}\dot{Y}_{(A)} = \left(\frac{\pi T}{2m}\right)^{1/2} r^2 \, [&Y^{EQ}(1)\,\{\exp\,[X[(A-1)^{2/3}-A^{2/3}]]\,A^{2/3-\tau}(A+1)^{\tau}Y(A+1)\\ &-\exp[X[(A-2)^{2/3}-(A-1)^{2/3}]](A-1)^{5/3-\tau}A^{\tau-1}Y(A)\}\\ &+3\,Y^2(1)/Y^{EQ}(1)\{\exp[X[(A-1)^{2/3}-(A-2)^{2/3}]]A^{2/3-\tau}\,(A-1)^{\tau}\,Y(A-1)\\ &-\exp[X[A^{2/3}-(A-1)^{2/3}]](A+1)^{5/3-\tau}A^{\tau-1}Y(A)\}-4Y(A)Y(1)A^{2/3}\\ &+Y(1)/A\{(A-1)^{5/3}Y(A-1) + 3(A+1)^{5/3}\,Y(A+1)\}\,]\end{aligned} \tag{18}$$

Here $X = \frac{-4\pi r^2\sigma}{T}$. For a volume V initially filled with N gas particles, there will exist N such equations, each coupled to the others through their dependence on droplet concentrations of one size larger and one size smaller. This system of equations was solved numerically using a FORTRAN program on the CRAY-1, with 10,000 incremental time steps of size 0.1 fm/c.

5. NUMERICAL RESULTS

Consider the following results at temperatures of 8, 12, and 15.5 MeV. (See Figs. 2-4.) The solid line in each figure shows the equilibrium configuration given by eq. (6). The three other lines show times of 10, 50, and 150 fm/c. Notice that for all three temperatures there is steady convergence to equilibrium as time progresses. Typically, reaction rates increase with temperature, so that the approach to equilibrium is more rapid as T increases. This is evident in the figures.

6. CONCLUSIONS AND EXPECTATIONS

One expects this model to simulate the time evolution of a nonequilibrium nuclear system as it approaches equilibrium. Given sufficient reaction time the equilibrium configuration is reached. Notice in Figs. 2-4 that the equilibrium distribution does describe the system for sufficiently long times. However, relevant times to consider for heavy-ion reactions are on the order of $10<t<50$ fm/c, and again from Figs. 2-4, one observes deviation from equilibrium during those times. It is therefore necessary to consider droplet distributions away from the equilibrium configuration, and therefore necessary to consider a rate equation which predicts what these distributions will be. One must recall, however, the assumptions made in solving this problem: constant temperature, $\mu_G = \mu_L$, and no cluster-cluster reactions. If one were to incorporate these effects into a more sophisticated equation, then perhaps equilibrium would be approached more quickly, and the distribution for times betweem $10<t<50$ fm/c might then show a configuration closer to equilibrium. Those refinements have yet to be made. Nevertheless, it is encouraging to see the results of the present model falling within the range of expectation, still leaving room for improvements, and more exciting results.

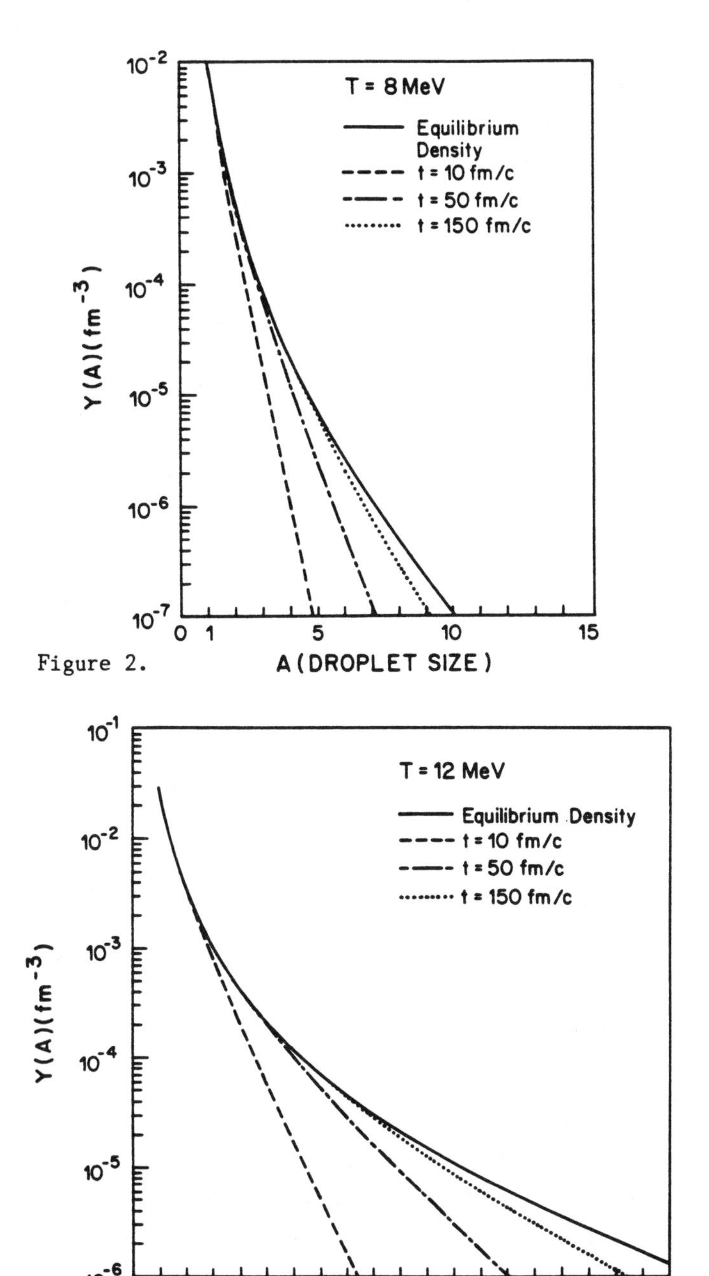

Figure 2.

Figure 3.

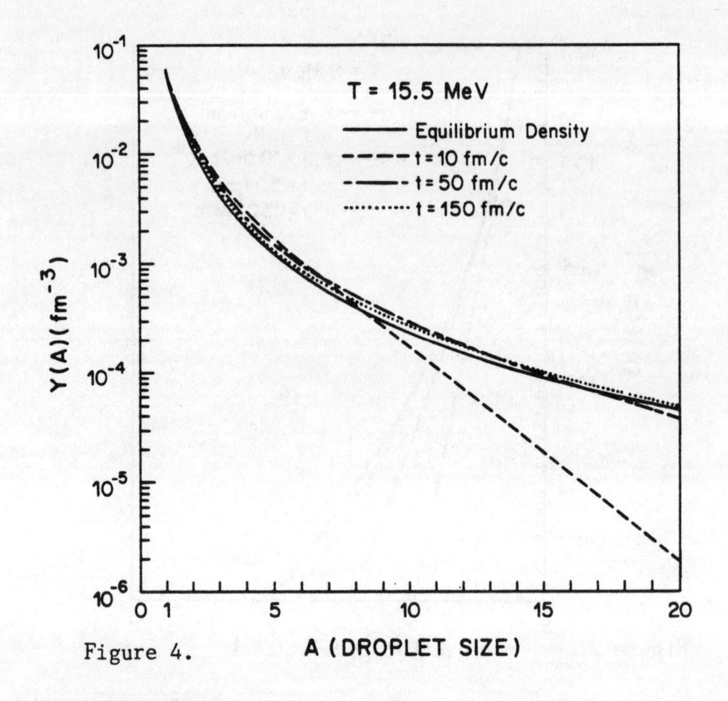

Figure 4.

ACKNOWLEDGEMENTS

The work presented here was done under the kind supervision of Joseph I. Kapusta. Special thanks also go to Laszlo P. Csernai, whose discussions and help with the computer code were greatly appreciated. This work was supported by the U.S. Department of Energy under contract DOE/DE-AC02-79ER-10364.

REFERENCES

1. For a recent review see: L.P. Csernai and J.I. Kapusta, Physics Reports 131, 225 (1986).
2. S.W. Koch, Lecture Notes in Physics, 207 (Springer-Verlag, 1984).
3. A.Z. Mekjian, Phys. Rev. C17, 1051 (1978).
4. D.H. Boal, Phys. Rev. C30, 749 (1984).

FRAGMENTATION OF HOT QUANTUM DROPS

Bruno Strack

Gesellschaft für Schwerionenforschung, Planckstr.1, Postf. 110541,
D-6100 Darmstadt 11, Federal Republic of Germany

Abstract

Static equation of state calculations are performed for infinite nuclear matter. Statistical and dynamical expansion calculations for finite and hot quantum systems lead to averaged evolution trajectories. Dynamicl calculations are performed and compared with static mean-field calculations. The break-up trajectories in the T,ρ plane fluctuate around the critical values of the EOS results. Small and large fragments are found near the critical temperature T_c. At critical temperatures fragments are found in U-shape spectra. Density distributions are studied and evidence for volume dependent multifragmentation processes is obtained for most break-up temperatures. Abcve the critical threshold fast and small fragments are counted. Mass spectra, multiplicities and momentum distributions are presented.

1. Introduction

Are there conditions leading to instabilities which themselves could be understood as a coexistence of nuclear vapor and fluid[1]? Are thermodynamical descriptions like the equation of state for infinite matter succesfully applicable to finite systems in energetic heavy ion reactions? In a recent paper[2], Vicentini et al. suggested the use of a classical model of molecular dynamics to study the fragmentation of small droplets. They showed that for hot liquid drops significant density inhomogenities and fragmentation takes place only if the central matter reaches the region of adiabatic instabilities, and if the matter expands far below the saturation density ρ_o. For initial densities of the order of equilibrium density matter does not fragment or develop large inhomogenities in the region enclosed by the isothermal and adiabatic spinodals. In this paper we like to show that dynamical and statistical

Monte-Carlo-Time-Dependent-Hartree-Fock[3, 4] (MCTDHF) calculations lead to evolution trajectories in the T,ρ plane that allow to generate small and large fragments in the critical region of the vapor-liquid phase transition. There are strong fluctuations near the critical point, which may provide a mechanism for entering the metastable and unstable regions of the phase diagram. Can these fluctuations wash out a first order liquid-gas phase transition for temperatures below T_c? The chosen initial conditions include density and temperatures wide enough to study a broad variety of fragmentation phenomena. For both the static equation of state (EOS) calculation as well as the dynamic quantum calculation for the fragmentation processes the same model hameltonian density is used. Thus, critical temperatures and densities are determined via static calculations for infinite matter and contrasted with results from finite and dynamical simulations of the nuclear expansion and disassembly stages. EOS calculations for infinite and homogeneous matter are presented. The dynamics of a hot nuclear drop is discussed afterwards within the MCTDHF theory. Finally, some averaged break-up trajectories are discussed in the thermodynamical T,ρ plane. The dependence of critical fragmentation properties like number of particles, compression, temperature and energy is mentioned. Excellent review papers can be found in ref.7 .

Fig. 1. Complex steam-water flows occur in a pot of boiling water. What occurs in heavy ion collisions?

2. *An Equation of State Calculation and the Critical Temperature*

A mean-field calculation of the equation of state for infinite and homogeneous nuclear matter is discussed. We wish to examine the coexistence problem[1] and to determine the critical density ρ_c and the associated critical temperature T_c. Furthermore the dependence of the fluid-vapor phase transition temperature on the nucleon number of the system is discussed. An effective nucleon-nucleon Skyrme type interaction is given by the following model Hamilton density

$$h(\rho,\tau) = h^2/2m\ \tau(p,n) + 3/8\ t_0\ \rho(p,n)^2 + 1/16\ t_3\ \rho(p,n)^3 \qquad (1)$$

where $\tau=\tau_p+\tau_n$ and $\rho=\rho_p+\rho_n$ are the kinetic energy and particle densities. To obtain the equation of state we assume the nuclear matter to be uniform and replace the single particle orbitals Ψ_λ of the physical system by plane waves. The parameters t_o and t_3 are determined from nuclear saturation properties. The Hamilton density h is used in this paper for both the static as well as the dynamic runs described later. For the infinite matter case protons and neutrons are not distinguished and spin iso-spin symmetry is fulfilled. To solve the equation of state the entropy density σ_q ($\sigma=\sigma_p+\sigma_n$) written for T>0 MeV via

$$\sigma_q = -k_B\ \Sigma_\lambda\ [{}_qf_\lambda \ln {}_qf_\lambda + (1-{}_qf_\lambda)\ln(1-{}_qf_\lambda)] \qquad (2)$$

has to be evaluated for the ${}_qf_\lambda$ occupation numbers of the s.p. orbitals Ψ_λ. The correlated free energy density ϕ is defined by

$$\phi(\rho,\tau) = h(\rho,\tau) - T\ \sigma(\rho,\tau) \qquad (3)$$

and leads to the final equation of state

$$P(\rho,\tau) = \rho\mu - \phi(\rho,\tau) \qquad (4)$$

where μ is the chemical potential.

In order to be able to contrast the static calculations for infinite matter with computational results from finite and quantum dynamical systems 2d calculations are performed. Microscopic equation of state calculations for homogeneous matter[5] - surface, curvature and Coulomb effects are ommitted - confirm that nuclei and bubbles do exist up to densities around (half) nuclear matter saturation densities. It is found that the critical densities are below $\frac{1}{2}\rho_0$. In our calculation the critical temperature T_c is around 15.5 MeV. The equations of state is illustrated in this figure for the case of infinite matter. Several regions within the plot can be distinguished. The hydrodynamical instability region (thin dotted area) is found where the compressibility $\kappa=\rho\ \partial_\rho P$ becomes negative. The region of coexistence is given for positive pressures and becomes smaller for increasing temperatures. This area is found in Fig.2 below the line of phase separation within the thin and thick dotted areas as well. The line of phase separation is constructed employing a Maxwell construction. A line at fixed T intersects this curve at two points. We note, that the influence of the finite size could shift the critical temperature T_c (not the associated density) by more than 30% to lower values[6].

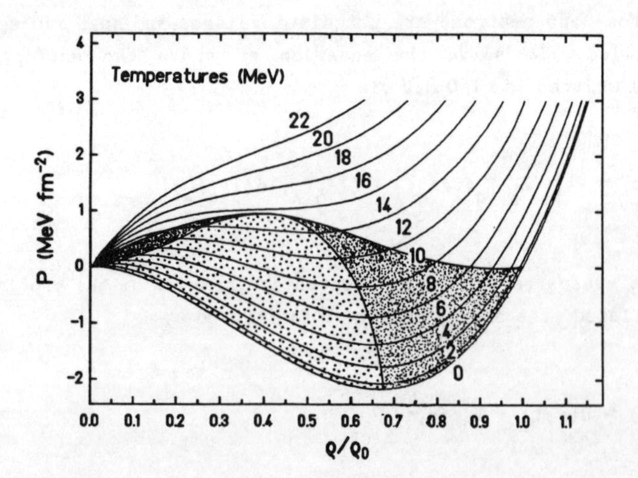

Fig. 2. Pressure curves shown as a function of density for the infinite matter case. The unstability area (thin dotted) and the upper line of phase separation are included besides the 12 isotherms: The coexistence region is defined for positive pressure and bounded by the line of phase separation.

3. Monte-Carlo Quantum Dynamics for Fragmentation

The selfconsistent and time-dependent Monte-Carlo-Hartree-Fock (MCTDHF) theory is sketched. The MCTDHF model simulates the expansion of energetic nuclear matter. Multy-body correlations are of importance and included in a Monte-Carlo description for the microscopic expansion processes. The physics of two-body correlations can be included in an effective interaction and be used in TDHF calculations with uncorrelated one-body wavefunctions. Many-body correlations may be simulated by Monte-Carlo techniques without an explicit treatment of collision terms. The hot hadronic matter may be defined at first by a high temperature T and at second by a high compression κ. The high temperature and therewith also the entropy of the excited system is introduced by a random procedure defining an ensemble of momentum distributions. From these a temperature parameter T can be extracted out of momentum distributions like in INC calculations. A temperature is not included in the model explicitely. The compression is chosen in such a way that the compressed matter occupies a physical reasonable coordinate space and reproduces an expected maximum density. And again, a random procedure defines an ensemble of density distributions. An averaged density distribution can be extracted.

The statistical ensemble ($M>\alpha>>1$) of Slater-determinants $_{\alpha}\Psi$ allows selfconsistent Monte-Carlo-Hartree-Fock calculations for all wavefunctions $_{\alpha}\Psi_{\lambda}$ and the evolution of strong density and momentum fluctuations at all times of the calculation. Thus, the high entropy ($S/A\leq5$) of the system is taken into account. It is pointed out that no coherent mean-field is underlying the calculation. For vanishing temperatures ($T\approx0$ MeV) or low energetic HIC the well known TDHF procedure is valid while the statistical ensemble of Slater-determinants $_{\alpha}\Psi$ is replaced by a single Slater-determinant ($\alpha=1$). In order to point out once more the statistical character of the MCTDHF[3] approach we give the relations underlying the definition of any observable O in cartesian coordinate space

$$O(r;t) \approx 1/M \Sigma_{\alpha} < {}_{\alpha}O(r_{\nu};t) > \qquad (5)$$

While discussing the MCTDHF method one may seperate the two essentials of the theory. The first is the use and the insight in the usefulness of statistical methods as it was described above. The second essential is the proposal to compute the expansion and break-up with the help of effective

Skyrme forces. In the dynamic calculations the same Hamilton density h is used as in the static EOS calculation. In the MCTDHF theory each Slater determinant ${}_{\alpha}\Psi$ carries its own field. No mean field is driving the dynamics because the ensemble of ${}_{\alpha}\Psi$ Slater determinants consists in an ensemble of different states (fields). The ensemble of determinants and the ensemble of observables ${}_{\alpha}O$ allow us to compute time-dependent fluctuations and instabilities.

With given Hamilton density ${}_{\alpha}h$ one derives the Monte-Carlo-Hartree-Fock equations for the s.p. wavefunctions by functional differentiation via the equation

$$ih\,\partial_t\,{}_{\alpha}\Psi_{\lambda} = \delta_{\,{}_{\alpha}\Psi^*_{\lambda}} < {}_{\alpha}h > \qquad (6)$$

For a local two-body potential for example and the one-body density matrix ${}_{\alpha}\rho(r,p)$ the equations can be written simply by a set of coupled and nonlinear equations

$$ih\,\partial_t\,{}_{\alpha}\Psi_{\lambda} = -h^2/2m^* \;\nabla^2\,{}_{\alpha}\Psi_{\lambda} + U_h\,{}_{\alpha}\Psi_{\lambda} - \int dr'\, U_f\,{}_{\alpha}\Psi_{\lambda} \qquad (7)$$

where U_h is the s.p. Hartree and U_f is the s.p. Fock potential derived via $U=\partial_{\rho}\,h$. The system is capable to form stable and unstable fragments. For high temperatures above the quasi-critical threshold the ratio of stable to unstable cluster is decreasing. Fragment species, mass distributions as well as multiplicities are easy to distinguish and to count.

We confine our study to a 2+1 dimensional model world driven by a Hartree force ($t_1=t_2=0$). Yukawa forces are neglected and Coulomb forces, because of their difficulties, in 2d calculations are not considered. Vector densities are omitted. A system with mass 80 amu is computed with an equivalent and equal number of 80 wavefunctions Ψ_{λ}. The initial phase-space distributions are assumed to be of Gauss type with random centroids in coordinate and random momenta in momentum space. In this paper we take only non orthonormal Gaussian wavefunctions. This injures the requirement of dealing with Fermi particles. But, if one looks in phase-space it turns out that there is a small overlap between the nucleons. So that actually we deal with nearly orthonormal sets of wavefunctions. In future work the wavefunctions will be orthonormalized. Each member of the ensemble numbered by α corresponds to one random set of wavefunctions. The initial wavefunctions are defined through the following set of wavefunctions at t_i=0 fm/c

$$^{0}{}_{\alpha}\Psi_{\lambda}(x,y) = \sqrt{\upsilon/\pi}\; {}_{\alpha}\exp_{\lambda}[-\tfrac{1}{2}\upsilon\{\Delta x^2 + \Delta y^2\} + i/h\,\{p_x x + p_y y\}] \qquad (8)$$

$1/\upsilon=4$ fm² is a standard value and connected with the standard deviation in space via $\Delta x=\sqrt{1/\upsilon}$ fm. The standart deviation of momenta is given through $\Delta p=43.3\,\sqrt{T}$ MeV/c. The momentum part is governed by a Maxwell distribution and defines the temperature parameter T. It is noted that T is not the temperature of the system as a Fermi gas implies a finite momentum spread even at zero temperature. The problem of determining a temperature is discussed in chapter 5.

Finally, we demonstrate the capability of the method and the computer code to discriminate between very small fragments. A composite particle or fragment is defined as any connected area in space where the density exceeds a certain threshold value; At present taken as $1/15\ \rho_0$.

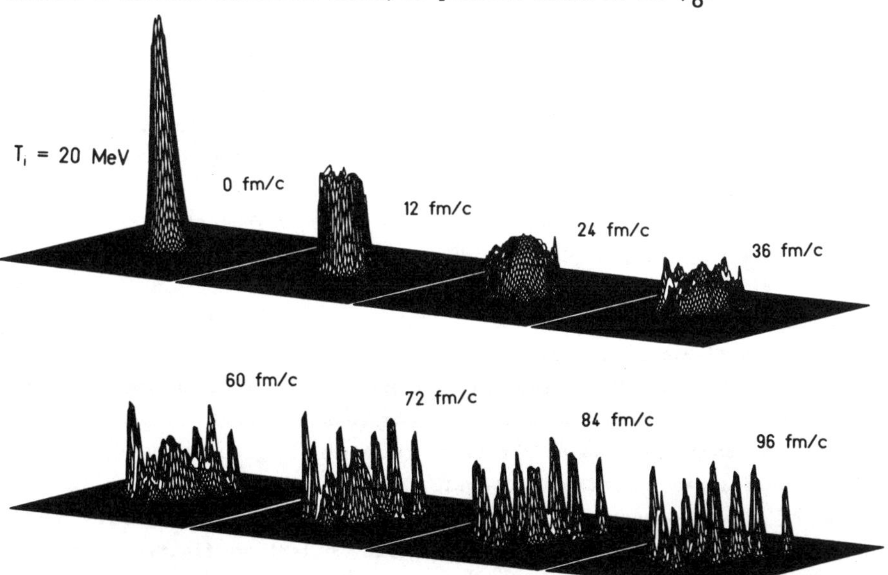

Figs. 3. Visualization of the onset of a nuclear disassembly. A sequence of density plots started at an initial temperature of Ti=20 MeV and a maximum density κi=3 ρo. A single Monte-Carlo event is displayed for a system containing 80 wavefunctions on a coordinate space of 70 times 70 fm.

4. Mass, Multiplicity and Momentum Spectra in Medium Systems

A hot and compressed system of 80 nucleons is studied in a 2+1 dimensional Hartree version. The initial phase-space distribution is of Gauss-type. The processes are driven by symmetric spin and iso-spin dependent forces and considered for two different initial temperatures T_i and a maximum density $\kappa_i=3\ \rho_o$. An extended study of finite size effects and a discussion for different forces can be found in ref.3 .

The expansion of a compressed nuclear system can be classified into three categories[4]:

- The system is hot and explodes while small fragments with great multiplicity are generated.

- The system is heated up such that small and heavy fragments are found with equal frequency.

- The system is cold and suffers (isoscalar mono) vibrations while compression and kinetic energy are steadily converted into each other. The multiplicity is equal to one. Some nucleons are evaporated.

For the first category we demonstrate how cluster are generated. For the second possibility we show spectra with U-shape. The fragment generation for the hot expansion (T_i=20 MeV) can be studied at plots of a single Monte-Carlo event in Figs.3. The density distribution tries to generate fragments in the interiour and the surface. A clear indication that volume effects are distinctive. This finding is in agreement with experimental data[8]. Corresponding spectra are found in Figs.5 in the lowest row. The great multipicity is characteristic. The collective evolution of averaged density distributions can not show single fragments. A hot system which disintegrates into small and fast cluster will flatten the density distribution very fast. A cold and compressed system can not spread steadily in coordinate space. It tryes to reach its groundstate density ρ_o via an overdamped vibration. Overall, the author wants to visualize the onset of the disassembly process (T_i=20 MeV >> T_c) with density plots averaged over 30 Monte-Carlo events. The collective flattening of the density distribution as function of time is illustrated in Figs.4 . It is a natural consequence of the used Monte-Carlo technique. The generation of fragments are much better studied at plots of single events like in Figs.3

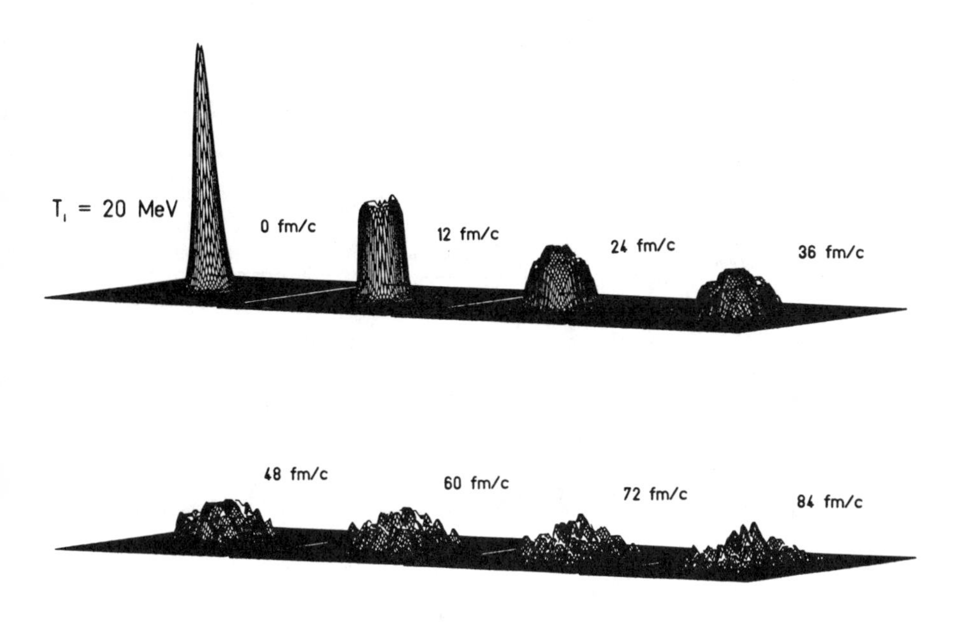

Figs. 4. Visualization of the onset of a nuclear disassembly. A sequence of density plots started at an initial temperature of Ti=20 MeV and a maximum density κi=3 ρo. The density is averaged over 30 Monte-Carlo events for a system containing 80 wavefunctions on a coordinate space of 70 times 70 fm.

Mass spectra, multiplicities and momentum distributions are are shown in Figs.5 for four different temperatures. The critical temperature (the mass yield dσ/dA is equal for all cluster) is around T_c=7.5 MeV. The effects of sequential decay[9] are not fully considered. Systems with increasing temperature will generate increasingly smaller cluster at increasingly greater multiplicity. The absolute momentum spectra show the same behaviour.

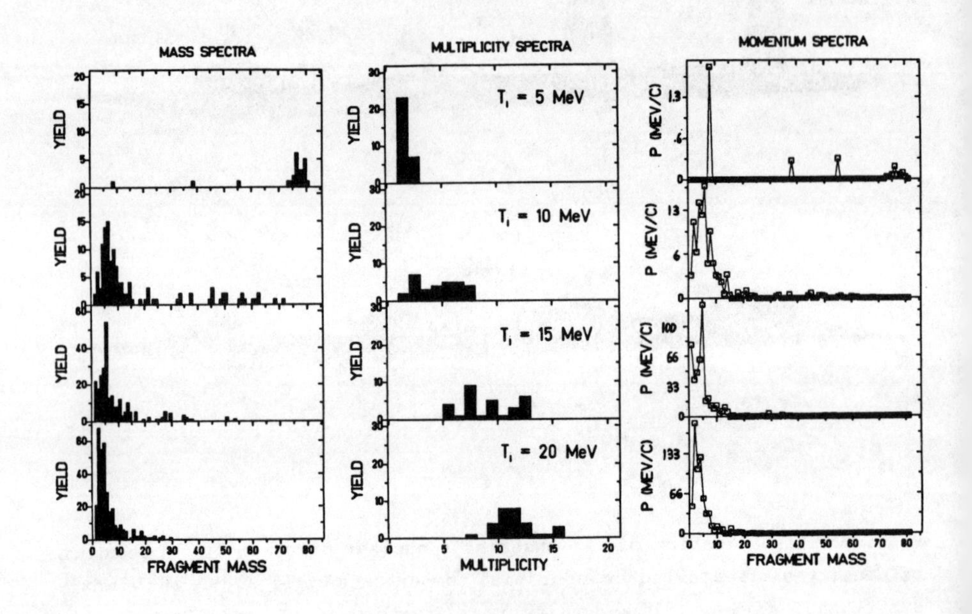

Figs. 5. Mass spectra(a), multiplicities(b) and momenta(c) distributions at four values of Ti and initial density of κi=3 ρo averaged over 30 runs for each temperature. Eighty wavefunctions are computed. A spin iso-spin symmetric force is used. The momenta are averaged.

5. Are There Condensation Phenomena in Finite Nuclear Systems ?

In the time-dependent approach one conveniently extracts critical temperatures T_c at which the average mass yield has U-shape or light and heavy masses are found with equal frequency. The temperature of the nucleus before break-up is given by the average nucleon energy E_{kin}. For a Maxwellian velocity distribution the temperature will be $T=E_{kin}$ in a 2d calculation. The present dynamic approach includes automatically in its description the finite size of the hot nuclear system and naturally also finite particle number effects. Dynamic calculations can lead to drastic deviations from expected mean-field results as already were discussed. Break-up densities are not much sensitive on different forces or initial conditions. All break-up densities are below $\frac{1}{2}\rho_o$.

Differ the break-up temperatures so strongly for different conditions during the thermalization stage that the static concept of liquid-gas phase transitions is useless? Or are dynamic results fluctuating around the static values such that their averaged values reproduce the mean-field values? In Fig.6 some averaged expansion trajectories for these scenarios are depicted within the thermodynamical T,ρ plane. The outer envelope is the Maxwell curve. The inner envelope is the isothermal spinodal. Inside the spinodal neither liquid nor gas is stable. The metastable region (dark pointed) and the instability region are easy to distinguish. The dynamical results of the evolution trajectories seem to fluctuate around the critical temperature values. Small and large fragments are found with equal frequency. Only part of the trajectories lead to total evaporization. Let us comment on the work of Vicentini et al.[2]. For studying fragmentation one has to consider that compression effects support the break-up processes. The hihger the matter is compressed the easier (less temperature) it is to break the matter into pieces. This is discussed extensively in ref. 3. If one studies the evolution with initial densities close to equilibrium density, as Vicentini et al. did, than one would expect high break-up temperatures. The MCTDHF calculations show them fluctuate around T_c.

Let us finally conclude with a critical remark. It is not clear if the reaction time is sufficiently long for a phase transition. Estimations based on the frequency of monopole oscillations[1] give times in the same range ($\sqrt{}10^{22}$s) and may support the transition assumption. In contradiction, there may be observable effects looking like transition effects but without any established phase transition or chemical equilibrium. In an unstable region the yield of light fragments might be

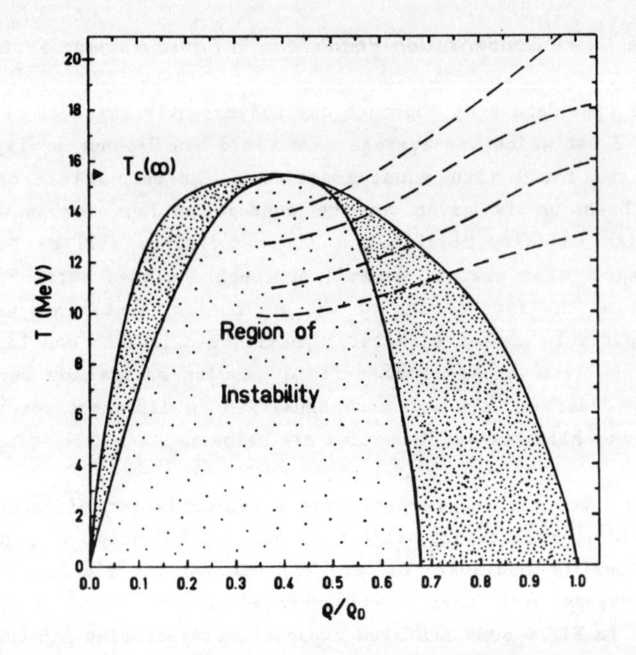

Fig. 6. Break-up densities (dashed lines) from dynamical calculations are shown in the T,ρ plane for infinite matter. The different trajectories correspond to (80,κi=3),(40,κi=3),(40,κi=3/2) and (8,κi=2) nucleon systems if looking from above.

enhanced. In infinite nuclear matter the theoretical evidence for liquid-vapor transition is obvious, in finite systems with dynamically defined short lifetime the reaction may go out of thermal equilibrium. The dynamical conditions may come close to those required for liquid-vapor phase transition but, equally important, there seems to be no physical need that they should do so[7]. Experimental data like energy dσ/dE or mass spectra dσ/dA are nearly insensitive on beam energy or the kind of reaction partners. To reach the conditions corresponding to a phase transition is rather difficult. Therefore, the author doubts whether the experimental data on multifragmentation have much to do with a phase transition[7 10]. But, this might be a too general statement .

The author thanks Professor V.R. Pandharipande for fruitful suggestions and Professor N.T. Porile for comments on the first draft of this paper.

References

1) J.E.Finn et al., Phys. Rev. Lett. 49(1982)1321;
R.W.Minich et al., Phys. Lett. 118B(1982)458;
M.W.Curtin, H.Toki, D.K.Scott, Phys. Lett. 123B(1983)289;
C.B.Chitwood et al., Phys. Lett. 131B(1983)289;
A.D.Panagiotou et al., Phys. Rev. Lett. 52(1984)49;
D.H.Boal, Phys. Rev. C30(1984)119.

2) A.Vincentini, G.Jacucci, V.R.Pandharipande,Phys. Rev. C31(1985)1783.

3) B.Strack, preprint, January (1985).

4) B.Strack, J.Knoll, Z. Phys. A315(1984)249;
J.Knoll, B.Strack, Phys. Lett. 149B(1984)45;
B.Strack, A.Rosenhauer, preprint, January (1985).

5) D.Q.Lamb et al., Nucl. Phys. A360(1981)459;
P.Bonche, D.Vautherin, Astron. Astro. 112(1982)268.

6) H.R.Jaqaman, A.Z.Mekjian, Z.Zamick, Phys. Rev. C29(1984)2067;
A.L.Goodman, J.I.Kapusta, A.R.Mekjian, Phys. Rev. C30(1984)851.

7) D.H.Boal, Adv.Nucl.Phys. Vol15, Negele Voigt eds., Plenum (1985)85;
J.Hüfner, Phys. Rep. 125(1985)131;
L.P.Csernai, J.I.Kapusta, Phys. Rep. 131(1986)223.

8) A.S.Hirsch et al., Phys. Rev. C29(1984)508.

9) C.Barbagallo, J.Richert, P.Wagner, preprint Strasbourg (1985);
D.Hahn, H.Stöcker, in preparation.

10) R.Trockel et al., GSI-85-45(1985).

Microcanonical Simulation of Nuclear Disassembly

Steven E. Koonin
W.K. Kellogg Radiation Laboratory
California Institute of Technology
Pasadena, CA 91125

and

Jorgen Randrup
Nuclear Science Division
Lawrence Berkeley Laboratory
Berkeley, CA 94720

"Counting phase space" is the natural response of any sensible physicist when faced with a complicated system. We know that if the system is sufficiently complicated and if our specification of its state is not too detailed (i.e., we stick to one- or few-body observables), then a statistical hypothesis is often a remarkably economical and accurate description.

In nuclear physics, the first counter of phase space was Bohr in his picture of the compound nucleus. Here, the energy of the initial configuration (say a proton incident on a heavy target nucleus) is assumed to "dissipate" into an equilibrated compound system, which then decays according to the phase space available. The description is based on a nearly degenerate Fermi gas whose Fermi kinetic energy is $\epsilon_F = 35$ MeV and whose density of states at an excitation energy ϵ is $\rho(\epsilon) \sim \exp(2\sqrt{a\epsilon})$, with $a \sim A/8$ MeV.[1]) This leads to a temperature $T \sim \sqrt{\epsilon/a}$, or $T = 2.8$ MeV for $\epsilon/A = 1$ MeV. Thus, if ϵ is not too large, $T \ll \epsilon_F$ and the picture of a single "pot" of Fermi gas is appropriate. Indeed, this "evaporation" picture is an accurate description of many low-energy nuclear reactions, provided adequate account is taken of effects like pairing, rotation of the nucleus, and shell structure.

With the advent of detailed measurements of collisions involving heavy ions, there is growing evidence that an "equilibrated" sub-system with much higher "temperatures" is formed, albeit transiently. While it remains to be established whether full equilibrium can be achieved during the short collision time in a small number of nucleons initially so far from equilibrium, it is clear that any extension of the conventional evaporation picture must incorporate new degrees of freedom (such as those allowing for the simultaneous emission of many large fragments).

This talk is a progress report on our attempts at a statistical description of small, highly excited nuclear systems. We imagine A nucleons with total energy E confined within a spherical volume Ω, as might describe the "freeze-out" stage of a heavy-ion collision. We would like to study how few-body observables vary with these parameters, how they are affected by interfragment interactions, how they differ from their values in the commonly assumed infinite and homogeous system, and what signatures there might be for the liquid-gas phase transition expected in nuclear matter. Our approach is based on the standard Monte-Carlo techniques used to describe equilibrium in condensed-matter systems. It is similar in spirit to work by Gross *et al.*[1]) and Bondorf *et al.*,[2]) but differs in important details of formulation and implementation.

We define a "fragmentation" F of the system as a partition of the A nucleons into N_F spherical fragments confined within Ω (equivalently, at a mean nucleon density $\rho = A/\Omega$), each fragment being specified by its mass number (A_n), its position ($\mathbf{r}_n$), its momentum ($\mathbf{p}_n$), and its internal excitation energy (ϵ_n). Note that we do not distinguish between neutrons and protons, an approximation that can be relaxed readily in more realistic studies. The total number of nucleons in F is

$$A_F = \sum_{n=1}^{N_F} A_n \ , \tag{1}$$

and we take the total energy to be

$$E_F = \sum_{n=1}^{N_F} \left[\frac{p_n^2}{2mA_n} - B_n + \epsilon_n + \frac{1}{2} \sum_{n \neq n'} V_{nn'} \right] \ , \tag{2}$$

where the binding energy is approximated by the semi-empirical mass formula

$$B_n = a_v A_n - a_s A_n^{2/3} - a_c Z_n^2 / A_n^{1/3} \ ; \quad Z_n = A_n/2 \ , \tag{3}$$

and the inter-fragment interaction has Coulomb and nuclear components:

$$V_{nn'} = \frac{e^2 Z_n Z_n'}{|\mathbf{r}_n - \mathbf{r}_{n'}|} + V_{\text{nuc}}(\mathbf{r}_n - \mathbf{r}_{n'}) \ . \tag{4}$$

We investigate several forms for $V_{\rm nuc}$ below.

Our fundamental statistical hypothesis is that all fragmentations consistent with the total nucleon number and energy are equally probable (we do not here impose constraints on the total linear and angular momenta, which are expected to be relatively unimportant). The microcanonical density of states is

$$\rho(\Omega, A, E) = \sum_F \delta(A_F - A)\delta(E_F - E) \,, \tag{5}$$

and from the entropy $S = \ln \rho$, we can calculate the temperature, chemical potential, and pressure as

$$\beta \equiv T^{-1} = \frac{\partial S}{\partial E} \quad ; \quad \eta = -T\frac{\partial S}{\partial A} \quad ; \quad P = T\frac{\partial S}{\partial \Omega} \,. \tag{6}$$

Other observables are given by averages over the microcanonical ensemble. For example, the average mass of fragment 1 (and, by symmetry, of any fragment) is

$$\bar{A} \equiv \langle A_1 \rangle = \frac{\sum_F \delta(A_F - A)\delta(E_F - E)A_1}{\sum_F \delta(A_F - A)\delta(E_F - E)} \,. \tag{7}$$

Our microcanonical formulation should be contrasted with the canonical or grand canonical approaches,[3)] where the density of states (5) is replaced by the partition functions

$$Z = \sum_F \delta(A_F - A)e^{-E_F/T} \,, \tag{8a}$$

or

$$Z = \sum_F e^{(\mu A_F - E_F)/T} \,. \tag{8b}$$

In the latter, μ and T are specified and E and A follow from the appropriate derivatives of Z, while μ and T for fixed E and A follow from (6) in the microcanonical formulation. Of course, for a large (thermodynamic) system, the two formulations are equivalent and the grand canonical formulation can be simplified greatly for non-interacting fragments. However, for the finite, interacting systems formed in heavy ion collisions, there may well be differences between the two approaches, the microcanonical being correct, of course.

The sum over fragmentations in (5) or (7) can be written more explicitly as

$$\rho(\Omega, A, E) = \sum_{N_F} \frac{1}{N_{F!}} \prod_{n=1}^{N_F} \left[\sum_{A_n=0}^{\infty} \int_\Omega \frac{d\mathbf{r}_n d\mathbf{p}_n}{(2\pi\hbar)^3} \int_0^\infty \rho_n(\epsilon_n) d\epsilon_n \right] \delta\left(\sum_n A_N - A\right) \delta(E_F - E) \tag{9}$$

where $\rho_n(\epsilon_n)$ is the density of excited states of fragment n. Because E_F depends quadratically on the fragment momenta, the $\mathbf{p}_n$ integrals can be done analytically, leading to:

$$\rho(\Omega, A, E) = \sum_N \prod_{n=1}^{\infty} \left[\sum_{A_n=0}^{\infty} \int \frac{d\mathbf{r}_n}{\Omega} \int_0^\infty d\epsilon_n \right] W \equiv \sum_C W(C) \tag{10}$$

Here, we have defined a "configuration" C as the specification of $\{A_n, \mathbf{r}_n, \text{and } \epsilon_n\}$ for each of the N fragments, and the weight of a given configuration is

$$W(C) = \frac{1}{N!} \frac{1}{\Gamma(3N/2)} \prod_{n=1}^{N} \left[\Omega \left(\frac{m A_n}{2\pi\hbar^2} \right)^{3/2} \rho_n(\epsilon_n) \right] \delta\left(\sum_{n=1}^{N} A_n - A \right) K^{3N/2-1} , \tag{11}$$

where

$$K = E - \sum_{n=1}^{N} [-B_n + \epsilon_n] + \frac{1}{2} \sum_{nn'} V_{nn'} \tag{12}$$

is the fragments' kinetic energy. Similar expressions can be derived for the canonical or grand canonical partition functions, Eqs. (8).

Observables can be expressed as sums over configurations, weighted by W. For example, from (6) and (10), we have

$$\beta = \frac{\sum_C W(C) \left[\left(\frac{3N}{2} - 1 \right) / K \right]}{\sum_C W(C)} , \tag{13}$$

while Eq. (7) becomes

$$\bar{A} = \frac{\sum_C W(C) A_1}{\sum_C W(C)} . \tag{14}$$

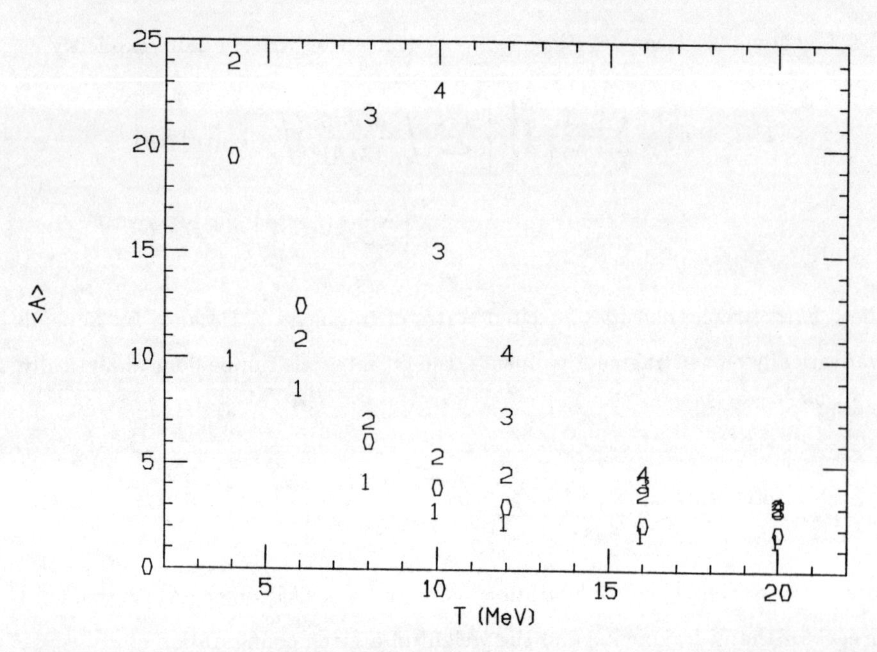

Figure 1. Average fragment mass as a function of temperature for a system of 40 nucleons. "0" denotes the "standard" run and "1"–"4" denote the respective cases, as described in the text. Statistical uncertainties are roughly of the size of the symbols plotted.

The large number of configurations that must be included in these sums preclude a direct evaluation in all but the smallest systems. We therefore evaluate observables of the form (13,14) as averages over a representative sample of configurations. In particular, if S configurations $\{C_s,\ s = 1,\ldots,S\}$ are distributed as $W(C)$ (i.e., the probability that any configuration C is contained in the $\{C_s\}$ is proportional to $W(C)$), then the average of any observables $\mathcal{O}$ can estimated by

$$\langle \mathcal{O} \rangle \equiv \frac{\sum_C W(C)\mathcal{O}(C)}{\sum_C W(C)} \approx \frac{1}{S}\sum_{s=1}^{S} \mathcal{O}(C_s)\,, \tag{15}$$

and the uncertainty of this estimate diminishes with increasing S as $S^{-1/2}$.

We generate the sample of configurations by the algorithm of Metropolis *et al.*,[4]), which provides a Markovian sequence of configuration $\{C_j, j = 1,\ldots,\}$ with the required properties. In particular, given any configuration C_j in the

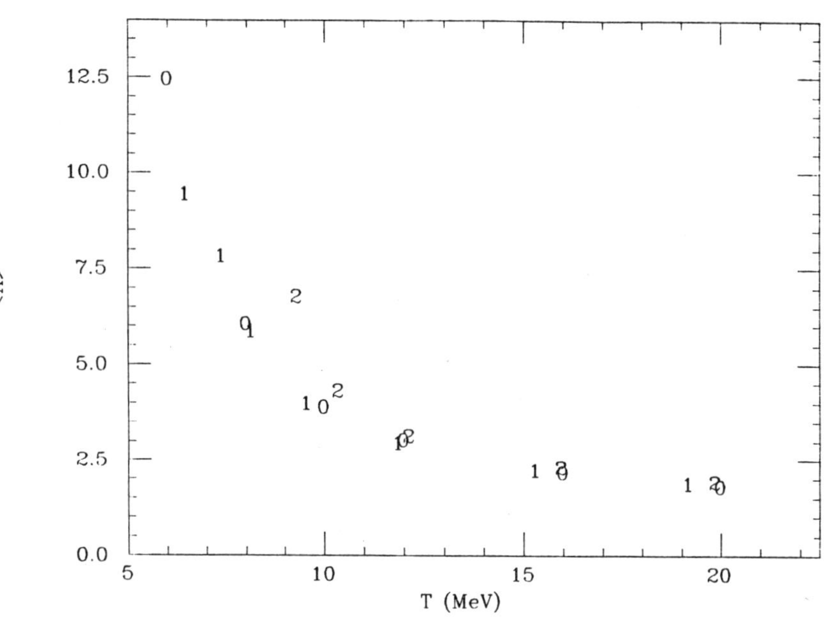

Figure 2. Average fragment mass as a function of temperature. "0" denotes the standard run of an $A = 40$ system, as in Figure 1, while "1" denotes a microcanonical simulation of the same situation and "2" denotes a microcanonical simulation when the total mass is reduced to $A = 10$.

sequence, we form a trial configuration, C', by making any one of several small "moves." These moves consist of changing the location or excitation energy of a fragment, exchanging mass and excitation energy between two fragments, fissioning a fragment, or fusing two fragments. If $W(C') > W(C_j)$ the next member of the sequence, C_{j+1}, is taken to be C', while if $W(C') < W(C_j)$, C_{j+1} is taken to be C' with probability $W(C')/W(C_j)$; otherwise $C_{j+1} = C_j$. This algorithm results in a diffusive exploration of phase space in a physically appealing way, and, if the $\{C_s\}$ are drawn from the $\{C_j\}$ at intervals large enough so that correlations between successive configurations vanish, the average required in Eq. (15) can be evaluated directly. It is important to note the flexibility and generality of this method; realistic mass formulas, level densities, and excitation energies can be used, as we need only be able to compute W for a given configuration.

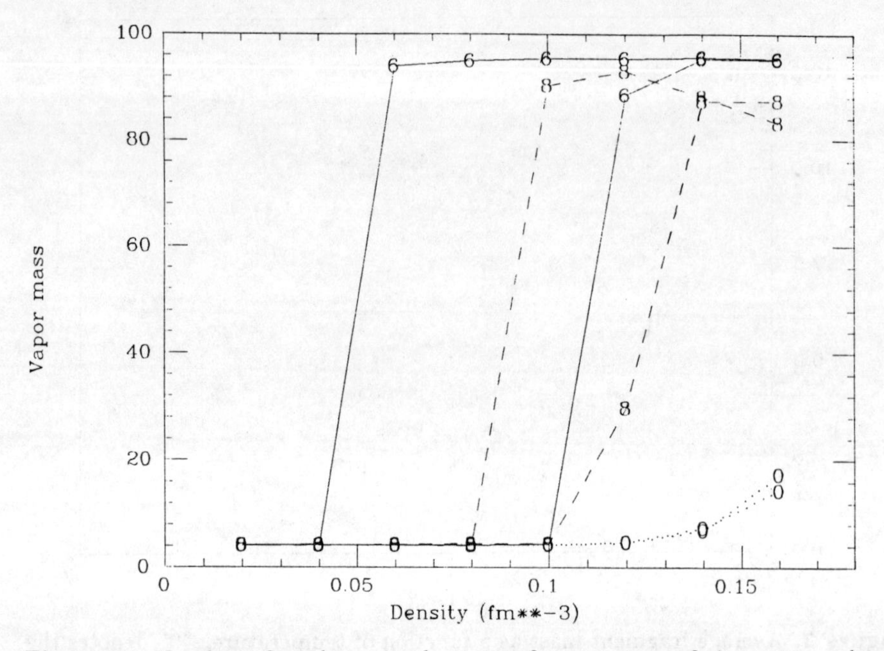

Figure 3. Number of nucleons in the vapor for a system of $A = 100$ nucleons forced through a compression and reexpansion at various temperatures. ____ $T = 6$ MeV; - - - - $T = 8$ MeV; $\cdots\cdot$ $T = 10$ MeV.

To illustrate the kind of results we obtain, we define our "standard" situation as a canonical simulation of an $A = 40$ system at 1/2 of nuclear matter density, an intrinsic level density $\rho(\epsilon) = \exp(2\sqrt{a\epsilon} - \epsilon/\tau_0)$, where $\tau_0 = 12$ MeV, and no interfragment interaction (viz., $V_{nn'} = 0$). The parameters of the semi-empirical mass formula, Eq. (3), have their usual values. Figure 1 shows the average fragment mass as a function of temperature for this case, together with the changes that result when 1) the density is halved to 1/4 of the saturation value; 2) intrinsic excitations of the fragments are forbidden; 3) the interfragment interaction is taken to be infinitely repulsive when fragments interact, and vanishing otherwise (hard sphere potential); and 4) the interfragment Coulomb repulsion is added to case 3). The latter two changes, by decreasing the effective volume available for each fragment, increase the effective density and hence the mean fragment size.

In Figure 2, we illustrate the differences between microcanonical and canonical results. Theses seem to be minimal, except for the smallest systems.

Our model also naturally allows for the incorporation of a vapor phase and the study of the associated phase transitions. We assume that nucleons can be in fragments or in a vapor that occupies the volume in Ω exterior to the fragments. The energy of the vapor is taken to be that of nuclear matter at the appropriate density, and the spectrum of its intrinsic excitations that of a Fermi gas. The Metropolis "moves" discussed above now must be generalized to include an exchange of nucleons and energy between a fragment and the vapor, as well as a change in the vapor's excitation energy.

Preliminary results are presented in Fig. 3. Here, a system of $A = 100$ non-interacting nucleons on the canonical formulation has been driven through a compressional cycle, begining with pure vapor at high density, expansion to low density, and then recompression. The average number of nucleons in the vapor is plotted. The hysteresis characteristic of a first order transition is clearly seen for $T = 6$ MeV and 8 MeV, but then vanishes at $T = 10$ MeV, presumably above the critical temperature. Similar results for a smaller system show a weaker hysteresis, implying a less well-developed phase structure.

The model and methods we have presented here are a well-founded formal and practical basis for quantitative studies of many aspects relating to hot, dilute nuclear matter. The results of these studies, now in progress, will be reported in a future publication.

This work was supported by the National Science Foundation Grants PHY85-05682 and PHY82-07732, and by the Director, Office of High Energy and Nuclear Physics of the Department of High Energy and Nuclear Physics of the Department of Energy under contract DE-AC03-76SF00098.

References

[1]) Gross, D.H.E., Phys. Lett. **161B**, 47 (1985).

[2]) Bondorf, J.P., *et al.*, Nucl. Phys. **A443**, 321 (1985); Nucl. Phys. **A444**, 460 (1985).

[3]) Koonin, S.E., and Randrup, J. Nucl. Phys. **A356**, 223 (1981); Fai, G., and Randrup, J., Nucl. Phys. **381**, 557 (1982).

[4]) Metropolis, N., *et al.*, J. Chem. Phys. **21**, 1087 (1953).

NUCLEUS-NUCLEUS REACTIONS

STATISTICAL AND COHERENT ASPECTS OF MEDIUM-ENERGY HEAVY ION COLLISIONS

Johanna Stachel

Department of Physics
State University of New York at Stony Brook, New York 11794

1. INTRODUCTION

Medium-energy heavy ion collisions have been studied extensively via light particle emission. These data are usually discussed in terms of phase space arguments and thermal characteristics of a (moving) hot source. However, such effects are obscured by e.g., spectator contributions or preequilibrium emission. Recently results have become available on the inclusive production of neutral pions[1,2]. Since pions creation consumes a sizeable fraction of the projectile's kinetic energy (at least 35% in the reactions discussed here for pions at rest in the c.m.) they are a "clean" probe for the statistical (thermal) aspects of such heavy ion reactions. On the other hand, if energetic pions are observed, one is approaching the limits of phase space and the pion production has to be governed by a coherent process. In the same reactions we have observed high energy (30-150 MeV) photon production. There the most prominent feature of the data is a strong dipolar anisotropy of the photons revealing a strong coupling to the projectile's direction of motion, which is unexpected in a statistical model.

2. PION PRODUCTION

We have studied π^0 production in reactions of 35 MeV/nucleon ^{14}N on targets of ^{27}Al, Ni and W at the MSU K=500 Superconducting Cyclotron and of 25 MeV/nucleon ^{16}O on ^{27}Al, Ni at the ORNL Holifield Heavy Ion Research Facility. Experimental details and a complete collection of the results can be found in refs. 1,3,4. Here only the main features of the data and the basic conclusions will be summarized.

(i) The pion kinetic energy spectra (in the lab frame) are peaked at T_{π}=10-15 MeV and are exponential for larger pion kinetic energies. The slope constants are large and, within the uncertainty, constant at 23±3 MeV for the systems were measured. This does not continue the decreasing trend observed[2] at beam energies between 84 and 60 MeV/ nucleon. Rather, our slope constants agree with the 60 MeV/nucleon result.

(ii) The shape of the angular distributions in the lab frame depends strongly on the target size (mass number). For the Al target the angular distribution is forward peaked at both beam energies while it is forward-backward symmetric with a minimum at 90° for the Ni target (the minimum is getting more pronounced with decreasing beam energy). Preliminary analysis for the W target exhibits backward peaking in the laboratory frame.

(iii) The integrated π^{0} production cross sections are relatively large: 115±15 nb and 2.3±0.52 nb for the Ni target at 35 and 25 MeV/nucleon, respectively.

The shape of the pion angular distributions as well as the centroid and shape of the rapidity distributions can be understood quantitatively as a combination of stopping of the projectile in the c.m. frame together with strong pion reabsorption effects. For details see again Ref. 3. The clue of the argumentation is the following: The projectile is stopped before traversing the whole target nucleus (as indicated e.g., in Monte Carlo calculations within the BUU formalism[5]) and pions emitted in forward direction have to traverse more nuclear matter and, hence, are absorbed more strongly. This effect is increasing as the target size is increasing. From pion absorption data[6] and optical model calculations[7] the pion mean free path λ_{abs} is known to be of the order of 2-3 fm, i.e., comparable or smaller to the relevant nuclear dimensions. The primary pion angular distributions have to exhibit a forward-backward peaking in the rest frame of the source. This is in accordance with data measured at higher beam energies[2].

The observed cross sections and also the slope constants of the pion kinetic energy spectra are by far too large to be explained in single-nucleon nucleon collision models employing a boost due to Fermi motion[8,9]. We, therefore, propose along the line of Refs. 10,11 a thermal model[12] combining the geometrical concepts of the "fireball" model[13] with statistical decay of the excited system[14]. The overlapping target and projectile matter form an excited system. Its thermal characteristics are calculated in the Fermi-gas model. It should be noted that this approach implies thermal but not chemical equilibrium in the composite system. Particle decay properties are inferred by applying Weisskopf theory for compound nucleus decay. Since the experimental inverse cross sections for pion absorption (and also photon absorption, see below) are known, this model does not have free parameters to be adjusted to the data. This model[12] can reproduce the experimental pion production cross sections within a factor of two. Only for the lowest beam energies (25 MeV/nucleon) and high pion kinetic energies the model tends to underpredict the data. We note that this thermal model starts to underpredict the data - independent on the beam energy and the colliding system - for reactions where more than 45%(±5%) of the total energy available in the c.m. is consumed to create a pion (with a given kinetic energy). We conclude that a thermal model applies for pion production in reactions where less than 40% of the total c.m. energy is carried away by one pion while processes, where the majority of the total c.m. energy is transferred to the pion are governed by a nonstatistical coherent process.

While such a process has not yet been worked out within theoretical models for the reactions and low beam energies discussed here, we want to show another experimental indication for the need of a coherent pion production mechanism. In Fig. 1 pion kinetic energy specta are shown as a function of the total energy available in the c.m. frame for pion production, i.e. $E_{cm}-m_{\pi}c^2-T_{\pi}^{cm}$. The zero on this scale corresponds to the phase space limit where all of the available c.m. energy is transferred to one energetic pion. From Fig. 1 it can be seen that, for the reaction 25 MeV/nucleon $^{16}O+^{27}Al$, the data reach this phase space limit within the experimental energy resolution. Such a process corresponding to pionic fusion (within the energy resolution) must be described as a coherent process. Furthermore, the data displayed in Fig. 1 indicate that, as the phase space limit is approached, the cross sections scale with the energy available in the c.m. while at higher beam energies the beam energy per nucleon seems to be the relevant quantity (see e.g., the cross sections for 84 and 74 MeV/nucleon in Fig. 1).

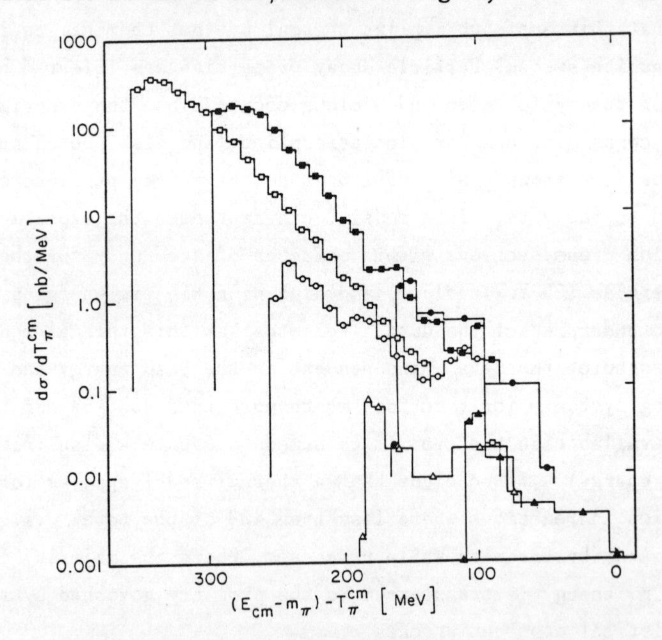

Fig. 1: Pion kinetic energy spectra as a function of the energy available in the cm system after pion emission (corrected for Q-values). Experimental values are shown for 25 MeV/nucleon $^{16}O+^{27}Al$,Ni (closed and open triangles, Refs. 3,4), 35 MeV/nucleon $^{14}N+^{27}Al$,Ni (closed and open dots, Ref. 3) 74 and 84 MeV/nucleon $^{12}C+^{12}C$ (closed and open squares, Ref. 2).

3. PHOTON PRODUCTION

To gain further insight into the above described heavy ion reactions where pion production was observed we also studied single photon production again using an array of lead glass Cerenkov detectors[15]. Inclusive photon production cross sections were measured in reactions of 35 MeV/nucleon ^{14}N+Ni at MSU and 25 MeV/nucleon ^{16}O, 21.8 MeV/nucleon ^{32}S, 15.8 MeV/nucleon ^{58}Ni at ORNL on targets of ^{27}Al, Ni, Cd, ^{197}Au, each. The rejection of cosmic and beam correlated background as well as the response of the set-up to single high energy photons are well understood and the main experimental results are the following:

(i) The integrated photon production cross sections for $E_\gamma \geq 35$ MeV are large, e.g. for 35 MeV/nucleon ^{14}N+Ni we obtain[15] σ=303 μb. Our measured cross sections agree well with an extrapolation from single high energy photon production data at higher beam energies of 48-84 MeV/nucleon[16].

(ii) The photon energy spectra are consistent with exponential decay. The slope constant for 35 MeV/nucleon ^{14}N+Ni is E_o=14.5 MeV (see Fig. 2). This slope constant is decreasing with decreasing projectile energy and increasing target mass number (E_o=7.8 MeV for 15.8 MeV/nucleon ^{58}Ni+^{197}Au).

(iii) The angular distributions in the lab-frame (again for $E_\gamma \geq 35$ MeV) are found to be very asymmetric and of dipolar shape. At 35 MeV/nucleon ^{14}N+Ni the anisotropy, expressed as the ratio of $d\sigma/d\Omega(90^o):d\sigma/d\Omega(30^o)$, is 1.7±0.2 as displayed in Fig. 3. This anisotropy is increasing drastically with increasing target and projectile mass (or charge). For 15.8 MeV/nucleon ^{58}Ni+^{197}Au we find an anisotropy $d\sigma/d\Omega(100^o):d\sigma/d\Omega(20^o)$ of 7.8.

To interpret these data we again applied the thermal model[12] introduced in section 2 using now experimental photon absorption cross sections[17]. The integrated photon cross sections are reproduced quite well, e.g., at 35 MeV/nucleon ^{14}N+Ni the model predicts σ=260 μb as compared at the experimental value of 303 μb. The same is true for the slope constant for the gamma energy spectra. For the above reaction at θ_γ=30° we calculate E_o=10.5 MeV (solid line in Fig. 2) as compared to

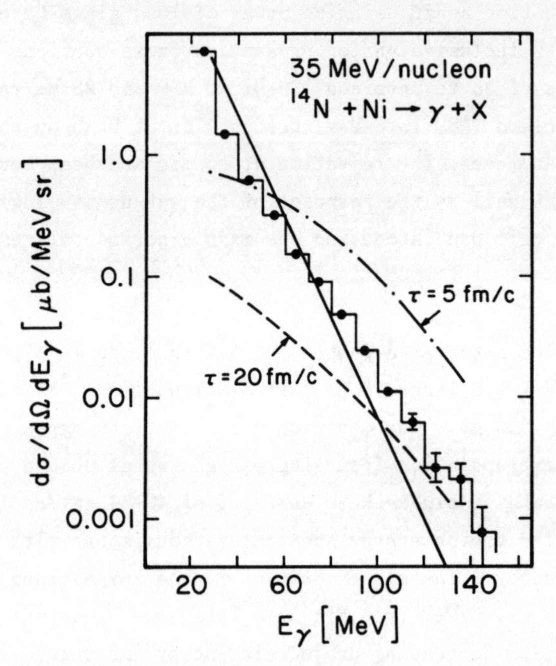

Fig. 2: Experimental γ-ray energy spectrum at θ_γ=30° together with the results of a statistical model (solid line) and and results of a nucleus-nucleus bremsstrahlung calculations for two different slowing down times τ (dashed and dashed-dotted lines).

the measured value of 14.5 MeV. However, in such a thermal model particle emission is isotropic in the respective source's rest frame i.e., forward peaked in the laboratory. The solid line in Fig. 3 displays the calculated photon angular distribution after transformation into the lab-frame. It clearly is at variance with the data.

Alternatively, high energy photons are predicted by models based on the bremsstrahlung mechanism. Nucleon-nucleon bremsstrahlung has been discussed in Refs. 18, 19. It is expected to be dominated by dipolar proton-neutron bremsstrahlung. However, for 35 MeV/nucleon the gamma spectrum has to terminate at 17.5 MeV/nucleon in the c.m. frame and in order to obtain the high photon energies that have been measured Fermi motion has to be employed. Although exact calculations for our systems have not been performed yet, we expect that, if the major momentum for high energy photon production stems from Fermi motion, the angular anisotropy will be reduced. This effect should be even more serious for lower beam energies where, however, the experimental anisotropy is increasing.

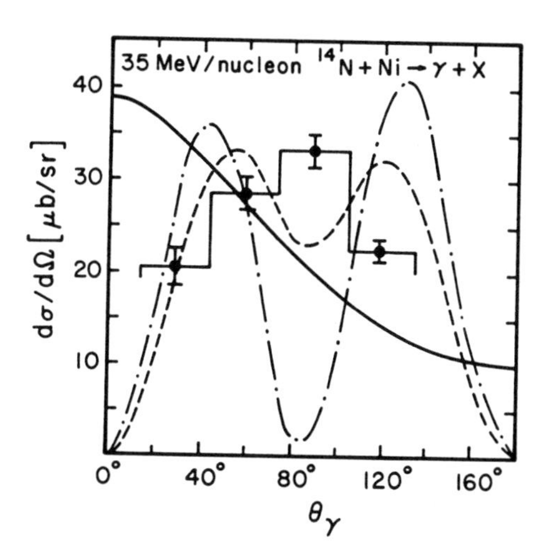

Fig. 3: Single photon angular distribtion in the laboratory frame for E_{γ}=35-150 MeV. The errors shown contain statistical and estimated systematic contributions. Also shown are results of a statistical model (solid line) and of nucleus-nucleus bremmstrahlung for $^{14}N+^{58}Ni$ (dashed-dotted line) and $^{14}Li+^{58}Ge$ (dashed line) with τ=5 fm/c.

High energy photons can also be produced as classical nucleus-nucleus bremsstrahlung. We have investigated[15] this process in the soft photon limit for 35 MeV/nucleon $^{14}N+Ni$ central collisions. Thereby, we include experimental form factors[20] for the colliding nuclei, and we assume a cross section of 500 mb contributing to this process (i.e., 50% of the geometrical cross section) and an exponential velocity profile $v=v_o \exp(-t/\tau)$. Figure 3 shows the resulting photon spectra for very rapid and more moderate slowing down times (τ=5 and 20 fm/c, respectively). The experimental spectrum is clearly steeper. It should be noted, that the shape of the resulting $1/E_\gamma$ spectra does not depend on the slowing down length (if τ>1 fm/c) while, of course, the overall cross section does. Due to the very similar Z/A ratios for the measured systems, the resulting angular distributions exhibit basically a quadrupole pattern as displayed by the dashed-dotted line in Fig. 3. Admitting non-zero impact parameters[21] and/or variations for the projectile and target charge during the initial phase of the collision could alleviate this deficiency. The dashed line in Fig. 3 demonstrates, that a strong dipole contributions is obtained if e.g., the projectile and target exchange 4 protons and neutrons corresponding, then, to a collsion of $^{14}Li+^{58}Ge$. Whether nucleon transfer which is known to occur at large internuclear distances, i.e., before substantial slowing down occurs, can cause such fluctuations is unclear. It seems, however, unlikely to exchange eight nucleons with large probability.

In summary, pion production has been observed in heavy ion reactions at beam energies per nucleon far below the free nucleon-nucleon threshold. In reactions where less than ~40% of the total energy available in the c.m. is consumed to produce a pion, this process can be described in a thermal (statistical) model. For higher pion kinetic energies and/or lower beam energy one is approaching the phase space limit. There data are found to scale with the total energy available in the c.m. and call for a coherent pion production mechanism. In the same reactions photon production for E_γ=35-150 MeV has been studied. The overall cross sections and slopes of the gamma energy spectra could again be reproduced fairly well within the same thermal model. The outstanding feature of these data, the very anisotropic dipolar angular distributions, however, calls for a mechanism that preserves the correlation between the direction of motion of the projectile and particle (photon) emission.

The results presented here have been obtained in collaboration with N. Alamanos, T.C. Awes, P. Braun-Munzinger, R.L. Ferguson, R.H. Freifelder, P. Paul, F. Plasil, M. Prakash, F.E. Obenshain, P. DeYoung, G.R. Young and P.H. Zhang. Financial support of the Alexander von Humboldt Foundation, the NSF and the US DOE (contract DEAC05-840R21400 with Martin Marietta Energy Systems) is gratefully acknowledged.

References

1. Braun-Munzinger, P. et al., Phys. Rev. Lett. *52*, 255 (1984).
2. Noll, H. et al., Phys. Rev. Lett. *52*, 1284 (1984).
3. Stachel, J. et al, Phys. Rec. *C33*, 1420 (1986).
4. Young, G.R., et al, Phys. Rev. *C33*, 742 (1986).
5. Aichelin, J. and Bertsch, G.F., Phys. Rev. *C31* 1730 (1985).
6. Nakai, K. et al., Phys. Rev. Lett *44*, 1446 (1980); D. Ashery et al., Nucl. Phys. *A335*, 385 (1980); Navon, I. et al., Phys. Rev. *C28*, 2548 (1982).
7. Huefner, J. and Thies, M., Phys. Rev. *C20*, 273 (1979); Mehrem, R.A.,
   Radi, H.M.A. and Rasmussen, J.O., Phys. Rev. *C30*, 301 (1984).
8. Shyam, R. and Knoll, J., Phys. Lett. *136B*, 221 (1984).
9. Guet, C. and Prakash, M., Nucl. Phys. *A428*, 119C (1984).
10. Aichelin, J., and Bertsch, G., Phys. Lett. *138B*, 350 (1984).
11. Aichelin, J., Phys. Rev. Lett. *52*, 2340 (1984).
12. Prakash, M, Braun-Munzinger, P., Stachel, J., Phys. Rev. *C33*, 937 (1986).
13. Westfall, G.D. et al., Phys. Rev. Lett. *37*, 1202 (1976); Gosset J. et al., Phys. Rev. *C16*, 629 (1977).
14. Weisskopf, V. Phys. Rev. *52*, 295 (1937).

15. Alamanos, N., et al Phys. Letters in print.

16. Grosse, E., et al, submitted to Europhysics Letters.

17. In the relevant energy range 25 MeV$\leq E_\gamma \leq$150 MeV photoabsorption cross sections were used as given by S. Ahrens et al, Nucl. Phys. *A251*, 479 (1975).

18 Ko, C.M., Bertsch G. and Aichelin, J., Phys. Rev. *231*, 2324 (1985).

19. Nifenecker, H., and Bondorf, J.P., Nucl. Phys. A*442*, 478 (1985).

20. Cycz, W., Lesniak L. and Maleski, A., Annals Phys. *42*, 97 (1967).

21. Vasak, D., preprint, April 1986.

SUBTHRESHOLD KAON PRODUCTION IN NUCLEUS-NUCLEUS COLLISIONS: A SENSITIVE PROBE FOR THE NUCLEAR EQUATION OF STATE?

Bernd Schürmann and Winfried Zwermann

Physik-Department, Technische Universität München,
D-8046 Garching,
FEDERAL REPUBLIC OF GERMANY

ABSTRACT

In principle, subthreshold kaon production provides an appropriate tool for testing the compressional part of the nuclear equation of state. Eventually, the experimental data have to be confronted with theoretical model predictions to extract the compressional energy. Due to the fact that the elementary kaon production cross section which is an essential input into the calculations, is poorly known in the energy regime needed, uncertainties of at least a factor of two arise in the calculated kaon yield; this makes the determination of the compressional energy from absolute kaon yields difficult. We find it more profitable to study kaon yield *ratios* of heavy to light colliding nuclei at fixed energy instead; in such ratios the mass number dependence is almost the same for the various parametrizations of the elementary kaon production cross section. With compression effects present, the mass number dependence will be largely reduced.

1. INTRODUCTION

As has often been emphasized, the main aim of nucleus-nucleus collisions at high energy (i.e. 400 MeV/nucleon to 4 GeV/nucleon beam energy) is to determine the nuclear equation of state under extreme conditions, i.e. to determine the internal energy per nucleon at high densities and/or high temperatures. Knowing in particular the compressional part of the equation of state is of utmost importance also in branches other than nuclear physics, such as astrophysics.

In the last few years, the attempt of Stock and collaborators[1] to extract the compressional part of the nuclear equation of state by utilizing the discrepancy between the measured pion production rate and the one calculated in the conventional numerical cascade model of Cugnon[2] has attracted a lot of attention. The basic point of this comparison is the assumption that the overestimate of the pion yield in these calculations is due the fact that in the cascade code no kinetic energy is converted into compression, and therefore, too much energy is available for pion production. Details of how one obtains from this overestimate the internal energy as a function of density at temperature T=0 can be found, e.g., in ref.[3]. Crucial for the method to work is that the number of pions should already be fixed at a quite early stage of the heavy ion collision, most preferably at the stage of maximum compression. The fact, however, that (i) pions are subject to strong absorption and (ii) that calculations with an appreciable number of pions still produced at late stages of the collision[4] yield inclusive differential cross sections which quantitatively agree with Cugnon's cascade calculations render the method of Stock et al. disputable. This, however, does not diminish the usefulness of their idea in general to use the particle production rate as a probe for the nuclear equation of state.

Whereas it thus seems that pions are less than ideal candidates to probe the equation of state, this does not apply for kaons because these are, due to strangeness conservation, hardly absorbed by the nuclear medium. Unfortunately, to date there exists experimental information on the kaon production cross section only at 2.1 GeV/nucleon[5]. Moreover, at this energy, heavy ion collisions are rather transparent, so this regime is not appropriate for a test of the equation of state as a function of density. As was pointed out recently by Aichelin and Ko[6] it is in this context much more profitable to consider kaon production at substantially lower energies where strong compression is possibly reached. This unavoidably implies to focus on <u>subthreshold</u> kaon production since the lowest energy for kaons to be produced in free nucleon-nucleon encounters is precisely in the unsuitable energy region, namely at 1.6 GeV/nucleon. Although the production cross sections naturally are considerably smaller below threshold than above, for sufficiently heavy colliding nuclei they can still be substantial, as has been pointed out by us in ref.[7]; for Pb on Pb at 800 MeV/nucleon we still obtain a total cross section of about 10 mb. In the coming years, such subthreshold strange particle cross sections will be measured with sufficient accuracy at GSI.

In this contribution, we discuss kaon inclusive yields in the incident energy range of 400 to 1000 MeV/nucleon. The calculation of the kaon excitation function on the basis of transport theory will be briefly reviewed in the next section. In section 3 we investigate the sensitivity of the excitation function to the (essentially unknown!) elementary kaon production cross section by employing various parametrizations of the latter. The inclusion of the nuclear mean field is briefly described in section 4. Its influence on the kaon inclusive yield is discussed in section 5, and compared to the uncertainties caused by the poor knowledge of the elementary kaon production cross section. Furthermore, in the same section we suggest to study the mass number dependence of <u>ratios</u> of the kaon inclusive yield.

2. KAON INCLUSIVE PRODUCTION IN THE MODEL OF TRANSPORT THEORY

We briefly describe how the kaon total production cross section can be evaluated in the model of transport theory which is an essentially analytic multiple collision approach. For details, we refer to ref.[7]. The kaons are assumed to be produced predominantly in elementary baryon-baryon (BB) collisions

$$B+B \rightarrow B+Y+K \tag{1}$$

where Y stands for either a Λ- or a Σ-hyperon. The baryons can be either nucleons or delta-resonances the latter of which we consider as stable during the collision. In the intermediate energy regime considered the production of kaons is subthreshold, and hence the elementary kaon production cross section will contribute only in the close vicinity of the threshold. There, this cross section is very small compared to the BB total scattering cross section. Therefore, kaon production may be treated perturbatively, i.e. we may neglect the influence of the kaons, once produced, on the baryon distributions.

We view a nucleus-nucleus collision as a superposition of free baryon-baryon encounters; furthermore we assume that, at least in the early stage of the collision, there exist two well distinguishable classes of participant baryons, one of which is moving in the nucleus-nucleus c.m. system on the average in the beam-(z-) direction, and the other in negative z-direction. We denote these two classes by "beamlike" ("B") and "targetlike" ("T"), respectively. Hence, the essential inputs are the elementary kaon production cross section $E_k d\sigma_{elem}/d^3k$ and the distributions of the colliding baryons with which the former has to be folded. The key quantities therefore are

$$J^{\mu\nu}_{mn} = \rho^{\mu}_{m-1}\rho^{\nu}_{n-1}\int\frac{d^3p_1}{E_1}\int\frac{d^3p_2}{E_2}\,Q^{(B)\mu}_{m-1}(\vec{p}_1)\,Q^{(T)\nu}_{n-1}(\vec{p}_2)\left(E_k\frac{d\sigma_{elem}}{d^3k}\right)^{\mu\nu} \tag{2}$$

where $Q^{(B)\mu}_{m-1}$ ($Q^{(T)\nu}_{n-1}$) denotes the momentum distribution of beam-(target-) like baryons after m-1 (n-1) collisions, occurring with probabilities ρ^{μ}_{m-1} (ρ^{ν}_{n-1}) as nucleons or delta-resonances (denoted by the superscripts μ,ν). To obtain the inclusive cross section for kaon production, $E_p d\sigma/d^3p$, the key quantities $J^{\mu\nu}_{mn}$ have to be weighted and summed in an appropriate way:

$$E_p \frac{d\sigma}{d^3p} = \sum_{i=1}^{\infty}\sum_{j=1}^{\infty} \frac{\sigma_A(i)}{\sigma_{NN}} \frac{\sigma_A(j)}{\sigma_{NN}} \sum_{m=1}^{i}\sum_{n=1}^{j}\sum_{\mu,\nu} J_{mn}^{\mu\nu} \, . \tag{3}$$

We note that in eq. (3) the inclusive invariant cross section is given in the nucleus-nucleus c.m. frame as a function of the momentum $\vec{p}$ whereas in eq. (2) the elementary cross section depends on the momentum $\vec{k}$ in the c.m. system of two colliding baryons. To a good approximation, the baryonic distributions in eq. (2) are given by relativistically generalized Gaussians with width $\sigma_n^{\nu 2}$ and average velocity $\vec{v}_n^{\nu}$ in the c.m. frame of the colliding nuclei,

$$Q_n^{(B)\nu}(\vec{p}) = N_n^{\nu} \gamma_n^{\nu} (E - \vec{v}_n^{\nu} \cdot \vec{p}) \exp\left[-m\gamma_n^{\nu}(E - \vec{v}_n^{\nu} \cdot \vec{p})/\sigma_n^{\nu 2}\right], \tag{4}$$

displayed here for beamlike baryons only.

In this work, we are interested in the kaon total production cross section only, $\sigma_K = \int d^3p \, d\sigma_K/d^3p$. This quantity can be calculated analytically up to a one-dimensional integration. From eq. (3) we then obtain

$$\sigma_K = \sum_{i=1}^{\infty}\sum_{j=1}^{\infty} \frac{\sigma_A(i)}{\sigma_{NN}} \frac{\sigma_A(j)}{\sigma_{NN}} \sum_{m=1}^{i}\sum_{n=1}^{j}\sum_{\mu,\nu} \sigma_{mn}^{\mu\nu} \tag{5}$$

with

$$\sigma_{mn}^{\mu\nu} = \int dS \, \sigma_{elem}(S) \, F(S;\, v_{mn}^{\mu\nu},\, \sigma_n^{\nu 2},\, \sigma_m^{\mu 2}) \tag{6}$$

where the function F is given in closed form and $\sigma_{elem}(S)$ is the elementary integrated kaon production cross section as a function of the invariant energy squared S.

3. THE ELEMENTARY KAON PRODUCTION CROSS SECTION

Virtually nothing is known experimentally about the elementary kaon inclusive production cross section in the vicinity of the lowest threshold $NN \to N\Lambda K$ at 1.6 GeV/nucleon; it is, however, precisely this vicinity which will enter into our subthreshold kaon production investigation. The situation is not quite as hopeless as it looks at first sight, as we shall demonstrate in this section. Instead of using the variable S in σ_{elem} of eq. (6), we rather use the quantity p_{max}, defined by

$$p^2_{max} = \frac{1}{4S}\left[S-(m_B+m_Y+m_K)^2\right]\left[S-(m_B+m_Y-m_K)^2\right] \qquad (7)$$

with m_K, m_B, m_Y being the kaon, baryon and hyperon masses in the exit channel, respectively. The quantity p_{max} is the maximum momentum the kaon can reach in an elementary BB collision in the c.m. system. Near threshold

$$p^2_{max} \approx \frac{2(m_B+m_Y)m_K}{m_B+m_Y+m_K}(E-E_{Thr}) \qquad (8)$$

where E_{Thr} is the c.m. threshold energy, $E_{Thr}=m_B+m_Y+m_K$, and E is the total c.m. energy, $E=\sqrt{S}$.

In fig. 1, the various parametrizations of the elementary kaon inclusive production cross section,used as inputs in eq. (5), are shown in comparison with the scarce experimental data as a function of p_{max} for the production channel with the lowest threshold, $NN \rightarrow N\Lambda K$. It is important to remark that at most the three lowest experimental points are of significance for the subthreshold kaon yield. The parametrizations are (i) the one of Randrup and Ko[8] (dotted line) which fits the data points best, (ii) one which depends linearly on p_{max} (full line), and (iii) one which is proportional to the fourth power of p_{max} (dashed line). The corresponding results for the kaon inclusive total production cross section from the reaction Nb on Nb are displayed as a function of the bombarding energy per nucleon in fig. 2. The cases (i) and (ii) lead to nearly identical curves in the entire energy range considered; this is not the case for the parametrization (iii). Important in this comparison are the <u>shapes</u> of the curves; the absolute magnitude is of less significance because we have taken the result obtained with case (i) as a reference curve and fitted the proportionality constants for the other elementary cross sections such that they lead to results close to this curve. With regard to

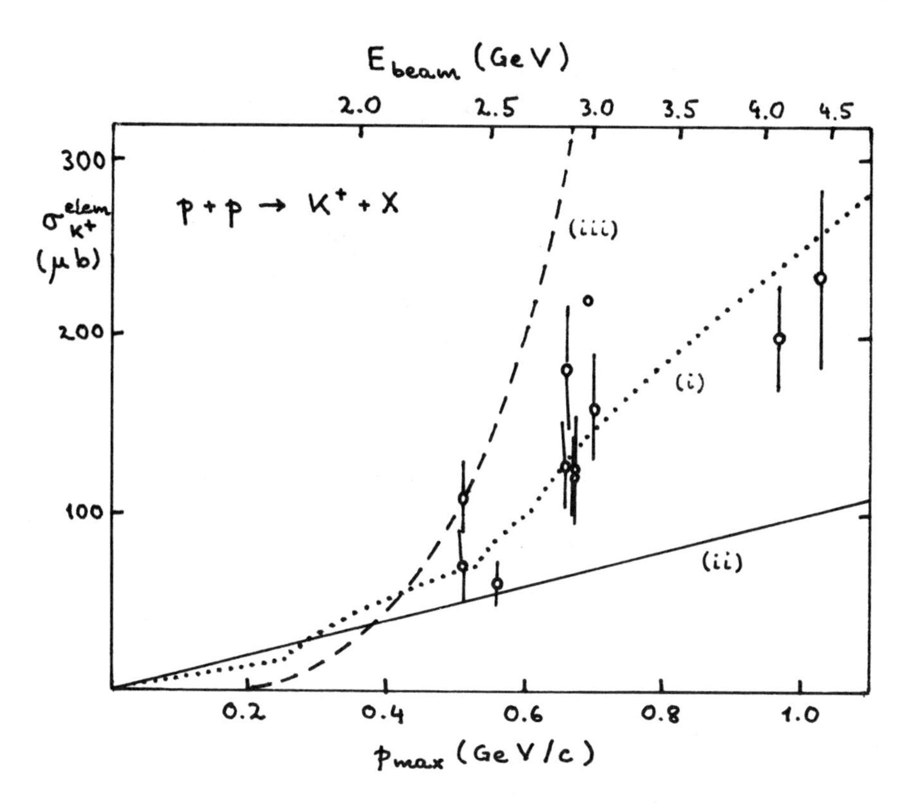

Fig. 1. The various parametrizations for the elementary kaon inclusive production cross section. See text for details.

parametrization (iii) which leads to deviations of about factors of two from the other kaon yields, we remark that the p_{max}^4-dependence arises near threshold from a pure phase space consideration (constant transition matrix elements). Furthermore, this dependence also results near threshold from a theoretical investigation[9)] based on the one-meson exchange process.

4. INCLUSION OF NUCLEAR MEAN FIELD EFFECTS

In this section we introduce the so far neglected nuclear mean field effects. It has become apparent from event per event measure-

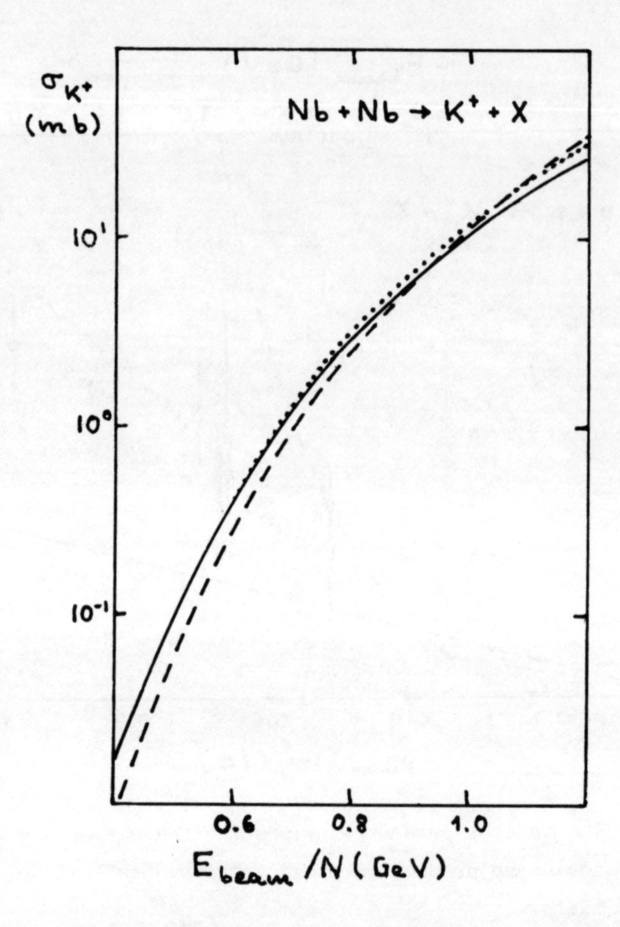

Fig. 2. The kaon inclusive yield obtained with the parametrizations of fig. 1.

ments that such effects have to be taken into account in the discussion of the experimentally observed kinetic energy sideward flow of the particles emitted in individual heavy ion collision events. Only how they have to be included in theoretical models is still subject to discussion. For comments on this point, we refer to ref. [10]. In the same paper, a mechanism is proposed which is based on the assumption of transparency of the heavy ion collision, i.e. compression effects

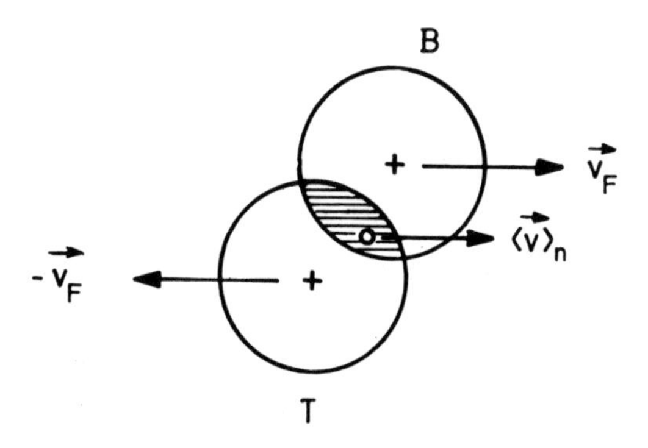

Fig. 3. Our view of a high-energy heavy ion collision.

are a priori absent. We now briefly explain this mechanism (cf. fig. 3). A beamlike participant nucleon (corresponding considerations apply for a targetlike participant) which after n binary collisions on the average moves in the nucleus-nucleus c.m. system in the beam-(z-) direction with velocity $\langle\vec{v}\rangle_n$, experiences the influence of two separate nuclear mean fields (beamlike and targetlike, respectively) which move in opposite directions along the z-axis with velocities $\vec{v}_F$ and $-\vec{v}_F$, respectively. The net effect on the participants is an attractive force which leads to a nonvanishing momentum transfer $\vec{\delta}_n$ given to the participant. If we choose the (x,z)-plane as the reaction plane, $\vec{\delta}_n$ will be in x-direction. The average velocity of a participant nucleon after n collisions is now not in z-direction anymore but possesses a transverse component,

$$\langle\tilde{\vec{v}}\rangle_n=(v_{nx},0,v_{nz}) \tag{9}$$

with $v_{nx}=\delta_{nx}/m$ and v_{nz} unchanged. The transverse component v_{nx} is here understood as being averaged over all impact parameters. As can be seen from eq. (6), in the kaon production cross section the modulus v_{nm} between the two colliding nucleons which produce the kaon, enters. This quantity is <u>enlarged</u> if the mean field effect according to the mechanism just described is included, as is obvious from eq.(9). Thus, there is more energy available to produce kaons than without mean field, so the kaon inclusive production cross section will be _enhanced_.

In contrast to the treatment of the mean field just discussed, in numerical simulations[11] of the physical situation described by the Boltzmann-Uehling-Uhlenbeck equation a common nuclear mean field is considered which is built up by all the participant nucleons and consequently depends on the density of these. For a "stiff" equation of state (i.e. large incompressibility) an appreciable amount of the interaction energy will be stored into compression; this energy is unavailable for kaon production, just like in the pion case discussed in the introduction. Of course the reduction of the kaon production rate is appreciable only if most of the kaons are produced at the stage of large compression which, according to the numerical calculations of ref.[6], is in fact the case.

5. SIGNATURES OF COMPRESSION EFFECTS

First we discuss how the kaon inclusive total production cross section will be altered if we include the mean field effects within the extended transport theory as described in section 4. The basic quantity here is the transverse momentum transfer δ_{nx}. An explicit expression for it is given in ref.[10] asa function of the impact parameter. We average δ_{nx} over all impact parameters, insert it in eq. (9), then calculate the modulus v_{nm} of the relative velocity, enter into eq. (6), and finally calculate the total production cross section. In the energy range considered, the curves of fig. 2 are changed by less than 10 per cent. We therefore can safely say that this mechanism has no visible influence on the kaon production rate , even though it leads to a sizeable sideward flow.

The situation is different in the numerical calculations of ref.[6]. Two nuclear equations of state have been considered there, a stiff and a soft one, with the parameters of ref.[11]. The system investigated is Nb on Nb. At an incident energy of 700 MeV/nucleon the kaon inclusive production rate with the stiff equation of state is reported to be roughly half of that with the soft equation of state. Unfortunately, no absolute values for the inclusive production rate are given at other bombarding energies. However, the reduction will presumably not become larger than a factor of two at lower energy; at higher energy, it will be smaller anyhow, because of the growing transparency of the collision. Hence, the differences in the calculations with a stiff and soft equation of state (or with no compression at all) will not exceed the uncertainties caused by the insufficient knowledge of the elementary kaon production cross section which are also about a factor of two (at least) as we have shown in section 3. It therefore appears that a comparison of measured with calculated absolute subthreshold kaon yields will not lead to reliable information on the compressional part of the nuclear equation of state.

A more promising way of getting hold of the compressional energy is to look instead at <u>ratios</u> of the kaon yields between heavy and light colliding identical nuclei, at fixed bombarding energy; in other words, one should study the mass number (A) dependence of kaon yield ratios. Our optimism is based on the results we have obtained for such

Table 1. The (Nb+Nb)/(Ne+Ne) kaon inclusive yield ratio with the parametrizations (i)-(iii).

E_{lab}(MeV)	(i)	(ii)	(iii)
400	22.7	22.6	24.0
500	21.9	21.8	23.2
600	21.1	20.9	22.6
700	20.4	20.2	22.0
800	19.7	19.5	21.5
900	19.1	18.8	20.9
1000	18.5	18.1	20.4
1100	18.0	17.6	20.0
1200	17.5	17.1	19.6

ratios within the transport model where compression effects are a priori absent. In table 1 we show such ratios for the systems Nb on Nb and Ne on Ne as a function of the incident energy, for the parametrizations (i)-(iii) of fig. 1 of the elementary production cross section. The numbers of table 1 exhibit two important features: (i) the kaon yield ratios differ by less than 10 per cent for the various parametrizations, and hence, much of the uncertainty caused by the elementary cross section in the kaon yields is eliminated in the ratios; (ii) the numbers differ from an A^2-dependence ($A^2_{Nb}/A^2_{Ne}=21.6$) by less than 20 per cent. We consider the approximate A^2-dependence of kaon inclusive yields to be a rather universal feature of models which lack compressional energy[7,8], and therefore expect such a behaviour also from corresponding future experimental results, if compression effects are absent; if they are present, the mass number dependence is expected to be largely reduced, for the following reason. For a light system like Ne on Ne, or, even better, C on C, compression effects should be small; for a heavy system like Nb on Nb, or, even better, Pb on Pb, compression effects, if they are present at all, should be large. A reduction factor of two as suggested by the calculations of ref.[6] would then reduce the A^2-dependence by a factor of two, an effect which should clearly be visible from the experimental data.

To summarize, in contrast to the sideward flow which can be explained by both the mean field mechanisms described above, subthreshold kaon production provides a unique tool to probe the compressional part of the nuclear matter equation of state, if ratios of the measured kaon yields for heavy and light colliding nuclei, taken at the same bombarding energy per nucleon, are considered. Complementary measurements of the elementary inclusive production cross section near the lowest threshold of 1.6 GeV would certainly be helpful in gaining additional information from a comparison between measured and calculated <u>absolute</u> kaon yields.

REFERENCES

1. Stock, R., Bock, R., Brockmann, R., Dacal, A., Harris, J.W., Maier, M., Ortiz, M.E., Pugh, H.G., Renfordt, R.E., Sandoval, A., Schroeder, L.S., Ströbele, H., and Wolf, K.L., Phys. Rev. Lett. 49, 1236 (1982).

2. Cugnon, J., Nucl. Phys. A387, 191c (1982).

3. Stock, R., Phys. Reports 135, 259 (1986).

4. Malfliet, R. and Schürmann, B., Phys. Rev. C31, 1275 (1985).

5. Schnetzer, S., Lemaire, M.-C., Lombard, R., Moeller, E., Nagamiya, S., Shapiro, G., Steiner, H., and Tanihata, I., Phys. Rev. Lett. 49, 989 (1982).

6. Aichelin, J. and Ko, C.M., Phys. Rev. Lett. 55, 2661 (1985).

7. Zwermann, W. and Schürmann, B., Nucl. Phys. A423, 525 (1984).

8. Randrup, J. and Ko, C.M., Nucl. Phys. A343, 519 (1980) and A411, 537 (1983).

9. Ferrari, E., Nuovo Cim. 15, 652 (1960).

10. Schürmann, B. and Zwermann, W., Phys. Lett. 158B, 366 (1985).

11. Bertsch, G.F., Kruse, H., and Das Gupta, S., Phys. Rev. C29, 673 (1984) and C33, 1107 (1986).

REACTION TRAJECTORIES IN THE HADRONIC PHASE TRANSITION REGION

David H. Boal[†]

Department of Physics

University of Illinois at Urbana-Champaign

1110 West Green Street

Urbana, IL 61801

ABSTRACT

Computer simulations are used to find the trajectories of nuclear reactions in the liquid-gas and quark-hadron phase transition regimes. It is found that proton induced and heavy ion induced reactions at intermediate energies enter the liquid-gas transition region, albeit in different ways. The simulations also show that the quark-hadron phase transition region should be accessable in heavy ion reactions at $\sqrt{s_{NN}} > 40$ GeV/c.

1. INTRODUCTION

One of the problems in accelerator based searches for signatures of hadronic phase transitions* is that the reactions are, by their very nature, time dependent: any possible phase transition (using the word loosely here) is a transient phenomenon whose effects may be obscured by the subsequent evolution of the reaction. One method of dealing with the time evolution of such finite systems is the use of computer simulations. Subject to the accuracy of the model, simulations may allow one to find out whether a particular reaction

†Permanent address: Dept. of Physics, Simon Fraser University, Burnaby, B.C. V5A 1S6 Canada.

*Since this is not meant to be a review paper, the interested reader is referred to the many other articles in this proceedings for further references. The references quoted in this paper are only those which pertain directly to the text.

accesses the phase transition region, what experimentally observable effects are generated by the transition and whether any of these effects survive the further evolution of the system. The finite size of the reaction region is obviously included as well. In this paper, some aspects of the nuclear liquid-gas transition will be investigated via computer simulation for both proton and heavy ion induced reactions. As well, ultra-relativistic heavy ion collisions will be examined to see if the conditions appropriate to the formation of a quark-gluon plasma can be achieved.

2. LIQUID-GAS TRANSITION.

2.1 Phase Diagram

Many calculations have been performed in the past decade or more delineating the liquid-gas coexistence regions of nuclear matter. One example[1] which used a zero range Skyrme interaction, is shown in Fig. 1. Curves of constant entropy per nucleon (S/A) are indicated by the dashed lines. The unstable regions are bounded by the

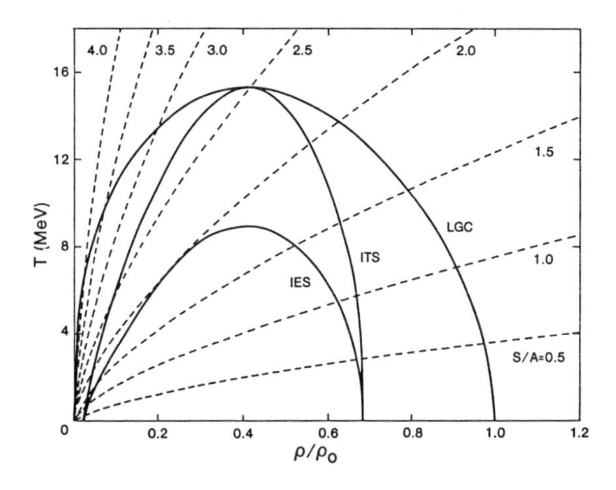

Figure 1. Phase diagram for nuclear matter from zero range Skyrme interaction of Ref. 1. The dashed curves are isentropes with the value of S/A indicated for each. Shown as well are the liquid-gas coexistence boundary (LGC), isothermal spinodal (ITS) and isentropic spinodal (IES).

isothermal spinodal (ITS) and isentropic spinodal (IES) curves. It is clear from the analysis of low and intermediate energy nuclear reaction data (for example, the two particle correlation measurements described elsewhere in these proceedings) that the trajectories of these reactions traverse the phase transition region near freeze out: typical temperatures at freeze-out are a few MeV and densities are of the order $1/2\ \rho_o$. Although the current data analysis methods are mainly concerned with the freeze-out region, the information they yield can be used as a check on the simulations. For example, in Ref. 2 the NN scattering rate was compared with the rate of expansion of an isotropic fireball to estimate its freeze-out characteristics. Such comparisons with data give one at least some confidence that the results of the simulations can be trusted for other parts of the reaction trajectory.

2.2 Proton - Induced Reactions

Proton (and presumably electron) induced reactions have the advantage of producing a system whose density is relatively uniform compared to the large scale inhomogeneities which may be present in a heavy ion reaction. Their disadvantage, for both experiments and simulations, is the low yield of fragment products. Hence, most of the discussion of fragmentation will be deferred to the following section on heavy ion reactions for the sake of providing better statistics. However, there are some simplifications which, while they would be unjustified for heavy ion reactions, are not too bad for the initial stages of a proton induced reaction. These simplifications allow a tremendous decrease in computer execution time per event and, in turn, allow calculation of some quantities which require good statistics. The model is as follows:

The simulation is performed for a single nucleon on a one hundred nucleon target. The motion of the particles is followed classically. The target nucleons are placed in a space-fixed square well potential (it can be verified by using the mean field generated potential model described in 2.3 that this is not a bad approximation in proton induced reactions as long as the reaction time frame is kept short, less than 60 fm/c). The Pauli principle is incorporated only crudely:

i) The momenta of the nucleons in the target obey a Fermi distribution.

ii) A NN collision is considered Pauli blocked unless the final momenta of both nucleons are lifted above the Fermi surface.

Further details can be found in Ref. 1. There are two results from this calculation which will be commented on. The first is the distribution of unbound nucleons in momentum space. Shown in Fig. 2 is the distribution 24 fm/c after the projectile nucleon has entered the target. One can see the semi-circular contours of constant probability characteristic of the thermal model analysis[3] of the (p,p') data show through clearly. The absolute normalization of the distribution is also in the right range. However, even though this distribution is similar to what one expects from a single equilibrated reaction region, in fact, the system is only barely in equilibrium. (In other words, the coordinate space density, and hence the collision rate which maintains equilibrium, is very low).

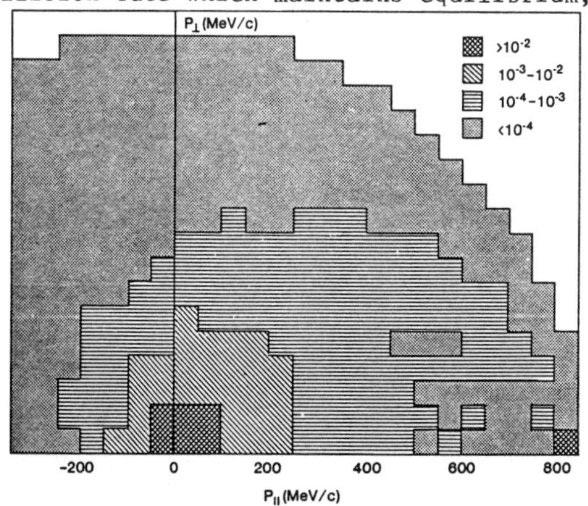

Figure 2. Model prediction for the free particle momentum space density in an impact parameter averaged 300 MeV p + (N = Z = 50) collision. The density is given in units of nucleons per $(50\ \mathrm{MeV}/c)^3$. Nucleons bound in the nucleus are not included in this plot. The distribution was sampled at 8×10^{-23} sec after the projectile has entered the nucleus. From Ref. 1.

However, what is relevant for the phase transition is the reaction trajectory. It has been proposed[4] that fragment formation in proton induced reactions at much higher energies than those considered here, may proceed via condensation near the critical point. Since pion production is not included in the simple model under consideration here, we have nothing to say about such high energy fragmentation. What we can say at lower energies based on the free nucleon multiplicity, is that events with enough nucleons emitted to form a sizeable fragment are very rare. It is more likely that fragmentation in this regime corresponds to breakup of the residual system. To find out if the system actually reaches the mechanical instability region, the numerically generated energy occupation probability f(ε) is integrated directly to give the entropy:

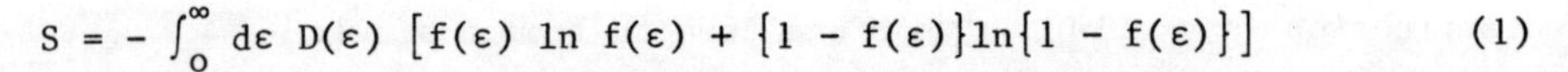

$$S = -\int_0^\infty d\varepsilon\, D(\varepsilon)\left[f(\varepsilon)\ln f(\varepsilon) + \{1 - f(\varepsilon)\}\ln\{1 - f(\varepsilon)\}\right] \qquad (1)$$

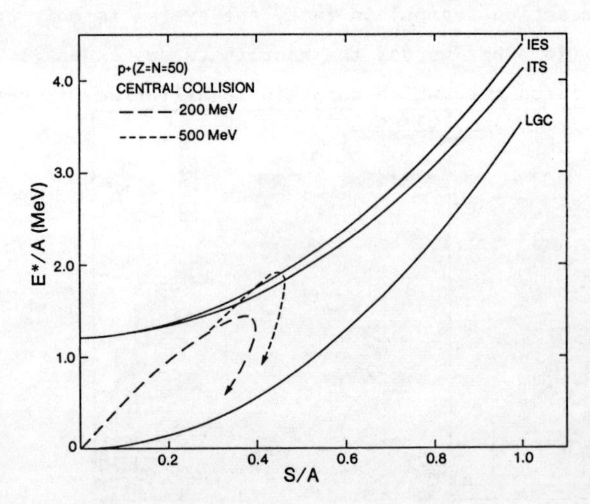

Figure 3. Isothermal spinodal (ITS), isentropic spinodal (IES) and liquid-gas coexistence curve (LGC) shown as a function of excitation energy and entropy per nucleon. The reaction trajectories for the excited nuclear system predicted by the model for 200 and 500 MeV p + (N = Z = 50) central collisions are shown by the dashed lines. From Ref. 1.

where $D(\varepsilon)$ is the density of states. The results of the calculation are illustrated in Fig. 3, where the reaction trajectory is shown as a function of E^*/A (excitation energy per nucleon) and S/A for a central collision at two bombarding energies. (What is shown is the entropy of the nucleons within the target radius, not the entropy of the system as a whole). One can see that the trajectory does reach the isentropic spinodal. One can assume a chemical equilibrium model at breakup to predict the fragment yields, with the parameters of the model being determined by where the trajectory intersects the is entropic spinodal. While the predicted yields are in at least qualitative agreement with the data, this does not constitute proof of spinodal decomposition.

2.3 Heavy Ion Reactions

The fixed target potential simplification which we used above for proton induced reactions is clearly inapplicable for heavy ion reactions and one must find a better way of handling the nucleus. One such approach, which has been used[5] in simulations of the Boltzmann equation, is to introduce a density dependent mean field $U(\rho)$ which allows the nucleons to bind together and form a nucleus. In the Boltzmann simulations, the density is an averaged quantity. Gale and Das Gupta[6] suggested that one could also use the mean field to propagate the fluctuations generated in an individual collision event. The model[7,8] which we wish to use here to investigate the phase transition region uses both a mean field and scattering terms, and, as such, is related to the Vlasov-Uehling-Uhlenbeck equation (VUU). However, the model is still essentially a classical equation of motion calculation, as can be seen from its ingredients:

i) Subject to certain constraints, a random initialization is made for the positions and momenta of the nucleons in a nucleus.

ii) The nucleon's position in phase space is smeared by means of a Gaussian of the form $\exp\{-[\alpha^2(\Delta r)^2] - [(\Delta p)^2/(\hbar\alpha)^2]\}$.

The smeared positions are used to calculate densities in both phase space and coordinate space.

iii) The coordinate space positions of the nucleons are propagated classically according to the variation in $U(\rho)$.

iv) A collision term is present, and the phase space occupancy is used to determine whether a given collision is allowed.

The only place where a quantum mechanical effect comes in (aside from the initial Fermi gas momentum distribution which is put in by hand) is in the Pauli blocking present in the collision term. For example, there is no variable de Broglie wavelength associated with each particle, and the only way that identical particles can be kept out of each other's phase space is through the collision term. Nevertheless, we feel that it is a useful first step on the way to constructing a quantum mechanical model.

Only a few of the results we have compiled will be touched on. For brevity, we will restrict ourselves to results for an equal mass collision of two ($Z = N = 20$) nuclei at 150 A•MeV in the lab. A scatter plot of the products of the reaction as a function of impact parameter (b) is shown in Fig. 4. As one would expect, collisions at large impact parameter do very little damage to the original nuclei: the greatest proportion of fragments come from more central collisions. Let's specialize to central collisions, then, and follow their temporal evolution.

One calculation which was performed was to follow the local coordinate and phase space density observed by nucleons which ultimately emerged as members of a cluster of given mass. The average coordinate space density is shown in Fig. 5 for several cluster masses: 1, 4 and 30. One can see that the average local density for each of the clusters starts out at about the same value at 20 fm/c, corresponding to maximum overlap of the target and projectile. The mass 1 systems observe a local density which simply decreases with time, as the free nucleons become further separated from nearby clusters. The mass 30 systems show very different behavior, which we observed for all heavy systems. The average

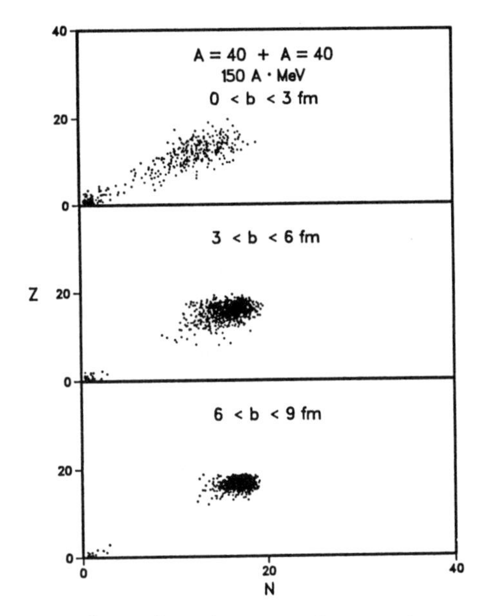

Figure 4. Scatter plots for the reaction products predicted by the VUU model for equal mass (Z = N = 20) collisions at 150 A•MeV. Three ranges of impact parameter are shown: 0-3 fm, 3-6 fm and 6-9 fm. From Ref. 7.

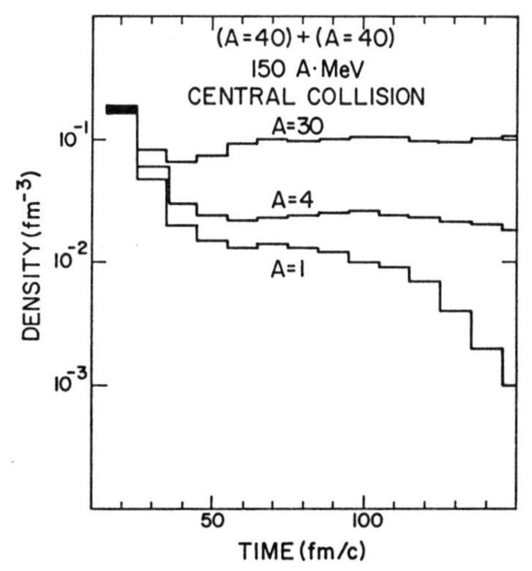

Figure 5. Average coordinate space density for several cluster masses (A = 1, 4, 30) shown as a function of time for a central collision of two (A = 40) nuclei at 150 A•MeV in the VUU model. From Ref. 7.

spatial density decreases, usually to about 0.06 fm^{-3}, then increases and oscillates about its asymptotic value (the average density of a light nucleus will be considerably less than the central value). This behavior is what one expects from the mechanical instability region: if a system has enough energy to enter the region at low density, it fragments; if it does not have enough energy, it oscillates. This effect can be seen even more dramatically in Fig. 6, where the average phase space density is shown for the same cluster masses. Even at t = 20 fm/c, where the coordinate space densities for all the clusters are similar, the phase space densities are very different. Nucleons which emerge as free particles already have been scattered into low density regions of phase space at this early stage of the reaction. The nucleons of heavy systems remain in relatively constant density regions throughout.

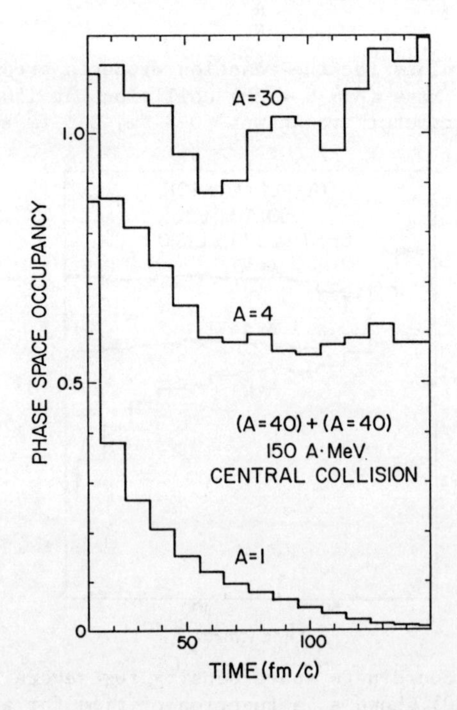

Figure 6. Average phase space occupancy for the same conditions as Fig. 5. From Ref. 7.

In summary, a picture of fragmentation at intermediate energy emerges in which the phase space fluctuations are the driving force behind the eventual outcome of a reaction. However, we have not yet found a way of inverting experimentally measured yields to determine where the instability boundary lies.

3. ULTRA-RELATIVISTIC HEAVY ION COLLISIONS

We now wish to shift our energy scale abruptly and work in the language of quarks and gluons. In this paper, we will limit ourselves to looking at the question of thermalization in an ultra-relativistic heavy ion collision. The method we adopt to look at this question is the parton cascade model,[9] which is certainly not applicable in the non-perturbative regime, but may be approximately valid for predicting certain quantities in the perturbative regime. A brief outline of the model's ingredients is:

i) Nucleons are represented by 3 quarks and 5 gluons each, the fractional momentum x carried by each massless parton being randomly assigned according to the distributions found from deep inelastic scattering.[10]

ii) The trajectories of the partons are followed classically and the parton collision cross sections are taken from lowest order QCD.[11].

iii) In a collision, the partons may be scattered off mass shell according to a m^{-2} distribution. They decay with a lifetime of $\hbar/\alpha_s m$. The decay products are two collinear massless partons with momentum fraction distributed according to the Altarelli-Parisi[12] splitting functions.

There are no antiquarks present in the initialization chosen; in this model they are produced only through collisions. Hence, the time evolution of their energy density provides a simple means of following the thermalization step of the collision. For example, in a central (A = 50) + (A = 50) collision at $\sqrt{s_{NN}}$ = 40 GeV the antiquark energy density reaches a maximum at 4×10^{-24} sec after the nuclei begins to interpenetrate. Similar results were found for other reactions. Thus, a time scale of ~ $1/2 \times 10^{-23}$ sec seems to be indicated for the thermalization step.

To see whether the gluons, which carry much of the momentum, are reasonably thermalized, we show their distribution in momentum space in Fig. 7. The reaction is (A = 50) + (A = 50) at $\sqrt{s_{NN}}$ = 40 GeV with the reaction stopped at 6×10^{-24} sec. The top part of Fig. 7 shows the distribution of all of the gluons (an average of 1500 per event;

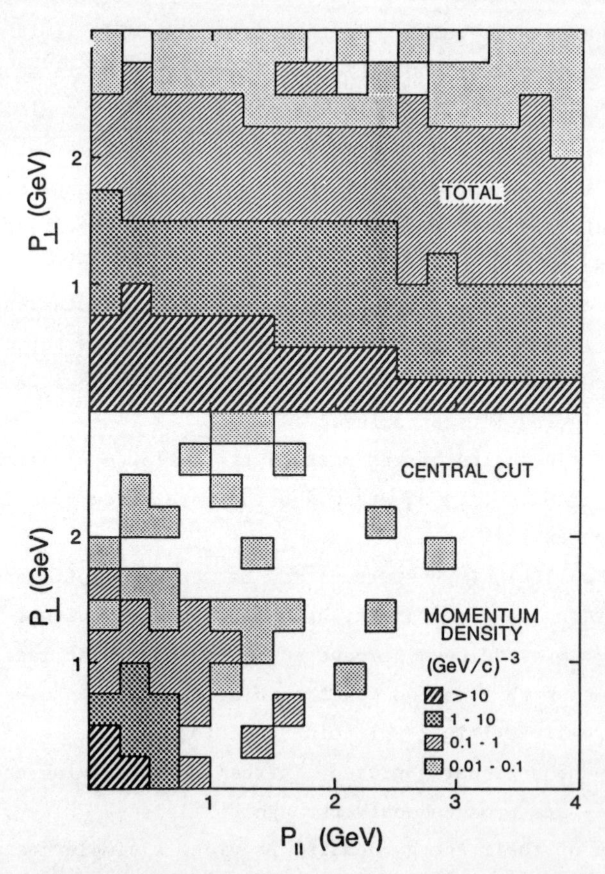

Figure 7. Momentum distribution of gluons in a central (A = 50) + (A = 50) collision at $\sqrt{s_{NN}}$ = 40 GeV observed at 6×10^{-24} sec after the nuclei begin to overlap. Distributions for all the gluons, and for those subject to a central cut in coordinate space, are shown. From Ref. 9.

40 events were summed over) in the directions perpendicular and parallel to the collision axis. One can see that there is a strong longitudinal component to the distribution. However, if one makes a central cut in coordinate space, the distribution is much more isotropic. The cut used in the lower part of Fig. 7 is a cylinder 2 fm in length and 1 fm in radius centered on the center of mass. Enlarging the coordinate space cut in the perpendicular direction introduces more of a perpendicular component to the distribution, corresponding to radial flow.

Having established that the central region is fairly thermalized, the major question is what is the energy density. For the reaction shown in Fig. 7, the energy density associated with the

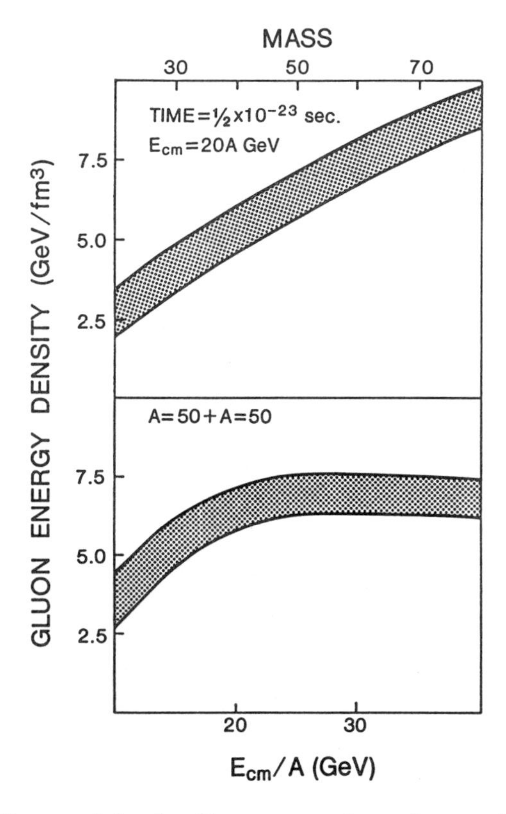

Figure 8. Mass and bombarding energy dependence of the gluon energy density in the central region from Ref. 9.

gluons in the central region is 5 GeV/fm^3. In this code, chemical equilibrium has not been established among the species by 5×10^{-24} sec, and most of the energy is carried by the gluons. However, this value is still more than the few GeV/fm^3 which, it is argued, is required for plasma formation. Larger energy densities can be achieved by increasing either the energy or mass. Examples are shown in Fig. 8 for the gluon energy density in a central region defined as a cylinder 1 fm in length, 3 fm in radius with the reaction stopped at 5×10^{-24} sec. The top part of the figure illustrates the mass dependence (equal projectile and target masses) for a (20 A•GeV) + (20 A•GeV) central collision. Calculations were carried out to (A = 100) + (A = 100), beyond which execution time became prohibitively long (3 CPU hours/event at 5 mips). The cross-hatched region represents the uncertainty caused by using only a finite number of events. By using large nuclei, energy densities in excess of 10 GeV/fm^3 can be achieved. In the calculations performed so far, the dependence of the energy density on bombarding energy is not so pronounced. For an (A = 50) + (A = 50) collision, the energy density appears to flatten out above 20 A•GeV in the center of mass, as shown in the lower portion of Fig. 8.

To summarize, using the parton cascade approach, we have demonstrated that the central region in coordinate space achieves approximate thermalization after $1/2 \times 10^{-23}$ sec in an ultra-relativistic collision. The region is baryon number depleted, particularly when comparing the net baryon number density with the gluon number density. The energy densities achievable appear to be quite adequate for the formation of the quark-gluon plasma state, particularly for projectile/target masses in excess of 100 and center of mass energies above $\sqrt{s_{NN}}$ = 40 GeV.

ACKNOWLEDGEMENTS

Alan Goodman (Tulane) and Glenn Beauvais (Simon Fraser University) were principal collaborators on the study of the liquid-gas phase transition question and their work on this problem is happily acknowledged. This work is supported in part by the Natural Sciences and Engineering Research Council of Canada.

REFERENCES

1. Boal, D. H. and Goodman, A. L., Phys. Rev. C (in press).
2. Boal, D. H. and Shillcock, J. C., Phys. Rev. C33, 549 (1986).
3. Boal, D. H. and Reid, J. H., Phys. Rev. C29, 973 (1984).
4. See Finn, J. E., et al., Phys. Rev. Lett. 49, 1321 (1982) and references therein.
5. Bertsch, G., Kruse, H. and Das Gupta, S., Phys. Rev. C29, 673 (1984).
6. Gale, C. and Das Gupta, S., Phys. Lett (in press).
7. Beauvais, G. E. and Boal, D. H., University of Illinois report P/86/2/26.
8. A model which is formally equivalent to Ref. 7, although it differs in its application, has been advanced independently by Aichelin, J. and Stocker, H., Max Planck Institute (Heidelberg) report MPI H-1986-V6.
9. Boal, D. H. in Proceedings of the Workshop for Experiments for a Relativistic Heavy Ion Collider, Haustein, P. and Woody, G., eds., BNL Report and submitted to Phys. Rev.
10. Abramowicz, H., et al., Z. Phys. C17, 283 (1983).
11. Cutler, R. and Sivers, D., Phys. Rev. D17, 196 (1978).
12. Altarelli, G. and Parisi, G., Nucl. Phys. B126, 298 (1977).

SOURCE RADII AND EMISSION TEMPERATURES IN INTERMEDIATE ENERGY NUCLEUS-NUCLEUS COLLISIONS

C.K. GELBKE

Department of Physics
and
National Superconducting Cyclotron Laboratory
Michigan State University, East Lansing, MI 48824, USA

Introduction

For intermediate energy nucleus-nucleus collisions, it has been clearly established that particle emission occurs prior to the attainment of full statistical equilibrium of the composite system.[1-7] In the absence of a complete dynamical treatment, recourse is often taken to models based on the assumption of statistical particle emission from highly excited subsets of nucleons[1-10] which are characterised by their average collective velocity, space-time extent, and excitation energy density or temperature. Clearly, the experimental characterisation of the collective and statistical properties of these subsets is important. In the following, I will discuss two-particle correlation measurements at small relative momenta which provide information about the space-time extent and temperature of the emitting system.

Most attempts to extract temperatures are based on analyses of the kinetic energy spectra of the emitted particles.[4] The interpretation of such spectra can, however, be complicated by sensitivities to the collective motion[11] and the temporal evolution of the emitting system.[6,12] Alternative determinations of the "emission temperature", i.e. the temperature at the point at which the particles leave the equilibrated subsystem, are based on the relative populations of states.[13-17] These "temperature measurements" can have large uncertainties whenever the primary population ratio is altered by secondary processes;[14,15,18] they should become insensitive to secondary processes if the level separation is much larger than the emission temperature.[15] These considerations suggest that accurate temperature determinations should be possible from the measurements of the relative populations of widely separated states. For the case of light nuclei, suitable states are generally particle unstable.

To first order within equilibrium thermodynamics, the relative populations of particle unstable states can be expressed as:

$$\frac{d\,N(E)}{d\,E} = N_0 \cdot e^{-E/T} \cdot \frac{1}{\pi} \sum_{J,\ell} (2J+1) \frac{\partial\, \delta_{J,\ell}}{\partial\, E} , \qquad (1)$$

where T denotes the temperature of the emitting source and $\delta_{J,\ell}$ is the scattering phase shift in the exit channel. If the energy dependence of the phase shifts is dominated by a series of resonances, one obtains:[15]

$$\frac{d\,N(E)}{d\,E} = N_0 \cdot e^{-E/T} \cdot \sum_i (2J_i+1) \frac{\Gamma_i/2\pi}{(E-E_i)^2 + \Gamma_i^2/4} . \qquad (2)$$

Information about the space-time extent of the emitting system can be obtained from the measurement of two-particle correlation functions[19-25] at small relative momenta for which interactions between the emitted particles are important. The two-particle correlation function, R(q), is defined in terms of the singles yields, $Y_1(\vec{p}_1)$ and $Y_2(\vec{p}_2)$, and the coincidence yield, $Y_{12}(\vec{p}_1,\vec{p}_2)$, of particles 1 and 2:

$$\sum Y_{12}(\vec{p}_1,\vec{p}_2) = C \cdot \left(1+R(q)\right) \cdot \sum Y_1(\vec{p}_1) Y_2(\vec{p}_2) . \qquad (3)$$

Here, $\vec{p}_1$ and $\vec{p}_2$ are the laboratory momenta of the two particles, q is the momentum of relative motion, and C is a normalization constant. Experimental correlation functions are obtained by inserting the measured yields into the equation and performing the summation over all energies and angles which satisfy a given gating condition and which correspond to a given relative momentum.

To first order within the framework of equilibrium thermodynamics, the two-particle correlation function for two non-identical particles of spins s_1 and s_2 can be approximated as[23]

$$R(q) = \frac{3}{(2s_1+1)(2s_2+1)\cdot 2r^3q^2} \sum_{J,\ell} (2J+1) \frac{\partial \delta_{J,\ell}}{\partial q} , \quad (4)$$

where r denotes the source radius. In this approximation, measurements of the two-particle correlation function and of the relative populations of states provide independent information on the temperature <u>and</u> the space-time evolution of the emitting system.[17,23]

Historically, the sensitivity of two-particle correlation functions to the space-time extent of the emitting system was derived from the modifications of the wave functions of relative motion due to final-state interactions[19] or quantum statistics.[24,25] If the time-dependence of the emission process is neglected, one can express the two-particle correlation function in terms of the single particle source function, $\rho(\vec{r})$, and the two-body wave function $\Psi(\vec{r}_1,\vec{r}_2)$:[19,23]

$$R(q) = \int d^3r_1 d^3r_2 \{ |\Psi(\vec{r}_1,\vec{r}_2)|^2 - 1 \} \cdot \rho(\vec{r}_1)\rho(\vec{r}_2) \times \left(\int d^3r \, \rho(\vec{r}) \right)^{-2} . \quad (5)$$

This formula has recently been shown[23] to be consistent with the thermal model, eq. 4. For our calculations, we have adopted the original formulation of Koonin and assumed a source of Gaussian spatial density, $\rho(r) \propto \exp(-r^2/r_0^2)$, and negligible lifetime.

Emission Radii

Particles with different reaction cross sections are expected to decouple from the emitting system at different average densities or source radii.[22] Detailed information about the breakup of highly excited nuclear systems may, therefore, be obtained from investigations of two-particle correlation functions for different particle pairs which are emitted in the same reaction.[21,22] Furthermore, particles of different energies may be emitted at different stages of the reaction.[17,20] The energy dependence of the correlation functions can, therefore, provide useful additional information about the space-time characteristics of the equilibration process.

Figures 1-3 show two-particle correlation functions[26] for several different pairs of light particles emitted in ^{14}N induced reactions on ^{197}Au at E/A=35 MeV. In order to explore the dependence of the correlation functions on the energy of the outgoing particles, the correlation functions were evaluated for different constraints on the sum energy, E_1+E_2, of the two coincident particles. These constraints are indicated in the figures.

Two-particle correlation functions for pairs of identical hydrogen isotopes are presented in Figure 1. For the two-proton correlation function, the attractive singlet S-wave interaction gives rise to a maximum at $q \approx 20$ MeV/c. The two-deuteron and two-triton correlation functions, on the other hand, do not exhibit maxima since the interactions between these pairs of particles are not resonant at low relative momenta.[21,27,28]

The p-α and d-α correlation functions are presented in Figures. 2 and 3, respectively. The p-α correlation function, Fig. 2, exhibits a broad maximum near $q \approx 50$ MeV/c which is due to the decay of the ground

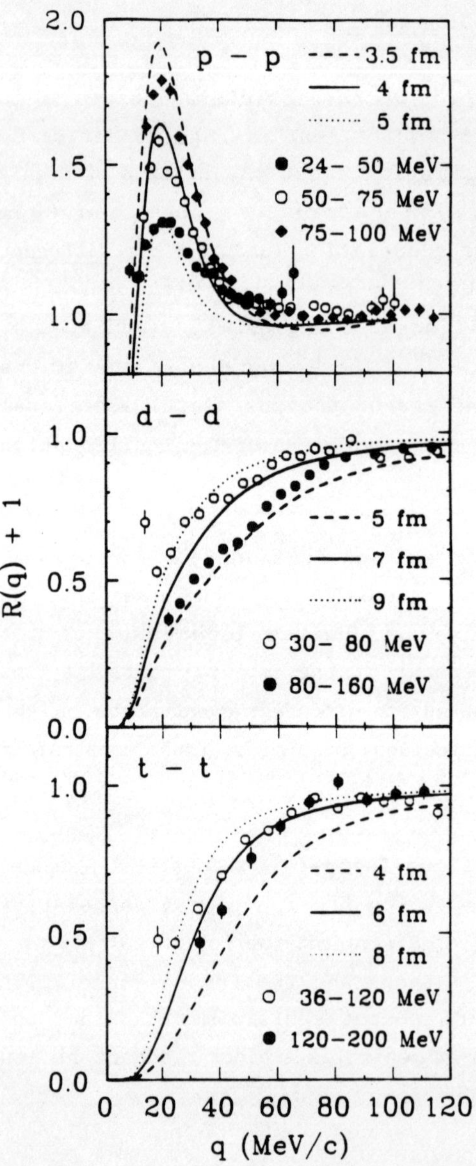

Fig. 1: Two-particle correlation functions for proton, deuteron, and triton pairs emitted at $\Theta_{av}=35°$ in ^{14}N induced reactions on ^{197}Au at E/A=35 MeV. (From ref. 26)

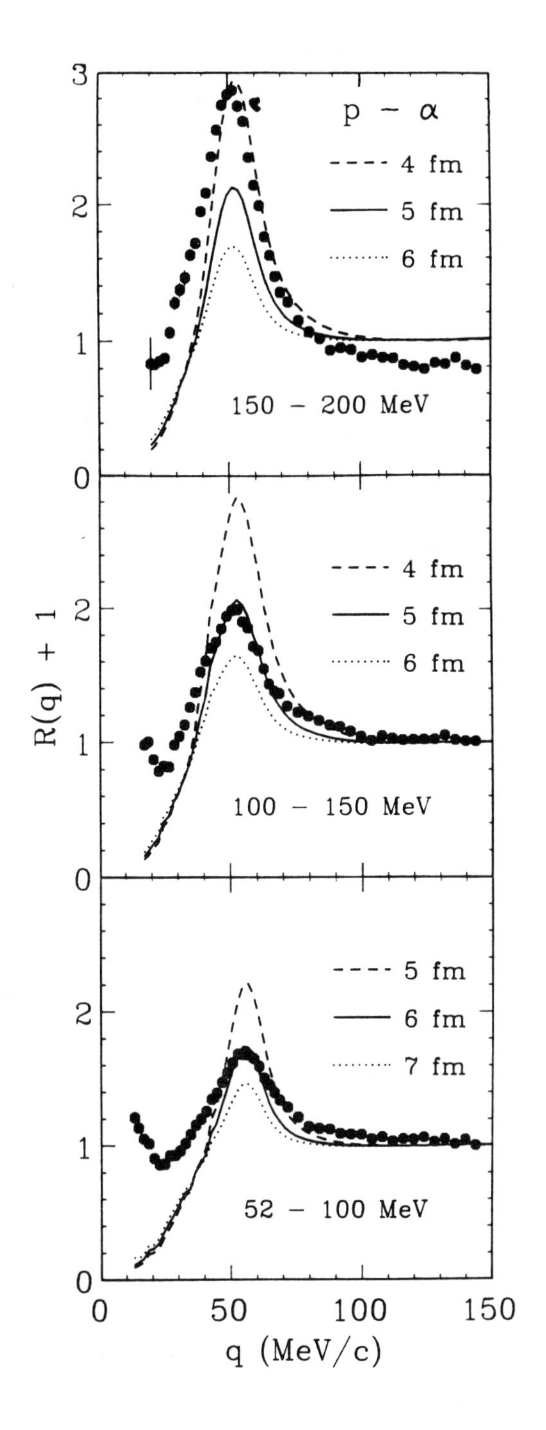

Fig. 2: Correlation functions between coincident protons and alpha particles emitted at $\Theta_{av}=35°$ in ^{14}N induced reactions on ^{197}Au at E/A=35 MeV. (From ref. 26)

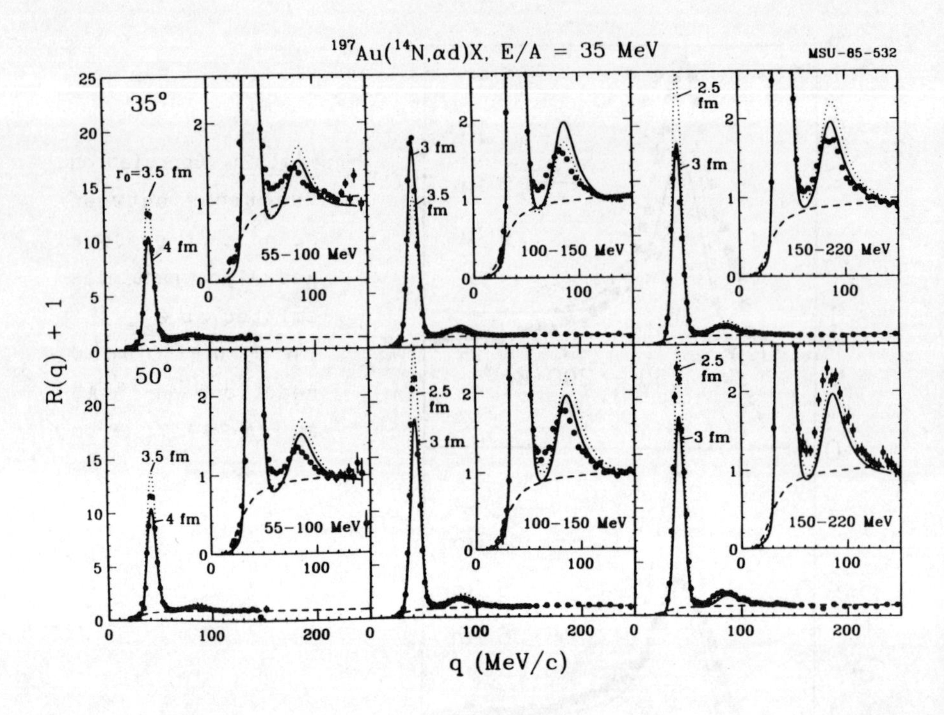

Fig. 3: Correlation functions between coincident deuterons and alpha particles emitted at $\Theta_{av} \approx 35°$ in ^{14}N induced reactions on ^{197}Au at E/A=35 MeV. (From ref. 17)

state of ^5Li ($J^\pi = \frac{3}{2}^-$, Γ=1.5 MeV) in the Coulomb field of the heavy residue.[16] The rise of the correlation function at small relative momenta, $q \lesssim 25$ MeV/c, is caused[29] by the decay of $^9B \to 2\alpha+p$. The α-d correlation function,[15,17] Fig. 3, exhibits two maxima corresponding to the T=0 state in ^6Li at 2.186 MeV ($J^\pi=3^+$, Γ=24 keV, $\Gamma_\alpha/\Gamma_{tot}$ = 1.00) and the overlapping T=0 states at 4.31 MeV ($J^\pi=2^+$, Γ=1.3 MeV, $\Gamma_\alpha/\Gamma_{tot}$= 0.97) and at 5.65 MeV ($J^\pi=1^+$, Γ=1.9 MeV, $\Gamma_\alpha/\Gamma_{tot}$= 0.74).

The solid, dashed and dotted curves shown in the figures represent theoretical correlation functions predicted by eq. 5. The two-particle wave functions were generated by numerical integration of Schroedinger's equation using potentials which reproduce the experimental phase shifts. The potential parameters used for the p+p, d+d, p+α, and d+α, wave functions are given in refs. 21 and 22. For the t+t system, only S-wave phase shifts were available;[28] these phase shifts were reproduced with a repulsive Woods-Saxon potential with the parameters V_0=30 MeV, R=4 fm, a=0.6 fm. For higher partial waves, only the Coulomb interaction was included. Because of this approximation, source radii extracted from t-t correlations are less reliable than the ones extracted for the other particle pairs. We have estimated the resulting uncertainties, by using the ℓ=0 potential parameters also for the ℓ = 1 and 2 partial waves. With this prescription, the extracted source radii were systematically larger by about 2 fm.

With the exception of the sharp peak of the α-d correlation function at q≈40 MeV/c, the measured correlations were not affected by the experimental resolution. The theoretical α-d correlations include corrections for the finite resolution of the experimental apparatus. Source radii were extracted from the peak at q≈40 MeV/c. (Radii extracted from the second peak are larger by about 0.5 fm. At present, the origin of this discrepancy is not clear.) It could be due to sequential feeding from highly excited primary reaction products, the temporal evolution of the reaction, or uncertainties in the resonance parameters. For the theoretical p-α correlations, the line shape was corrected, to first order,[29] for the acceleration of the emitted particles in the Coulomb field of the heavy reaction residue.

For all cases investigated, the measured correlations become more pronounced with increasing energy of the two coincident particles indicating that more energetic light particles are emitted from sources which are more localized in space-time. This feature is quantified by the estimated source radii summarized in Table 1. Also included in the table are source radii extracted from the measurements at $\Theta_{av}=50°$. No significant dependence of the correlation functions on angle is established. The observation of considerable correlations at angles significantly larger than the grazing angle renders interpretations[30)] in terms of the sequential decay of projectile fragments unlikely.

Table 1: Source radii, r_0, for a source of negligible lifetime and Gaussian density, $\rho(r)=\rho_0 \cdot \exp(-r^2/r_0^2)$, extracted from two-particle correlation functions; the corresponding rms radii are given by: $r_{rms} = \sqrt{3/2} \cdot r_0$; equivalent sharp sphere radii are given by: $R=\sqrt{5/2} \cdot r_0$.

correlation	E_1+E_2 [MeV]	$r_0(35°)$ [fm]	$r_0(50°)$ [fm]
p+p	24 - 50	4.9 ± 0.5	5.2 ± 0.5
	50 - 75	4.3 ± 0.3	4.0 ± 0.3
	75 - 100	3.8 ± 0.2	3.6 ± 0.2
d+d	30 - 80	8 ± 2	-
	80 - 160	5.5 ± 1	-
t+t	36 - 120	6.5 ± 1	-
	120 - 200	5.5 ± 1	-
p+α	52 - 100	6 ± 0.5	6 ± 0.5
	100 - 150	5 ± 0.5	4.8 ± 0.3
	150 - 200	4 ± 0.5	3.7 ± 0.3
d+α	55 - 100	3.8 ± 0.2	3.9 ± 0.2
	100 - 150	3.0 ± 0.2	2.8 ± 0.2
	150 - 220	3.0 ± 0.2	2.7 ± 0.2

The source sizes extracted for the emission of energetic p-p and α-d pairs are smaller than the size of the target nucleus [$r_0(Au)$ = $\sqrt{2/3}\cdot r_{rms}(Au) \approx 4.3$ fm]. For the other particle pairs, larger source dimensions are obtained. Since the inclusion of the temporal evolution of the emitting system is expected to reduce the calculated two-particle correlations, source radii extracted under the assumption of negligible lifetime represent upper limits for the spatial extent of the emitting system. The extracted source radii may be ordered approximately as follows: $r_0(\alpha+d) \leq r_0(p+p) \leqq r_0(\alpha+p) \leqq r_0(t+t)$, $r_0(d+d)$. For an interpretation of these results, several questions should be addressed: (i) Different reaction products may have different impact parameter weightings; protons may have larger contributions from larger impact parameters than composite particles.[31] (ii) The problem of sequential feeding from highly excited primary reaction products may alter the two-particle correlation functions. (iii) Different particle species may go out of equilibrium at different densities, depending on their interaction cross sections.[22] A consistent treatment of the temporal evolution of the emitting source is not yet available. (iv) The calculation of the theoretical correlation functions involve several uncertainties. The effects of the Coulomb interaction with the residual nuclear matter have not yet been treated reliably. In addition, there are still uncertainties in the low energy phase shifts (particularly for the t+t system).

External Coulomb Distortions

Correlations between two particles which experience non-resonant final-state interactions and which have different mass-to-charge ratios are sensitive to the Coulomb field of the residual nuclear system at the point of freeze-out. This sensitivity is derived from the facts that no long-lived intermediate states are formed if the final-state interactions are non-resonant and that particles with different charge-to-mass ratios experience different accelerations in the Coulomb field of the residual nuclear system.[29] These effects have been observed for correlations between coincident protons and deuterons.[26]

Because of the repulsive and non-resonant[33] proton - deuteron interaction, the coincidence yields between protons and deuterons exhibit a clear minimum at small relative velocities, $v_p \approx v_d$. However, the exact location of the minimum is displaced from the line of minimum relative velocity because of the accelaration of the two particles in the Coulomb field of the residual nuclear system.[26]

In order to analyze this effect, it is convenient to introduce new coordinates, $S = (2E_p - E_d)/\sqrt{5}$ and $T = (E_p + 2E_d)/\sqrt{5}$, along the directions parallel and perpendicular to the line of minimum relative velocity, $E_d \approx 2E_p$. In terms of these coordinates, the line $E_d = 2 \cdot E_p$ corresponds to S=0 and acceleration in the Coulomb field of the residual system shifts the minimum of the correlation function, R(S), to S>0. In analogy to eq. (1), this correlation function is defined as:

$$\sum Y_{pd}(S,T) = C \cdot \big(1+R(S)\big) \cdot \sum Y_p[E_p(S,T)] \cdot Y_d[E_d(S,T)].$$

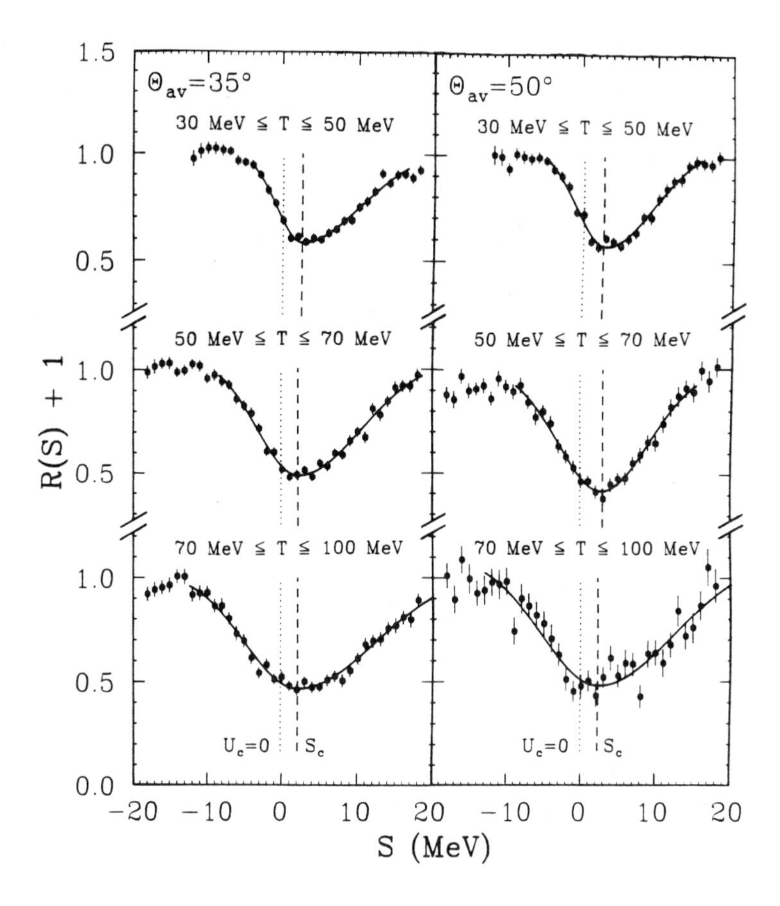

Fig. 4: Correlation function for coincident protons and deuterons as a function of the coordinate S defined in the text.

Figure 4 shows correlation functions R(S) measured for Θ_{av}=35° and 50°. The summation was performed over all detector pairs for which $\Delta\Theta$=6.1° . Apart from the energy thresholds, $E_p \geq 12$ MeV and $E_d \geq 15$ MeV, the indicated constraints on T were applied. The values, S_0, corresponding to minimum relative velocity for U_c=0 are marked by the dotted lines; the locations of the observed minima, S_c, are marked by the dashed lines. No significant energy or angle dependence is observed; the average value is $\bar{S}_c$= 2.6±0.8 MeV. Three-body Coulomb trajectory calculations[26] indicate that this value is consistent with a distance of emission of $d < 15$ fm; it is consistent with emission from the surface of the composite system. Significant contributions to the emission of coincident protons and deuterons from the sequential decay of long-lived projectile fragments[30] can be excluded. It would clearly be interesting to have a more quantitative treatment of the p-d correlation function which includes Coulomb and nuclear interactions in a consistent manner.

Emission Temperatures

Emission temperatures can be obtained by comparing the experimental yield of particle unstable nuclei with thermal calculations.[15-17] In order to extract the coincidence yield, Y_c, resulting from the decay of particle unstable nuclei, we have assumed that the coincidence yield, Y_{12}, is given by $Y_c = Y_{12} - C \cdot Y_1 Y_1 [1+R_b(q)]$, where $R_b(q)$ denotes the background correlation function.

In Figure 3, the background correlation functions are shown by the dashed lines. The resulting yields from particle unstable $^6Li^*$ nuclei are shown in Figure 5 as a function of the kinetic energy, $T_{c.m.}$, in the ^6Li rest frame. The curves shown in the figure correspond to the theoretical coincidence yields resulting from the decays of thermally emitted particle unstable ^6Li nuclei. The calculations of the decay yields incorporate the appropriate decay branching ratios as well as the efficiency and resolution of the hodoscope.[15-17] The calculated yields are normalised to reproduce the experimental yield integrated over the energy range of $T_{c.m.} = 0.3 - 1.2$ MeV. In order to extract emission temperatures, we have integrated the decay yields over the energy ranges of $T_{c.m.} = 0.25-1.45$ and $1.5-6.25$ MeV and compared the ratio of these yields to the corresponding theoretical ratio. The results are summarized in Table 2.

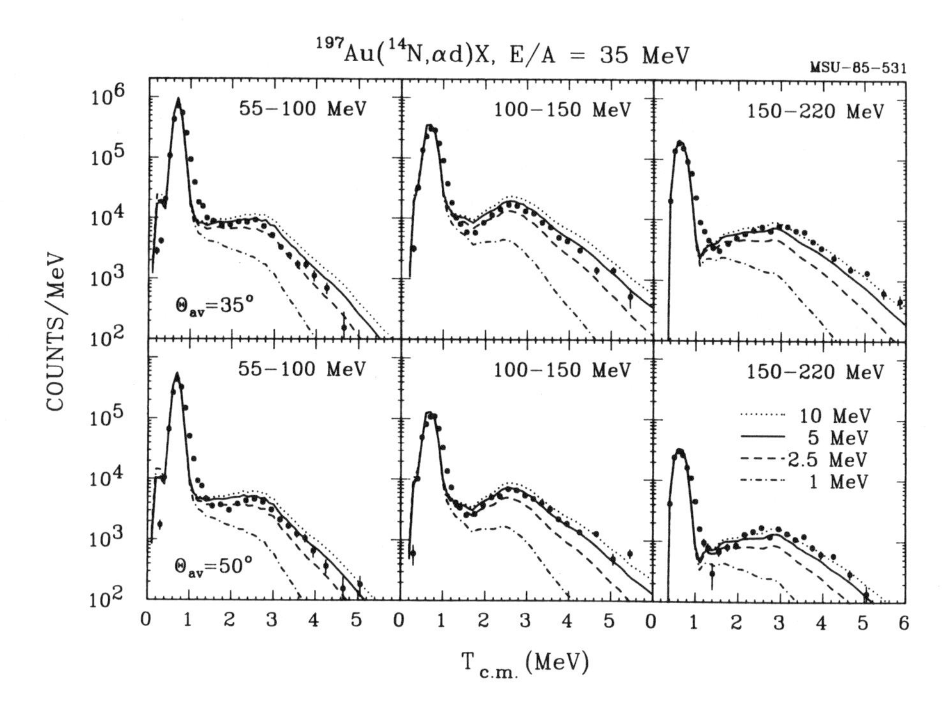

Fig. 5: Energy spectra resulting from the decay $^6Li^* \rightarrow \alpha+d$. For a detailed discussion of the figure see text.

Table 2: Emission temperatures extracted from the decay $^6Li^* \rightarrow \alpha+d$.

$E_1 + E_2$	T(35°)	T(50°)
55 - 220 MeV	4 MeV	4 MeV
55 - 100 MeV	4 MeV	3 MeV
100 - 150 MeV	4 MeV	5 MeV
150 - 220 MeV	7 MeV	9 MeV

Higher emission temperatures and smaller source radii are extracted for higher kinetic energies of the emitted particles. These findings are consistent with particle emission from a subsystem which is in the process of cooling and expanding. Cooling and expanding subsystems of high excitation could arise from the equilibration of participant matter with the surrounding cold target nuclear matter[6] or from an isentropic expansion as expected from intranuclear cascade calculations.[32] Although these experimental results are highly suggestive, one should realize that the temperatures extracted from the α-d coincidence measurements shown in Figs. 2 and 3 have considerable uncertainties (≈25%, see ref. 17). These uncertainties are due to uncertainties in the α-d background correlation function, uncertainties concerning the importance of feeding from the sequential decay of heavier mass primary fragments[15-17] and due to the saturation of the coincidence yields at higher temperatures. It is also conceivable that the observed energy dependence of the relative populations of particle unstable states of ^{6}Li is caused by an energy dependence of the feeding from higher lying particle-unstable states.

More accurate temperature determinations can be made by measuring the relative populations of states separated by significantly larger energy intervals;[16] e.g.: $^{5}Li_{0.0} \rightarrow \alpha+p$, $^{5}Li_{16.7} \rightarrow d+{}^{3}He$; $^{8}Be_{3.0} \rightarrow \alpha+\alpha$ and $^{8}Be_{17.6} \rightarrow {}^{7}Li+p$. As an example, Figures 6 and 7 show correlations[16] which can be attributed to states in ^{5}Li; these correlations were measured by the MSU-GANIL collaboration for ^{40}Ar induced reactions on ^{197}Au at E/A=60 MeV. Figure 5 shows the α-p correlation function. The broad peak near q≈50 MeV/c is due to the decay of $^{5}Li_{gs} \rightarrow \alpha+p$ in the Coulomb field of the heavy target residue.[28] (The sharp peak near q≈15 MeV/c is not related to a resonance in the mass five system:[29] it is caused by the three particle decay of $^{9}B \rightarrow \alpha+\alpha+p$.) Figure 7 shows the d-$^{3}He$ correlation function. The most pronounced structure in this correlation function is due to the decay $^{5}Li^{*}_{16.7} \rightarrow d+{}^{3}He$. From the relative populations of these two states an average emission temperature of T≈4-5 MeV was extracted.[16] This value is considerably lower than the temperature parameters of T≈20 MeV which characterise the kinetic energy spectra of the emitted particles.[5,15]

Very similar emission temperatures were measured for ^{40}Ar induced reactions[15,16] at E/A=60 MeV and for ^{14}N induced reactions[17] at E/A=35 MeV. At present, this similarity between emission temperatures obtained for two very different reactions is not understood. One possibility is that the two measurements contain very different contributions from peripheral reactions. To eliminate such contributions, measurements must be performed in coincidence with central collision triggers. Clearly, more measurements are needed.

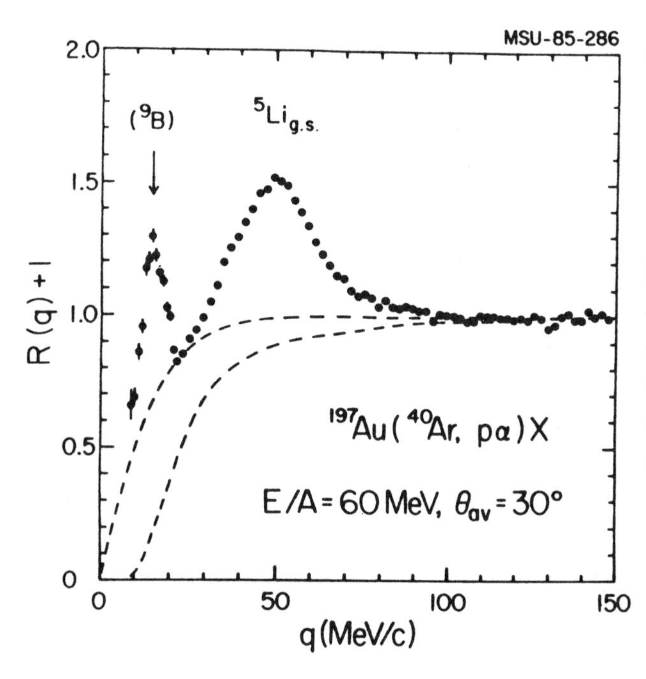

Fig.6: Correlation function for coincident protons and alpha particles for the reaction ^{40}Ar+^{197}Au at E/A=60 MeV. The dashed lines indicate the extremes within which the background correlation function was assumed to lie. (From ref. 16)

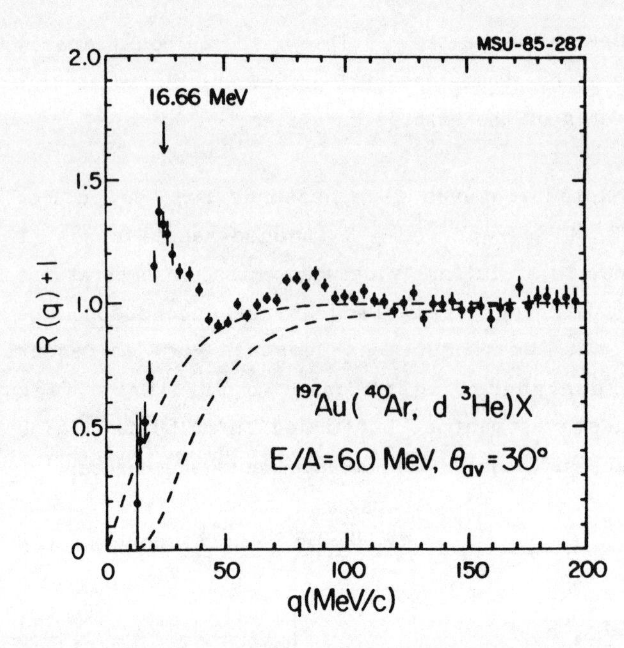

Fig.7: Correlation function for coincident deuterons and ^{3}He nuclei for the reaction ^{40}Ar+^{197}Au^{197}Au at E/A=60 MeV. The dashed lines indicate the extremes within which the background correlation function was assumed to lie. (From ref. 16)

Conclusion

Detailed investigations of two-particle correlations at small relative momenta have only been performed very recently. The technique bears significant promise as a diagnostic tool for studies of nucleus-nucleus collisions at intermediate energies. Two-particle correlation functions contain information about the space-time characteristics of the emitting system. Since particles with different reaction cross sections are expected to decouple from the emitting system at different average densities, one may be able to obtain a detailed understanding about the freeze out phase of the reaction by comparing correlations between different pairs of particles emitted in the same class of reactions. For particles with non-resonant final state interactions and different mass-to-charge ratios the two-particle correlation functions exhibit sensitivities to the Coulomb field of the residual system at the point of freeze out. The consistent quantum mechanical formulation of this problem is not yet available. Emission temperatures can be obtained from the relative populations of widely separated states. An important problem to be addressed by future experimental and theoretical work concerns the importance of sequential decay contributions from particle unbound states to the measured particle cross sections.

Nearly identical emission temperatures of T≈4-5 MeV were measured for the reactions $^{14}N+^{197}Au$ at E/A=35 MeV and $^{40}Ar+^{197}Au$ at E/A=60 MeV. This result raises the question to which extent the assumption of partial (or local) equilibrium will remain to be useful for the interpretation of more detailed measurements. Clearly the use of statistical concepts which were originally developed for the discussion of bulk matter presents an over-simplification. Of paramount importance will be a detailed understanding of fluctuations (in density, "temperature", etc.) which must be large for the systems which are experimentally accessible.

I would like to acknowledge many stimulating discussions with D.H. Boal, W.A. Friedman, W.G. Lynch, and J. Pochodzalla. The results reported here were obtained in collaboration with: D. Ardouin, D. Boal, G. Bizard, Z. Chen, C.B. Chitwood, H. Delagrange, H. Doubre, D.J. Fields, W.A. Friedman, C. Grègoire, A. Kyanowski, F. Lefèbvres, W.G. Lynch, M. Maier, W. Mittig, A. Pèghaire, J. Pèter, J. Pochdzalla, F. Saint-Laurent, J.C. Shillcock, B. Tamain, M.B. Tsang, Y.P. Viyogi, J. Quèbert, and B. Zwieglinski. This work was supported by the National Science Foundation under grants PHY 83-12245 and PHY 84-01845.

References:

1) Awes, T.C., Poggi, G., Gelbke, C.K., Back, B.B., Glagola, B.G., Breuer, H., and Viola, V.E. Jr., Phys. Rev. C24, 89 (1981)
2) Awes, T.C., Poggi, G., Saini, S., Gelbke, C.K., Legrain, R., and Westfall, G.D., Phys. Lett. 103B, 417 (1981)
3) Awes, T.C., Saini, S., Poggi, G., Gelbke, C.K., Cha, D., Legrain, R., Westfall, G.D., Phys. Rev. C25, 2361 (1982)
4) Westfall, G.D., Jacak, B.V., Anantaraman, N., Curtin, M.W., Crawley, G.M., Gelbke, C.K., Hasselquist, B., Lynch, W.L., Scott, D.K., Tsang, M.B., Murphy, M.J., Symons, T.J.M., Legrain, R., Majors, T.J., Phys. Lett. 116B, 118 (1982)
5) Jacak, B.V., Westfall, G.D., Gelbke, C.K., Harwood, L.H., Lynch, W.G., Scott, D.K., Stöcker, H., Tsang, M.B., and Symons, T.J.M., Phys. Rev. Lett. 51, 1846 (1983)
6) Fields, D.J., Lynch, W.G., Gelbke, C.K., Tsang, M.B., Utsunomiya, H., Aichelin, J., Phys. Rev. C30, 1912 (1984)
7) Westfall, G.D., Gosset, J., Johansen, P.J., Poskanzer, A.M., Meyer, W.G., Gutbrod, H.H., Sandoval, A., Stock, R., Phys. Rev. Lett. 37, 1202 (1976)
8) Gosset, J., Kapusta, J.I., Westfall, G.D., Phys. Rev. C18, 844 (1978)
9) Knoll, J., Phys. Rev. C20, 773 (1979)
10) Friedman, W.A. and Lynch, W.G., Phys. Rev. C28, 16 (1983)
11) Siemens, P.J. and Rasmussen, J.O., Phys. Rev. Lett. 42, 880 (1979)
12) Stöcker, H., Ogloblin, A.A., Greiner, W., Z. Phys. A303, 259 (1981)
13) Morrissey, D.J., Benenson, W., Kashy, E., Sherrill, B., Panagiotou, A.D., Blue, R.A., Ronningen, R.M., van der Plicht, J., and Utsunomiya, H., Phys. Lett. B148, 423 (1984)

14) Morrissey, D.J., Benenson, W., Kashy, E., Bloch, C., Lowe, M., Blue, R.A., Ronningen, R.M., Sherrill, B., Utsunomiya, H., and Kelson, I., Phys. Rev. C32, 877 (1985)

15) Pochodzalla, J., Friedman, W.A., Gelbke, C.K., Lynch, W.G., Maier, M., Ardouin, D., Delagrange, H., Doubre, H., Grégoire, C., Kyanowski, A., Mittig, W., Péghaire, A., Péter, J., Saint-Laurent, F., Viyogi, Y.P., Zwieglinski, B., Bizard, G., Lefèbvres, F., Tamain, B., and Québert, J., Phys. Rev. Lett. 55, 177 (1985)

16) Pochodzalla, J., Friedman, W.A., Gelbke, C.K., Lynch, W.G., Maier, M., Ardouin, D., Delagrange, H., Doubre, H., Grégoire, C., Kyanowski, A., Mittig, W., Péghaire, A., Péter, J., Saint-Laurent, F., Viyogi, Y.P., Zwieglinski, B., Bizard, G., Lefèbvres, F., Tamain, B., and Québert, J., Phys. Lett. 161B, 275 (1985)

17) Chitwood, C.B., Gelbke, C.K., Pochodzalla, J., Chen, Z., Fields, D.J., Lynch, W.G., Morse, R., Tsang, M.B., Boal, D.H., and Shillcock, J.C., Phys. Lett. B (in press)

18) Boal, D.H., Phys. Rev. C30, 749 (1984)

19) Koonin, S.E., Phys. Lett. 70B, 43 (1977)

20) Lynch, W.G. , Chitwood, C.B., Tsang, M.B., Fields, D.J., Klesch, D.R., Gelbke, C.K., Young, G.R., Awes, T.C., Ferguson, R.L., Obenshain, F.E., Plasil, F., Robinson, R.L., and Panagiotou, A.D., Phys. Rev. Lett. 51, 1850 (1983)

21) Chitwood, C.B., Aichelin, J., Boal, D.H., Bertsch, G., Fields, Gelbke, C.K., Lynch, W.G.,,Tsang, M.B., Awes, T.C., Ferguson, R.L., Obenshain, F.E., Plasil, F., Robinson, R.L., and Young, G.R., Phys. Rev. Lett. 54, 302 (1985)

22) Boal, D.H. and Shillcock, J.C., Phys. Rev. C33, 549 (1986)

23) Jennings, B.K., Boal, D.H., and Shillcock, J.C., Phys. Rev. C (in press)

24) Kopylov, G.I. and Podgoretskii, M.I., Yad. Fiz. 18 (1973) 656 [Sov. J. Nucl. Phys. 18, 336 (1974)

25) Kopylov, G.I., Phys. Lett. 50B, 472 (1974)

26) Pochodzalla, J., Chitwood, C.B., Fields, D.J., Gelbke, C.K., Lynch, W.G., Tsang, M.B., Boal, D.H., and Shillcock, J.C., to be published

27) Hale, G.M. and Dodder, B.C., Few-Body Problems in Physics, edited by Zeidnitz, B., (Elsevier, Amsterdam, 1984), Vol. 2, p. 433

28) Hale, G.M., private communication

29) Pochodzalla, J., Friedman, W.A., Gelbke, C.K., Lynch, W.G., Maier, M., Ardouin, D., Delagrange, H., Doubre, H., Grégoire, C., Kyanowski, A., Mittig, W., Péghaire, A., Péter, J., Saint-Laurent, F., Viyogi, Y.P., Zwieglinski, B., Bizard, G., Lefèbvres, F., Tamain, B., and Québert, J., Phys. Lett. 161B, 256 (1985)

30) Bond, P.D and de Meijer, R.J., Phys. Rev. Lett. 52, 2301 (1984)

31) Beauvais, G.E. and Boal, D.H., University of Illinois preprint, 1986, and to be published

32) Bertsch, G. and Cugnon, J., Phys. Rev. C24, 2514 (1981)

TREATMENT OF HEAVY ION COLLISIONS AT INTERMEDIATE ENERGIES

J.Aichelin
Institut für Theoretische Physik der Universität Heidelberg
and
Max-Planck-Institut für Kernphysik, Heidelberg, F.R.G.

Abstract - Microscopic calculations of intermediate energy heavy ion collision are performed employing the Boltzmann Uehling Uhlenbeck (BUU) formalism. We see that at beam energies around the Fermi energy the system does not equilibrize although the spectra show an exponential form. Preequilibrium particles are preferably emitted to the opposite site of the impact parameter as suggested by the Fermi jet model. We discuss furthermore the linear momentum transfer,the nuclear mean free path and the formation of a third cluster at midrapidity which reflects nicely the importance of both - two body collisions and mean field - at these beam energies.

I - INTRODUCTION

At low as well as at high beam energies (Ekin/N < 10 MeV and > 200 MeV) heavy ion reactions can be understood in simple models. At low beam energy with increasing impact parameters we see the formation of compound nuclei, deep inelastic processes and peripheral reactions. At high beam energies the single particle inclusive data can be well described assuming that the geometrically overlapping zones form an equilibrated source of particle emission. So at low and at high beam energies we observe equilibration, however, the mechanism which drives the system towards equilibrium is completely different.

At low energies only very few particles can escape from the combined system of projectile and target. The attractive mean field binds the nucleons together long enough to equilibrize although the residual two-body interaction is highly suppressed due to the Pauli principle which hinders the scattering in already occupied states. At high energies the mean field plays no role anymore being negligible compared to the average kinetic energy of the nucleons. At these energies however the Pauli principle is less severe and consequently the mean free path for nucleon nucleon collisions decreases. Therefore mutual two-body interactions which occur while projectile and target move through each other are sufficient to equilibrize the interaction zone.

When a while ago heavy ion beams between Ekin = 10 MeV/N and 200 MeV/N became available first experiments indicated [1] that single particle inclusive spectra also show the exponential forms which one would expect for particle emission from an equilibrated source. However the value of the slope parameter cannot easily be reconciled with the expected temperature of the total system or a geometrically defined subsystem. So it remained a question to theory whether also at intermediate energies a thermal system can be identified.

Another topic of current interest is the question which mechanism leads to the multifragmentation of the target into several fragments of charges larger than 2. This process has first been established at high beam energies but even at energies as low as 84 MeV/N [2] we see the same mechanism at work. At even lower energies the situation is more puzzling. Multifragmentation occurs in central collisions and is therefore associated with a large multiplicity of nucleons. Although it seems that multifragmentation is still observed at 30 MeV/N [3], at this energy also coincidences between projectile like fragments and light clusters are reported [4]. A theoretical explanation of the underlying mechanism of this kind of peripheral reactions is not at hand.

At energies below 10 MeV/N the linear momentum transfer between projectile and target is complete. At higher energies less and less momentum can obviously be transferred [5]. Is this due to peripheral reactions in which the projectile is cleaved and only part of it transfer momentum whereas the rest moved almost uninfluenced, or is it due to the fact that the nuclear stopping power is limited and therefore even in central collisions the projectile cannot completely be stopped at this energy ?

Already at moderate beam energies the single particle inclusive data can be described as a sum of two components, both exponential in form but with different slope parameters [6]. The low energy component can be well understood assuming that the particles stem from compound decay. The hard component was phenomenologically described as resulting from a Fermi jet mechanism which emerges from the interaction region prior to equilibration [7].

All these points intensified the demands for a comprehensive theory of heavy ion reactions which includes both, mean field dynamics which governs nuclear reactions at low energies as well as two-body collisions which dominates at high energies. The first results of such a theory are obtained by now. I will concentrate here on the above mentioned questions : a) Is the Fermi Jet mechanism the true source of preequilibrium particle emissions at energies around 25 MeV/N ? b) Why are three large clusters formed at beam energies around the Fermi energy ? c) Does thermalization occur in medium energy heavy ion reaction ? d) How much momentum can be transferred in this reaction and does there exist an upper limit as indicated by experiments ?

II - BUU THEORY.

In principal one would like to know the time evolution of the A_t+A_p body density matrix which is formally expressed by the von Neumann equation. However neither do we know the initial n body density matrix nor do we know how to solve this equation. Furthermore we are not interested in all informations contained in this equation but rather in that part which dominates the time evolution.

Rather than to introduce approximations into the quantum mechanical von Neumann equation we choose a different ansatz. We start with the classical analogon - the Liouville equation - and put in the dominating quantum mechanical features by hand. It will be shown later that on the one-body level, the time evolution of the classical phase space density and its quantummechanical counterpart is almost identical.

In our calculations [8,9] the nucleons are classical points like particles in coordinate and momentum space. Initially we assign to each nucleon a position in a sphere of R = 1.2 a**1/3. and a momentum between o and p_f randomly. The Fermi momentum p_f is determined by a local Fermi gas approximation. This procedure guarantees that on the average each nucleon occupies initially a phase space volume of h**3/4. The nucleons interact via stochastic two-body scattering with an isotropic cross section of 40 mb as well as via a selfconsistent density dependent mean potential :

$$U(\rho) = a*\rho + b*\rho**c$$

For the calculations presented here we use a = -353 MeV, b = 303 MeV, c = 7/6 which gives a compressibility of 200

MeV. When two nucleons come closer than d = $\sqrt{(\sigma/\pi)}$ the nucleons may scatter. We check whether the scattering violates the Pauli principle by examining the phase space around the phase space coordinates of the scattered particles. We calculate that part of this phase space volume (Ω) which is already occupied by other particles ($\Delta\Omega$) . Finally we determine the scattering probability by p = max [1 - $\Delta\Omega/\Omega$,0]. The total scattering probability is the product of that of both collision partners.

On the one-body level the equation we solve is identical with the Boltzmann equation with an Uehling Uhlenbeck scattering term. In contrary to the Boltzmann scattering term this scattering term takes care of the Pauli blocking and can be considered as a first order approximation to an exact quantal scattering term.In its underlying equation, however not in the numerical methods applied, it is identical with the so called VUU calculations [9].

Since our approach is not a straightforward derivation of the underlying quantum mechanical n body equation we have first of all to make sure that that it contains the essential ingredients for the time evolution of the system of interest. The best method to prove this is a comparison between the classical mean field calculation and its quantal equivalent, the Time Dependent Hartree Fock calculation [10]. This comparison is shown in figure 1 for the system 84 MeV/N C+C at b = 1 fm. Column 1 and 2 show the density at different times for the quantal and the classical mean field calculation. We see an almost complete agreement of the quantities of interest, the linear momentum transfer and the deflection angle. We can therefore conclude that at these energies the detailed form of the wavefunction is irrelevant for the time evolution of the system. Choosing the coordinates of the nucleons randomly, as done in the classical calculation, yields the same result as determining the Slater determinants properly by solving the Schroedinger equation. In column 3 we see the time evolution of the system if we allow Pauli corrected collision. We observe a completely different time evolution : Besides remnants we see a midrapidity source of particle emission. These nucleons have suffered at least one collision. This midrapidity source is essential to describe the relatively mild angular dependence of the inclusive proton spectrum in the nucleus nucleus center of mass system [1].

III - PREEQUILIBRIUM EMISSION OF FAST PARTICLES - FERMIJETS

The spectra of protons emitted in heavy ion reactions for incident energies ranging from a few MeV up to 20 MeV/n above the Coulomb barrier cannot be completely explained in terms of emission from a thermalized system [6]. At high

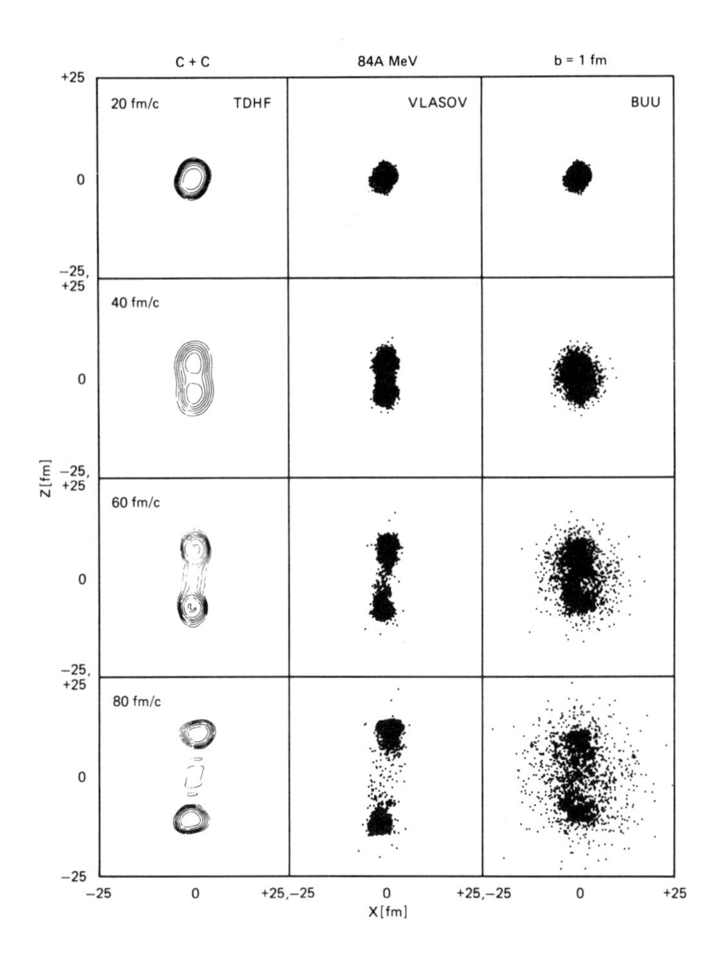

Fig. 1 : Time evolution of the single particle distribution function for the reaction C (85 MeV/N) + C at b=1 fm as predicted by the quantal mean field theory without collision term (TDHF, left-hand side column), the classical mean field theory (Vlasov's equation without collision term, center column) and the Vlasov equation with a Uehling-Uhlenbeck collision term included (right-hand side column). The transparency is evident in both mean field results without collision term. In contrast, a rapid deceleration of 60% of the incident nuclei is observed once the collision term is included into the Vlasov equation.

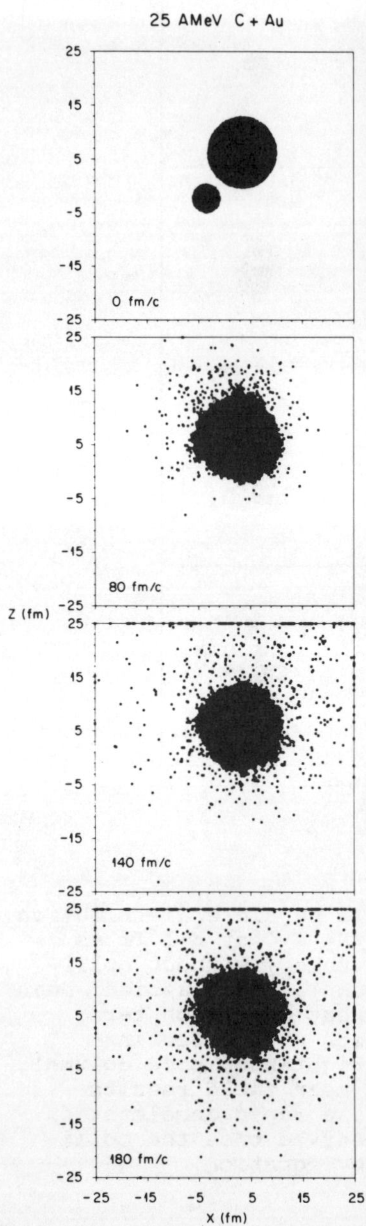

energies and in forward direction there is a considerable enhancement indicating an emission prior to equilibrium. This nonequilibrium component has been phenomenologically explained by a jet mechanism [7]. It is assumed that projectile nucleons from the overlap region of projectile and target can travel through the target without suffering collisions and are only little affected by the nuclear mean field. For some of these projectile nucleons the relative momentum with respect to the target surface is enhanced by the Fermi motion. Those are assumed to be able to surmount the potential barrier at the target surface and escape.

In order to verify this reaction mechanism we performed calculation for the system 25 MeV/N C+Au at b= 7 fm. The time evolution of this system is displayed in figure 2. For four different times (t = 0,80, 140,180 fm/c) we see the projected density distribution. At 80 fm/c we see the emission of many particles in forward direction in the nucleus nucleus center of mass system whereas only very few are seen in the backward direction. There is no large projectile remnant. The attractive mean field is sufficiently strong to absorb the projectile although initially there was less than 50% overlap. At 140 fm/c we see a strong enhancement of particle

Fig. 2 : Survey of the reaction 25A MeV C+Au at b = 7fm. Here the coordinates of all particles of 100 simulations of the reaction are projected onto the xz plane, were z is the beam direction and x the direction of the impact parameter. The density distribution is plotted at t = 0,80,140,180 fm/c.

emission opposite to the direction of the impact parameter. Only in the very late stage of the reaction the emission pattern is fairly isotropic as expected from the decay of a compound nucleus.

The left-right asymmetry of the emission get even more pronounced tagging the particles according their kinetic energy in the nucleus nucleus center of mass system. Particles with Ekin < 25 MeV show an almost isotropic distribution : 3.07 (3.53) have negative (positive) p_x values. The energetic particles, however, are most probably emitted on the opposite site to the impact parameter. We see 1.59 (1.13) particles with $p_x > 0$ ($p_x < 0$). This findings were recently confirmed by experiment [11].

Where were these energetic particles located at the beginning of the reaction ? Since our theory deals with classical particles we can trace back the momentum and position of the particles in time. In figure 3a we display the position at time t = 20fm/c of those particles which were finally emitted with energies larger than 25 MeV, in figure 3b we display the same quantity for all emitted particles. We see that the energetic particles come with a high probability from the overlap region. The targetlike nucleons are predominantly scattering partners of projectile nucleons. They have exchanged the momentum with the projectile nucleons due to the isotropic cross sections. The projectile nucleons of figure 3a have also a 30% higher momentum compared to the average of all projectile nucleons. As already suggested in the phenomenological models, additional momentum from the Fermi motion helps to overcome the potential barrier at the surface of the target. However, there is one mechanism present which is not counted for in these models:

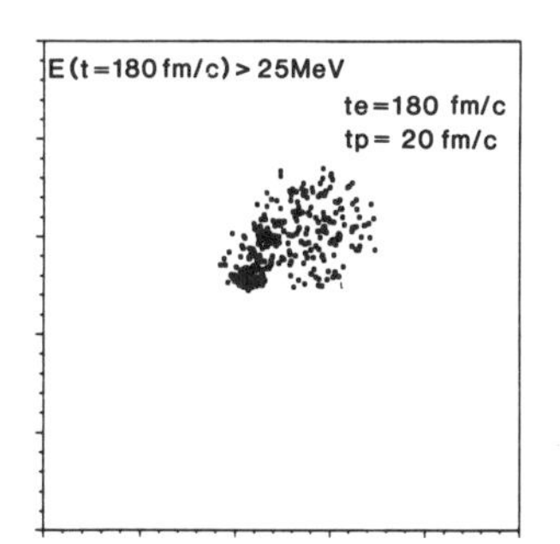

3a

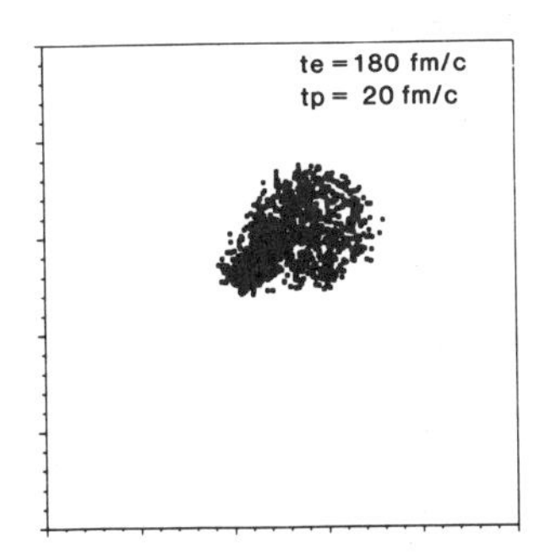

3b

Fig. 3 : Density profile of the emitted particles. The position at time tp is displayed of all those particles which are emitted at time te. Fig. 3a shows those particles whose final Ekin is larger than 25 MeV. Fig. 3b displays all emitted particles.

The absorption of the projectile generates a zone of higher density. The mean field tries to lower the density in this zone by accelerating particles in the direction of low density regions. Hence also the energetic projectile nucleons get an extra push which makes them even faster compared to the velocity of the target, thus enhancing their probability of escaping from the system. The lower energy particles are much more isotropically distributed over projectile and target as may be seen from figure 3b. They reflect almost an emission from an equilibrated system.

IV - FORMATION OF THREE HEAVY CLUSTERS

The formation of three large cluster is another phenomenon which is predicted by the BUU theory for energies around the Fermi energy. It is formed by a complicated interplay between two-body collisions and the nuclear mean field. Figure 4 displays the density profile for the reaction 44 MeV/N Ar+ Ni for four different impact parameters. The coordinates of all particles of 100 simulations are projected onto the xz plane where z is the beam direction and x is the direction of the impact parameter. The density profile is plotted at t = 0,40,80,120,180 fm/c from top to bottom. At low impact parameters the system fuses. At b = 4.3 fm we see at t = 180fm/c three large stable clusters which are well separated from each other. At larger and smaller impact parameters we see finally two clusters only.

This is first of all a completely unexpected phenomenon considering the condition under which three clusters may be observed in BUU calculations. The BUU calculation is essentially a mean field calculation in which all fluctuations are strongly damped. If multifragmentation is induced by density fluctuations we would not expect to see it in a BUU calculation. Here only reaction mechanism survive which are very close to the average behaviour of the time evolution of the system. Observing three clusters means therefore that this is the dominant process and not caused by fluctuations.

What is the driving mechanism behind this process ? This can be inferred from figure 5 where the position of the center of mass of different types of nucleons are plotted as a function of time. Depending where the nucleons were at t = 0 and t = 180 fm/c, we distinguish six classes i.e all possible combinations of origins (target, projectile), and final destinations (target remnant, projectile remnant and midrapidity cluster). The thickness of the lines is proportional to the number of class members. For b = 2.5fm we see only four classes because there exists no third cluster as can be seen from figure 4. Here some of the nucleons from the geometrical overlap move from projectile to target or

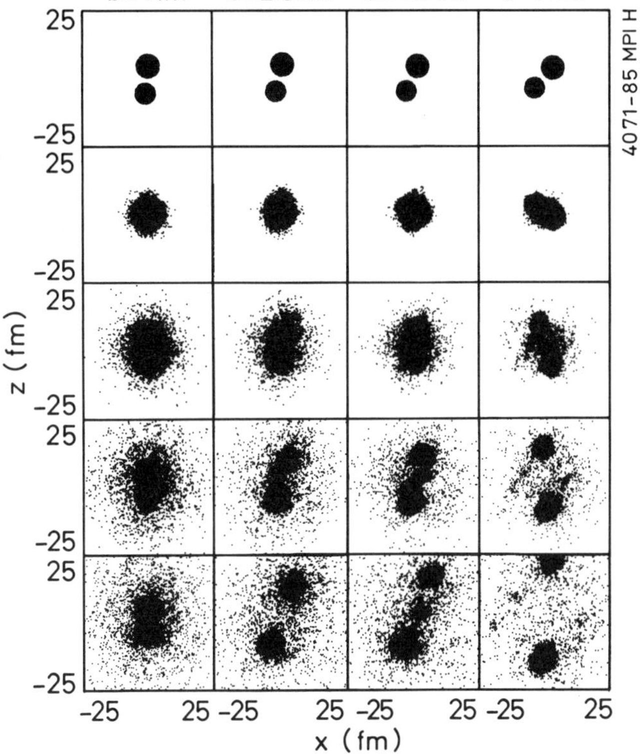

Fig. 4 : Survey of the reaction 44 A MeV Ar+Ni at different impact parameters. Here the coordinates of all particles of 100 simulations of the reaction are projected onto the xz plane, where z is the beam direction and x the direction of the impact parameter. The density distribution is plotted at t = 0,40,80,120,180 fm/c.

vice versa and get dragged along with the host nucleus. At b = 4.3 fm this classes of nucleons exist as well, but in addition a third cluster is formed. Initially these nucleons are placed between the both aforementioned classes. They get less accelerated by the mean field - which wants to lower the density in the overlap region - than those finally absorbed by the host nucleus, but much more than those which keep their identity as target or projectile nucleons. Because the overlap is smaller in size, there are less collisions,the total linear momentum transfer between projectile and target is smaller and consequently projectile and target remnant separate much faster at b = 4.3fm compared to b = 2.5 fm. So at b = 2.5 fm the remnants have still the possibility to reabsorb those nucleons which potentially would form the third cluster. At b = 4.3fm the remnants have lost this ability. Rather the system disrupts at the surface of the remnants, forming a cluster almost at rest in the nucleus nucleus center of mass.

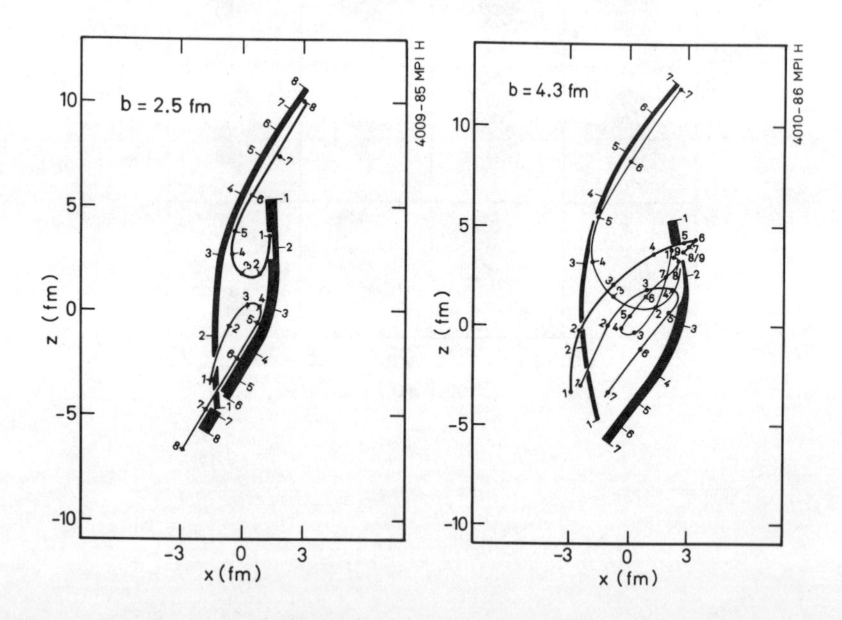

Fig. 5 : Mean positions of different classes of nucleons for two impact parameters (b = 2.5 fm and b = 4.3 fm). We see the mean values of those nucleons which belong finally to projectile remnant, target remnant or to the third cluster. The thickness of the lines is proportional to the number of nucleons which belong to each class. The numbers mark the center of mass of each class in timesteps of 20 fm/c.

Summarizing these observations we see in the reaction 44A MeV Ar+Ni a third heavy cluster for a range of intermediate impact parameters.It is created by a cooperation of the mean field and collisions. Collisions cause a rapid deceleration which is necessary to initialize the formation of the third clusters.Therefore this process is not seen in time dependent Hartree Fock (TDHF) calculations. The mean field however is responsible for the acceleration of those nucleons in transverse direction which directs the nucleons in different regions in space. A neck is formed. At low impact parameters resp. larger linear momentum transfer the potential field of the remnants is strong enough to accelerate the neck nucleons considerably towards the center of one of the remnants. So the neck nucleons can catch up and get absorbed. The systems disrupts in the middle of the neck. For smaller momentum transfer resp. larger impact parameters, projectile and target remnants emerge faster from the interaction region and consequently their ability to absorb the neck nucleons get reduced. Furthermore the neck nucleons have a larger transverse momentum which enlarge the momentum transfer necessary to catch up with the remnants. Both effects together hinder an absorption of the neck nucleons. The system disrupts at the surface of the remnants. The interacting region contains many nucleons at reasonable high density. It can therefore stabilize and forms a third cluster.However, even at the same energy per nucleon we do not see a third cluster in case of a carbon-projectile. For the formation of the third cluster we need two different classes of nucleons : those which are only little decelerated and those whose longitudinal momentum is almost zero in the nucleus nucleus center of mass frame but have gained transverse momentum. Only if both classes can be formed simultaneously the system disrupts at the surface of projectile and target remnant. Small projectiles are insufficient for this process for two reasons. First of all they cannot simultaneously provide enough nucleons for a remnant, large enough to be a stable cluster, and for a dens neck region, large enough to stabilize. In addition also the viscosity is too large. In small projectiles the possible velocity difference between the leftmost and rightmost nucleons is limited. The outermost nucleons are influenced if the innermost nucleons get decelerated whereas here (at t = 40 fm/c and b = 4.3 fm) the most left resp. right nucleons move with projectile resp. target velocity with an almost linear velocity profile in between.

V - THERMALIZATION AT INTERMEDIATE ENERGIES.

At low (Ekin < 5 MeV/N) as well as at very high energies (Ekin > 200 MeV/N) the inclusive single particle energy spectra are exponential in shape and the angular distribution is almost isotropic in the rest system of the compound system resp. the fireball. This has been interpreted in terms of an emission from an equilibrated system. Figure 6 shows the proton spectra in the reaction 84 MeV/N C+Au comparing the theory with results of Jakobsson et al [1]. We see that also at intermediate energies the high energy part of the spectra looks quite exponential. However, the slope parameter is larger than expected from a compound nucleus and smaller than expected from a fireball model. In order to connect this exponential form of the spectrum with a temperature we have to find the emitting source. Jakobsson has shown that the spectrum can be well described using three sources with different temperatures and different velocities. In order to check whether the situation is that simple and lacks any sign of preequilibrium we calculate the first and second moment of the momentum distribution of the emitted particles as a function of their emission time. These quantities are displayed in figure 7. We see that the average momentum of the emitted particles with respect to the nucleus nucleus center of mass system as well as the width of the momentum distribution are strongly dependent

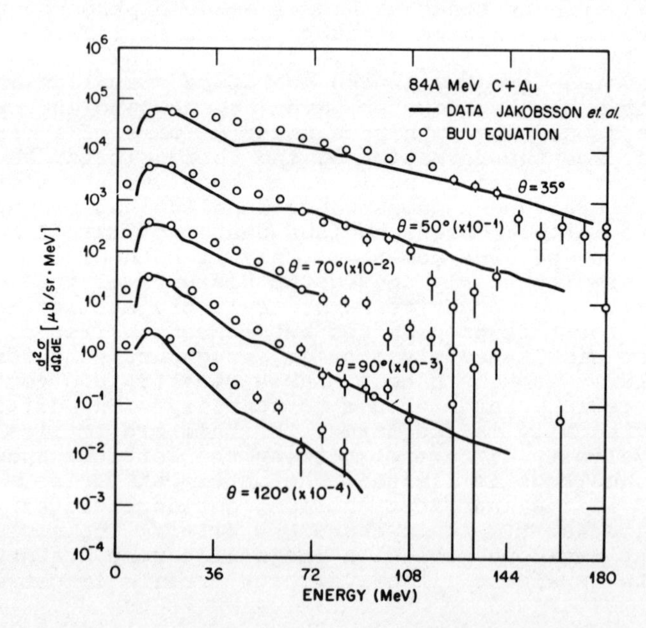

Fig. 6 : The proton spectrum in the reaction 84 Mev/N C+Au comparing the theory with results of Jakobsson et al.[1]. The curves show the energy spectra at angles of 35°, 50°, 65°, 110°, 135° with curves displaced by a factor of 10 for clarity in the figure.

on the emission time. At the beginning of the reaction the emitted particles have in the average half the beam energy and a very large width in momentum space. These particles predominantly suffered one scattering. With increasing time, the average momentum as well as the width decrease, but only at a very late stage the average momentum coincides with the nucleus nucleus center of mass momentum as expected for particle emission from a compound nucleus. At that stage the system has only little excitation energy left. From this observation we can conclude that the single particle spectra - although exponential in shape - are actually a quite complicated superposition of distributions of particles emitted at different times. This is not surprising : Projectile and target nucleus initially have a Fermi distribution. After one collision between a projectile and a target nucleon we expect for an isotropic cross section a center of mass velocity of the collision partner of half the beam momentum and the momentum distribution being the convolution of the Fermi distribution of projectile and target which comes close to a Gaussian. Particles emitted later have more collisions in the average. Therefore their width is reduced and the average momentum has decreased. Only at the very end we see the formation of a compound nucleus, however with small excitation energy. Only very few of the observed particles come from this final stage of the reaction.

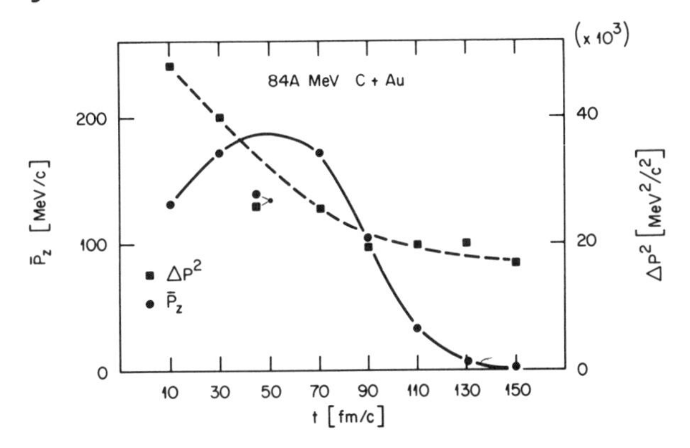

Fig. 7 : Mean momentum $\bar{p}_z$ and $\Delta p^2 = \Sigma(p_i - \bar{p}_z)^2/N$ of the particles emitted in the reaction 84 MeV/N C+Au as a function of the emission time.

Summarizing these observations we see that at intermediate energies the observed particles - although their spectrum is exponential in shape - do not come from an equilibrated source as at high and at low beam energies. Almost all of them are emitted prior to equilibrium, having average velocities between nucleon nucleon and nucleus nucleus center of mass velocities, although the width decreases because the excitation energy left in the remnant decreases. So neither is the mean field strong enough to bind projectile and target nucleons long enough for the formation of a compound nucleus nor is the mean free path sufficiently short that equilibration is reached while the projectile travels through the target. The latter is a consequence of the Pauli blocking.

VI - LINEAR MOMENTUM TRANSFER

One of the primary challenges in intermediate energy heavy ion physics is to determine the linear momentum which can be transferred between nuclei. This quantity determines the excitation energy which can be stored in a nucleus and hence the maximal temperature assuming that a global equilibrium can be obtained. Recently models were advanced [12,13] which rely on very large momentum transfer. We have shown [8] that linear momentum transfer predicted in the BUU approach agrees reasonably with the value measured by Galin et al. [5] for the case 60 MeV/N C+Au. Here we report on an extension of this calculation to determine the target mass and projectile energy dependence of this quantity for central collisions. Figure 8 shows our prediction for four projectile target combinations at three different energies: 60 MeV/N, 84 MeV/N, 250 MeV/N C+Ni, C+Sm, Ne+Ni, Ne+Sm. We see in all cases the momentum transfer being incomplete. Hence the incomplete momentum transfer in heavy ion reactions is not a geometrical effect resulting from the partial overlap of projectile and target. At these energies the momentum transfer is proportional to A**1/3 indicating that two-body collisions are the dominant mechanism for transferring momentum. At all three energies the total momentum transferred is proportional to the projectile mass, indicating that each projectile nucleon transfers on the average a momentum which is independent of the size of the projectile.

Investigating the momentum transfer as a function of the projectile energy we find that there is a maximal momentum which can be transferred to the system. A further increase of the projectile energy changes this quantity only slightly; for small targets we find even a decrease. This finding agrees with the observations of Galin et al. [5], however, our absolute values are somewhat higher. At 84 MeV/N we find for central collisions a momentum transfer of 2.2 (3.0) GeV for C+Ni (Sm) and 3.3 (5.1) for Ne+Ni (Sm). Hence already at this energy the momentum transfer is quite

ergy up to energies around 200 MeV/N is correlated with a

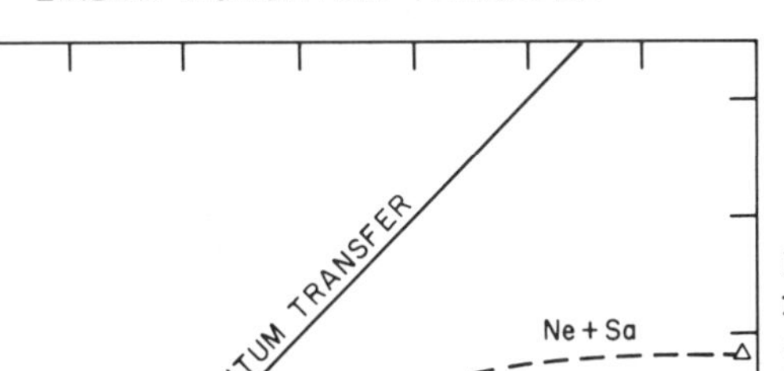

Fig. 8 : Linear momentum transfer as a function of the projectile energy for different systems : Ne+Ni, Ne+Sa, C+Sa.

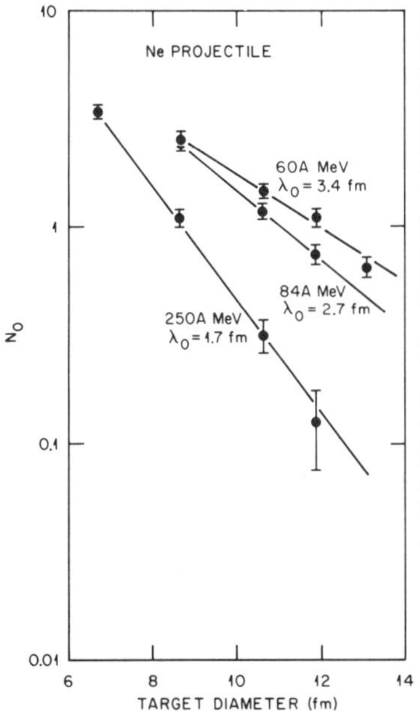

decrease of the mean free path. This can be seen from figure 9. Here the mean free path of the projectile nucleons is calculated for three different beam energies by counting the number of projectile nucleons (N) which have passed the target without a collision at central impact parameters. We see that at 60 MeV/N (84 MeV/N) the Pauli blocking is still quite important. It enlarges the expected mean free path for unblocked collision $\lambda = \surd(1/\sigma/\rho)$ by a factor of 2.0 (1.6). At 250 MeV/N the Pauli blocking is unimportant concerning the mean free path, i.e. the probability to have at least one collision.

Fig. 9 : Number of unscattered projectile nucleons as a function of the diameter of the target for three different beam energies. λ denotes the extracted mean free path.

SUMMARY

Heavy ion reactions at beam energies between 10 MeV/N and 200 MeV/N show a variety of phenomena which is neither known from high energy nor from low energy physics. This is due to the equal importance of the effect of the nuclear mean field and that of two-body collisions. In this transition region we still see signs of the familiar low energy behaviour like formation of a compound nucleus in the late stage of the reaction but also the first hints of typical high energy phenomena like a midrapidity sources. In addition to that new phenomena - like the formation of three cluster - appear which are specific to this energy interval.

Hence one should not wonder that the simple models, which describe high or low energy physics so well, do not work at intermediate energies. We see that the assumptions of a thermalized source of particle emission cannot be verified in contradistinction to reactions at lower or higher beam energies. Neither is the mean free path sufficiently short to thermalize projectile and target matter while travelling through each other nor is the mean field strong enough to force the nucleons together until complete thermalization is reached. But exactly this is the reason why intermediate energies are so interesting. They offer the unique possibility to study how a many-body system comes to equilibrium which is sufficient in number to have interesting features but contains little enough particles that it can be studied theoretically on a microscopic level.

REFERENCES

[1] B. Jakobsson et al., Proceedings of the 4th Bergen Workshop in Nucl. Physics (1982), A. Oskarsson, Lund University Report 8303.

[2] H. Rabe, Thesis, Univ. Heidelberg (1986).

[3] D.J. Fields W.G. Lynch, C.B. Chitwood, C.K. Gelbke, M.B. Tsang, H. Utsunomiya and J. Aichelin, Phys. Rev. C30 (1984) 1912.

[4] G. Bizard et al. Proceedings of the XXIII International Winter Meeting on Nuclear Physics, Bormio, 1985.

[5] J. Galin et al., Phys. Rev. Lett. 48 (1982) 1787; S. Leray, Proceedings, this Conference.

[6] L. Westerberg, D.G. Sarantites, D.C. Hensley, R.A. Dayras, M.L. Halbert and J.H. Barker, Phys. Rev. C18 (1978) 796.

[7] J.P. Bondorf, J.N. De, G. Fai, A.O.T. Karvinen, B. Jakobsson, and J. Randrup, Nucl. Phys. A333 (1980) 285.
[8] J. Aichelin and G. Bertsch, Phys. Rev. C31 (1985) 1730;
J. Aichelin, Phys. Rev. C33 (1986) 537;
Proceedings of "Phase space approach to nuclear dynamics", Trieste 1985, World Scientific Publishing Co., Singapore.
[9] H. Kruse, B.V. Jacak J.J. Molitoris, G.D. Westfall and H. Stöcker Phys. Rev. C31 (1985) 1770.
[10] J. Aichelin and H. Stöcker, Phys. Lett. 163B (1985) 59.
[11] M.B. Tsang, Proceedings this Conference.
[12] D. Vasak et al., Phys. Lett. 93B (1980) 243; Nucl. Phys. A428 (1984) 291c.
[13] A.D. Panagiotou et al., Phys. Rev. C31 (1985) 55.

QUANTUM KINETIC THEORY AND HYDRODYNAMICS

APPROACHES TO THE THERMALIZATION PROBLEM

Rudolph C. Hwa
Institute of Theoretical Science and Department of Physics
University of Oregon, Eugene, Oregon 97403

Abstract

The two approaches to the study of the thermalization problem in ultra-relativistic nuclear collisions in the flux-tube model and in the parton model are reviewed. The deficiencies of the models are pointed out. The similarities of the results as well as their differences are compared.

One of the important theoretical problems associated with the experimental attempt to create quark-gluon plasma in the laboratory is the determination of the initial temperature T_i and the time τ_i needed to form such a plasma. For if T_i is too low there would not be a plasma phase. Lattice QCD calculations have determined the deconfinement temperature to be in the neighborhood of 200 MeV, but the calculations are done under the assumption of thermal equilibrium for the system. Can that condition be achieved in a high-energy nuclear collision? More specifically, is the thermalization time short enough so that the system can attain thermal equilibrium before it disintegrates?

At present our knowledge on strong interaction is insufficient to render a determination of T_i and τ_i from first principles. Although rough estimates have been made on the basis of extrapolations back from the final states observed in hadron-hadron and hadron-nucleus collisions, some recent studies have been directed toward the thermalization problem from the initial state. In the absence of a basic theory useful for treating soft processes, such studies have relied on models that have proven successful in describing certain dynamical problems in particle physics. There are two basic approaches that are from the outset very different in their simplest forms: the string model and the parton model. These models have been modified and improved so as to describe more complicated systems. In order to understand better the similarities and differences in the estimates made for plasma formation, it is useful to recall the origins of these models and then to review the changes that have been made to adapt to the nuclear case.

1-A. String Model. The string model[1] is most succesful in describing jet characteristics in e^+e^- annihilation. In that process the ends of the string are precisely known, namely: a quark-antiquark pair; their initial relative momentum is also not controversial, being a matter under experimental control. As the ends of the string move apart, a gluon emission becomes a kink, and a $q\bar{q}$ pair can be produced by tunneling to break the string (the Schwinger mechanism). Although the details of color neutralization by pair creation and the formation of hadrons are model dependent, the essence of the physics involved has evidently been captured by the string model.

1-B. Parton Model. The parton model[2] is most succesful in describing lepton-nucleon deep inelastic scattering. The measured strucuture functions provide information on the quark distributions in the nucleon. Those partons (quarks and gluons) are regarded as free during the course of scattering in the Bjorken limit, $\nu \to \infty$, $Q^2 \to \infty$, x fixed. The parton model is useful when the nucleon is regarded as a collection of weakly interacting partons in a frame in which the nucleon moves fast and the partons have momentum-fraction distributions determined by lepton-production.

2. pp Collision. The two models have been applied to hadronic collisions with varying degrees of sophistication, summarized briefly below.

2-A. Naive String Model. In an early version, initiated by Low and Nussinov,[3] the two incident hadrons exchange a gluon so a string is stretched by two octet ends. The color flux-tube model[4] is based on such a dynamical picture and focuses on the pair-production process in the tube. What the color charges are at the ends of the tube has not been investigated with care, so the rate of color neutralization is uncertain to the exent of the uncertainty in the initial color field. That is a problem that still persists in later generalization to nuclear collisions. In that sense the string model for hadronic collision is very different from that for e^+e^- annihilation. The Lund model[5] is not successful when applied to pp collisions,[6] because it fails (a) in giving too narrow a multiplicity distribution, (b) in having negligible forward-backward multiplicity correlation, and (c) in not exhibiting KNO scaling for energies below those of the S$\bar{p}$pS collider.

2-B. Naive Parton Model. When applied to pp collisions, the parton model is most useful for hard processes, such as massive lepton-pair production and large-p_T inclusive reactions, for which the model serves as a link between the hadronic initial particles and the initial partons in the hard subprocesses, calculable in perturbation theory. In extending the application to soft processes at low-p_T there is an uneasy feeling about its validity, despite Feynman's origin paper[7] on the parton model being precisely for such processes. Gross features of multiparticle production at ISR energies seemed reproducible assuming short-range interaction in rapidity among the partons. Inclusive distributions of hadrons in the fragmentation region can well be described by the parton model, when supplemented by the recombination model for hadronization.[8] However, in the central region there are difficulties at the naive level, the same as those mentioned in the naive string model above.

2-C. Sophisticated String Model. The uncertainty in the color charges at the ends of the naive string is resolved in very specific terms in the dual-topological-unitarization (DTU) scheme.[9] A proton is split up into two components consisting of a quark and a diquark, so that between the two protons there are two strings each being stretched by quark of one proton and the diquark of the other. Although these quarks are not the partons in the parton model (which has no diquarks), taking the hadron constituents into consideration represents a step closer to the parton model. At higher energies more strings can be added in the central region between quarks and antiquarks in the sea.[10] Larger fluctuation in multiplicity is thereby attainable, but KNO scaling is not forced by any first principles in DTU. One question that can be raised concerns the relevance of DTU to the thermalization problem, since unitarity puts constraints on the physical multiparticle states which occur far later in time relative to the time scale for thermalization. Yet, without DTU there is no guidance in determining the color charges at the ends of the strings for pp collision. The dilemma is rooted in the S-matrix heritage of DTU, which does not specify the space-time development of the scattering process of the hadron constituents.

2-D. Sophisticated Parton Model. To have KNO scaling the soft partons in the central region must interact and cluster in non-trivial ways. Through the study of branching equations[11] that guarantee $D = (\overline{n^2} - \bar{n}^2)^{1/2}$ being proportional to $\bar{n}$ at high energy, a dynamical branching model[12] can be constructed that is compatible with the parton model through duality.[13] It is found that the partons must cluster with non-planar topology. Thus the naive parton model must be modified in a way that brings it closer to the sophisticated string models. Concerning thermalization it is clear that space-time development at early times must also be added to the consideration.

2-E. A Mid-course Comparison. While the final theory for soft hadronic interactions remains unclear, a picture is emerging that appears to be approachable from either ends. The essence of the original string model is that all quark pairs and gluon pairs are produced, while that of the parton model is that they are already in the sea of the incident hadron. A more realistic picture for hadron-hadron collision is somewhere in the middle: the partons in the hadrons produce string-like clusters. It may be just as hard to ascertain the charging process from the string approach as it is to determine the production process from the parton approach. It is with this duality in mind that we embark on the nucleus-nucleus collision problem.

3. AA Collision. We consider here the problem of thermalization in the central rapidity region. The incident energy is assumed to be high enough so that such a central region is well-developed and separates the two fragmentation regions. The two colliding nuclei are sufficiently large so that at small impact parameters we may ignore transverse motion during the course of thermalization.

3-A. Colored Rope. The string is now thickened to a rope. The initial color electric field E_0 in the rope is proportional to the color charges Q_0 at the ends. The charging process is regarded as a random walk in color space[14], so[15]

$$Q_0 \propto N^{1/2} \tag{1}$$

where N is the number of steps in the walk. There are a number of uncertainties in (1). Firstly, workers in the subject [14,16,17] have not considered the complications discussed in Sec. 2-C. Charging is assumed to take place in units of effective nucleon collisions; for proton-nucleus collision N is taken to be ν, the number of interactions that the projectile proton suffers while traversing the target nucleus. Thus for AA collision it is assumed that[17]

$$N \propto A^{1/3} \times A^{1/3} \tag{2}$$

I have for some time expressed the view [8,18] that in pA collision at high energy the proton does not maintain its identity during collision and that it is a stream of partons that propagate through the nuclear target. This view is, of course, a natural development from the parton model. Recently, however, a similar view has been expressed starting from the DTU approach.[19] The second uncertainty in (1) concerns the proportionality factor, which, of course, is related to the question on N. One gluon exchange at every step is an assumption based mainly on simplicity. Even the more specific prescription of DTU is subject to question: a quark-diquark separation may be granted for pp collision, but a diquark staying permanent bound as it transverses a nucleus is an unreasonably strong assumption. A recent work on color neutralization by Gyulassy and Iwazaki[20] alludes to color charging by partons taking random walk. That seems to be a more realistic picture of the process, although quantitative estimate would be hard. What one can only say at this stage is that

$$Q_0 \propto A^{1/3} \tag{3}$$

which is consistent with (1) and (2), and bears the same degree of uncertainty. The color neutralization time τ_0 or the decay time of E_0 follows essentially from dimensionality[15]

$$\tau_0 \sim E_0^{-1/2} \propto A^{-1/6} \tag{4}$$

The first part of the relationship has been studied with care,[20] but the precision gained in the proportionality factor is overwhelmed by the uncertainty in the second part. Similar general conclusion can, of course, also be obtained, as in Ref. 17, using dimensional analysis. Using a model kinetic equation the evolution of matter during the pre-equilibrium regime has been investigated[21] so that the thermalization time can be related in numerical solutions to the input parameters τ_o and the collision time τ_c. It is noteworthy that despite the uncertainties in (4) it has been estimated[20] that τ_o (AA) is an order of magnitude smaller than τ_o (pp) because the screening effect of the produced pairs on the color electric field while important for pp[4] is negligible for AA collision.[16] Assuming that τ_o (pp) is of the order of 1 fm/c, one would infer that τ_o (AA) is only of order 0.1 fm/c, a surprisingly small value which we shall return to later.

We conclude this short sketch of the color (frayed) rope approach to thermalization by giving the A-dependence of the initial temperature T_i. Since the energy density ε of the plasma should initially be proportional to $E_0{}^2$ when the field energy is totally converted to the thermal energy of the quarks and gluons, we expect[15]

$$T_i \sim \varepsilon^{1/4} \sim E_0{}^{1/2} \sim Q_0{}^{1/2} \propto A^{1/6} \qquad (5)$$

where, again, the last step following from (3) is most uncertain.

3-B. Thermalization in the Parton Model. For the most part of what remains in this review I shall describe briefly the determination of T_i and τ_i in the parton model.[22] Our method is independent of the details of how the parton model is modified to treat the AA collision. The essence remains that the partons preexist in the nuclei before the collision takes place. The two basic properties that they have are (a) nuclear transparency and (b) distributed contraction. The former refers to negligible nuclear stopping power on fast quarks,[18] which underlies the phenomenological fact that in $a + b \rightarrow c + X$ the inclusive distribution of c in the fragmentation region of a at high energy is independent of b, if b is a hadron, or only mildly dependent on b if b is a nucleus, A. Distributed contraction refers to the quantum mechanical property that the uncertainty in longitudinal position Δz of the partons is contracted in accordance to their rapidities; thus $\Delta z \approx \ell/\cosh y$ where ℓ is the hadronic length scale of about 1 fm. This implies that only the fast partons are contracted to a thin disk; the wee partons are not. Combining these two properties we see that the thermal energy density cannot increase indefinitely, even as $s \to \infty$, because by (a) the fast partons from the two nuclei, though contracted, do not interact effectively, while by (b) the slow partons that do interact strongly are not densely packed. In Ref. 22 we were able to make a quantitative estimate of T_i starting from these tenets of the parton model.

An important part of the procedure is to recognize the global change of the rapidity distribution in a space-time cell as a function of the proper time τ. Let the location of the cell be specified by (τ, η), where η is the spatial rapidity, $\eta = 1/2 \ln (t + z)/ (t - z)$. Fixing η at any value and increasing τ, we know in the no-interaction case that the parton rapidity distribution $P(\tau, \eta, y)$ for $\tau \approx 0$ is very broad, from $-Y$ to $+Y$, and decreases in width until it approaches $\delta(y - \eta)$ as $\tau \to \infty$, at which point the partons have hadronized and end up in a detector located at η. Property (b) mentioned above leads to the independence of the width 2Δ of $P(\tau, \eta, y)$ on η for fixed τ. The dependence of Δ on τ is calculable.

In the realistic situation parton interactions change $P(\tau, \eta, y)$ to the true distribution $F(\tau, \eta, y)$. The determination of how $P(\tau, \eta, y)$ becomes $F(\tau, \eta, y)$ is hard, and is the heart of the thermalization problem. Only partial attacks on the problem have so far been attempted.[23-26] Fortunately, for our purpose of calculating T_i and τ_i, the full solution is not needed, provided that high accuracy is not demanded. In the following discussion we need only regard $F(\tau, \eta, y)$ as the thermal distribution, characterized by the temperature $T(\tau)$ at every cell.

Now, in the free-streaming case we have $P(\tau, \eta, y)$ from which the energy density $\varepsilon'(\tau)$ can be calculated from kinematics and the proton density of the colliding nuclei, assuming that each nucleon has partons in the sea (i.e. central rapidity region) as prescribed in the parton model. In the thermalized system we have $F(\tau, \eta, y)$ from which the energy density $\varepsilon(\tau)$ can also be calculated. The key point in our approach is to identify $\varepsilon'(\tau)$ with $\varepsilon(\tau)$ at $\tau = \tau_i$ as a first order requirement in the approximation of short collision time. Our reasoning is that the change $P \to F$ is due to local interaction in rapidity. There are exchanges of partons among neighboring cells and creation of partons within the cells. So long as partons in the fragmentation region do not contribute to $\varepsilon(\tau)$ in the central region (due to nuclear transparency), energy conservation and Lorentz invariance along a fixed τ hyperbola within the central region require that

$$\varepsilon'(\tau_i) = \varepsilon(\tau_i) \tag{6}$$

One may even take this condition as a definition of τ_i in the "ideal" case.

Independent of the normalizations of the distribution functions we can also separately obtain

$$\varepsilon'/\rho' = f(\tau), \qquad \varepsilon/\rho = 3T \sim \tau^{-1/3} \tag{7}$$

where ρ' and ρ are the parton densities in the free-streaming and thermal cases. $f(\tau)$ is completely calculable and behaves as $m_T'\ell/4\tau$ at small τ, where m'_T is the transverse mass of the partons. We shall set $m'_T\ell \approx 3$. At large τ, $f(\tau)$ approaches m'_T. In Fig. 1 we show both parts of (7). The first intercept at the smaller τ value corresponds to the solution we seek because in that case we have

$$T_i\,\tau_i \approx 1/4 \tag{8}$$

which exhibits the correct inverse relationship between T_i and τ_i. The second intercept at higher τ has no physical significance. (8) is to be compared to a similar relationship in the flux-tube approach where $\tau_o \sim E_o^{-1/2} \sim T^{-1}$. The realistic solution for the physical system presumably follows a smooth curve shown in Fig.1; its distance from the asymptotes depends on the detailed properties of soft interaction. If the collision time τ_c is not too large, the intercept of the asymptotic lines of (7) provides a good approximation for the true values of T_i and τ_i. As this point there is no good estimate of τ_c. The conservative view that it is on the order of 1fm/c is not pursuasive, since the pion mass or the perturbative-QCD scale parameter Λ is not necessarily relevant for a highly turbulent and high-density system whose natural length scale may well be far less than the usual hadronic scale. The fact that τ_i is around 0.15 fm/c attests to the possibility that τ_c can be equally small at early times when the energy and parton densities are high.

Since $\varepsilon'(\tau) \sim \tau^{-2}$ and $\varepsilon \sim T^4$, we obtain from (6) and (8) a value for T_i^2 that depends only on the properties of the colliding nuclei[15]

$$T_i^2 \sim \int_0^L dz\rho'(z) \sim A^{1/3} \tag{9}$$

where the integration is over the average longitudinal nuclear distance, $L = 4R_A/3$. Thus T_i^2 is a measure of the integrated parton density per unit transverse area. In principle, in an A + A' collision (A right-moving, A' left-moving) T_i for $y > 0$ depends on $A^{1/6}$ while for $y < 0$ the dependence is $A'^{1/6}$, but in practice it may be hard to verify the difference experimentally, since, apart from other complications, thermal conductivity in the plasma tends to equalize the temperature in the system. Putting in typical numbers for nuclear and quark densities, we get for AA collision

$$T_i = 180\ A^{1/6}\ \text{MeV} \tag{10}$$

$$\tau_i = 0.27\ A^{-1/6}\ \text{fm/c} \tag{11}$$

Amazingly, the A dependences are the same as in (4) and (5). Even the order of magnitude of τ_i is similar to that estimated in Ref. 20, as already mentioned following (4).

Note that (6) and (7) implies $\rho' = \rho$. This may seem to imply that there is no creation of new partons during the collision, a scenario that is totally opposite to the string model where all pairs are created after the collision. However, since our input for ρ' is based on the quark and gluon distribution $F(x)$ at $x = 0$ as determined by leptoproduction [$F(x = 0) = 5$], it actually includes some produced partons due to collisional effect. For our usage in the present problem, the condition $\rho' = \rho$ should be regarded as a nomalization condition on $P(\tau,\eta,y)$ to render the proper parton density at τ_i compared to that of the thermal system, not one chosen for a non-interacting nucleus. In Ref. 22, $F(x = 0)$ is related to the particle density in rapidity in the central region. Indeed, if one were concerned, for example, with the slow increase of dN/dy for the pions produced at higher energy, one would have to let $F(x = 0)$ increase accordingly at the present level of our treatment; that is, the production of partons can be accommodated by a posteriori connection with the produced hadrons, but not in any systematic way from $\tau = 0$. The inadequacy of the treatment here is only a part of a bigger problem to be discussed below.

It is important to understand why the results from the two approaches are so similar. The origin of (4) and (5) is clearly different from that of (10) and (11). The former is based on random walk in color charging, while the latter is rooted in the parton density expressed in (9). Nevertheless, (3) seems to indicate that the two are not very far apart despite the very different languages used in the two approaches. Future studies should focus on the relationship between Q_0 in the frayed rope and the sea quarks and gluons in the parton model.

4. Speculations

As a way to conclude this mini-review let me put the subject in a perspective that reflects my impression of what has occurred at his Workshop. It is my observation that despite the small size of this Workshop there are various subcultures that do not communicate with one another. In particular, the subject matter discussed on the first day concerns the final-state characteristics of pp and $\bar{p}p$ collisions, namely: multiplicity distributions, forward-backward correlations, negative binomials, fluctuations, branching, etc. Near the end of the meeting we heard talks on AA collisions, phase transition, transport theory, hydrodynamical flow, etc. The two parts could have been held simultaneously in separate quarters, and there would not have been, it seems, any lack of stimulation due to the absence of the complementary parts. Yet is there anyone who is willing to predict that the complications associated with pp collisions will go away for AA collisions at comparable energies per nucleon? What we need is more integration between the two subcultures. It is toward that end that I conclude this talk.

The major difference between pp and AA collisions is that there is a plasma phase (if we are lucky) between the initial and final states of the latter type of collision, and presumably not so for the former at energies that have been explored. If the fluctuations observed for pp are associated with the color charging and neutralization processes of the strings, it is possible that similar fluctuations that can occur for AA collisions would get damped out by the plasma state, since the thermal and color conductivities of the quark matter produced would serve to "short" out the irregularities (spikes[27]) and non-uniformities (fluctuations in rapidity density). Total multiplicity fluctuations from event to event may or may not survive, but forward-backward correlation is likely to be suppressed since the plasma state erases any memory of the multi-string structure of the pre-equilibrium regime, the strucutre that has been held responsible for such correlations. These considerations are partially based on the reasoning suggested by the dual parton model,[10] which, however, has no clear statements about the space-time evolution of the system, nor about the formation of the quark matter. Indeed, DPM, being based on both the initial parton distributions of the incident nuclei and the final physical multiparticle state due to unitarity, would seem to encounter major difficulties if quark matter intervenes in the middle and invalidates the DTU diagrams whose quark and diquark lines connect the initial and final states. There is, however, another possible scenario. The fluctuations observed in pp collisions could well be due to the hadronization process rather than during the thermalization process. Applied to the AA collision case, it means that they would occur during and after the phase transition to hadronic clusters and would be independent of the mechanism of plasma formation. Such a possibility would be supported by the observation of similar fluctuations in AA collisions as in the pp case. The parton model for thermalization would then seem more appropriate. The description of time evolution through phase transition and on to hadron emission by means of hydrodynamical flow would be unreliable, since hydrodynamics is based on the assumption of small fluctuations in a macroscopic dynamical system. The formation of hadronic clusters which successively branch into smaller clusters can introduce large derivations from the mean in the multiplicity of hadrons produced. The stochastic nature of the problem is outside the scope of hydrodynamics and would present quite a new challenge to the study of ultrarelativistic AA collisions.

I want to thank M. Gyulassy for helpful discussions. The work was supported in part by the U. S. Department of Energy under contract number DE-FG06-85ER40224.

REFERENCES

1. B. Anderson, G. Gustafson, G. Ingelman, and T. Sjostrand, Phys. Rep. 97, 33 (1983).
2. R. P. Feynman, Photon-Hadron Interactions (W. A. Benjamin, New York, (1972).
3. F. E. Low, Phys. Rev. D12, 163 (1975); S. Nussinov, Phys. Rev. Lett. 34, 1286 (1975).
4. N. K. Glendenning and T. Matsui, Phys. Rev. D28, 2890 (1983).
5. B. Anderson, G. Gustafson, I. Holgerson, and O. Mansson, Nucl. Phys. B178, 242 (1981).
6. T. Sjöstrand, Fermilab-Pub-85/119-T (Aug. 1985).
7. R. P. Feynman, "High Energy Collisions", ed. by C. N. Yang (Gordon and Breach, N.Y., 1969) p. 237; Phys. Rev. Lett. 23, 1415 (1969).
8. R. C. Hwa, "Partons in Soft-Hadronic Processes", Proc. of the Europhysics Study Conference, Erice, March 1981, ed. by R. T. Van de Walle (World Scientific, Singapore, 1981), p. 137, and other references cited therein.
9. G. Veneziano, Nucl. Phys. B74, 365 (1974); B 117, 519 (1976); H. M. Chan et al., Nucl. Phys. B86, 470 (1975); B92, 13 (1975); G. Chew and C. Rosenzweig, Nucl. Phys. B104, 290 (1976) and Phys. Rep. 41 C, 263 (1978).
10. A. Capella et al, Phys. Lett. 81B, 68 (1979); Z. Phys. C3, 329 (1980); H. Minakata, Phys. Rev. D20, 1656 (1979); G. Cohen-Tannoudji et at, Phys. Rev. D212689 (1980); A. Capella and J. Tran Thanh Van, Z. Phys. C10 249 (1981); A. Capella, C. Pajares and A. V. Ramallo, Nucl. Phys. B241, 75 (1984).
11. S. D. Ellis, SLAC Report No. 267 (184), p. 1; B. Durand and S. D. Ellis, Snowmass Proceedings, ed. R. Donaldson and J. G. Morfin (1984), p. 234; C. S. Lam and M. A. Walton, Phys. Lett. 140B, 246 (1984).
12. D. C. Hinz and C. S. Lam, McGill University preprint (1985).
13. R. C. Hwa and C. S. Lam, Oregon preprint OITS-312 (1985).
14. H. Ehtamo, J. Lindfors, and L. McLerran, Z. Phys. C 18, 341 (1983); T. S. Biro, H. B. Nielsen, and J. Knoll, Nucl. Phys. B245, 449 (1984).

15. We use the notations that $\propto$ denotes "proportional to" with unknown proportionality factor, while ~ denotes the same but with known factors not shown explicitly.
16. A. Bialas and W. Czyz, Phys. Rev. D31, 198 (1985).
17. A. K. Kerman, T. Matsui, and B. Svetitsky, MIT preprint CTP 1299 (1985).
18. R. C. Hwa and M. S. Zahir, Phys. Rev. D31, 499 (1985).
19. A. Capella et al., LPTHE Orsay 86/10 (1986).
20. M. Gyulassy and A. Iwazaki, Phys. Lett. 165B, 157 (1985).
21. K. Kajantie and T. Matsui, Phys. Lett. 164B, 373 (1985).
22. R. C. Hwa and K. Kajantie, Phys. Rev. Lett. 56, 696 (1986).
23. G. Baym, Phys. Lett. 138B, 18 (1984).
24. U. Heinz, Nucl. Phys. A 418, 603c (1984).
25. R. C. Hwa, Nucl. Phys. A418, 559c (1984), and Phys. Rev. D32, 637 (1985).
26. P. Carruthers and F. Zachariasen, Rev. Mod. Phys. 55, 245 (1983).
27. G. Ekspong, these proceedings.

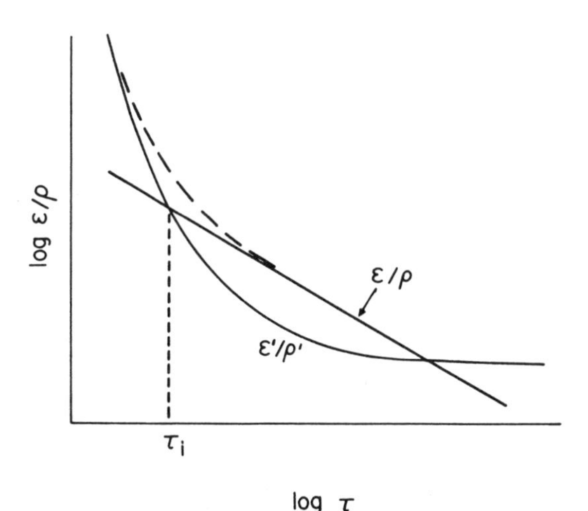

Fig. 1. Sketch of log-log plot of ε'/ρ' and ε/ρ versus τ. The broken curve represents a possible real solution to the problem.

PHASE TRANSITIONS AND DOUBLE SHOCK CREATION

László P. Csernai+

School of Physics and Astronomy, University of Minnesota
Minneapolis, Minnesota 55455

ABSTRACT

The nuclear equation of state is discussed in the phase transition region into QCD plasma. In case of a strong 1st order phase transition two possibly observable effets on the reaction dynamics are pointed out.

1. INTRODUCTION

In recent years considerable effort has been invested into studies which aim to predict the signatures and the formation mechanism of the QCD plasma [1-10]. Lacking experimental verifiability the interest in these studies diminished somewhat, but we hope that this will change as soon as the first data become available from BNL and CERN.

The theoretical studies fall into three major categories: i) the reaction mechanism, ii) the equation of state at the transition and iii) the search for experimental signals. Nuclear stopping power experiments led to the conclusion that in the ultrarelativistic region two basically different reaction mechanisms can be expected: Up to about 15 - 35 GeV/nucleon laboratory beam energy the stopping power of the heaviest nuclei is sufficiently strong to keep the baryons in the C.M. system [11], while at higher energies the colliding nuclei become transparent. On the equation of state side most of the theoretical work is invested in the study of pure SU(N) Yang-Mills theory on the lattice. Thus, these calculations, were restricted so far to zero net baryon density or zero chemical potential. To form such a low or "zero" baryon density matter in the deconfined phase one may need extremely high energy: $E_{Lab} \cong 100$ GeV/nucleon or more.

Relatively simple "phenomenological" theories, on the other hand, are able to provide us with an equation of state (EOS) in the phase transition region for cold matter [4], for zero baryon charge at

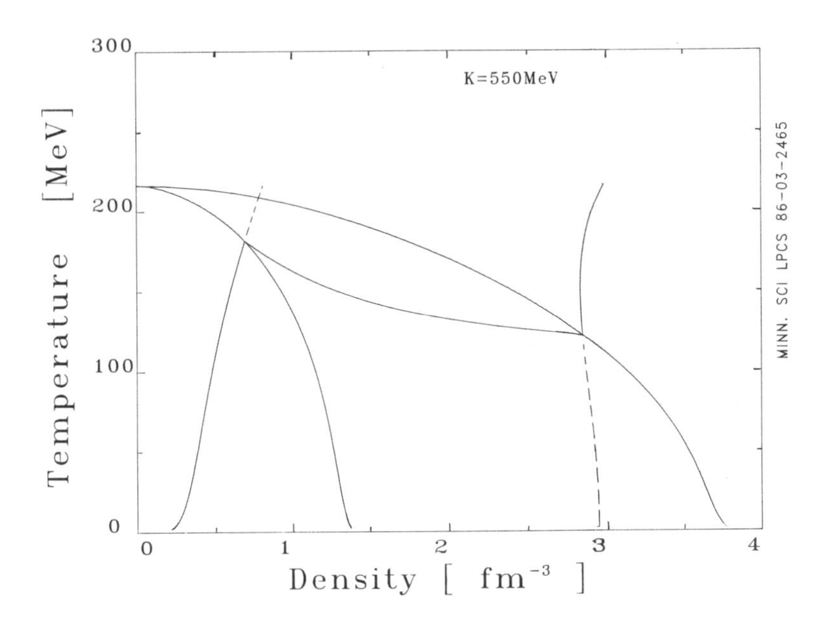

Fig.1 The nuclear phase diagram on the density (n) temperature (T) plane. The hadronic phase is described by the Sierk-Nix EOS (K=550MeV) with masless pions, the QCD plasma phase is parametrized by the MIT bag model EOS with a bag constant $B^{1/4}$ = 300MeV. The two curves connecting the critical point at T_c =216MeV with the critical density points (n_{CH}=1.37 fm^{-3}, n_{CQ}=3.77 fm^{-3}) separate the hadronic mixed and quark phases. The curve starting at the ground state nuclear mater n_o= 0.145 fm^{-3} is the compressional Taub adiabat. When it reaches the mixed phase the temperature decreases due to the latent heat that should be invested into the transition.

finite temperatures [5] and in the complete phase space for finite density and temperture too [2,3,8,10,12,19]. These phenomenological equation of state studies can yield a good qualitative insight into the phase transition problem until apriori QCD calculations become available in the complete density-temperature domain. There are even some advantages in the phenomonological approach: The nuclear equation of state in the BEVALAC domain is under extensive experimental and theoretical study right now via pion multiplicities and energy flow meaurements. The results of these studies can easily be incorporated in the phase transition studies.

The equation of state studies provide an equilibrium equation of state of course. Thus, this can be exploited for reaction studies in the lower part of the ultrarealtivistic energy domain where thermalization is still possible in the baryon rich matter. As transparency sets in with increasing energy, transport properties of the matter become more and more important [14]. At extremely high energies where baryon free plasma is formed a new, different type of rapid thermalization is possible in the local rest frame within 1 fm/c, but baryons or their valence quarks do not take part in this thermalization.

2. EQUATION OF STATE

2.1 Hadronic Equation of State

We have some information about the properties of nuclear matter from conventional nuclear physics. There is a stable equilibrium state at $n_o = 0.145\ fm^{-3}$ with a binding energy of 16 MeV/nucleon and a compressibility of K=180-240MeV. The value of K is of importance, as we will see later the high density behavior of matter is very sensitive to K, but also on the form of the equation of state! So, the value of K does not tell us too much about the equation of state at $n>3n_o$.

The first phenomenological equations of states were constructed

[2-5] by using a very simple functional form. The energy density e_H in terms of density n and temperature T was parametized as

$$e_H = n \{ m_N + W_o + K (n/n_o - 1)^2/18 + 3T/2 \}, \qquad (1)$$

where m_N is the nucleon rest mass, W_o the binding energy, the 3rd term is the compressinal energy e_{comp}, usually called "quadratic", and the last term is the thermal energy described as that of a Boltzmann ideal gas. (This approximation is adequate at high temperatures). At high temperatures pion pairs should also be taken into account. Neglectiong their rest mass the pion contribution is

$$e_{\pi} = \pi^2 T^4/10 , \qquad p_{\pi} = \pi^2 T^4 /30. \qquad (2)$$

At high temperatures some nucleons can be exited. The most important contribution is coming from delta resonances. Since the total baryon charge is conserved

$$n = n_N + n_\Delta , \qquad (3)$$

in the Boltzmann approximation the delta to nucleon ratio is given by

$$n_\Delta /n_N = 4 (m_\Delta /m_N)^{3/2} \exp(-[m_\Delta - m_N]/T) . \qquad (4)$$

This causes a change in the energy density where the rest mass term is now $n_\Delta m_\Delta + n_N m_N$, and also influences the other thermodynamical variables.

At high densities there is another constraint on the EOS, the requirement of causality. Eg.(1) yields a sound speed larger than the speed of light at large densities [20] and thus it is acausal. This can be remedied by using another functional form for the compressional energy: the "Sierk-Nix" parameterization

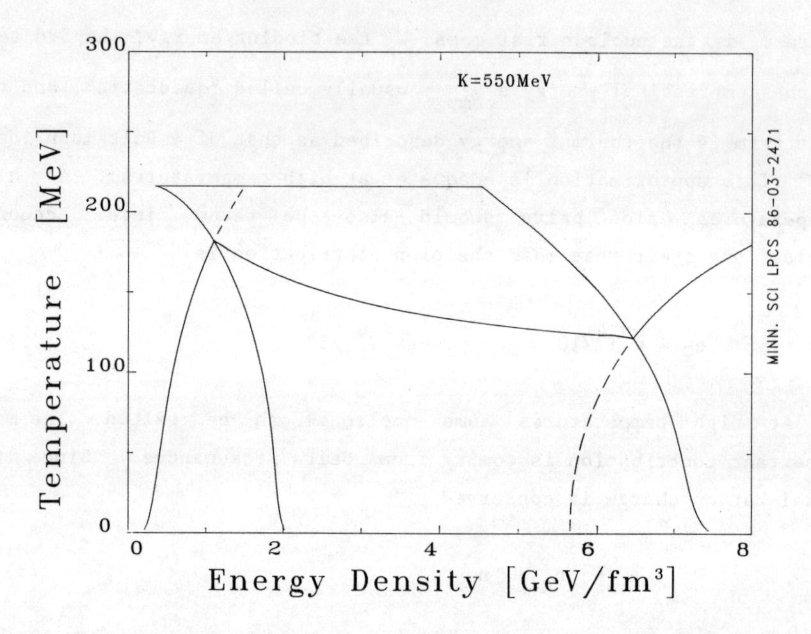

Fig. 2 The phase diagram on the energy density, temperature plane for the same EOS as Fig. 1.

$$e_{comp} = n \{ 2K [(n/n_o)^{1/2} - 1]^2/9 \}, \tag{5}$$

or the so called "linear" form:

$$e_{comp} = K n_o (n/n_o - 1)^2 / 18 . \tag{6}$$

These three parameterziations provide almost identical behavior [20] at $n = 3\text{-}4\ n_o$ if the compressibility is choosen for the

"Quadratic": K=275MeV, for the Sierk-Nix: K=550MeV, and for the "Linear": K=1800MeV.

It is important to emphasize that the nuclear EOS strongly influences the phase transition and the phase diagram. The compressional energy is particularly important. When it is neglected [19] the resulting phase diagram may lead to pathological behavior, the matter at n_o , and T=0 being in the mixed phase. (Also the possibility of a first order phase transition is restricted to a very small range of bag constants: $B^{1/4}$= 149-154MeV [19]. One way of including the compressional energy into the hadronic phases is the excluded volume approximation [8], which is a standard way of treating nuclear matter in relativistic nuclear collisions, first introduced in ref. [21]. This approximation leads to a phase diagram [8] similar to the nuclear EOS-s with explicit compressional energy (Fig. 1).

2.2 QCD - Plasma Equation of State

Usually for applications in heavy ion physics we can restrict ourselves to two flavors (u and d) in the quark-gluon phases. In zeroth order of perturbation theory then the pressure p_Q in terms of T and μ is

$$p_Q = 37\pi^2 T^4/30 + \mu^2 T^2 /9 + \mu^4/162\pi^2 - B, \qquad (7)$$

where B is the bag constant. This is also called as the M.I.T. bag model EOS. For phase transition studies this simple form is frequently used. [2,3,19]. In some cases 1-loop [8] or 2-loop [10] perturbative corrections are also included. These perturbative corrections contain a running coupling constant, which leads to the introduction of a new parameter Λ , fixing the renormalization scale. The introduction of these perturbative terms leads to a 10-20%

increase of the critical temperature and to a similar decrease of the critical densities [8,10]. n_{CH} and n_{CQ}.

Having defined both the Hadronic and QCD plasma EOS one can create a complete equation of state by Maxwell construction, containing pure phases and a region where the above two phases coexist. In this region Gibb's criteria

$$p_Q = p_H , \quad T_Q = T_H , \quad \mu_Q = \mu_H , \tag{8}$$

should be satisfied. This can be done relatively easily since $p_Q(\mu,T)$ is given and the chemical potental of the hadronic phase is al. o well known. A typical phase diagram arising from this construction is shown on Figs. 1-3.

The phase diagram of the first order phase transition is sensitive to both the nuclear and plasma parameters. An increase of compressibility K leads to a decrease of both critical densities, n_{CH} and n_{CQ} while increase of the bag constant B leads to the increase of the critical temperature, and densities. The inclusion of hadronic resonances has negligible effect on the phase diagram at T = 0 or μ = 0 but it pushes the phase boundaries to higher n and T values in the intermediate region, leading to an increase of the pure nuclear matter domain in the (μ,T) plane. The equilibrium pressure at fixed intermediate chemical potentials also increases due to the inclusion of hadronic resonances. The equilibrium pressure is higher at T=0 than at T_C which is an interesting feature first observed in ref. [10].

The energy density where the mixed phases formation becomes possible is about 1-2 GeV/fm^3 at finite densities, and it decreases below 1 GeV/fm^3 when the density tends to zero. To form pure QCD plasma one should, however, reach 2-6 GeV/fm^3 energy density at finite densities and 1-4 GeV/fm^3 at n=0 . This of course does not mean that n=0 QCD plasma can be formed with less beam energy!

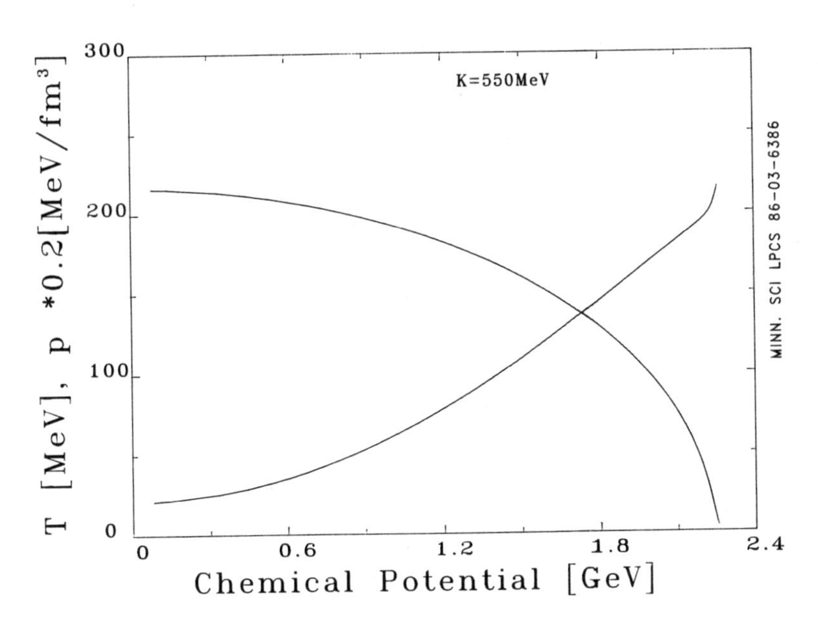

Fig. 3 The equilibrium temperature and the equilibrium pressure of the mixed phase versus the common chemical potential μ. The equilibrium temperature decreases from T_c to zero as μ increases, but the pressure increases unlike the pressure of thr familiar Van der Waals gas.

3. PLASMA FORMATION IN THE STOPPING REGION

3.1 Thermalization Considerations

In order to make use of the EOS one needs a system at least close to equilibrium. We do not know whether such a system can be formed in an ultrarelativistic neclear collisions or not. The sucess of the global flow analysis indicates that in the BEVALAC energy region the assumption of a locally near equilibrium matter is acceptable and the use of an EOS in the analysis of BEVALAC fragmention and pion production data is also accepted. We also know that non-equilibrium features are present already at the BEVALAC especially in peripheral collisions and in collisions of smaller nuclei. To describe these non-equilibrium features cascade and two and more component fluid dynamical models are also developped. Comparing the results of different theoretical models [12] one can conclude that the one-fluid dynamical model is one, still acceptable, extreme which leads to the highest density increases and produces the smallest overall dissipation.

Based on the experience we gained at the BEVALAC and on theoretical expectations, the one-fluid model is probably still acceptable to predict some basic quantities in the stopping energy region. The plasma phase transition might even increase the energy range of it's applicability because of the large increase of degrees of freedom. In any case, the predictions of fluid dynamical model should be considered as one extreme, and the development of a transport theoretical model for the baryon rich region capable of describing the phases transition would be welcome.

3.2 Fluid - Dynamical Predictions

There are relatively few calculations describing the baryon rich matter in the ultrarelativistic energy region [2,3,7,8,10,13]. Three dimensional calculations [13,22] did not consider a phase transition into QCD plasma in the EOS. One dimensional calulations yielded some interesting results arising from the phase transition.

The simplest way to present these results is to make use of Rankine - Hugoniot - Taub relations. These are simply the conservation laws of of relativistic flow in a supersonic compression.

Given the EOS of the matter the energy-momentum tensor of a perfect fluid is

$$T^{\mu\nu} = (e + p)\, u^{\mu} u^{\nu} - p\, g^{\mu\nu}, \tag{9}$$

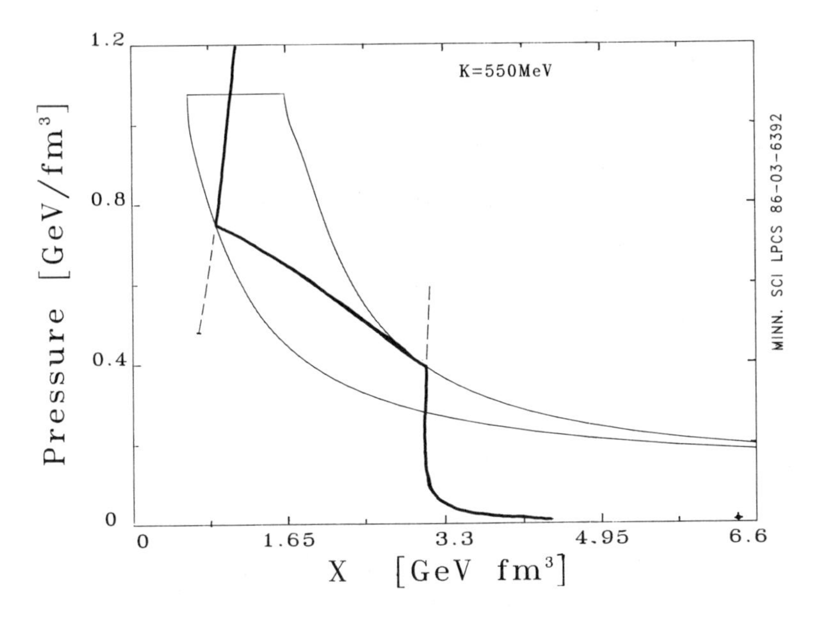

Fig. 4 The Taub adiabat in the pressure, p, generalized specific volume, $X = (e+p)/n^2$, plane. It starts from the nuclear ground state (p=0, X=6.5GeV fm), then as the pressure increases it goes through the hadronic, mixed and plasma phases. The change of X is not monotonic along the Taub adiabat. The thin lines indicate the phase boundaries.

where $u^{\mu} = (\gamma, \gamma\bar{v})$ is the four-velocity and $g^{\mu\nu}$ is the metric tensor. In a central collision of large nuclei the matter is stopped in the middle while the outer edges of the nuclei still propogate inward undisturbed. Here we assumed that the compression takes place in a front which is narrower than the nuclei, it should not be infinitely narrow. Inside this front there is a strong velocity gradient and large dissipation takes place due to viscous transport processes. But before the compression, stage 1, the matter flows inwards unformly, so

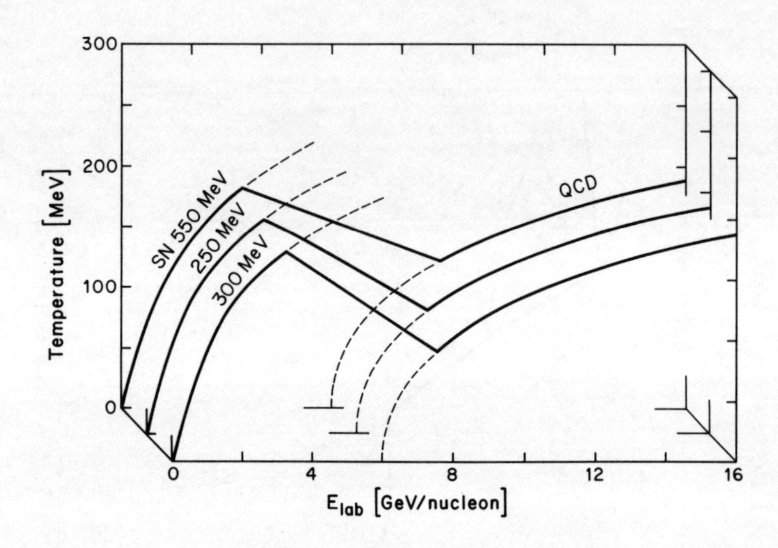

Fig. 5 Temperature of the maximally compressed state reached at a given beam energy. The 3 EOS's presented are the Sierk-Nix (SN) and the Quadratic EOS with two different compressibilities K. In the mixed phase the temperature decreases for all EOS-s.

the perfect fluid approximation is alright and after full compression, stage 2, also because here the matter stands still. Viewing the process from the moving front the energy, momentum and baryon conservation requires the equality of the incoming and outgoing currents:

$$T^{\mu\nu}_{(1)} \Lambda_\nu = T^{\mu\nu}_{(2)} \Lambda_\nu , \tag{10}$$

and

$$N^{\mu}_{(1)} \Lambda_\mu = N^{\mu}_{(2)} \Lambda_\mu , \tag{11}$$

where Λ^{μ} is the normal vector of the front where the compression takes place and N^{μ} is the baryon four current $N^{\mu} = n\, u^{\mu}$. Eliminating the four-velocity from Eqs.(10-11) yields the Taub adiabat

$$(p_{(2)} - p_{(1)})\,(X_{(2)} + X_{(1)}) = (h_{(2)}X_{(2)} - h_{(1)}X_{(1)}), \tag{12}$$

where h is the entralpy density h=e+p and X is the generalized specific volume $X = h/n^2$. The Taub adiabat connects the initial incoming state (1), which is the ground state nuclear matter, with the compressed state, which can be in hadron or quark phase or even a mixture of the two. The possible final states reachable in a supersonic compression are shown in Figs. 1 and 2 on the phases diagram, for the Sierk-Nix plus M.I.T. bag EOS. Deviations from the one-fluid dynamical behavior would lead to density decrease and entropy increase.

The compression in the hadronic phase is coupled to rapid temperature increase (Fig. 1). If the formation of phase mixture is inhibited this increase would continue above the critical temperature (dashed line). The formation of plasma droplets in the mixture region leads to temperature decrease, because of the large latent heat of the phase transition. This should be true in any model with a large latent heat. The temperature can rise further with compression if the matter is completely converted into quark phase.

It is convenient to discuss this supersonic compression process on the p,X plane. On this plane the slope of the straight line connecting the initial and final states is proportional with the baryon current across the front:

$$dp/dX = - (n_{(1)}v_{(1)})^2 = - (n_{(2)}v_{(2)})^2, \tag{13}$$

where (1) is the initial state, $n_{(1)} = n_o$!

In Fig.4 the phase boundaries and the Taub adiabat are plotted on the p,X plane. The phases boundaries start at a high equilibrium

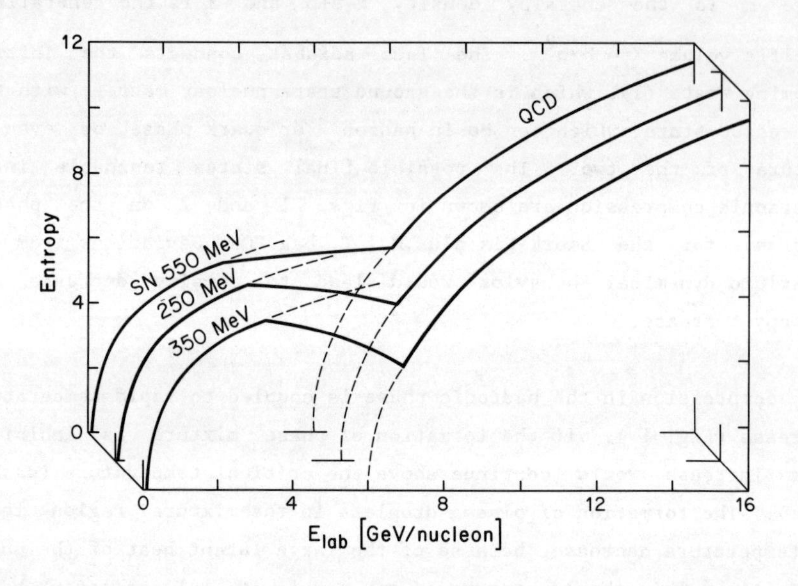

Fig. 6 Entropy of the compressed state for the same EOS's as in Fig.5. The entropy decreases in the mixed phase for the "Quadratic" EOS indicating the instability of the front.

pressure decreasing gradually as X increases (at the same time n decreases and T increases). The shock adiabat starts at the nuclear ground state (p=0, X=6.5GeV fm^3) and first slowly increases with decreasing X (compression) then goes over into a sharp rise and bends back. The sharp rise lasts until the adiabat reaches the boundary of the mixed phase. The formation of the mixed phase results in a sharp kink and the pressure increases once more with decreasing X until the pure quark phase is reached and the pressure rises rapidly again. Eq. (13) establishes a connection between the beam energy, so with small beam energy ,small slope, only hadronic final states can reached while large energy, large slope, leads to plasma final state.

The most striking feature is that the transition is not gradual around the threshold. A very small slope increase leads to a tremendous change in volume. This might lead to some observable phenomena just around the phase transition threshold.

This was first pointed out in refs. [1-2]. A slightly stiffer nuclear EOS, like Eq.(1), leads to a Taub adiabat where the straight line (13) can cross the adiabat in 3 points. This situation is discussed in detail [2] and it is shown that it can result in shock instability or double shock formation.

3.3 Beam Energy Dependence

Without going into the details of the complicated discussion [2] of double shock formation we can show that something unusual should happen at the phase transition threshold. Let us calculate the final state depending on the beam energy by using Eq. (13). Fig. 5 shows the final temperature for 3 different EOS's. The high energy end, the QCD plasma, does not change, there is a samll change at low beam energies where the final state is in the hadronic phase. The most sensitive is the mixed phase where the temperature drops with increasing energy.

More enlightening is to study the final entropy. It should monotonicly increase with beam energy. Fig. 6 shows the final entropy versus beam energy for 3 EOS's. The Sierk-Nix EOS yields monotonic entropy increase showing a very strong structure. The other two curves, corresponding to the "Quadratic" EOS, show a decrease in the mixed phase! This solution cannot be physical and coincides with the cases where shock instabilities were predicted [2].

It is a good first approximation that the expansion after the maximum conpression is adiabatic [8,12]. In this case the entropy excitation function of Fig. 6 should be observable. To measure the entropy experimentally is a difficult task [12]. One can make a rough estimate of the connection between the pion multiplicity and the entropy:

$$S \sim a \ln (n_{\pi} / n) + b. \tag{14}$$

Although accelerator experiments are not yet available in this energy region, very recently cosmic ray emulsion data were analysed and published in the phase transition region [18]. The n_{π}/n multiplicities show a structure which resembles somewhat the structure we expect to see in the entropy (Fig. 7). Although this resemblence obviously does not prove or indicate the existence of the phase transition, it does suggest that there is a possibility to find the phase transition by similar studies.

3.4 Flow Analysis

It was shown [2] that there exist the possibility of a shock instability in the presence of a strong 1st order phase transition into quark-gluon plasma especially for a stiff nuclear EOS. In the one fluid picture this appears as the splitting of the front into two. In the first front which propagates faster the matter is precompressed, usually still in the hadronic phase. From this precompressed state a second slower front leads to the plasma [1]. In 3 dimensions not only the speed but also the direction of the propagation is different. The nuclear size is, however, probably not sufficient to form two separate fronts. Thus we can expect an essential broadening of the front at the threshold instead. This would smooth out the sharply peaked flow angle distributions observed at the BEVALAC.

Two and three dimensional fluid dynamical studies are in progress [17] to give quantitative answer on the changes of the flow around the phase transition threshold.

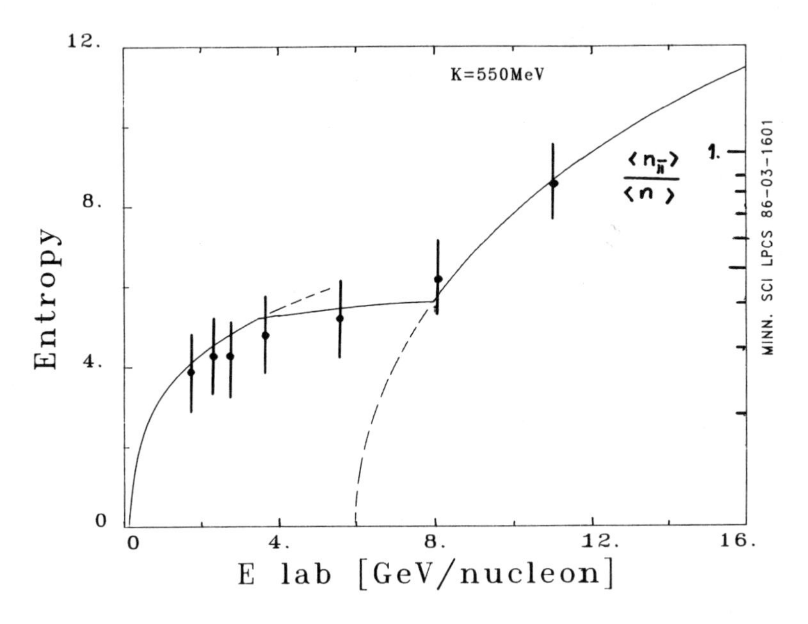

Fig. 7 Entropy of the final state calculated with the Sierk-Nix EOS and experimental pion multiplicity data for heavy projectiles Z>20 from ref. [18] plotted versus the incident nucleus energy.

4. SUMMARY

The features of several phenomenological EOS's were discussed which describe the QCD plasma phase transition. The EOS depends sensitively on nuclear parameters like compressibility and plasma parameters like the bag constant. Most realistic hadronic and QCD plasma EOS-s yield a 1st order phase transition with large latent heat.

In the stopping energy region which is expected to include the phase transition threshold one can make use of the equation of state if the matter is not too far from local thermal equilibrium. The fluid dynamical model as one extreme, assuming local thermalization, yields strong changes both in the final entropy and in the flow pattern just at the phase transition threshold. The first cosmic ray experiments [18] in this energy region indicate that to discover a significant structure in the pion multiplicity excitation function caused by the phase transition is not impossible.

5. REFERENCES

+ On leave of absence from the CRIP, Budapest, Hungary

1. G.Chapline; Proc. of 7th High Energy Heavy Ion Study, Darmstadt, Oct. (1984) p. 45.
2. H.W.Barz, L.P.Csernai, B.Kämpfer, B.Lukács; Phys.Rev. C32, 115 (1985).
3. H.W.Barz, et al.; Phys.Rev. C31, 268 (1985) and Phys.Lett. 143B, 334 (1985); L.P.Csernai, B.Lukács; Phys.Lett. 132B, 295 (1983).
4. G.Baym, S.A.Chin; Phys.Lett. 62B, 241 (1976). G.Chapline, M.Nauenberg; Phys.Rev. D16, 456 (1977). B.Freedman, L.McLerran; Phys.Rev. D17, 1109 (1978).
5. T.Celik, H.Satz; Z.Phys. C1, 163 (1979).
6. M.Gyulassy, K.Kajantie, H.Kurki-Suonio, L.McLerran; Nucl.Phys. B237, 477 (1984).
7. H.Stöcker, G.Graebner, J.Maruhn, W.Greiner; Z.Phys. A295, 401 (1980).
8. P.R.Subramanian, H.Stöcker, W.Greiner; Frankfurt UFTP-169/86 rep.
9. V.M.Galitskij, I.N.Mishustin; Phys. Lett. 72B,285 (1978); E.V.Shuryak; Phys.Rep. 61, 71 (1980).
10. J.Kuti, B.Lukács, J.Polónyi, K. Szlacsányi; Phys.Lett. 95B, 75 (1980); S.A.Chin; Phys.Lett. 78B, 552 (1978).
11. L.P.Csernai, J.I.Kapusta; Phys.Rev. D29,2664(1984); D31,2795(1985)
12. L.P.Csernai, J.I.Kapusta; Phys.Rep. 131, 223 (1986).
13. R.B.Clare, D.Strottman; Phys.Rep. in press
14. T.Elze, M.Gyulassy, D.Vasak; LBL-21137 (1986) rep.
16. A.K.Kerman, T.Matsui, B.Svetitsky; Phys.Rev.Lett. 56, 219 (1986).
17. D.Strottman, L.P.Csernai; in preparation
18. T.W.Atwater, P.S.Freier; Phys.Rev.Lett. 56, 1350 (1986).
19. J.Cleymans, R.Gavai, E.Suhonen; Phys.Rep. 130, 217 (1986).
20. C.Grant, J.I.Kapusta; Phys.Rev. C32, 663 (1985).
21. P.R.Subramanian, et al.; J.Phys. G7, L241 (1981).
22. H.Stöcker, W.Greiner, Phys.Rep. in press

Large Transverse Momenta and the Hydrodynamical Model of Multiparticle Production

K. Wehrberger
Institut für Kernphysik, TH Darmstadt
D- 6100 Darmstadt
Federal Republic of Germany

The goal of the present work is to investigate whether the range of applicability of a hydrodynamical model of multiparticle production extends beyond the region of very small transverse momenta. To be specific, we use the Landau model [1] with cylindrical initial conditions and ideal relativistic hydrodynamics, applied to pp- and p$\bar{p}$- collisions at ISR- and SPS- energies. We choose an effective velocity of sound $c_o = 1/\sqrt{3}$ [2], i.e. pressure p and energydensity ε are related by the equation of state

$$p = \frac{1}{3}\varepsilon \qquad (1)$$

This model has been applied successfully to many features of high energy collisions at small transverse momenta [1,3] ($p_t <$ some m_π). There are two sources of transverse momentum: the thermal distribution at the (energyindependent) break-up temperature $T_f = m_\pi$ and hydrodynamical flow. The limitation and approximate constancy of the average p_t observed in multiparticle events is naturally explained by the dominance of the thermal source at small p_t. However, with increasing p_t this contribution drops exponentially and hydrodynamical flow is the main source of intermediate transverse momentum (some $m_\pi < p_t < 5\text{-}10$ GeV). On the other hand in the region of large p_t ($p_t > 5\text{-}10$ GeV) we have events with very large momentum transfer, perturbative QCD is applicable and we do not expect a hydrodynamical description to work. In this study therefore we solved the equations of the threedimensional Landau model in an approximation which makes sense in the region of intermediate transverse momentum.

We visualize the evolution of the central region of the fireball as follows. Longitudinal shock waves and rarefaction waves pass through the medium, until the nontrivial region develops. We neglect transverse expansion during this very brief first stage. Then matter starts flowing out also transversally as a simple wave the front of which moves with the velocity of light. Its rear end propagates inward with the velocity of sound, and our interest concentrates on the fluid in between, which is the source of intermediate p_t. With

$$\Delta = \frac{r_\circ}{\gamma} \quad , \quad \gamma = \frac{K\sqrt{s}}{2m_p} \tag{2}$$

($r_\circ$, m_p: radius, mass of the proton; $\sqrt{s}$ cms-energy; K: inelasticity) and

$$\tau = \frac{\sqrt{t^2 - z^2}}{\Delta} \, , \quad \rho = \frac{r}{\Delta} \tag{3}$$

(r,z transverse, longitudinal coordinate) the boundary between transverse rarefaction wave and the nontrivial region is given by

$$\rho = \rho_\circ - c_\circ \tau \tag{4}$$

This is where we differ from [4] where an outward moving boundary was considered which is in our opinion not adequate. Introducing transverse rapidity ξ and logarithmic temperature ω (T_o: initial temperature of the hot fireball)

$$\tanh\xi = \frac{p_t}{\sqrt{p_t^2+m_\pi^2}} \quad , \quad \omega = \ln(T/T_o) \qquad (5)$$

the boundary conditions, obtained from the asymptotic form of the exact solution of the onedimensional model [5] , read

$$\omega = - c_o^2 \ln\tau \quad , \quad \xi = 0 \qquad (6)$$

Now we make the following approximations. First, we neglect the thermal distribution to the transverse momentum, as was discussed above. Second, we assume a flat distribution in longitudinal rapidity which is justified in the central region. Third, we assume $\xi >> 1$ which is consistent with the first approximation. Then the equations of motion simplify

$$\begin{aligned} \frac{\partial\omega}{\partial\tau} + \frac{\partial\omega}{\partial\rho} + \frac{c_o^2}{1-c_o^2}\left[\frac{1}{\tau}+\frac{1}{\rho}\right] &= 0 \\ \frac{\partial\xi}{\partial\tau} + \frac{\partial\xi}{\partial\rho} - \frac{c_o^2}{1-c_o^2}\left[\frac{1}{\tau}+\frac{1}{\rho}\right] &= 0 \end{aligned} \qquad (7)$$

and may be solved analytically with the boundary conditions (4) on (6). An upper limit (p_t < 5-10 GeV) of applicability of the approximation follows from the observation that we cannot go too close to the outer edge because the particles there are already frozen out. We obtain the spectrum of transverse momenta from the usual freeze-out condition.

Our results may be summarized as follows. The average transverse momentum is a slowly growing function of the available energy

$$\langle p_t \rangle \approx m_\pi \left(\frac{K \sqrt{s}}{m_p} \right)^{1/6} \tag{8}$$

which agrees with earlier calculations [6,7] but does not say anything specific about the spectrum. At large p_t the behaviour of the cross section is

$$\frac{dN}{dp_t^2} \sim p_t^{-5} , \tag{9}$$

at smaller p_t the falloff is sharper. This is in qualitative agreement with data and should also be compared to the prediction of a p_t^{-4}- spectrum at large p_t from perturbative QCD.

Our conclusion is that the flattening of the p_t- distribution at large p_t does not rule out hydrodynamics as a model for production of transverse momenta, but rather confirms its validity.

This report is based on work done in collaboration with D. Syam and S. Raha [8].

References

1) for a review, see E.I.Daibog et al., Fortschr. d. Phys. 27(1979)313
2) K.Wehrberger and R.M.Weiner, Phys.Rev. D31(1985)222
3) Proceedings of the First International Workshop on Local Equilibrium in Strong Interaction Physics, Singapore 1985
4) I.Yotsuyanagi, Prog.Theor.Phys. 55(1976)539
5) I.M.Khalatnikov, Zhur.eksp.teor.fiz. 27(1954)529
6) G.Milekhin, Sov.Phys.JETP 8(1959)829
7) M.Chaichian et al., Phys.Lett. 50B(1974)362
8) K.Wehrberger, D.Syam and S.Raha, Phys.Rev. D, to be published

The QCD Quark Wigner Operator and Semiclassical Transport Equations

H.-Th. Elze, M. Gyulassy, and D. Vasak [1]

Nuclear Science Division, Lawrence Berkeley Laboratory
University of California, Berkeley, California 94720

Abstract: Using exact gauge covariant equations of motion for the SU(N) quantum chromodynamic Wigner operator, we derive the semiclassical limit of the generalized on-shell constraint and the transport equation respectively. The transport equation reduces to a Vlasov type equation with non-Abelian and spin-dependent terms. We show how quantum corrections can be calculated systematically. Under suitable conditions the transport equation reduces to a particularly simple set of Abelian equations in the Cartan basis of SU(N).

1. Introduction

A central problem in the study of quark-gluon plasma properties [1,2] is to determine the formation time of the plasma and its evolution towards chemical and thermal equilibrium (cf. the contribution by R. Hwa in this volume). For that purpose it is necessary to derive transport equations for quarks and gluons that can be applied even if equilibrium is never achieved. While there has been a considerable amount of work on deriving relativistic Abelian transport equations in the past [3]-[8], work on gauge covariant non-Abelian transport equations [9,10,11] has only recently been begun. In a recent paper we contributed to that development by deriving constraint and transport equations for the *relativistic gauge covariant Wigner operator* for fermions interacting via SU(N) gauge fields [12].

Based on investigations of the external field problem in QCD and of particle production by covariant constant fields in particular [13], one is led to conjecture that at least in certain limiting situations the transport equations for quarks and gluons may reduce to a particularly simple set of Abelian equations. The simplification of the equations of motion (Dirac's equation together with Yang-Mills' equations) in the semiclassical limit suggests that at least for slowly varying fields, the transport equations for quarks and gluons may also reduce to a set of Abelian Vlasov type equations:

$$(p_\mu \partial^\mu_x + g\vec{q} \cdot \vec{F}_{\mu\nu}(x) p^\nu \partial^\mu_p) f_{\vec{q}}(x,p) = C_{\vec{q}} + S_{\vec{q}} \ , \tag{1.1}$$

where $f_{\vec{q}}(x,p)$ is the phase space density for particles with effective charge $\vec{q}$, and $C_{\vec{q}}$ and $S_{\vec{q}}$ are collision and source terms. In this picture the self consistent fields

[1] Work supported by the Director, Office of High Energy and Nuclear Physics of the Department of Energy under Contract DE-AC03-76SF00098. H.-Th. Elze and D. Vasak gratefully acknowledge support by DAAD-NATO Postdoctoral Fellowships.

$\vec{F}^{\mu\nu}$, $\vec{F}^{\mu\nu} \equiv \partial^\mu \vec{A}^\nu - \partial^\nu \vec{A}^\mu$, satisfy a set of Maxwell equations

$$\partial_\mu \vec{F}^{\mu\nu} = g\vec{J}^\nu_{\text{ext}} + \sum_{\vec{q}} g\vec{q} \int d^4p\, p^\nu f_{\vec{q}}(x,p) \quad , \tag{1.2}$$

where $\vec{J}^\mu_{\text{ext}}$ is an external source current and where the second term is the induced current in the plasma.

A new feature of the above equations is the expected occurrence of source terms that result from pair production in the external fields. In QED such terms are never important in practical applications. However, in QCD the assumption of confinement requires that such terms appear to neutralize the color fields. In ref.[13] the total pair creation rate of quanta with charge $\vec{q}$ was calculated for covariant constant color electric fields, $\vec{F}^{30} = \vec{E}$ to be

$$\int d^4p\, S_{\vec{q}}(x,p) = \frac{\gamma g^2}{24\pi}(\vec{q}\cdot\vec{E})^2 \quad , \tag{1.3}$$

where $\gamma = 1$ $(1/2)$ for fermions (vector bosons) respectively. While the transverse momentum dependence of $S_{\vec{q}}$ is understood, there is considerable uncertainty as yet about the longitudinal momentum dependence [14].

The above chromo-transport equations represent the simplest and most natural generalization of relativistic QED plasma equations. However, it is far from clear whether and under what special conditions they may actually apply.

Usually, the operator that is expected to have the closest connection with the classical distribution functions is the Wigner operator [5,6,7]

$$\hat{W}(x,p) \equiv \int \frac{d^4y}{(2\pi)^4} e^{-ip\cdot y} \bar{\psi}(x+\tfrac{1}{2}y) \otimes \psi(x-\tfrac{1}{2}y) = \int \frac{d^4y}{(2\pi)^4} e^{-ip\cdot y} \bar{\psi}(x) e^{\frac{1}{2}y\cdot\partial_x^\dagger} \otimes e^{-\frac{1}{2}y\cdot\partial_x} \psi(x) \; , \tag{1.4}$$

where $\partial_x^\dagger \equiv \overleftarrow{\partial/\partial x^\mu}$ and $\partial_x \equiv \overrightarrow{\partial/\partial x^\mu}$ are the generators of translations acting to the left and to the right respectively. Unfortunately, the above definition cannot be correct for a gauge theory, since $\hat{W}$ does not transform covariantly under a gauge transformation. A gauge covariant definition can, however, be constructed by substituting the covariant derivative, $D^\mu \equiv \partial^\mu + igA^\mu$, and its adjoint in place of ∂^μ and its adjoint. Applying this minimal substitution rule leads to the following definition of the relativistic gauge covariant Wigner operator for spin-1/2 particles:

$$\hat{W}(x,p) \equiv \int \frac{d^4y}{(2\pi)^4} e^{-ip\cdot y} \bar{\psi}(x) e^{\frac{1}{2}y\cdot D_x^\dagger} \otimes e^{-\frac{1}{2}y\cdot D_x} \psi(x) \; . \tag{1.5}$$

The tensor product in eq.(1.5) implies that $\hat{W}$ is a matrix in spinor (4×4 components) as well as in color space ($N \times N$ components). Under a local gauge transformation, $S(x) \equiv \exp[i\theta_a(x)t_a]$, $\psi(x) \longrightarrow S(x)\psi(x)$ and $D^\mu \rightarrow S(x)D^\mu S^{-1}$, and the Wigner operator, eq.(1.5), transforms covariantly:

$$\hat{W}(x,p) \longrightarrow S(x)\hat{W}(x,p)S^{-1}(x) \; . \tag{1.6}$$

As we showed previously [12] this definition of $\hat{W}$ is equivalent to that proposed in ref.[10]. Note, however, that the covariant derivative is the operator that corresponds to the ordinary *kinetic* momentum, $\hat{\pi}^\mu \equiv P^\mu - gA^\mu = iD^\mu$, where $P^\mu = i\partial_x^\mu$ is the canonical momentum conjugate to the position coordinate. The on-shell condition $\hat{\pi} \cdot \hat{\pi} = m^2$ applies to the kinetic rather than the conjugate momentum in the classical limit. Therefore, it is natural that the covariant rather than the ordinary derivative appears in the definition of the Wigner function in eq.(1.5).

Our objective in this paper is to derive and discuss the proper semiclassical limit of the non-Abelian equations of motion for $\hat{W}$ derived from Dirac's field equations in ref.[12]. We show that in the classical limit our equations reduce to those proposed by Heinz in ref.[10] and also how quantum and spin-dependent corrections to the classical equations can be calculated systematically. We note that a non-Abelian transport theory for scalar particles was studied by Winter [11]. A thorough study of scalar QED transport theory has been presented by Remler [3]. We reproduce the semiclassical transport equation for scalar particles derived in ref.[3] as a special case of our general transport equation. Extensive lists of references dealing with non-gauge-covariant Wigner functions can be found in refs.[5,6]. We also refer the reader to contributions by U. Heinz and M. Rhoades-Brown to these proceedings.

The plan of our paper is as follows. In Sec.2 we list definitions and main results derived in ref.[12] which we employ afterwards. In Sec.3 we perform the operator expansion of the transport equations which yields the semiclassical limit. We find that only for "slowly" varying fields and Wigner functions (in a generalized sense derived there) the transport equations can be reduced to Vlasov's equation. Finally, in Sec.4 we show how at least the Vlasov part of the chromo-transport equation, eq.(1.1), arises from our general results in a model where the expectation value of the Wigner operator is assumed to be diagonal in the gauge that diagonalizes the field tensor. We close by pointing out directions for further studies and applications.

2. The Quantum Transport Equations

We use the metric $g^{\mu\nu} \equiv diag(1,-1,-1,-1)$, $a \cdot b \equiv a_\mu b^\mu$, and choose units such that $\hbar = c = 1$. The gauge field potential is a $N \times N$ matrix in color space defined by

$$A_\mu \equiv A_\mu^a t_a \ , \tag{2.1}$$

with the $N^2 - 1$ Hermitian generators of SU(N) in the fundamental representation satisfying $Tr\ t_a = 0$, $Tr\ t_a t_b = \delta_{ab}/2$, and $[t_a, t_b] = i f_{abc} t_c$. The covariant derivative,

$$D_\mu \equiv \partial_\mu + igA_\mu \ , \tag{2.2}$$

is thus an $N \times N$ matrix in color space as is the field strength tensor, $F_{\mu\nu} \equiv [D_\mu, D_\nu]/(ig)$, which obeys the field equation

$$[D_\mu, F^{\mu\nu}] = gJ^\nu \ . \tag{2.3}$$

The color current operator is given by

$$J^\mu \equiv \hat{\jmath}^\mu_a t_a \equiv t_a \bar{\psi}\gamma^\mu t_a \psi = \int d^4p\, t_a\, Tr\, \gamma^\mu t_a \hat{W}(x,p) \ , \tag{2.4}$$

where the trace refers to spinor and color indices. Under a local gauge transformation S not only D_μ and $F^{\mu\nu}$ transform covariantly but also J^μ, i.e., $\hat{J} \to S(x)\hat{J}S^{-1}(x)$, because $t_a \cdot Tr(t_a S\hat{W}S^{-1}) = St_aS^{-1} \cdot Tr(t_a\hat{W})$.

Our derivation of the equations of motion for the Wigner operator defined in eq.(1.5) was based on the field equations for the Heisenberg operators ψ and $\bar{\psi}$,

$$(i\gamma^\mu D_\mu - m)\psi = 0 = \bar{\psi}(i\gamma^\mu D^\dagger_\mu + m) \ , \tag{2.5}$$

where henceforth $D^\dagger$ is defined to act always to the left, and $\bar{\psi} \equiv \psi^\dagger\gamma^0$, $\gamma^{\mu\dagger} \equiv \gamma^0\gamma^\mu\gamma^0$, with $\gamma^0 \equiv diag(1,1,-1,-1)$. We suppress quark flavor indices. We also used the quadratic form of Dirac's equation and its adjoint

$$\begin{aligned}(D_\mu D^\mu + \tfrac{1}{2}g\sigma^{\mu\nu}F_{\mu\nu} + m^2)\psi &= 0 \ , \\ \bar{\psi}(D^\dagger_\mu D^{\dagger\mu} + \tfrac{1}{2}g\sigma^{\mu\nu}F_{\mu\nu} + m^2) &= 0 \ ,\end{aligned} \tag{2.6}$$

where $\sigma^{\mu\nu} \equiv \frac{1}{2}i[\gamma^\mu,\gamma^\nu]$ and $g^{\mu\nu} = \frac{1}{2}\{\gamma^\mu,\gamma^\nu\}$.

Eq.(2.5) implies the following generalized *gauge covariant operator constraint equation* for $\hat{W}$ [12]:

$$\begin{aligned}&\left(\gamma^\mu p_\mu - m + \tfrac{1}{2}i\gamma^\mu \mathcal{D}_\mu(x)\right)\hat{W}(x,p) \\ &= -\tfrac{1}{2}ig\partial^\nu_p\gamma^\mu \left(\int_0^1 ds\, (1-\tfrac{1}{2}s)\, [e^{-\frac{1}{2}i(1-s)\partial_p\cdot\mathcal{D}(x)}\, F_{\nu\mu}(x)]\, \hat{W}(x,p) \right. \\ &\left. \quad + \hat{W}(x,p)\int_0^1 ds\, \tfrac{1}{2}(1-s)\, [e^{\frac{1}{2}is\partial_p\cdot\mathcal{D}(x)}\, F_{\nu\mu}(x)] \right) \ ,\end{aligned} \tag{2.7}$$

where ∂_p always acts on $\hat{W}$ and where the covariant derivative of a second-rank tensor $\mathcal{T}$ is defined by

$$\mathcal{D}(x)\mathcal{T}(x) \equiv \partial_x\mathcal{T}(x) + ig[A(x),\mathcal{T}(x)] \ . \tag{2.8}$$

Henceforth and in eq.(2.7) in particular the multiplication of $\hat{W}$ by any operator from the left or right has to be understood in the sense of matrix multiplication. An operator acting on $\hat{W}$ from the left (right) actually is meant to be inserted immediately to the right (left) of the $\otimes$-sign in the definition, eq.(1.5). We remark here that this convention is quite essential for keeping track of the operator ordering, since for interacting quantum fields e.g. $[(F_{\mu\nu})_{ab}, U_{cd}] \neq 0$. For the special case of a purely classical external field ordinary matrix multiplication rules suffice, since we must only keep track of the sequence of matrix indices, while the order of matrix elements is irrelevant. In general those matrices are field operators and their ordering is important.

As was shown in ref.[12], eq.(2.7) can be separated into its Hermitian and anti-Hermitian parts which yield a proper *generalized on−shell constraint equation* and a second equation, which can be recognized as a transport equation in the simple case of a constant Abelian field $F_{\mu\nu}$. In the general case, however, of interacting matter and gauge fields we found it more appropriate, although tedious, to derive the *quantum transport equation for the QCD Wigner operator* from eqs.(2.6). The result of the lengthy calculation is:

$$
\begin{aligned}
0 \; &= \; p\cdot\mathcal{D}(x)\,\hat{W}(x,p) \\
&+ \; \tfrac{1}{2}gp^{\mu}\partial_p^{\nu}\int_0^1 ds\,\Big(\,[e^{-\frac{1}{2}i(1-s)\Delta}F_{\nu\mu}(x)]\,\hat{W}(x,p)+\hat{W}(x,p)\,[e^{\frac{1}{2}is\Delta}F_{\nu\mu}(x)]\,\Big) \\
&- \; \tfrac{1}{4}ig\mathcal{D}^{\mu}(x)\partial_p^{\nu}\int_0^1 ds\,\Big(\,(s-1)[e^{-\frac{1}{2}i(1-s)\Delta}F_{\nu\mu}(x)]\,\hat{W}(x,p)+s\,\hat{W}(x,p)\,[e^{\frac{1}{2}is\Delta}F_{\nu\mu}(x)]\,\Big) \\
&+ \; \tfrac{1}{4}ig\partial_p^{\nu}\int_0^1 ds\,\Big(\,(s-1)[e^{-\frac{1}{2}i(1-s)\Delta}\mathcal{D}^{\mu}(x)F_{\nu\mu}(x)]\,\hat{W}(x,p)+s\,\hat{W}(x,p)\,[e^{\frac{1}{2}is\Delta}\mathcal{D}^{\mu}F_{\nu\mu}(x)]\,\Big) \\
&- \; \tfrac{1}{8}ig^2\partial_p^{\mu}\partial_p^{\nu}\int_0^1 ds\int_0^1 d\tilde{s}\;(s-1+\tilde{s})[e^{-\frac{1}{2}i(1-s)\Delta}F_{\mu}{}^{\lambda}(x)]\,\hat{W}(x,p)\,[e^{\frac{1}{2}i\tilde{s}\Delta}F_{\nu\lambda}(x)] \\
&- \; \tfrac{1}{16}ig^2\partial_p^{\mu}\partial_p^{\nu}\int_0^1 ds\int_0^1 d\tilde{s}\,\Big(\,(s^2(1+\tilde{s})-2s)[e^{-\frac{1}{2}i(1-s)\Delta}F_{\mu}{}^{\lambda}(x)]\,[e^{-\frac{1}{2}i(1-s\tilde{s})\Delta}F_{\nu\lambda}(x)]\,\hat{W}(x,p) \\
&\qquad\qquad + (1-s)^2(\tilde{s}-2)[e^{-\frac{1}{2}i(1-s)(1-\tilde{s})\Delta}F_{\mu}{}^{\lambda}(x)]\,[e^{-\frac{1}{2}i(1-s)\Delta}F_{\nu\lambda}(x)]\,\hat{W}(x,p) \\
&\qquad\qquad + s^2(1+\tilde{s})\,\hat{W}(x,p)\,[e^{\frac{1}{2}is\Delta}F_{\mu}{}^{\lambda}(x)]\,[e^{\frac{1}{2}is\tilde{s}\Delta}F_{\nu\lambda}(x)] \\
&\qquad\qquad + \;(1-s)(2s+(1-s)\tilde{s})\,\hat{W}(x,p)\,[e^{\frac{1}{2}i(s+(1-s)\tilde{s})\Delta}F_{\mu}{}^{\lambda}(x)]\,[e^{\frac{1}{2}is\Delta}F_{\nu\lambda}(x)]\,\Big) \\
&+ \; \tfrac{1}{4}ig\,\Big(\,\sigma^{\mu\nu}[e^{-\frac{1}{2}i\Delta}F_{\mu\nu}(x)]\hat{W}(x,p) \;-\; \hat{W}(x,p)\sigma^{\mu\nu}[e^{\frac{1}{2}i\Delta}F_{\mu\nu}(x)]\,\Big)\;,
\end{aligned}
\tag{2.9}
$$

where $\Delta \equiv \partial_p\cdot\mathcal{D}(x)$ and the small brackets, [...], delimit where the $\mathcal{D}$-derivative from Δ acts; ∂_p always acts on $\hat{W}$.

At this point it should be noticed that one purpose among others of our previous derivation of the gauge covariant quantum transport equations, eqs.(2.7, 2.9), was to provide completely general results from which semiclassical approximations can be obtained in a systematical and transparent way. The semiclassical limit will be studied in the next section.

We finally remark here that $\hat{W}$ as defined in eq.(1.5) can be reexpressed [12] in the form proposed in ref.[10],

$$
\hat{W}(x,p)=\int\frac{d^4y}{(2\pi)^4}e^{-ip\cdot y}\;\bar{\psi}(x+\tfrac{1}{2}y)U(x+\tfrac{1}{2}y,x)\otimes U(x,x-\tfrac{1}{2}y)\psi(x-\frac{1}{2}y)\;, \tag{2.10}
$$

where the *link operator* U is given by the path ordered exponential of a line integral [15,16],

$$U(b,a) \equiv P\exp\left(-ig\int_a^b dz^\mu A_\mu^a(z)t_a\right) \quad , \tag{2.11}$$

and the path of integration is the *straight line* between the end points,

$$z(s) \equiv z(b,a,s) = a + (b-a)s \quad , \; 0 \leq s \leq 1 \; . \tag{2.12}$$

The formal properties of link operators studied in ref.[12] formed the backbone of our derivation of eqs.(2.7, 2.9).

The advantage of our definition, eq.(1.5), is that the path defining $U(b,a)$ automatically turns out to be the straight line. Starting with eq.(2.10) instead raises the question of the choice of the path. A unique definition of the path is important, however, since the value of $U(b,a)$ depends on the path except in the trivial case, when $A^\mu = S\partial^\mu S^{-1}$ is a pure gauge field with $F_{\mu\nu} = 0$. Therefore, the path ambiguity must be removed. The ambiguity is removed by requiring that the variable p in $\hat{W}(x,p)$ corresponds to the kinetic momentum in the classical limit. That requirement is what led to our definition of $\hat{W}$ given in eq.(1.5). We showed previously that for any other choice of the path the interpretation of p would have to differ from what we want on physical grounds [12].

3. The Semiclassical Limit

In our units where all quantities are measured in terms of a length scale L, $\hbar$ did not appear in the equations. However, we see from eqs.(2.7, 2.9) that $F_{\mu\nu}$ is always acted on by an operator of the form

$$\hat{O}(\triangle) = \int ds\; f(s)\; e^{g(s)\,\triangle} = \sum_{n=0}^{\infty} a_n\, \triangle^n \quad , \tag{3.1}$$

where a_n are dimensionless constants and $\triangle = \partial_p \cdot \mathcal{D}$ as before. If we would have chosen ordinary classical units for p in terms of some momentum scale, while retaining length units for x, then obviously $\triangle$ would have to be replaced by

$$\triangle \longrightarrow \hbar\triangle \; . \tag{3.2}$$

In the semiclassical limit, we therefore retain only the leading powers of $\triangle$ in eq.(3.1). Thus we perform an expansion in powers of the space gradients of $F_{\mu\nu}$ and in powers of momentum gradients of $\hat{W}$ simultaneously.

Carrying out the above expansion in the constraint equation (2.7) leads to

$$\begin{aligned}(\gamma^\mu p_\mu - m)\,\hat{W}(x,p) \; &= \\ -\tfrac{1}{2}i\gamma^\mu \mathcal{D}_\mu(x)\,\hat{W}(x,p) \; &+ \; \tfrac{1}{2}ig\partial_p^\mu\gamma^\nu\left(\,F_{\mu\nu}(x)\,\hat{W}(x,p) \; - \; \tfrac{1}{4}\left[\,F_{\mu\nu}(x),\hat{W}(x,p)\,\right]\,\right) \\ &+ O(\partial_p\cdot\triangle F\cdot\gamma\hat{W}) \quad . \end{aligned} \tag{3.3}$$

In the case of an *external Abelian gauge potential* eq.(3.3) reduces to

$$(\gamma^\mu p_\mu - m)\,\hat{W}(x,p) \;=$$

$$-\,\tfrac{1}{2}i\gamma_\mu\partial_x^\mu\,\hat{W}(x,p) + \tfrac{1}{2}ig\partial_p^\mu\gamma^\nu F_{\mu\nu}(x)\,\hat{W}(x,p) \;+\; O(\partial_p\cdot\triangle F\cdot\gamma\hat{W}) \quad . \tag{3.4}$$

We see that the on-shell condition, $p^2 = m^2$, can hold for quarks only, when the *covariant derivative* of $\hat{W}$ is small compared to p (in the reference frame of a specified ensemble) *and* the *field strength* is also small. By small we mean that the ensemble averaged terms in eq.(3.3) satisfy:

$$\langle p\hat{W}\rangle \;\gg\; \langle \mathcal{D}\hat{W}\rangle \ , \tag{3.5}$$

$$\langle p\hat{W}\rangle \;\gg\; \langle gF\,\partial_p\hat{W}\rangle \ . \tag{3.6}$$

These conditions constrain the possible ensembles for which a simple semiclassical picture may hold. We return to this point in the next section. Note that for non-Abelian fields, the smallness of the covariant derivative means not only that the spatial gradients in the system are sufficiently small, but also that A and $\hat{W}$ approximately commute, i.e.

$$\langle p\hat{W}\rangle \gg \langle [gA,\hat{W}]\rangle \ . \tag{3.7}$$

When the above conditions are not satisfied, then significant off-shell corrections must be applied to the transport theory of quarks.

For covariant constant fields [13] satisfying

$$\mathcal{D}_\lambda F_{\mu\nu} = 0 \ , \tag{3.8}$$

the higher order corrections in $\triangle F$ vanish. However, in general $O(\triangle^n F)$ corrections must be calculated. Only for slowly varying fields, in the sense that

$$\langle F\hat{W}\rangle \gg \langle \mathcal{D}F\,\partial_p\hat{W}\rangle \ , \tag{3.9}$$

can an expansion in powers of $\triangle$ possibly converge rapidly.

In the general case for strong or rapidly varying fields the full quantum equation, eq.(2.7), must be solved. This would be equivalent to solving the field equations, eqs.(2.5), i.e. hopeless at this time. Only under the rather restrictive conditions, eqs.(3.5, 3.6, 3.9), can we expect that the transport theory for quarks reduces to a simpler, more manageable form.

Turning now to the transport equation, eq.(2.9), we obtain by the $\triangle$ expansion the *semiclassical transport equation for the QCD Wigner operator*:

$$p\cdot\mathcal{D}(x)\,\hat{W}(x,p) \;+\; \tfrac{1}{2}gp^\mu\partial_p^\nu\left\{\,F_{\nu\mu}(x),\hat{W}(x,p)\,\right\} \;+\; \tfrac{1}{4}ig\left[\,\sigma^{\mu\nu}F_{\mu\nu}(x),\hat{W}(x,p)\,\right]$$

$$=\,-\,\tfrac{1}{8}ig\partial_p^\mu\left[\,F_{\mu\nu}(x),(\mathcal{D}^\nu(x)\,\hat{W}(x,p))\,\right] \;-\; \tfrac{1}{16}ig^2\partial_p^\mu\partial_p^\nu\left[\,F_\mu{}^\lambda(x)F_{\nu\lambda}(x),\hat{W}(x,p)\,\right]$$

$$+\,O(\triangle F) \ . \tag{3.10}$$

For example, the lowest order corrections to the Vlasov and spin terms on the l.h.s. of eq.(3.10) appearing among the $O(\triangle F)$ terms are given by

$$+ \tfrac{1}{8} i g p^{\mu} \partial_p^{\nu} \partial_p^{\lambda} \Big[(\mathcal{D}_{\lambda}(x) F_{\nu\mu}(x)), \hat{W}(x,p) \Big] - \tfrac{1}{8} g \partial_p^{\lambda} \Big\{ \sigma^{\mu\nu} (\mathcal{D}_{\lambda}(x) F_{\mu\nu}(x)), \hat{W}(x,p) \Big\} \ . \tag{3.11}$$

Eq.(3.10) clearly generalizes Vlasov's equation in the following sense: i) it is still an operator equation; ii) the first and second term on the l.h.s. of eq.(3.10) present the usual combination of phase space variables and derivatives, however, modified by (anti)commutators, which can only be simplified for the external Abelian field problem (cf. eq.(3.12)); iii) explicit spin-dependent corrections arise; iv) quantum corrections can be systematically calculated via expansion of eq.(2.9) in powers of the $\triangle$ operator.

We remark here that the purely classical transport equation for quarks given in ref.[10] corresponds to the sum of the first two terms on the l.h.s. of eq.(3.10) being set equal to zero and decomposing them in color space. Furthermore, adding the corrections from (3.11) to the r.h.s. of eq.(3.10) and dropping all $\sigma^{\mu\nu}$-dependent terms, we reproduce the result for scalar quarks obtained in ref.[11], where, however, there is a sign error in the Vlasov term and correspondingly in the $O(\triangle)$ correction to it.

As a final limiting form, we consider the transport equation which follows from eq.(2.9) for the case of an *external Abelian gauge potential*:

$$\begin{aligned} & p \cdot \partial_x \hat{W}(x,p) + g p^{\mu} \partial_p^{\nu} F_{\nu\mu}(x) \hat{W}(x,p) + \tfrac{1}{4} i g F_{\mu\nu}(x) \Big[\sigma^{\mu\nu}, \hat{W}(x,p) \Big] \\ & = -\tfrac{1}{12} \Big(g \partial_p \cdot [\partial_x F_{\mu\nu}(x)] \partial_p^{\mu} \partial_x^{\nu} + g^2 \partial_p \cdot [\partial_x F_{\mu}{}^{\lambda}(x)] F_{\nu\lambda}(x) \partial_p^{\mu} \partial_p^{\nu} \Big) \hat{W}(x,p) \\ & \quad - \tfrac{1}{8} g \partial_p \cdot [\partial_x F_{\mu\nu}(x)] \Big\{ \sigma^{\mu\nu}, \hat{W}(x,p) \Big\} + O(\partial_x^2 F \partial_p^2 \hat{W}) \ . \end{aligned} \tag{3.12}$$

This is Vlasov's equation for the Wigner operator including spin-dependent (quantum) corrections. It generalizes the result obtained by Remler [3] for the case of scalar QED to spinor QED.

Finally, we remark again that the expressions *classical* or *semiclassical* employed in this section only refer to the approximate expansion of operators in powers of $\triangle$ acting on F. To relate the operator equations to the transport equations presented in the introduction, expectation values of the equations with respect to a suitable ensemble must be calculated as we discuss in the next section.

4. Discussion

The constraint and transport operator equations are completely equivalent to the original Dirac's field equation, eq.(2.5). They provide, however, a systematic means to study the semiclassical limit of QCD as expressed in terms of the Wigner function $\langle \hat{W} \rangle$. The choice of the ensemble over which $\hat{W}$ is to be averaged is, of course, dictated by the particular physical situation under study. Unfortunately, at present it is not known which field configurations are most relevant in actual physical processes such as nuclear collisions. This uncertainty is due to the unsolved confinement problem in QCD. To make further progress it is necessary to make an *ansatz*, i.e. a bold and unjustified guess, about the characteristics of that ensemble. We thus proceed in the spirit of the MIT Bag model and the color flux tube models [13,14] and assume that it is sensible to talk about quarks, as if they obeyed some effective plasma transport equation at least in some finite volume $V = L^3$.

Because of the asymptotic freedom property of QCD, we expect that the effective coupling, $g^2/4\pi \sim 1/\log(1/L\Lambda)$, where $\Lambda \sim 200$ MeV is the scale of QCD, vanishes as $L \to 0$. Thus in the limit of very small confined plasmas the transport equations are expected to reduce to the trivial free gas equation, $p\partial_x\langle\hat{W}\rangle = 0$. From the renormalization group approach we expect [1,2] furthermore that for high temperature and/or density plasmas the running coupling becomes small for any L. In fact, the strength of the coupling is always expected to be controlled by the largest of the parameters $M = T, \mu, L^{-1}$, or$(F^{0i})^{1/2}$ with dimensions of momentum that pertain to the problem via $g^2/4\pi \sim 1/\log(M/\Lambda)$. This inherent dependence of the coupling on the physical situation greatly complicates the analysis of the transport equations for quarks. Therfore we must treat g as a phenomenological parameter.

In general, we found that the transport equations can only be simplified if rather restrictive conditions, eqs.(3.5, 3.6, 3.9), apply for the relevant physical ensemble. As a final exercise we now show that at least in the context of the color flux tube models the reduction can be carried further and that the plasma transport equations become the simple intuitive ones given in the introduction.

With the above reservations clearly in mind, we consider an ensemble that corresponds to a quark plasma in a finite flux tube in which quarks are subject to a covariant constant or at most slowly varying mean field, $\langle F^{\mu\nu}(x)\rangle$. There exists a gauge where $\langle F^{\mu\nu}(x)\rangle$ is diagonal. Since every traceless $N \times N$ matrix can be expanded in terms of the $N-1$ commuting generators, h_i, of SU(N) defined by [13]

$$h_j \equiv (2j(j+1))^{-\frac{1}{2}} diag(1,\cdots,1,-j,0,\cdots,0) \ , \tag{4.1}$$

with $-j$ appearing in the $j+1$ column, we can always write

$$\langle F_{\mu\nu}(x)\rangle \equiv S(x)\vec{F}_{\mu\nu}(x)\cdot\vec{h}S^{-1}(x) \ , \tag{4.2}$$

where $S(x)$ is a particular gauge transformation.

If we now make the *ansatz* that the ensemble in the flux tube is such that $\hat{W}$ is diagonal in the *same gauge* where $\langle F_{\mu\nu}(x)\rangle$ is diagonal, then we can express

$$\langle\hat{W}\rangle \equiv S(x)\left(\sum_{j=1}^{N-1} W^j h_j + W^0 \underline{1}\right) S^{-1}(x) \ , \tag{4.3}$$

i.e. decompose $\langle\hat{W}\rangle$ in terms of N Wigner *functions* depending on x, p. We suppress presently irrelevant spinor indices, but a similar decomposition in spinor space in terms of the Clifford algebra can be performed [5]. Note that the property $\hat{W}^\dagger =$

$\gamma^0 \hat{W} \gamma^0$ insures that the N diagonal elements in color space $\langle \hat{W} \rangle^{ii}$ are real numbers after taking appropriate traces in spinor space.

Eq.(4.3) is a strong model assumption. It is, however, plausible that for slowly varying fields it is satisfied if the initial conditions for the plasma are assumed to correspond to $\langle \hat{W} \rangle = 0$, as in flux tube models [13,14]. The present assumptions are consistent as long as the gluon field can be treated as a classical external field, which implies that color exchanging scattering processes in the plasma must be negligible. Therefore our model may well be applicable to the description of the initial stages of plasma formation, if the flux tube picture is valid.

If eq.(4.3) holds, then it is obvious that the most convenient gauge to work in is the $S(x)$ gauge. In that gauge by assumption

$$\begin{aligned} \langle \hat{W} \rangle_{ij} &= (\vec{W} \cdot \vec{h} + W^0 \underline{1})_{ij} \\ &= \delta_{ij} (\vec{W} \cdot \vec{\epsilon}_j + W^0) \; \equiv \; \delta_{ij} f_j \;, \end{aligned} \tag{4.4}$$

with the "charges" $\vec{\epsilon}_j$ given by

$$\vec{\epsilon}_j \equiv (\vec{h})_{jj} = ((h_1)_{jj}, \cdots, (h_{N-1})_{jj}) \;. \tag{4.5}$$

These are just the elementary weight vectors of SU(N). Eq.(4.4) provides the model dependent relation between the Wigner operator and the classical quark distribution functions discussed in the introduction.

The semiclassical transport equations for the $f_j(x,p)$ in this model are obtained by taking the expectation value of eq.(3.10) in this ensemble. Using eqs.(4.2-4.5), the color structure of the semiclassical transport equation, eq.(3.10), is simplified considerably:

$$\begin{aligned} &(p \cdot \partial_x + g \vec{\epsilon}_j \cdot \vec{F}_{\mu\nu} p^\nu \partial_p^\mu) \; f_j(x,p) \\ &\quad = -\tfrac{1}{4} i g \vec{\epsilon}_j \cdot \vec{F}_{\mu\nu} \, [\, \sigma^{\mu\nu}, f_j(x,p) \,] \; + \; \tilde{C}_j(x,p) \;, \end{aligned} \tag{4.6}$$

where $\tilde{C}_j$ represents correlation terms of the form

$$\tilde{C}_j(x,p) = -\tfrac{1}{2} g p^\mu \partial_p^\nu \left(\langle \{ F_{\nu\mu}, \hat{W}(x,p) \} \rangle - 2 \langle F_{\nu\mu} \rangle \langle \hat{W}(x,p) \rangle \right)_j + \cdots \;. \tag{4.7}$$

We shall make no attempt in this paper to decompose the important correlation terms. Rather, we stop here with having shown how the Vlasov terms arise in a particular phenomenological model from the underlying quantum theory. Note that all the non-Abelian commutator terms dropped out for the model *ansatz*, eq.(4.3).

The model transport equation, eq.(4.6), confirms some of the expectations discussed in Sec.1. In particular, we observe the effective coupling constant $g\vec{\epsilon}_j$ entering the above set of N Abelian-like equations for the components f_j of $\langle \hat{W} \rangle$ (cf. eq.(4.4)). In the absence of correlations the equations would simply decouple. This equation seems to be as close as one can get to the classical Vlasov's equation starting from the Wigner operator defined in eq.(1.5). It is valid for fermions with positive or negative energy. For $p^0 < 0$ we obtain the equation for antiquarks by replacing p^μ by $-p^\mu$ everywhere, which effectively reverses the sign of g in eq.(4.6) as expected. We see the explicit spin-dependence as well. The important quantum corrections, especially pair production, are buried in the uncalculated correlation terms.

We remark here that another kind of model could be set up in the spirit of the MIT Bag model. Starting from an Abelianized version of our transport theory one would include MIT Bag boundary conditions for the quark and gluon fields. Thus the influence of the non-perturbative QCD vacuum could be incorporated while keeping the kinetic description of the confined plasma as simple as possible.

For an electron-positron plasma the transport equations derived in ref.[12] may be useful as well. To our knowledge spin and systematic quantum corrections to that equation had not been derived previously. For a U(1) gauge symmetry one only has to replace $\vec{\epsilon}_j \to 1$ and the matrix transport equation becomes a single Vlasov's equation including corrections.

Obviously a great deal of work remains in the development of the transport theory for quarks and gluons. The extraction of Debye screened collision terms and color neutralizing pair production terms from the correlation terms needs to be performed. Presently we are working on a gauge covariant Wigner operator for gluons and the derivation of the gluon transport equation from the field equations. The question of the proper physical basis for the ensemble averaging also is a crucial area for study. These and other problems will have to be solved in order to connect upcoming nuclear collision data with the properties of quark-gluon plasmas.

Bibliography

[1] H. Satz, ed., Thermodynamics of Quarks and Hadrons, North-Holland, Amsterdam 1981; T. W. Ludlam and H. E. Wegner, eds., Quark Matter '83, Nucl.Phys. A418 (1984); K. Kajantie, Quark Matter '84, Lecture Notes in Physics 221 (Springer-Verlag, Berlin, 1985).

[2] E. V. Shuryak, Phys.Rep. 115 (1984) 151; B. Mueller, The Physics of the Quark-Gluon Plasma, Springer lecture notes in physics, Springer, Berlin 1985; J. Cleymans, R. V. Gavai, and E. Suhonen, Phys.Rep. 130 (1986) 217.

[3] E. A. Remler, Phys.Rev. D16 (1977) 3464.

[4] I. Bialynicki-Birula, Acta Phys. Austriaca, Suppl. XVIII (1977) 111.

[5] R. Hakim, Riv. Nuovo Cim. 1 no. 6 (1978).

[6] S. R. de Groot, W. A. van Leeuwen, and Ch. G. van Weert, Relativistic Kinetic Theory, (North-Holland, Amsterdam 1980).

[7] P. Carruthers and F. Zachariasen, Rev. Mod. Phys. 55 (1983) 245.

[8] S.P. Li and L. McLerran, Nucl. Phys. B214 (1983) 417.

[9] K. Kajantie and C. Montonen, Physica Scripta 22 (1980) 255.

[10] U. Heinz, Phys.Rev.Lett. 51 (1983) 351; 56 (1986) 93;
Ann. Phys. (N.Y.) 161 (1985) 48; Phys. Lett. 144B (1984) 228.

[11] J. Winter, J. de Phys. 45 C6 (1984) 53.

[12] H.-Th. Elze, M. Gyulassy, D. Vasak, LBL-preprint 21137 (March 1986), submitted for publication.

[13] M. Gyulassy and A. Iwazaki, Phys. Lett. 165B (1985) 157.

[14] K. Kajantie and T. Matsui, Phys. Lett. 164B (1985) 373.

[15] C. Itzykson and J.-B. Zuber, Quantum Field Theory, McGraw-Hill, New York 1980.

[16] P. Becher, M. Bohm, and H. Joos, Eichtheorien der starken und elektroschwachen Wechselwirkung, Teubner, Stuttgart 1981.

COLOR RESPONSE AND COLOR TRANSPORT IN A QUARK-GLUON PLASMA *

Ulrich Heinz
Physics Department
Brookhaven National Laboratory
Upton, NY 11973
U S A

ABSTRACT

Using color kinetic theory, we discuss color conduction and color response in a quark-gluon plasma. Collective color oscillations and their damping rates are investigated. An instability of the thermal equilibrium state in high T QCD is discovered.

*Invited talk presented at the
Second International Workshop
Local Equilibrium in Strong Interaction Physics
Santa Fe, New Mexico
April 9-12, 1986

The submitted manuscript has been authored under contract DE-AC02-76CH00016 with the U.S. Department of Energy. Accordingly, the U.S. Government retains a nonexclusive, royalty-free license to publish or reproduce the published form of this contribution, or allow others to do so, for U.S. Government purposes.

1. INTRODUCTION

One of the basic differences between normal nuclear matter and quark matter is that the latter is a color conductor, whereas the former is a color insulator[1]. While in nuclear matter color cannot leak out appreciably from those little bags ($V \sim 2\ fm^3$) called nucleons, in the quark-gluon plasma it is free to move anywhere within the 4-volume occupied by the plasma state. This opens the possibility of having large-scale collective color oscillations in the plasma, generating effective (massive) degrees of freedom in the plasma phase[2] which substitute the massless gluons.

In the generally accepted picture of the quark-gluon plasma its ground state is given by a thermal distribution of quarks, antiquarks, and gluons, filling an otherwise empty (so-called perturbative) vacuum in such a way that the colors on the particles add up to local color neutrality. Any color disbalance in the plasma is expected to eventually disappear; the dominant mechanism to equilibrate color and thermalize the energy contained in the perturbation would be provided by those above-mentioned collective color modes since they show up as peaks in the dissipative part of the plasma's color response function.

As an example, let us consider the color-string breaking model for quark matter formation in heavy ion collisions[4-6], where the quarks and gluons to form the plasma are assumed to be created from the vacuum in the strong color electric fields of strings connecting the struck baryons from the two colliding nuclei. In this model color transport has to be an important ingredient in understanding equilibration of the plasma and developing a picture of the so far elusive pre-equilibrium stage of quark matter forming in heavy-ion collisions. Furthermore, estimates of the initial energy density in the collision based on hydrodynamic back-extrapolation of measured rapidity distributions[7] are only useful if the concept of local thermal equilibrium underlying the hydrodynamic framework is applicable to sufficiently small times; a theory for pre-equilibrium dynamics will help extend the extrapolation backwards in time[5] if the equilibration time scale is not small enough to apply Bjorken's arguments.

In this talk I review our attempts to construct and analyze a kinetic theory for plasmas with non-Abelian interactions. After presenting the basic equations and their justification on the basis of QCD, I will show you results for the coefficient of color conductivity and the dispersion relation for colored plasma oscillations obtained from a linear response analysis of a near-equilibrium plasma. I will comment on the damping mechanisms for color oscillations and point out several basic differences between a QED and a QCD plasma. In particular I will show that in QCD there is a difficulty with the sign of the imaginary part of the plasmon frequency which appears to indicate instability rather than damping of color oscillations. I will demonstrate the origin of this sign and speculate on its physical implications, pointing to a possible need to reorganise our thinking about the physical vacuum in high-temperature QCD.

2. THE COLOR TRANSPORT EQUATIONS

The essential quantity in a non-equilibrium description of a plasma is the particle distribution in phase space. The quantum mechanical version of the phase-space distribution function is given by the Wigner-function, which is the Weyl-transform of the density matrix[8].

Its well known non-relativistic definition can be extended to relativistic gauge field theories if proper care is taken to preserve gauge invariance[9,10]. The quark-antiquark distribution is determined by the ensemble average of the following operator[9],

$$\hat{F}(x,p) = - \int \frac{d^4y}{(2\pi)^4} e^{-ip\cdot y} U(x,x-y/2) \psi(x-y/2) \bar{\psi}(x+y/2) U(x+y/2,x), \quad (1)$$

where $\psi, \bar{\psi}$ are the (Dirac-spinor) quark field operators, and U is given in terms of the gluon field operator as the path ordered line integral

$$U(x,x-y/2) \equiv P \exp\left(i\frac{g}{2}\lambda_a \int_{x-y/2}^{x} A_\mu^a(z)dz^\mu\right) . \quad (2)$$

[Because of the Gell-Mann matrices λ_a in the exponent U is a 3 × 3 matrix.] In general U, and therefore $\hat{F}$, depends on the path chosen to get from x-y/2 via x to x+y/2; although this path dependence turns out to disappear in the semiclassical limit[9,10], we have to specify a particular path to define the quantum theory uniquely. Let us take a straight line; in this case it can be shown[10] that (1) is identical to

$$\hat{F}(x,p) = - \int \frac{d^4y}{(2\pi)^4} e^{-ip\cdot y} e^{-y\cdot D_x/2} \psi(x) \bar{\psi}(x) e^{y\cdot D_x^{\dagger}/2} , \quad (3)$$

where $D_{x,\mu} \equiv \partial/\partial x^\mu - i(g/2)[\lambda_a A_\mu^a,\cdot]$ is the covariant derivative operator, and $D_x^{\dagger}$ is its adjoint with the partial derivative acting to the left. Point-splitting the quark density matrix with the covariant derivative representing the quark <u>kinetic</u> momentum, rather than their <u>canonical</u> momentum, renders the role of the variable p in the Weyl-transform (3) as the classical kinetic momentum manifest in general (and not only in the semiclassical limit, as would be true for non-straight paths in (2)). This is a good guiding principle[10] for specifying the path in the more general definition (1).

Recently, Elze, Gyulassy, and Vasak[10] derived an <u>exact</u> equation of motion for the quark Wigner-operator (3) from the Dirac equation. The result is not yet in a very intuitive form, and much more work is needed to analyze and interpret their equation. However, in the semiclassical limit (obtained by considering only the lowest order in momentum derivatives of and color commutators with $\hat{F}$) things simplify considerably, and one obtains[9,10]

$$p \cdot D_x \hat{F}(x,p) + \frac{g}{4} p^\mu \partial_p^\nu \{F^a_{\mu\nu}\lambda_a, \hat{F}(x,p)\} + \frac{ig}{8}[\sigma^{\mu\nu}F_{\mu\nu}\lambda_a, \hat{F}(x,p)] = \hat{C}(x,p) \tag{4}$$

where the "collision term" $\hat{C}$ contains all higher order terms. If we neglect the spin effects contained in the third term on the left and decompose the equation into its color trace (singlet component) and the trace with the 8 λ-matrices (octet components), and finally split $\hat{F}$ into its positive (quark) and negative frequency (antiquark) components, we arrive at the classical transport equations[9,11]:

$$p^\mu \partial_\mu \overset{(-)}{f}(x,p) = \pm g p^\mu F^a_{\mu\nu}(x) \partial_p^\nu \overset{(-)}{f}_a(x,p) + \overset{(-)}{C}(x,p) \tag{5}$$

$$p^\mu[\partial_\mu \delta_{ac} + g f_{abc} A^b_\mu(x) \pm \frac{g}{2} d_{abc} F^a_{\mu\nu}(x)\partial_p^\nu] \overset{(-)}{f}_c(x,p) = \\ = \frac{g}{6} p^\mu F^a_{\mu\nu}(x) \partial_p^\nu \overset{(-)}{f}(x,p) + \overset{(-)}{C^a}(x,p) \quad . \tag{6}$$

Here, bars denote the antiparticle distribution functions and collision terms, and the upper (lower) sign is for quarks (antiquarks). These equations can also be derived as color singlet and octet moments of an underlying transport equation for a classical, color dependent distribution function defined on an extended color-phase-space[9,11].

Self-consistency of the theory is achieved by requiring the color mean fields A^a_μ, $F^a_{\mu\nu}$ in (5,6) to be generated by the color current of the quarks and antiquarks in the plasma:

$$\partial_\mu F_a^{\mu\nu}(x) + g f_{abc} A^b_\mu(x) F_c^{\mu\nu}(x) = g j^\nu_a(x) \equiv g \int p^\nu [f_a(x,p) - \bar{f}_a(x,p)] dP \quad . \tag{7}$$

Here dP denotes the momentum space integration measure, including the mass-shell condition.

In addition to eqs. (5,6) there should also be a kinetic equation for the phase-space distribution of the thermally excited gluons, and a gluonic contribution to the color current in eq. (7). For gluons the problem of deriving a gauge invariant transport equation has not yet been solved; therefore they are here omitted. In most of the applications of the theory below we will be able to correct for the gluonic contribution <u>a posteriori</u>, such that this omission is not too embarassing. However, already the damping rates for color oscillations turn out to be dominated by thermal gluons, and the kinetic theory based on eqs. (5-7) alone fails to yield reasonable results. Therefore, clearly more work based on the recent progress in the quantum transport theory has to be done.

3. EXTERNAL COLOR FIELDS

3.1 Color Conductivity

Noting that the collision terms C, $\bar{C}$ conserve energy-momentum, baryon number and color, it is easy to derive macroscopic conservation laws for the baryon current, color current, and energy momentum tensor[11]. Performing a near-equilibrium expansion of the conserved currents, one derives a set of "chromohydrodynamic" equations for a viscous, colored fluid and defines the transport coefficients for shear and bulk viscosity as well as heat and color conduction[11]. For example, the color conductivity tensor is defined by the response of a locally color neutral plasma to an external color field $F^a_{\mu\nu}$ via an induced color current

$$j^\mu_a(x) = \sigma^{\mu\nu}_{ab} F^b_{\nu\lambda}(x)\, u^\lambda(x) \equiv \sigma^{\mu\nu}_{ab} \tilde{E}^b_\nu(x) \ , \tag{8}$$

where u^λ is the local 4-velocity of the plasma, and $\tilde{E}^b_\nu$ is the "color electric field 4-vector" which in the local rest frame reduces to the usual electric field vector $\vec{E}^b$. Due to the antisymmetry of $F^b_{\nu\lambda}$, the components of $\sigma^{\mu\nu}_{ab}$ parallel to u^μ are irrelevant, and in the local rest frame the conductivity tensor has only spacial components σ^{ij}_{ab}. For an isotropic medium it reduces to a positive Lorentz-scalar σ.

We will compute σ in the relaxation time model, writing the collision term as[12]

$$\overset{(-)}{C}(x,p) = - p^\mu u_\mu(x) \left(\overset{(-)}{f} - \overset{(-)}{f}_0\right)/\tau_0 \tag{9}$$

where f_0 is the equilibrium distribution function. All the physics of equilibration is put into the relaxation time τ_0 which is assumed to be independent of the energy transferred in those collisions. For the purpose of computing the plasma viscosity the relevant τ_0 is argued to be dominated by quark-quark and quark-gluon scattering[13-15] and found to be $\tau_0 \sim 1/\left(T\alpha_s^2 \ell n(1/\alpha_s)\right)$. We will see later that this expression is not applicable in the case of color conduction, and that actually fundamental problems occur when we try to extract τ_0 in (9) from thermal QCD. But for the moment, let us simply assume we knew what value to take for τ_0. [Note that without a collision term color conduction vanishes due to entropy conservation by the collisionless transport equations[11].]

We linearize the kinetic equations around a color neutral thermal equilibrium configuration, in which the quark singlet distribution f(x,p) is given by a Fermi-Dirac distribution f_0, and the octet distributions f_a and all color fields vanish. In momentum space it is easy to express the induced octet distribution functions in terms of the external color electric field, which can then be inserted into the right hand side of eq. (7) to obtain the induced color current. One reads off[16]

$$\sigma^{\mu\nu}_{ab}(k) = \delta_{ab} \frac{ig^2}{2T} \int \frac{d^3p}{(2\pi)^3E} \frac{p^\mu p^\nu}{k\cdot p + iu\cdot p/\tau_0} [f_0(1-f_0) + \bar{f}_0(1-\bar{f}_0)] \qquad (10)$$

for an external perturbation with an arbitrary momentum spectrum $\tilde{E}^b_\nu(k)$. (Note the diagonality in color!) We will see in Section 4 that this expression is related to the transverse part of the color response function of the quark-gluon plasma. The coefficient of color conductivity is defined as the static field limit of eq. (10), i.e., $k,\omega \to 0$. After splitting off [16] the Lorentz-tensor structure of eq. (10), we are left in the static limit with a simple integral over Fermi distribution functions which far massless quarks can be evaluated analytically. We find

$$\sigma = \tau_0\, \Omega_p^2 \qquad (11)$$

where $\Omega_p^2 = m_{el}^2/3$ is the well-known contribution from one fermionic spin degree of freedom to the electric screening mass (plasma frequency) in thermal QCD. Since several spin and flavor degrees of freedom and also the gluons contribute additively to the induced color current, the full conduction coefficient is obtained by inserting in (11) the complete one-loop result for the plasma frequency,

$$\Omega_p^2 = g^2T^2/3\ \left[1 + 1/6 \sum_f \left(1 + 3(\mu_f/\pi T)^2\right)\right] . \qquad (12)$$

This procedure will be better justified below when we look at the complete momentum dependence of the color response function.

3.2 Color String Breaking

The collision term (9) neglects one crucial aspect of a system subjected to a constant external color electric field, namely the possibility to produce $q\bar{q}$ pairs and gluons from the vacuum by the Schwinger mechanism[4-6,18]. In a finite temperature plasma the just computed color conduction will permit the system to dynamically screen the external color electric field, thereby reducing particle production. During the initial phases of a heavy-ion collision, however, and in particular in the central rapidity region, particle production will occur first, and equilibration via a collision term like eq. (9) will follow. In the flux tube model[4-6] particle creation occurs from the color electric field generated between the leading projectile and target nucleons by color exchange. The differential rate per unit 4-volume and transverse energy interval dE_T^2 for this process is given by

$$\frac{d\omega}{dE_T^2} = \pm \frac{\sigma}{4\pi^2} \ln[1 \pm \exp(-\pi E_T^2/\sigma)] \tag{13}$$

where the +(-) sign applies for gluon (quark-antiquark) pair creation $\sigma = g|E_{eff}|$ is the effective electric color field the produced pair sees and will be discussed below.

Eq. (13) does not give any information on the longitudinal momentum distribution of the produced pairs. Since a constant color electric field is invariant under boost transformations along the direction of the field, any function of $\xi = y-\eta$ (where y and η are the momentum and space-time rapidity) that integrates up to eq. (13), generates a possible longitudinal momentum dependence of the pair creation rate[5]. A particularly simple and still reasonable assumption takes $\delta(\xi)$ as the rapidity dependence[5], which implies that in any frame particles appear at point z with a longitudinal velocity $v_L/c = z/t$ due to the acceleration in the color electric field.

Kajantie and Matsui[5] investigated this model in the Abelian approximation where color degrees of freedom are neglected. In this case $\sigma = |gE|$ where E is the Abelian external electric field. They omitted the Vlasov-term in the kinetic equation (5) (which would lead to acceleration of the produced particles in the electric field) arguing that all particles are produced with some transverse momentum and that the orientation of the electric field switches randomly in this direction. Therefore, they solved

$$p^\mu \partial_\mu f = - p^\mu u_\mu \frac{f-f_0}{\tau_C} + \frac{|gE|}{4\pi^3} \exp\left(-\pi p_T^2/|gE|\right) \delta(y-\eta) \tag{14}$$

(massless quarks) where the source term was expanded in lowest order of

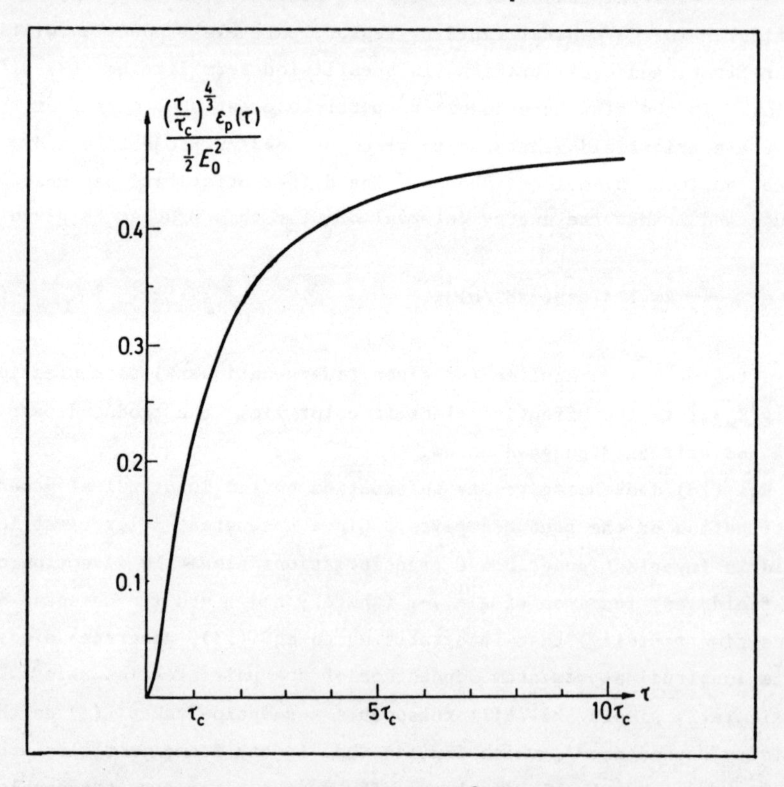

Fig. 1 Fraction of field energy density $E_0{}^2/2$ converted into particle energy density, scaled by a factor $(\tau/\tau_C)^{4/3}$. Boost-invariant hydrodynamic behavior would show up as a constant line and is approximately achieved for $\tau \gtrsim 10\tau_C$. From Ref. 5.

the exponential, and boost-invariance of the solution was imposed in order to comply with the symmetries of the external field. Introducing a particle formation time $\tau_0 \sim E^{-1/2}$, the only parameter of their solutions turned out to be the ratio τ_c/τ_0.

Fig. 1 shows the energy density contained in the produced quark-antiquark pairs as a function of proper time for the case $\tau_c = \tau_0$. The energy axis is scaled by $(\tau/\tau_0)^{4/3}$ such that boost-invariant hydrodynamic behavior[7] would show up as a constant line. One sees that the hydrodynamic region (local equilibrium) is only attained at proper times about an order of magnitude longer than the basic formation and equilibration times τ_0, τ_c. The particle energy density never exceeds about 1/4 of the initially present field energy density[5], due to rapid expansion. Extrapolating the measured energy density (from rapidity distributions) back in time along the Bjorken scaling curve overestimates the energy density near τ_0 by a factor 2-5.

Obviously the Abelian model is somewhat oversimplified because it neglects gluon production, which may enhance the particle energy density at small times considerably. Recently Gyulassy and Iwazaki[6] were able to properly include the color degrees of freedom into Schwinger's pair-creation formula. They noted that a constant color electric field can always by a global gauge choice be rotated such that it has only components in the 3 and 8 direction of color space. To the extent that higher order non-Abelian interactions between the produced particles can be neglected, the Wigner function of the produced quark-antiquark pairs will also be diagonal in color, i.e., a superposition of the unit matrix and λ_3 and λ_8. In this sense the problem looks Abelian, but still the effect of a, say, $\vec{E}_8$ field on red quarks is different from that on blue or green quarks, implying different pair creation rates.

Writing the color components of the quark distribution function as diag(f_r, f_g, f_b) [rather than as singlet and octet components as in eqs. (5,6)] and introducing the three root vectors $\vec{\varepsilon}_r = (1/2, 1/2\sqrt{3})$, $\vec{\varepsilon}_g = (-1/2, 1/2\sqrt{3})$, $\vec{\varepsilon}_b = (0, -1/\sqrt{3})$ they showed that the effective interaction of quarks of color i (i = r,b,g) with the color electric field $\vec{E} = (E_3, E_8)$ is given by the product $g\vec{\varepsilon}_i \cdot \vec{E}$. In the formula for the pair creation rate this effective field has still to be corrected for the interaction between the produced pair which leads to partial screening of the external field[6]:

$$\sigma(\vec{\varepsilon}_i,\vec{E}) = \left|g\vec{\varepsilon}_i\cdot\vec{E} - \frac{g^2}{2A_T}\vec{\varepsilon}_i\cdot\vec{\varepsilon}_i\right| = \left|g\vec{\varepsilon}_i\cdot\vec{E} - \frac{1}{3}\sigma_A\right| . \tag{15}$$

Here A_T is the transverse area of the color string formed between the produced pair, and $\sigma_A \equiv g^2/2A_T$ is the string tension of the adjoint string (which has a single color octet charge on each end).

Inserting everything into eqs. (5,6), we obtain the generalization of eq. (14) for the quarks (i = r,b,g)

$$\begin{aligned} p^\mu[\partial_\mu + g\vec{\varepsilon}_i\cdot\vec{F}_{\mu\nu}\,\partial_p^\nu]f_i &= -\,p^\mu u_\mu \frac{f_i - f_{i,eq}}{\tau_i} \\ &\quad - \frac{\sigma(\vec{\varepsilon}_i,\vec{E})}{4\pi^3}\,\ell n\left[1 - \exp\left(-\pi E_T^2/\sigma(\vec{\varepsilon}_i,\vec{E})\right)\right]\delta(y-\eta) . \end{aligned} \tag{16}$$

Following Gyulassy and Iwazaki[6], I note that no A_μ^3 or A_μ^8 gluons will be produced from the external field, and that for the remaining six "charged" gluons we can define root vectors $\vec{\eta}_{ij} = \vec{\varepsilon}_i - \vec{\varepsilon}_j$ in terms of which the effective field in the gluon pair creation rate becomes $\sigma(\vec{\eta}_{ij},\vec{E})$ as defined in (15). I therefore expect[31] for the gluons an equation similar in structure to (16):

$$\begin{aligned} p^\mu[\partial_\mu + g\,\vec{\eta}_{ij}\cdot\vec{F}_{\mu\nu}\,\partial_p^\nu]f_{ij} &= -\,p^\mu u_\mu \left(\frac{f - f_{eq}}{\tau}\right)_{ij} \\ &\quad + \frac{\sigma(\vec{\eta}_{ij},\vec{E})}{4\pi^3}\,\ell n[1 + \exp(-\pi p_T^2)/\sigma(\vec{\eta}_{ij},\vec{E})]\delta(y-\eta) . \end{aligned} \tag{17}$$

This rather wild guess may serve as another hint as to the possible color structure of the so far missing kinetic equation for the gluons.

4. COLOR OSCILLATIONS

In section 3.1 we computed the color current induced in an equilibrium plasma by an external color perturbation. Inserting the result into the YM equation and keeping self-consistently all terms linear in the perturbations, we can determine the color fields induced by the perturbation. We find[16]

$$\delta A^a_\lambda = \left\{ \frac{M_L}{k^2 - M_L} Q^{\lambda\rho} + \frac{M_T}{k^2 - M_T} P^{\lambda\rho} \right\} A^a_{\rho,ext} \quad . \tag{18}$$

The curly bracket defines the color response function, separated into a longitudinal ($\sim Q^{\lambda\rho}$) and a transverse ($\sim P^{\lambda\rho}$) part; it is diagonal in color. The scalar functions $M_{T,L}$ depend on the momentum k^μ of the external perturbation and are given by integrals similar in structure to eq. (10); for massless quarks they are easily calculated analytically. Let's take the collisionless limit ($\tau_0 \to \infty$); we then find[19]

$$\begin{aligned} M_L(\eta) &= m^2_{el} \left[1 - \eta^2 + \frac{1}{2}\eta(\eta^2-1)\ \ln\frac{\eta+1}{\eta-1}\right] ; \\ M_T(\eta) &= \left[m^2_{el} - M_L(\eta)\right]/2 \ . \end{aligned} \tag{19}$$

Here $\eta = \Omega/K$ is the ratio of frequency and momentum of the color perturbation evaluated in the plasma center of mass frame. m^2_{el} is again the contribution to the screening mass from a single spinless quark flavor; surprisingly, the η-dependence of $M_{L,T}$ agrees with the high temperature limit of the one-quark-loop contribution to the thermal gluon polarization operator[20]. Furthermore, in that limit the gluon loops are known[20] to have exactly the same η dependence, such that their only contribution is by changing the value of the electric screening mass to eq. (12). This justifies our procedure to correct the color response function for thermal gluon effects by simply adjusting the value for m^2_{el}.

The poles in (18) determine the dispersion relation for collective color oscillations. They are shown in Fig. 2. There is one longitudinal and one transverse optical mode, both starting at the plasma frequency $\Omega_p = m_{el}/\sqrt{3}$. There are no acoustical modes starting at $\Omega = 0$. The effect of a baryonic chemical potential is rather weak as it turns out that the gluons dominate the dispersion relation. Note that both modes always stay in the timelike region $\Omega^2 > K^2$, in contrast to electrodynamic plasmas where the longitudinal mode becomes spacelike for sufficiently large K[21,22]. This is due to the masslessness of the gluons contributing to the collective oscillations, whereas in QED only the massive electrons take part.

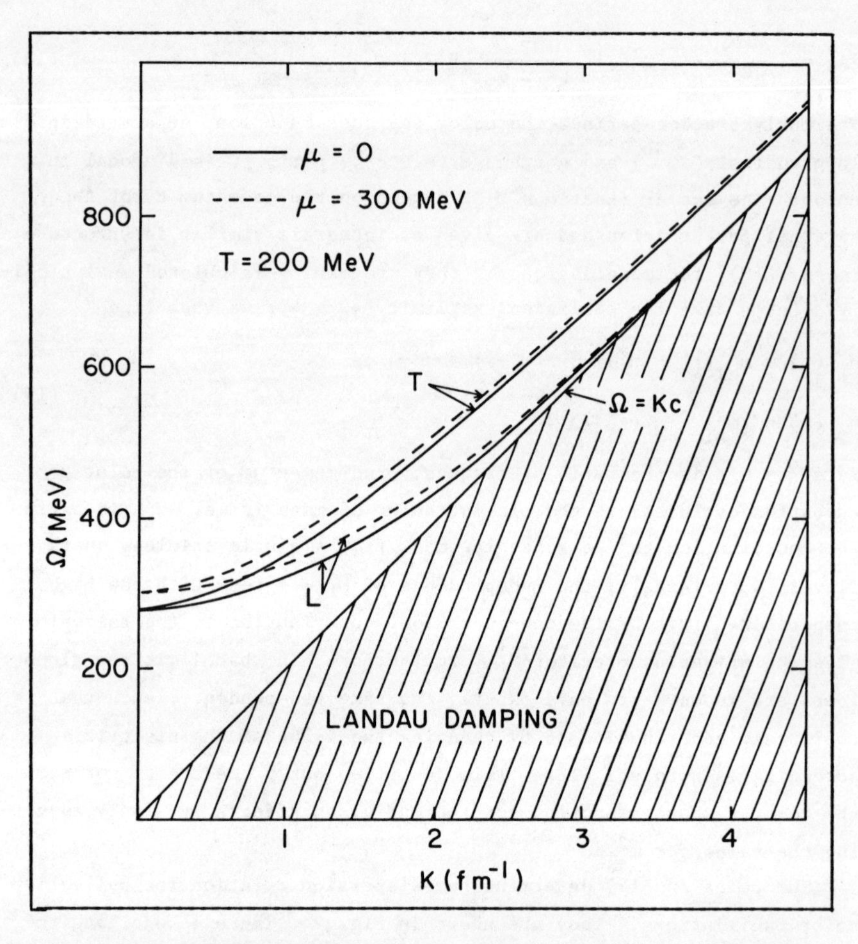

Fig. 2 Dispersion relation for longitudinal (L) and transverse (T) colored plasma oscillations. Parameters: g=2.0 (α_s=0.32); $m_{u,d}$=10 MeV, m_s=150 MeV; T=200 MeV. The effect of a baryonic chemical potential associated with u and d quarks (dashed vs. solid lines) is minimal. From Ref. 19.

5. DAMPING RATES - THE POSSIBLE INSTABILITY OF THE PERTURBATIVE VACUUM EVEN AT HIGH T

For $\eta < 1$ the logarithm in (19) develops an imaginary part. In the QED case therefore the frequency obtains a negative imaginary part wherever the dispersion relation is spacelike; this results in the so-called collisionless or Landau damping for longitudinal plasma modes. The underlying mechanism is that electrons in the plasma get scattered by the potential of the collective wave, or in Feynman diagram language, quanta of the collective mode get absorbed on the thermally distributed electrons (or quarks and gluons in the QCD case, Fig. 3a,b).

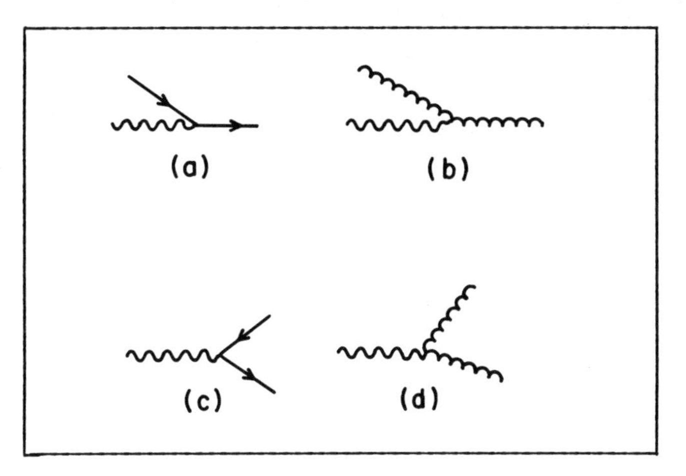

Fig. 3 Damping mechanisms for colored plasma oscillations: Landau damping due to absorption of collective quanta (wavy lines) by a thermal quark (a) or a (massless) gluon (b); decay of the collective mode into thermal $q\bar{q}$ pairs (c) or gluon pairs (d).

However, this process is kinematically forbidden if the collective momentum is timelike, and therefore the QCD plasma oscillations of Fig. 2 never experience Landau damping. In that sense, a QCD plasma is quite different from electrodynamic plasmas. However, this does not imply that QCD plasmons are undamped; the collective modes can decay by gluon or quark pair creation (Fig. 3c,d). The pair decay processes are only possible in the timelike region; the mass-threshold $\Omega^2 - K^2 > 4m^2$ effectively suppresses them in usual electrodynamic plasmas, but due to the masslessness of gluons is hardly effective in the QCD plasma. For finite baryon density the $q\bar{q}$ decay gets Pauli suppressed, but the gluon pair decay is present for all values of μ and, due to Bose enhancement and the large number of final state degrees of freedom, always dominant.

Obviously, these decay processes change the number of particles and can therefore not be described without a collision term in the kinetic equation. Weldon[23)] has shown how the imaginary part of the color polarization operator (which describes these decay processes) can be brought into a form that resembles a collision term of the Boltzmann or Uehling-Uhlenbeck type, or even the relaxation time model of eq. (9). Taking the dominant decay into a pair of gluons, the imaginary part of the thermal gluon-loop contribution yields for the rate in Landau gauge and in the zero momentum limit[24)]

$$\Gamma_{L,2g}(\Omega,K) \equiv \frac{1}{\tau_0(\Omega,K)} \xrightarrow[K\to 0]{} -\frac{5g^2}{8\pi}\,\Omega_P \coth(\Omega_P/4T) \ . \tag{20}$$

Here we tried to interpret Γ as an inverse relaxation time[23)], but it came out with a negative sign! The $q\bar{q}$ decay contributes a positive, but smaller number so that the net result is still negative. What happened?

This question was recently attacked in a somewhat different approach by Jorge Lopez et al.[25)]. They computed the dissipative part of the color response function (which is related to the real part of the gluon polarization operator by a fluctuation-dissipation theorem[16)]) directly from thermal QCD in the one-loop approximation, using real-time Feynman rules[20,26)]. In Fig. 4 we show the imaginary part of the (color) dielectric function which is related to the longitudinal part of the color response function (polarization operator)[16,20)]:

$$\mathrm{Im}\,\varepsilon = -\frac{1}{k^2}\,\mathrm{Im}\,\pi_L = -\coth\frac{\Omega}{2T}\,\mathrm{Im}\,R_L \ . \tag{21}$$

We see that both at zero temperature (where the imaginary part vanishes for $\Omega^2 < K^2$ since there are no thermally excited particles present to allow the processes in Fig. 3a,b) and at finite T the sign of the imaginary part stays the same throughout momentum space for all T, but switches from QED to QCD. At T = 0 the sign change is easily identified; the imaginary part stems from the logarithm in

$$\pi^{\mu\nu}(k^2) = (g^{\mu\nu} - k^\mu k^\nu/k^2)\, g^2\, \frac{2N_f - 5N}{48\pi^2}\, k^2 \ln\left(-k^2/M^2\right) \ , \tag{23}$$

via $\ln(-k^2/M^2) = \ln|k^2/M^2| - i\pi\,\Theta(k^2)$. The sign switch of the imaginary

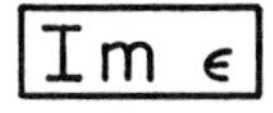

QED:

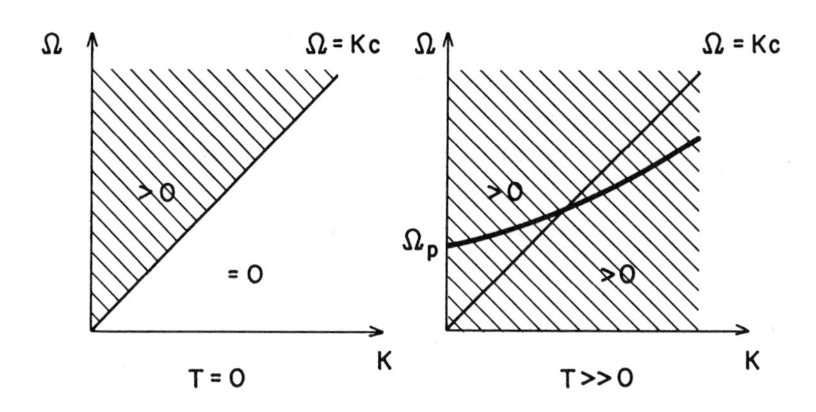

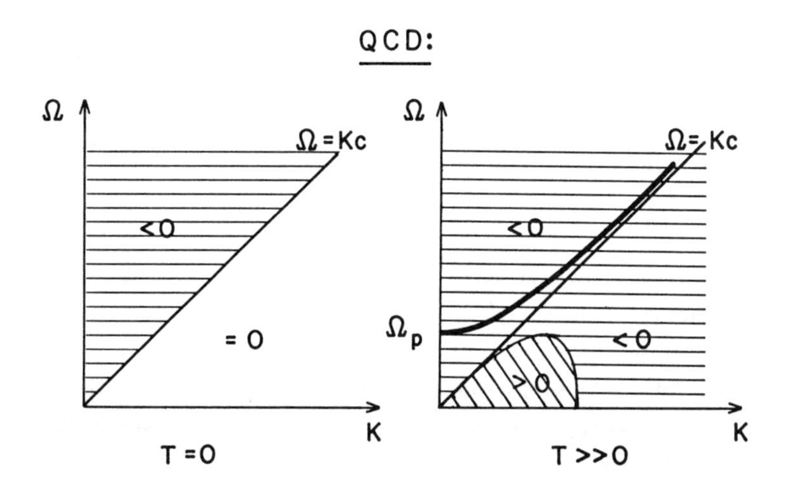

Fig. 4 Sign of the imaginary part of the (color) dielectric function in QED and QCD at zero and finite temperatire. Im ε peaks along the longitudinal collective mode as indicated.

part is the same as that of the real part which changes the direction of running of the effective coupling constant, i.e., infrared freedom (QED) into asymptotic freedom (QCD). Finite temperature effects do not influence the sign but rather "enhance" it because now Im ε develops a strong peak along the longitudinal collective mode[25]. The positive sign of Im ε in QED is usually interpreted[27] as being necessary for stability of the system against electromagnetic perturbations: energy dissipates from the perturbation into the system (most effectively through the collective mode), and the perturbation is damped away.

Figure 4 indicates the opposite in QCD[25]: except in a small stable region of space like momenta (K less than several times T), color perturbations receive spontaneously energy from the system (irrespective of temperature) and get amplified, most strongly so along the longitudinal dispersion relation; the decay of the perturbative vacuum (thermal ground state) into gluon pairs seems to be the doorway channel for this instability. The instability does not seem to care about deconfinement at high temperature (as indicated by Debye screening of color electric fields) or the smallness of the coupling constant. This indicates a problem in the magnetic sector. Indeed, at T = 0 it is known that the perturbative vacuum is unstable against spontaneously developing a constant color magnetic field[28]; however, the minimum in the effective potential at finite values of $\vec{B}$ evaporates at large T, apparently indicating restoration of the perturbative vacuum[29]. Fortunately, there is a hole in this argument: in the calculation of the effective potential an unstable model (leading to an imaginary part of the potential) is usually neglected. At T = 0 this mode was shown to lead to a complicated structure of electric vortex lines in the QCD vacuum[30]. This mode keeps contributing to the effective potential at high temperatures, and my guess is that it is also responsible for the instability found above.

If this turns out to be really true, a major revision of our basic picture of the quark-gluon plasma seems indicated. Starting from the above string-breaking picture to generate quarks and gluons, the chances that they ever approach that local equilibrium state of thermally distributed quarks and gluons in an otherwise emply vacuum, that we like so much, are essentially zero. Entirely different degrees of freedom will have to be used if, for example, the assumption of local equilibrium and a hydrodynamic treatment is expected to make sense.

This work was supported by the U.S. Department of Energy under contract DE-AC02-76CH00016.

REFERENCES

1) Baym, G., Physica 96A, 131 (1979).
2) Carruthers, P., Phys. Rev. Lett. 50, 1179 (1983); Nucl. Phys. A418, 501c (1984).
3) For recent overviews, see B. Müller, "The Physics of the Quark-Gluon Plasma", Lecture Notes in Physics 225, Springer Verlag 1985; E.V. Shuryak, Phys. Rep. 115, 151 (1984); K. Kajantie, ed., "Quark Matter '84", Lecture Notes in Physics 221, Springer Verlag 1985.
4) Casher, A., Neuberger, H., and Nussinov, S., Phys. Rev. D20, 179 (1979);
Glendenning, N.K. and Matsui, T., Phys. Rev. D28, 2890 (1983);
Bialas, A. and Czyz, W., Phys. Rev. D31, 198 (1985);
Kerman, A.K., Matsui, T., and Svetitsky, B., Phys. Rev. Lett. 56, 219 (1986).
5) Kajantie, K. and Matsui, T., Phys. Lett. 164B, 373 (1985).
6) Gyulassy, M. and Iwazaki, A., Phys. Lett. 165B, 157 (1985).
7) Bjorken, J.D., Phys. Rev. D27, 140 (1983).
8) Balescu, R., "Equilibrium and Non-equilibrium Statistical Mechanics", Wiley, New York 1975;
de Groot, S.R., van Leeuwen, W.A., and van Weert, Ch.G.,"Relativistic Kinetic Theory", North Holland, Amsterdam 1980.
9) Heinz, U., Phys. Rev. Lett. 51, 351 (1983).
10) Elze, H.-Th., Gyulassy, M., and Vasak, D., LBL-21137, to appear in Nucl. Phys. B (1986).
11) Heinz, U., Ann. Phys. 161, 48 (1985).
12) Anderson, J.L. and Witting, H.R., Physica 74, 466 and 489 (1974).

13) Hosoya, A. and Kajantie, K., Nucl. Phys. B250, 666 (1985).
14) Danielewicz, P. and Gyulassy, M., Phys. Rev. D31, 53 (1985).
15) Gavin, S., Nucl. Phys. A435, 826 (1985).
16) Heinz, U., Ann. Phys. 167, (1986).
17) Gross, D.J., Pisarski, R.D., and Yaffe, L.G., Rev. Mod. Phys. 53, 43 (1981);
Nadkarni, Phys. Rev. D27, 917 (1983).
18) Schwinger, J., Phys. Rev. 82, 664 (1951).
19) Heinz, U. and Siemens, P.J., Phys. Lett. 158B, 11 (1985).
20) Weldon, H.A., Phys. Rev. D26, 1394 (1983).
21) Krall, N.A. and Trivelpiece, A.W., "Principles of Plasma Physics", McGraw Hill, New York 1973.
22) Heinz, U., Phys. Lett. 109A, 385 (1985).
23) Weldon, H.A., Phys. Rev. D28, 2007 (1983).
24) Heinz, U. and Weldon, H. A., to be published.
25) Lopez, J., Parikh, J., and Siemens, P. J., "Instability of the QCD Plasma", Texas A&M preprint, 1986. This reference contains a slightly different arrangement of signs in QCD for spacelike momenta; the signs shown in Fig. 4 are based on my own recalculation of Im ε.
26) Dolan, L. and Jackiw, R., Phys. Rev. D9, 3320 (1974).
27) Jackson, J.D., "Classical Electrodynamics", chpt. 7.5, Wiley & Sons, New York, 1975.
28) Savvidy, G.K., Phys. Lett. 71B, 133 (1977);
Nielsen, N.K. and Olesen, P., Nucl. Phys. B144, 376 (1978).
29) Müller, B. and Rafelski, J., Phys. Lett. 101B, 111 (1981);
Kapusta, J., Nucl. Phys. B190, 425 (1981).
30) Nielsen, N.K. and Olesen, P., Phys. Lett. 79B, 304 (1978).
31) I thank M. Gyulassy for a discussion of this point.

Quantum Kinetic Theories of Heavy-Ion Collisions

M. J. Rhoades-Brown
Physics Department
State University of New York
Stony Brook
N.Y. 11794

M. Ploszajczak
Institute of Nuclear Physics
PL-31-342, Krakow
Poland

With the promise of new heavy-ion facilities capable of accelerating nuclei up to several GeV/A or beyond, there will be a tremendous energy range available for the experimental and theoretical study of heavy-ion collisions. Traditional heavy-ion physics i.e. the study of heavy-ion collisions at a few MeV/A appears to be reasonably well described by the Schrödinger equation approximated with a mean-field potential for the many particle dynamics. Indeed, it appears for energies below the Coulomb barrier, that the energy dependant competition between the fusion and quasi-elastic channels can be qualitatively understood[1)] within the coupled-channel Schrödinger picture plus a simple and local phenomenological potential.

At intermediate bombarding energies i.e. $\sim$ 65 MeV/A the choice of dynamical equation is not so clear, for there is now evidence that the experimental observables i.e. mass distribution or energy spectra of ejected particles is strongly influenced by either individual nucleon-nucleon collisions[2)] or coherent processes[3)] of unknown origin. At intermediate energies the theoretical questions appear to be; what are the dynamical limits of the mean-field approximation? what is the relative importance of individual nucleon-nucleon collisions? (i.e. going beyond the mean-field approximation) and are there coherent mechanisms present at intermediate energies? Of course, a detailed study of nucleon-nucleon collisions also allows the simplifying concept of local thermal equilibrium within a finite fermion system to be studied.

Perhaps the most sophisticated kinetic equation applied to date, that allows for both a mean-field $V(\rho)$ and for the influence of nucleon-nucleon collisions, is the so-called Uehling-Uhlenbeck equation[2,4],

$$\frac{\partial}{\partial t} f(x,p,t) + \frac{p}{m}\frac{\partial}{\partial x} + \frac{\partial}{\partial x} V(\rho) \frac{\partial}{\partial p} f(x,p,t) = \int d^3p_1 d^3p_2 \frac{d\Omega_3}{(2\pi)^6} v \frac{d\sigma}{d\Omega}$$

$$[f(x,p_1,t)f(x,p_2,t)(1-f(x,p_3,t))(1-f(x,p,t)) - f(x,p,t)f(x,p_3,t) \\ (1-f(x,p_1,t))(1-f(x,p_2,t))] \cdot (2\pi)^3 \delta^{(3)}(p_1+p_2-p_3-p) \tag{1}$$

where $f(x,p,t)$ is a *classical* single-particle phase-space distribution function. The right hand side of this equation is a so-called collision integral which simulates the influence of the Pauli exclusion principle on the nucleon-nucleon collision. In this formula v is the relative velocity of the two colliding nucleons and $d\sigma/d\Omega$ is the nucleon-nucleon differential cross-section. We note that this form of the collision integral requires the modification of the mean-field dynamics, which is based on the self-consistent mean-field potential $V(\rho)$, to be perturbative.

Of particular interest to us, for the intermediate energy domain, is the mathematical structure of the Uehling-Uhlenbeck equation relative to more general kinetic equations. The homogeneous part of this equation is essentially the classical Boltzmann equation. Indeed, equation (1) is solved assuming the homogeneous part is exactly equivalent to Hamilton's classical equations of motion[4]. The classical nature of the distribution function requires $f(x,p,t)$ to be positive for all points in classical phase space. Quantum mechanically this condition has to be relaxed; the quantum mechanical distribution function allows for negative values for f(x,p,t) in the classically forbidden regions of phase space. As part of our general studies and developments of kinetic theories we have addressed the questions of how to introduce and approximate quantum mechanical features.

At the relativistic or ultra-relativistic energy domain there is much speculation in the literature on the formation of a quark-gluon plasma[5]. To date, much of our theoretical understanding of this problem is based on hydrodynamic calculations supplemented with equations of state that assume local thermal equilibrium[6]. Of course, hydrodynamic flow is a reduction of the general kinetic theory that is based on the concept of local equilibrium. We note that studies of nuclear stopping power[7] strongly indicate that at bombarding energies above a few GeV/A the interpenetrating nuclei may become transparent or essentially pass through each other as the number of available nucleon-nucleon collisions is insufficient to dissipate the bombarding energy. This transparency question has raised serious questions as to the validity of conventional hydrodynamics for describing colliding nuclei at these energies. In addition, the concept of local thermal equilibrium needs to be addressed again[8]. Assuming that a quark-gluon plasma has been formed, classical hydrodynamics has been used to investigate the lifetime and geometry of this matter[5,6]. Although these studies have provided tremendous insight into this problem, we believe it is important to develop a quantum hydrodynamic flow theory for the Q.C.D. plasma that respects the non-abelian interaction character of the matter, and also has its roots and connections with a more formal kinetic theory which is based on the Q.C.D. Lagrangian.

In this paper we discuss the nature of our self-consistent truncation scheme for kinetic theories applied to the simplest possible quantum mechanical problem i.e. a single particle in a local field[9] $V(x)$. This simple problem allows the clearest explanation of our general truncation scheme. In addition to this problem, we discuss generalizations to the many-body quantum mechanical system and relate our findings to possible extensions of the Uehling-Uhlenbeck model. The connection with, and extensions of the classical hydrodynamic limit are also briefly mentioned here. For the relativistic problem we discuss the scalar plasma and the Q.E.D. plasma (abelian interactions) in the mean-field limit[10]. For the quark-gluon plasma we briefly show more recent work on the form of quantum hydrodynamic flow for non-abelian interactions[11]. We note that the concept of local thermal equilibrium is not required in our approach.

As discussed in some detail by Carruthers and Zachariasen[12] the quantum mechanical generalization of the one particle distribution function $f(x,p,t)$ is the so-called Wigner function[13],

$$f(x,p,t) = (2\pi\hbar)^{-1} \int dy\, e^{ipy/\hbar} \rho(x - y/2, x + y/2, t), \tag{2}$$

where, as mentioned before, we restrict ourselves to one dimension for clarity. In the Hartree approximation, the dynamical equation satisfied by f(x,p,t) may be derived from the Liouville equation for the density matrix[9] $\rho(x, x', t)$,

$$\frac{\partial f}{\partial t} + \frac{p}{m}\frac{\partial f}{\partial x} + \frac{2}{\hbar} sin\left(\frac{\hbar}{2}\frac{\partial^{(V)}}{\partial x}\frac{\partial^{(f)}}{\partial p}\right) V(x) f = 0, \tag{3}$$

or,

$$\frac{\partial f}{\partial t} + \frac{p}{m}\frac{\partial f}{\partial x} + \frac{\partial V}{\partial x}\frac{\partial f}{\partial p} - \frac{\hbar^2}{4}\frac{\partial^3 V}{\partial x^3}\frac{\partial^3 f}{\partial p^3} + ... = 0. \tag{4}$$

We use the Hartree approximation only to illustrate our approximation technique for a simple but non-trivial model.

Equations (3) and (4) show that the lowest expansion of the *sin* operator corresponds to the self-consistent field component of the classical Eq. (1). This implies that the quantum correction to the homogeneous part of (1) can be incorporated simply by taking the next higher terms in an $\hbar$ expansion as shown in Eq(4). However, this idea is too simplistic, for the definition of f(x,p,t) itself contains $\hbar$ to all orders. The questions remain, where is the physics of the mixed representation? how do we extract this physics for the quantum domain?

The simplistic view of the $\hbar$ or semi-classical expansion of (4) is more easily seen by defining a time-independent distribution function $f(p', p, E)$ as[14]

$$f(p', p, E) = \int dxdt e^{iEt/\hbar} e^{ip'x/\hbar} f(x,p,t) \tag{5}$$

and looking at its *exact* dynamical equation i.e.

$$f(p', p, E) = f_0(p', p, E) + (E + i\epsilon - p'v_p)^{-1} \int \frac{dp''}{(2\pi)} V(p'')$$

$$\left(f(p' - \frac{1}{2}p'', p - p'', E) - f(p' + \frac{1}{2}p'', p - p'', E) \right), \tag{6}$$

where $f_0(p', p, E)$ is the free particle distribution function and v_p is the particle velocity. This form, analogous to the Lippmann-Schwinger equation, contains no visible $\hbar$ expansion, for it is buried in the definition of $f(p', p, E)$.

In order to illustrate how we extract physical information from the Wigner phase space representation we simply rewrite Eq.(2) as

$$\rho(x, x', t) = \int dp e^{-ip(x'-x)/\hbar} f(\frac{1}{2}(x + x'), p, t), \tag{7}$$

and introduce the moment functions[15] $< p^n(x,t) >$ as

$$\rho_0(x,t) < p^n(x,t) > = \left(\frac{\hbar}{2i}\right)^n \left(\left(\frac{\partial}{\partial x_i} - \frac{\partial}{\partial x_k} \right)^n \rho(x_k, x_i, t) \right)_{x_i = x_k = x} \tag{8}$$

$$= \int f(x,p) p^n dp. \tag{9}$$

where $\rho_0(x,t)$ is the diagonal density matrix. In this way (7) can be written as,

$$\rho(x, x') = \int dp \left(1 - \frac{ip}{\hbar}(x' - x) - \frac{p^2}{2\hbar^2}(x' - x)^2 + ... \right) f\left(\frac{x' + x}{2}, p \right) dp \tag{10}$$

or

$$\rho(x, x') = \rho_0 \left(\frac{x' + x}{2} \right) \left(1 - \frac{i}{\hbar}(x' - x) < p > - \frac{< p^2 >}{2\hbar^2}(x' - x)^2 + ... \right). \tag{11}$$

The expansion in Eq. (11) is not terribly useful, for convergence is likely to be too slow. However, we have obtained a more useful expansion for the density matrix by introducing a phase-amplitude representation for the Schrödinger amplitude $\psi(x,t)$ in $\rho(x_k.x_i;t)$ and then proceded to calculate (8) to a high order in n. For this representation $\left(\psi(x,t) = \rho_o^{1/2}(x,t)e^{iS(x,t)/\hbar}\right)$ we find the following general expressions for the *deviations* of the moments $< p(x,t) >$ from the mean momentum value i.e. $< (p- < p(x,t) >)^n >$ of the form,

$$\left(-\frac{i}{\hbar}\right)^n \frac{(x'-x)^n}{n!} < (p- < p >)^n >= -\frac{1}{\hbar}\left(\left(sin\left(\frac{(x'-x)}{2}\frac{\partial}{\partial x}\right)\right.\right.$$

$$\left.+sin\left(\frac{(x'-x)}{2}\frac{\partial}{\partial x'}\right)\right)e^{iS}\Big)_{x'=x=q} + \left(\left(\frac{(x'-x)}{2}\frac{\partial}{\partial x} + \frac{(x'-x)}{2}\frac{\partial}{\partial x'}\right)e^{iS}\right)_{x'=x=q}, \quad (12)$$

for odd n, and

$$\left(-\frac{i}{\hbar}\right)^n \frac{(x'-x)^n}{n!} < (p- < p >)^n >$$

$$= \left(cos\left(\left(\frac{(x'-x)}{2}\right)\left(i\frac{\partial}{\partial x'} - i\frac{\partial}{\partial x}\right)\right)\rho(x,x')\right)_{x=x'=q}, \quad (13)$$

for even n respectively. We note that even powers of n contain only derivatives of ρ_0 and odd powers of n contain only derivatives of S. It is important to observe that S(x,t) is not the classical action variable. Making use of equations (12) and (13) and rewriting the expansion (11) we obtain the following *identity* for the density matrix that is independent of the phase-amplitude representation utilized in its derivation;

$$\rho(x,x',t) = \rho_0(\frac{1}{2}(x'+x);t)e^{-\frac{i}{\hbar}(x'-x)<p>} < e^{\frac{-i}{\hbar}(x'-x)(p-<p>)} >, \quad (14)$$

where[9],

$$< e^{\frac{-i}{\hbar}(x'-x)(p-<p>)} >= 1 - \frac{i}{\hbar^2}(x'-x)(< p > - < p >)$$

$$-\frac{1}{2\hbar}(x'-x)^2 < (p- < p >)^2 > + \ldots. \tag{15}$$

Equation (14) forms the basis of our truncation scheme and relatively simple extensions of its structure allow the truncation scheme to be applied to both the many-particle problem and various kinds of plasma based on field representations. Equation (14) is an expansion of the density matrix in powers of $(x' - x)^n < (p- < p >)^n > /\hbar^n$ for n=0,1,2... . Near the classical domain, or more important, for physical problems with large numbers of degrees of freedom this expansion should converge faster than one based on the moments $< p^n(x,t) >$. For the simple one-body problem the quantum corrections begin at the level of $< (p- < p >)^2 (x'-x)^2/\hbar^2$.

In order to apply Eq. (14) we rewrite Eq.(3) using the definition (9). The resulting infinite set of equations for the moments is given by[9],

$$\frac{\partial}{\partial t}(\rho_o < p^n >) = -\frac{1}{m}\frac{\partial}{\partial x}(\rho_o < p^{n+1} >)$$

$$+\frac{4\rho_0}{\hbar^2}\sum_{k=1,3,5..}^{n}\left(-\frac{\hbar}{2i}\right)^{k+1}\binom{n}{k} < p^{n-k} > \frac{\partial^k}{\partial x^k}V, \tag{16}$$

where, n=0,1,2.... These equations that can be cut by applying the expansion (14). For the classical domain we may assume the relation $< p^n >= (< p >)^n$ is valid for all n and Eq.(14) reduces to the *classical* diagonal density. For this condition (16) also reduces to two equations of the form,

$$\frac{\partial}{\partial t}\rho_0 = -\frac{1}{m}\frac{\partial}{\partial x}(\rho_0 < p >) \qquad n = 0 \tag{17}$$

$$\frac{\partial}{\partial t} < p >= -\frac{\partial}{\partial x}\left(\frac{1}{2}\frac{(< p >)^2}{m} + V\right) \qquad n = 1. \tag{18}$$

These are simply Hamilton's classical equations of motion because for a point particle $\rho_0 = \delta(x - x')$and $< p >$ is the momentum. The n=2 equation from the set

(16) gives identically 0=0 and thus breaks the chain. The lowest order quantum corrections to the dynamics can be obtained by saying $< (p- < p >)^2 > \neq 0$ and in addition truncate (14) at the level $< (p- < p >)^3 > = < (p- < p >)^4 > = 0$. In this way the quantum mechanical equations for ρ, $< p >$ and $< (p- < p >)^2 >$ can be derived[9)]. We do not write them down again here but note the n=0 and n=1 equations remain unaltered. The n=4 equation from the set (16) gives identically 0=0. The n=3 equation gives an error measure for truncating at this order in n. We have derived the appropriate dynamical equations for truncations up to high n.

In order to include quantum corrections to the Uehling-Uhlenbeck equation a truncation scheme of this type needs to be incorporated into the homogeneous part and the collision integral. This is currently under investigation.

We note that the infinite set of equations (16) for the quantum one-body problem bear a striking resemblance to the coupled hierarchy of equations normally encountered in classical hydrodynamics. For instance n=0 and n=1 have the same *form* as the continuity equation and the Euler equation of classical hydrodynamics. This analogy was noticed a long time ago by Madelung[16]. However, we have exploited this analogy and derived a closed form for the quantum modifications to classical hydrodynamics that does not require the concept of local thermal equilibrium to achieve a closed set of equations[17].

As an example, if the equation for the higher moments is cut at n=2 and we introduce the concept of a pressure P_{ij}, now defined in three dimensions as,

$$P_{ij} = m\rho_0 < (p_i - < p_i >) > (p_j - < p_j >) > . \tag{19}$$

The n=0 continuity equation remains unchanged as does the n=1 equation for the conservation of momentum. However the n=2 equation reduces to a quantum analogue of the equation of state[17)] i.e.

$$\frac{\partial}{\partial t} P_{\beta\gamma} + \sum_\alpha < u_\alpha > \frac{\partial}{\partial x_\alpha} P_{\beta\gamma} = - \sum_\alpha \Big(P_{\beta\gamma} \frac{\partial}{\partial x_\alpha} < u_\alpha >$$

$$+ P_{\alpha\beta} \frac{\partial}{\partial x_\alpha} < u_\gamma > + P_{\alpha\gamma} \frac{\partial}{\partial x_\alpha} < u_\beta > \Big) \tag{20}$$

where $< u_\alpha >=< p_\alpha > /m$ and the greek indices run from 1 to 3. The n=3 equation of the infinite coupled set, in exact analogy to the quantum one-body problem, becomes a time-independent consistency condition[17],

$$P_{\beta\gamma}\frac{\partial}{\partial x_\delta}P_{\alpha\delta}+P_{\alpha\beta}\frac{\partial}{\partial x_\delta}P_{\gamma\delta}+P_{\alpha\gamma}\frac{\partial}{\partial x_\delta}P_{\beta\delta}=-\frac{1}{4}\frac{\rho_0^2\hbar^2}{m}\frac{\partial^3 V(\rho)}{\partial x_\alpha x_\beta x_\gamma} \tag{21}$$

where V is the mean-field potential. Hence it is possible to relate directly the mean-field potential to the pressure P_{ij} in a self-consistent way such that the quantum mechanical fluctuations are included to a given order and the assumptions about thermal equilibrium can be relaxed.

In the relativistic domain a problem of great interest to nuclear physicists is the formation and evolution of a quark-gluon plasma[5]. As part of a continuing program aimed specifically at this problem we have first applied our truncation approach to the simpler scalar plasma and electron plasma in a constant background magnetic field[10]. The constant background field simplifies the gauge invariance of the wavefunction. We do not go into details of the formalism here, for this has been published in some detail elsewhere[10], however, some characteristic differences between the field and particle problem are outlined.

The relativistic Liouville equation for charged scalar field densities is of the form,

$$(\Box_2-\Box_1)<\phi^\dagger(x_1)\phi(x_2)>=<\phi^\dagger(x_1)j(x_2)>-<j^\dagger(x_1)\phi(x_2)> \tag{22}$$

where $\Box$ is the D'Alembertian and $j(x)$ is a source function. The reduction of this equation to a closed equation for a N-point Wigner function requires a postulated form for $j(x)$ and in addition an approximation for the resulting product of field amplitudes. In general this latter step may be difficult to justify and may introduce unwanted infinities. If these problems are not addressed at the formal level, then in the mean-field limit the dynamical equation for a one-point Wigner function of a scalar plasma is identical in form to Eq. (3). This simplified equation, rewritten as moment equations, can be cut successfully using an expansion of the form (14).

Because of the four vector nature of the relativistic problem the cutting condition has to be modified to read[10],

$$< (p_1 - < p_1 >)^{k_1} (p_2 - < p_2 >)^{k_2} (p_3 - < p_3 >)^{k_3} (p_4 - < p_4 >)^{k_4} = 0 \qquad (23)$$

where $\sum_{i=1}^{4} k_i = n$ for cutting order n.

The electron plasma with abelian interactions offers an opportunity to study a multi-component Wigner function. These complications, introduced by the spinor nature of the plasma, requires the sixteen component Wigner function to be expanded in a basis of sixteen independent combinations of the Dirac γ matrices. In the mean-field limit each component satisfies a dynamical equation of the form[10] (14). This reduction will also be important for the quark-gluon plasma.

For the quark-gluon plasma it is necessary to consider the non-abelian nature of the interactions. For a general gluon field the *gauge invariant* Wigner function has been introduced by Heinz[18] i.e.

$$F(x,p) = (2\pi)^{-4} \int d^4 y e^{-ip_\alpha y^\alpha} <: \overline{\psi}(x+y/2) \Big(Pexp\Big(-iQ_a \int_x^{x+y/2} A^a_\mu(z) dz^\mu \Big)\Big)$$

$$\Big(Pexp\Big(-iQ_b \int_{x-y/2}^{x} A^b_\mu(z) dz^\mu \Big)\Big) \psi(x-y/2) :> \qquad (24)$$

where ψ is a solution of the Dirac equation $i\gamma^\mu(\partial_\mu + iQ_a A^a_\mu(x))\psi(x) = m\psi(x)$ and P denotes a path ordered exponential for the gluon field A^a_μ. Color is included by the usual color vectors $Q_a (a = 1, ...8)$. The fully quantum and gauge independent transport equation of which (24) is a solution has only recently been derived[19] and is somewhat lengthy and complicated. Heinz derived[18] the classical limit of the kinetic equation satisfied by $F(x,p)$.

We prefer an alternative viewpoint for the transport equation which has its roots in the derivation of the simple one-dimensional, one-body quantum problem. Rather than work with (24) we define directly the *gauge- independent* moments of the problem in the form[11],

$$M^{\mu}_{\alpha^n} = Tr \int d^4p \Big(p_\alpha - Q_b A^b_\alpha(x) \Big)^n \gamma^\mu f(x,p) \tag{25}$$

$$M^{\mu}_{[\alpha^n,a]\pm} = Tr \int d^4p \Big(\Big(p_\alpha - Q_b A^b_\alpha(x) \Big)^n Q^a \pm Q^a \Big(p_\alpha - Q_b A^b_\alpha(x) \Big)^n \Big) \gamma^\mu f(x,p). \tag{26}$$

In these equations $f(x,p)$ is a *gauge-dependant* Wigner function,

$$f(x,p) = -(2\pi)^4 \int d^4y e^{-ip_\alpha y^\alpha} <: \bar{\psi}(x - y/2)\psi(x + y/2 :> . \tag{27}$$

The transport equation satisfied by $f(x,p)$ is much simpler than the equation satisfied by $F(x,p)$ i.e.[11],

$$\frac{i}{2}\gamma^\mu \partial_\mu f(x,p) = -(\gamma^\mu p_\mu - m) f(x,p) + \int d^4y e^{-ip_\alpha y^\alpha} < \gamma^\mu Q_a A^a_\mu(x) \psi(x) \bar{\psi}(x') > \tag{28}$$

This equation is gauge-dependant and therefore not terribly useful; however a *gauge-dependant* set of coupled equations can be derived by combining (25) and (28). Just like the non-relativistic problem this set of coupled equations is fully equivalent to the kinetic equations satisfied by $F(x,p)$. We derive this set for the moments in (25) to be

$$i\partial_\mu M^{\mu}_{\alpha^n} = i\partial^{(P)}_\mu M^{\mu}_{\alpha^n} - Tr \sum_{t=0,2,4}^{n} \left(\frac{i}{2}\right)^t \binom{n}{k} \Big(\partial^{(t)}_\alpha A^a_\mu(x) \Big) M^{\mu}_{[a,\alpha^{n-t}]-}$$

$$+ Tr \sum_{t=1,3,5}^{n} \left(\frac{i}{2}\right)^t \binom{n}{t} \Big(\partial^{(t)}_\alpha A^a_\mu(x) \Big) M^{\mu}_{[a,\alpha^{n-t}]+} \qquad n = 0,1,2,\ldots\ldots \tag{29}$$

where we have taken the mean field-limit[11] and $\partial_\mu^{(P)}$ means the derivative acts only on the quantity $(p_\alpha - Q_a A_\alpha^a(x))$. To investigate the validity of our approach we have checked[11] our coupled set of equations and have shown that for n=0 we obtain the baryon number continuity equation and that for n=1 we get the well known[18] dynamical equation for the energy-momentum tensor. We are confident our approach to the complicated quark-gluon plasma problem[11,20] represents a considerable simplification and enables quantum hydrodynamic flow to be extracted from the Q.C.D. field theory.

In conclusion, we belive heavy-ion collisions represent a natural medium within which to develop and apply new kinetic theories or transport equations. The truncation scheme discussed here is applicable to simple one-body problems, quantum hydrodynamics and various kinds of plasma including those requiring non-abelian interactions. Practical implementation of the above scheme for nuclear physics, particle physics and astrophysics remains an interesting challenge for future studies.

This work was supported by U.S. Department of Energy under contract No. DE-AC02-76ER13001 with the State University of New York at Stony Brook.

References

1) S. Pieper,M. Rhoades-Brown,S. Landowne, Phys. Lett. 162B(1985)43

2) G. Bertsch, Invited paper presented at School in Heavy-Ion Physics, Erice, Sicily, July 17-23, 1984.

3) J. Stachel, contribution to this conference

4) G. Bertsch, H. Kruse and S. das Gupta, Phys. Rev. C29(1984)674;

J.Aichelin and G. Bertsch, Phys. Lett. 138B(1984)350;

J. Aichelin, contribution to this conference.

5) Quark Matter '84, Proceedings of the Fourth International Conference on Ultra-Relativistic Nucleus-Nucleus Collisions. Ed. K. Kajantie,

Lecture Notes in Physics 221 (Springer-Verlag).

6) J.D. Bjorken, Phys. Rev. D27(1983)140;

G. Baym, B. Friman, J.-P. Blaizot, M. Soyeur and W. Czyz, Nucl. Phys. A407(1983)541

7) R. Ledoux, contribution to this conference.

8) R. Hwa, contribution to this conference.

9) M. Ploszajczak and M.J. Rhoades-Brown, Phys. Rev. Letts. 55(1985)147.

10) M. Ploszajczak and M.J. Rhoades-Brown, Phys. Rev. D33(1986) in press

11) M. Carrington and M.J. Rhoades-Brown to be published

12) P. Carruthers and F. Zachariasen, Phys. Rev. D13(1976)950

13) E.P. Wigner, Phys. Rev. 40(1932)749

14) P. Carruthers and F. Zachariasen, Rev. Mod. Phys. 55(1983)345

15) E. Moyal, Proc. Cambridge Philos. Soc. 45(1949)99

16) E. Madelung, Z. Phys. 40(1926)322

17) M. Ploszajczak and M.J. Rhoades-Brown to be published

18) U. Heinz, Phys. Rev. Lett. 51(1983)351;Ann. of Phys. 161(1985)48

19) H.-Th. Elze, M. Gyulassy and D. Vasak to be published, (Berkeley preprint No. LBL-21137,1986)

20) M. Ploszajczak, M.J. Rhoades-Brown and M. Carrington to be published

PHASE TRANSITIONS

PHASE TRANSITIONS: THE LATTICE QCD APPROACH

R. V. Gavai[1]
Physics Department
Brookhaven National Laboratory
Upton, NY 11973
U S A

Abstract

Recent results in the field of finite temperature lattice quantum chromodynamics (QCD) are presented with special emphasis on comparison of the different methods used to incorporate the dynamical fermions. Attempts to obtain a nonperturbative estimate of the velocity of sound in both the hadronic and quark-gluon phase are summarised along with the results.

1. INTRODUCTION

Quantum Chromo Dynamics (QCD) is now widely believed to be the theory of strong interactions. Many of its predictions, especially those stemming from its property called asymptotic freedom, have been verified experimentally. With the advent of lattice gauge theories, and the subsequent use of Monte Carlo techniques to simulate them on the powerful supercomputers of today, we may now be on our way to understand the low energy (low momentum transfer) strong interactions as well. Already, static properties, such as the masses of the stable hadrons or the chiral condensate $\langle\bar{\psi}\psi\rangle$, have been reliably obtained from QCD using these techniques, confirming thus a long-held view that chiral symmetry is broken dynamically in our world. A natural question to ask is whether the theory can tell us anything about what we should expect under extreme conditions such as high temperatures or densities. Such conditions could have occurred in the very early

---

[1]Address after 1 October 1986, and permanent address: Theory Group, Tata Institute of Fundamental Research, Homi Bhabha Road, Bombay 400005 India

stages of our Universe and, interestingly enough, can perhaps be attained in the proposed heavy-ion experiments at Brookhaven National Laboratory and at CERN.

Lattice QCD at finite temperatures and finite densities offers us a way to obtain an answer to the question above. It may be emphasized that unlike many other approaches which too attempt to answer the same question, the lattice QCD approach is free of any arbitrary assumptions and does not need essentially any free parameters. The only parameters that enter are the quark masses and the scale of the theory Λ_{QCD}, all of which can be fixed by calculating hadron masses using the same methods.

Since the entire temperature domain of interest can be investigated using these methods, it is not a surprise that the very question of the existence of a phase transition has been pursued rather hotly in this field in the recent past. After a bit of controversy in the beginning, a consistent picture seems to be emerging now about the predictions of QCD with light, dynamical fermions, as I hope to show you later. Another interesting development in the past year or so has been the application of these techniques to obtain quantities such as the velocity of sound near the phase transition. Attempts to obtain a detailed space-time description of the evolution of the quark-gluon plasma would find such information quite useful. Before I present these results though, let me briefly review how one obtains them starting from first principles. I intend to give here only a broad idea; the interested reader may find it more rewarding to fill in the details from the literature[1)] elsewhere.

1.1 The Three Steps of Lattice Approach

The thermodynamic observables of a theory can be obtained in the canonical fashion from its partition function:

$$Z = \mathrm{Tr}\, \exp(-H/T) \,, \tag{1}$$

where H is the Hamiltonian of the theory and T is the temperature of the system. Tr denotes sum over all physical states of the theory. Usual thermodynamic formulae can be employed to obtain from Z the various physical quantities of interest, e.g., the energy density $\varepsilon = V^{-1}T^2\ \partial \ln Z/\partial T$. What we wish to do is to substitute the Hamiltonian of QCD in (1) and evaluate various physical quantities and order parameters as a function of temperature: a phase transition will show up as a nonanalytic behaviour. The lattice approach to do this consists of three major steps. First one rewrites the partition function in eq. (1) as a functional integral of the exponential of the euclidean action over all the fields of the theory. This essentially amounts to summing over all possible classical paths with a given boundary condition. Using a complete set of states $|i\rangle$ and dividing 1/T in n equal segments of length ε ($1/T=n\varepsilon$), one can rewrite eq. (1) as below:

$$Z = \sum_{i,i_1,i_2\ldots} \langle i|e^{-\varepsilon H}|i_1\rangle \langle i_1|e^{-\varepsilon H}|i_2\rangle \ \ldots\ \langle i_{n-1}|e^{-\varepsilon H}|i\rangle\ . \qquad (2)$$

Note that $\exp(-\varepsilon H)$ is the time evolution operator in the euclidean space-time. Thus each term in the sum in eq. (2) can be thought as one corresponding to a path which takes the system from the state i to i_1, to i_2 ... and so on, and back to i. In the limit $\varepsilon \to o$ such that $1/T$ = constant, eq. (2) becomes

$$Z = \int_{bc} D\phi \exp\left(-\ L\ \left(\phi(x),\ \partial\mu\phi(x)\right)\right)\ . \qquad (3)$$

In the case of QCD, ϕ in eq. (3) denotes the gluon fields and the (anticommuting) quark, antiquark fields; L is the usual QCD Lagrangian, and bc denotes periodic boundary conditions for the gluon fields and antiperiodic ones for the fermions.

One way to handle the complicated integrals in eq. (3) is to introduce a space-time lattice. Then the lattice spacing a acts as a regulator for the theory. The lattice theory can be made to respect gauge invariance by appropriately choosing the field variables and the action on the lattice. A popular choice[1)] is to place the fermion fields $\psi(n)$ on the lattice sites $n = (n_1,n_2,n_3,n_4)$ and the gauge fields U_n^μ are then associated with the (oriented) bonds of the lattice. In terms of these variables eq. (3) takes the form

$$Z = \int_{bc} \prod_{\substack{n\\ \mu}} dU_n^\mu \prod_n d\psi(n)\ d\bar{\psi}(n)\ \exp\left(-\ S(\psi,\bar{\psi},U_n^\mu)\right) \qquad (4)$$

where the lattice action S in (4) is chosen by demanding a) gauge invariance, and b) proper classical continuum limit, i.e., $\lim S(\psi,\psi,U_n^\mu) = L_{QCD}$. The gauge variables $U_n^\mu \sim \exp(ia\ g\ A^\mu(n))$ in this limit, where $A^\mu(n)$ is the continuum gluon field which is an analogue of the photon field in QED. If the lattice has N_β sites in the temporal direction, and a_β is the lattice spacing in that direction, then the temperature $T = 1/N_\beta a_\beta$. Analoguous quantities in the three spatial direction determine the volume: $V = N_\sigma^3 a_\sigma^3$. In practice all the final expressions are evaluated for $a_\sigma = a_\beta$ for simplicity, making it necessary that $N_\sigma \gg N_\beta$. $(N_\sigma = N_\beta \to \infty$ would correspond to $T = 0)$. The expression in (4) looks very similar to those used in statistical mechanics, e.g. the partition function of the Ising model. It is thus natural to expect that the methods to obtain the expectation values of various observables from Z above are borrowed from those areas of physics. Monte Carlo simulations is one such technique. Its popular use is dictated by the third step we have to take.

Introduction of the lattice above was merely for calculational convenience. One must remove the lattice finally by taking the limit of vanishing lattice spacng a. Only those answers which are obtained in this limit are relevant to our original problem. Employing the Monte Carlo technique, Creutz[2)] showed how one can take this limit numerically. Asymptotic freedom of QCD tells us how the bare coupling g^2 must change as $a \to 0$. It goes to zero according to the following relation:

$$a\Lambda_L = (b_0 g^2)^{-b_1/2b_0^2} \exp(-1/2b_0^2 g^2)\left[1 + O(g^2)\right], \tag{5}$$

with $b_0 = 33\text{-}2N_f/48\pi^2$ (6)

and $b_1 = 153\text{-}19N_f/384\pi^2$. (7)

Here N_f is number of massless flavours in the theory. Creutz showed that for $N_f = 0$ eq. (5) holds true for rather small lattices and rather large couplings g^2. In the (asymptotic) scaling region, where above equation is satisfied, one can obtain continuum results for any physical quantity of interest by using eq. (5).

2. PROBLEMS WITH DYNAMICAL FERMIONS

The anticommuting nature of the fermion variables $\psi,\bar{\psi}$ in eq. (4) makes it difficult to apply the above procedure in a straightforward manner to obtain the thermodynamics of QCD. One finds it usually convenient to carry out the fermionic integrals explicitly since S in (4) is typically $S = S_G(U) + \sum_{n,n'} \psi(n)Q_{nn'}\psi_n$. This leads to the following expression for Z:

$$Z = \int_{bc} \prod_{n,\mu} dU^{\mu} \exp\left(-S_G(U)\right) \cdot \det Q(U) \quad . \tag{8}$$

In a typical calculation, Q is a square matrix of dimension ~ 6000, and one needs to evaluate det Q about 1-10 million times. ($N_\sigma = 8$, $N_\beta = 4$ was assumed). Even with clever tricks which make use of the properties of Q such a calculation would need about three years on even a CRAY-XMP.

One therefore needs good approximation schemes which succeed in getting the essence of fermion loops contained in det Q with minimum computer time. The early calculations[3)] were done by dropping the determinant altogether, which can be thought of as the heavy quark limit of our world. There it was found that QCD had a strong first order deconfinement phase transition with a latent heat of about 1 GeV/fm^3. Research efforts in the past couple of years or so have

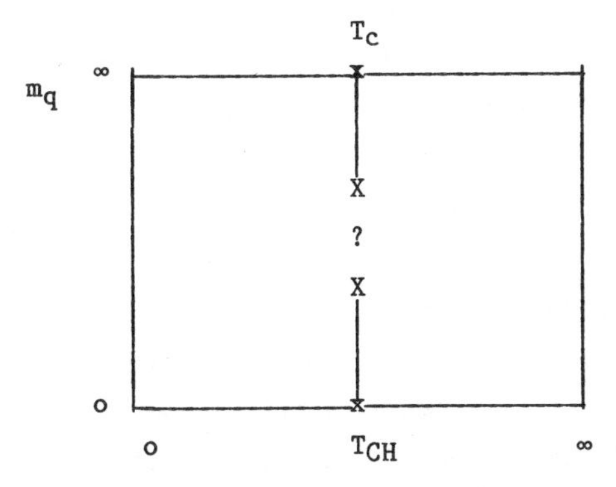

Fig. 1 Expected phase diagram of QCD.

been concentrated on the question of making those calculations more realistic by considering lighter quarks. Based on simple models which exploit only the symmetry aspects, one can argue what one would expect as the quark mass, m_q, is gradually lowered. These expectations are summarised in Fig. 1. The first order phase transition in the quenched or quarkless QCD is denoted there by T_c on the $m_q = \infty$ line. As m_q is lowered, one expects a line of first order phase transitions along which the latent heat decreases. The end of this line will be marked by a point where latent heat vanishes. It is not clear whether a second order line continues beyond this point. Starting from the other end, $m_q = 0$, one expects a chiral symmetry restoring phase transition there: we believe that chiral symmetry is broken in our world $\left(\langle\bar{\psi}\psi\rangle \neq 0,\ m_\pi^2 \to 0 \text{ as } m_q \to 0 \text{ etc.}\right)$ and it can be shown[4)] that at sufficiently high temperatures it must be restored. If this phase transition is also of first order, then as m_q is increased from zero, one expects an analogous scenario as that for the deconfinement phase transition.

The interesting question, of course, is about the positions of the two end points. Are they close to each other? or perhaps overlapping? Could one have two types of phase transitions for some value of m_q? These and other similar questions of details can only be answered after a good scheme to approximate the fermion determinant det Q is found. There are lots of proposals for such schemes in the literature, quite a few of which have been already used to study the full QCD thermodynamics, often leading to unfortunately confusing, sometimes even contradictory, results. In many cases the source of such a confusion is the method used to incorporate the fermion effects. Thus, one needs to test the methods rather thoroughly before drawing any firm conclusions. As I see, there are at least three necessary checks: i) one must study how stable the results are under variations of the parameters which keep the physics (i.e., temperature, number of flavours, quark mass, etc.) same, ii) one should compare results obtained by just varying the approximation schemes, and iii) one should compare the approximation with an exact numerical evaluation of the determinant in simpler situations.

Using the so-called pseudo-fermion method[5)], I have studied the full QCD from the standpoint of these checks. Along with Karsch[6)], I

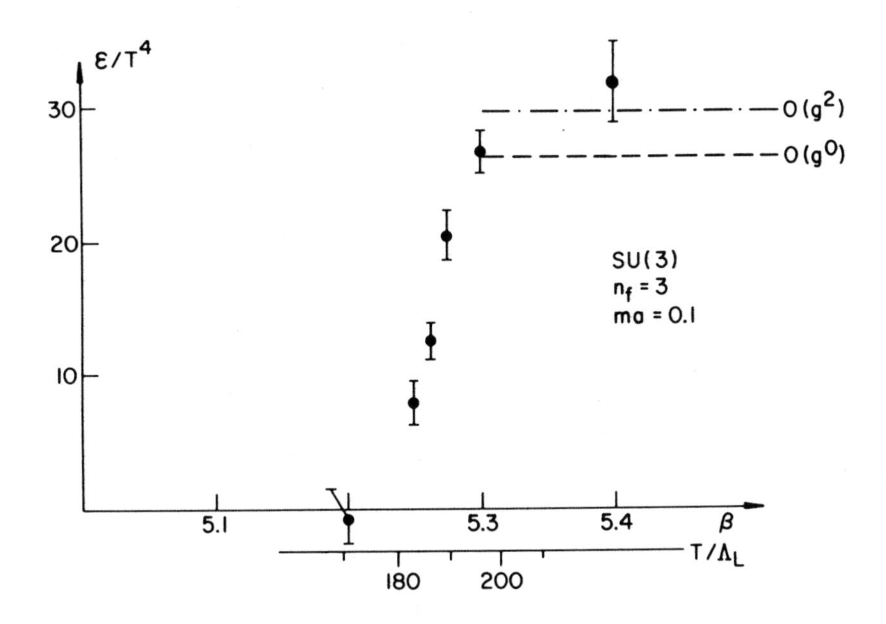

Fig. 2 Energy density as a function of β ($=6/g^2$) for QCD with three flavours of mass 0.1 on an $8^3 \times 4$ lattice. The temperature scale has been obtained by assuming the validity of eq. (5).

simulated the theory with three light dynamical flavours. Figures 2 and 3 show the energy density ε, and the order parameters $\langle\bar{\psi}\psi\rangle$ and $\langle L\rangle$ as a function of $\beta = 6/g^2$ or equivalently temperature on an $8^3 \times 4$ lattice. $\langle\bar{\psi}\psi\rangle$, the chiral condensate, can be thought of as a direct measure of the constituent mass of the quarks while $\langle L\rangle$ can be loosely described as the deconfinement order parameter: $\langle L\rangle \simeq 0$ corresponds to a confining phase, and $\langle L\rangle \neq 0$ to a deconfined phase.

One sees clearly that all these quantities undergo a rapid variation in a small range of temperature. The energy density jumps from a small value (~ 0.) to a value corresponding to that of an ideal gas of quarks and gluons. The constituent mass of the quarks becomes very small at the phase transition, and deconfinement seems to take place coincident with chiral symmetry restoration. We made attempts to look for the characteristic two state signal of a first order phase transi-

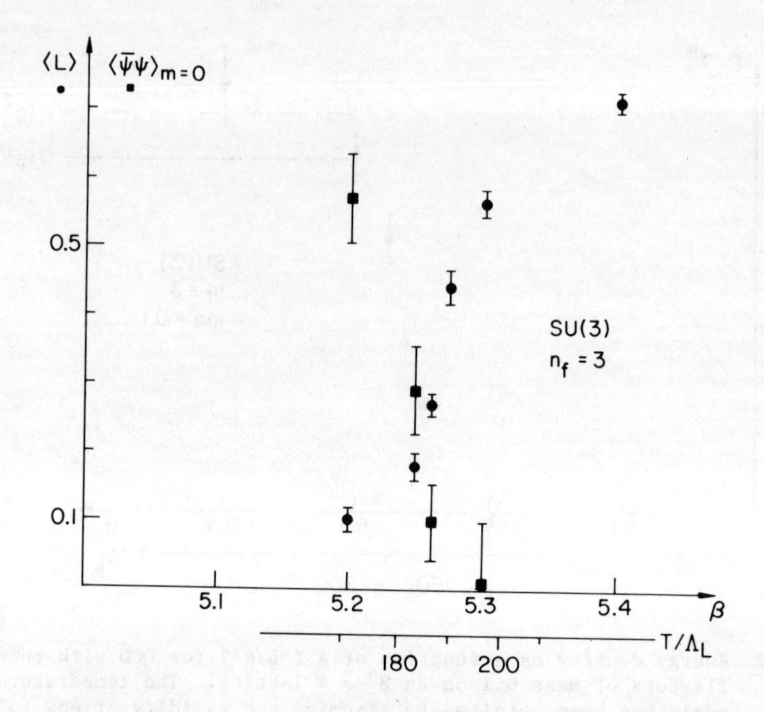

Fig. 3 The order parameters $\langle L\rangle$ and $\langle\bar{\psi}\psi\rangle_{m=0}$ versus $\beta(=6/g^2)$ and T/Λ_L. All the input parameters are the same as in Fig. 2 except that an additional quark mass of 0.075 was used to obtain a linear extrapolation to $\langle\bar{\psi}\psi\rangle_{m=0}$.

tion with negative results. While our results agree with the previous results[7] obtained on smaller lattices and with lesser statistics quantitatively, both sets of authors obtained β_c ~5.25, ref. 7 did find a first order phase transition. Our study suggests that the nature of the phase transition is sensitively dependent on the choice of parameters pertaining to the method. In particular, the signal observed in ref. 7 was washed out in higher statistics studies. Nonetheless, a safe conclusion would perhaps be that even in the phase transition in the full theory is indeed of first order, the latent heat (and similar discontinuities) is much smaller than was estimated in the earlier studies of quenched (or heavy quark) QCD. Phenomeno-

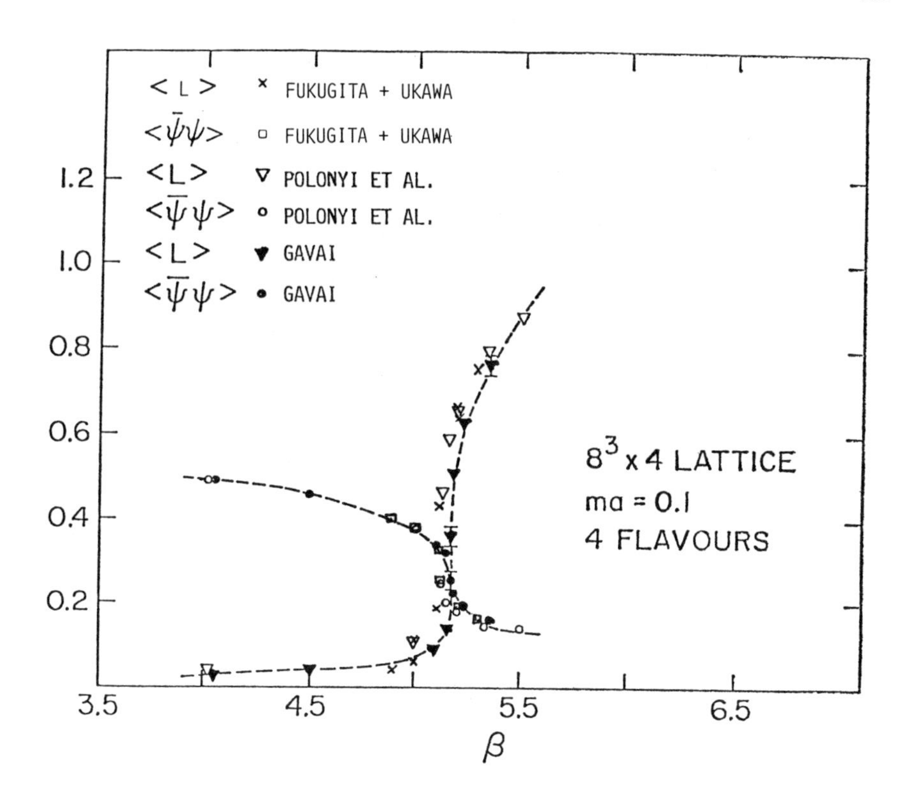

Fig. 4 $\langle L\rangle$ and $\langle\bar{\psi}\psi\rangle_{ma=0.1}$ as a function of β. All the data have been obtained for QCD with four light flavours of mass 0.1 on an $8^3 \times 4$ lattice. The dashed curves are drawn to guide the eye. Data from ref. 9, 10, and 11.

logical implications of this conclusion could be significant, especially in the studies of space-time evolution of the plasma and the experimental signatures to detect the plasma. Our lattice was perhaps not big enough to allow a good estimate of T_c in MeV. A rough estimate can, however, be obtained by using the recent spectroscopic calculations[8)] to set the scale Λ_L: $T_c \sim 200\text{-}250$ MeV.

Figures 4 and 5 exhibit the same quantities as in Figs. 2 and 3

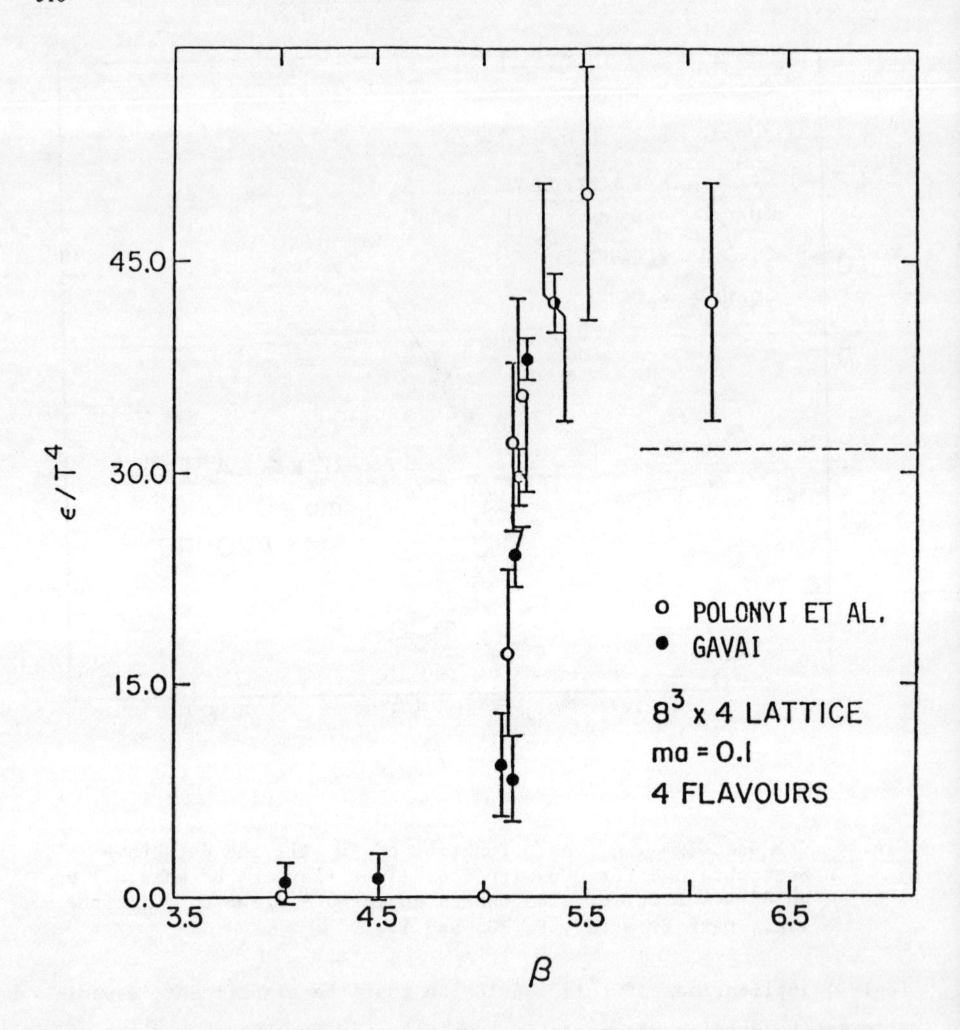

Fig. 5 The energy density ε versus β for the same theory as in Fig. 4. The dashed line represents the value for a corresponding ideal gas of 8 gluons and 12 quarks.

but calculated[9)] for four light flavours, again by using the pseudo-fermion method, in order to compare with a very different scheme to include the fermions, namely the microcanonical method[10)]. All the parameters which govern the physics, such as m_q or N_β were chosen to be the same for the two cases. A good quantitative agreement is evident in each of the three observables over the entire temperature range studied. I received a preprint[11] very recently where yet another method, called Langevin method, was used to study the same problem.

Table I: COMPARISON OF THE PSEUDO-FERMION METHOD WITH AN EXACT METHOD
Physics Parameters: Lattice size = 4^4, $N_f = 4$, $m_q a = 0.1$

	Observables						
METHOD	$\langle\bar{\psi}\psi\rangle$	W(1,1)	W(1,2)	W(1,3)	W(2,2)	W(2,3)	W(3,3)
Exact	0.402 ±0.006	0.415 ±0.003	0.177 ±0.003	0.074 ±0.003	0.035 ±0.003	0.008 ±0.002	-0.001 ±0.002
Pseudo-fermion	0.413 ±0.009	0.404 ±0.002	0.168 ±0.002	0.070 ±0.001	0.030 ±0.001	0.006 ±0.001	0.001 ±0.001

Their results also agree quantitatively with those discussed above, as one can see in Fig. 4. This is indeed very encouraging, and leads one to believe that these results are perhaps stable and reliable. Finally, Table I shows the comparison of the pseudo-fermion method with a method[12] using an exact numerical evaluation of the fermion determinant, but on a smaller lattice[13]. The physical observables labelled by W are relevant in the determination of hadron masses or the heavy quark potential. Once again, one finds an impressive agreement.

3. VELOCITY OF SOUND IN LATTICE GAUGE THEORIES

A popular approach to obtain a detailed space-time picture of the quark-gluon plasma produced in the relativistic heavy ion collisions is to employ the equations of relativistic hydrodynamics. One needs an equation of state to solve these equations. It is therefore of some interest to know the velocity of sound in the quark-gluon "fluid" in the non-perturbative region around the phase transition. Since $V_s^2 = (\partial P/\partial \varepsilon)_S$, one can use the lattice approach to obtain it. Unfortunately both ε and P are hard to obtain since both are dominated by two terms of about the same magnitude but opposite sign. It turns out[14], however, that one can relate it to another quantity which involves less of these uncertainties, and some preliminary results[14,15] have thus been obtained.

First, let us consider what one expects from simple considerations. Approximating the confined phase at low temperatures by a non-relativistic ideal gas of hadrons, it is simple to obtain $V_s^2 = \gamma\, T/m_H$ where m_H is the (effective) hadron mass. At large temperatures, one ought to have an ideal relativistic gas of quarks and gluons, and hence $V_s^2 \rightarrow 1/3$. In the absence of a phase transition, one expects thus the dashed line to represent V_s^2 as a function of T in fig. 6. At the phase transition point V_s^2 goes to zero, and the solid line in fig. 6 then would depict the behaviour of V_s^2.

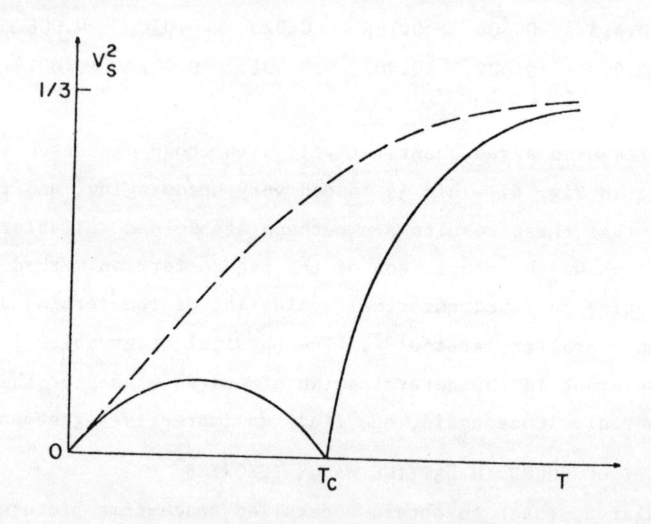

Fig. 6 A schematic representation of theoretical expectations for V_s^2 as a function of T.

Figure 7 shows the lattice evaluation of V_s^2 in the case of quenched QCD which is consistent with these naive expectations. Estimating the glueball mass from the low temperature relation above, one obtains m_G = 900 MeV which is certainly in the right ballpark. This calculation has now been performed with the dynamical fermions, and one again

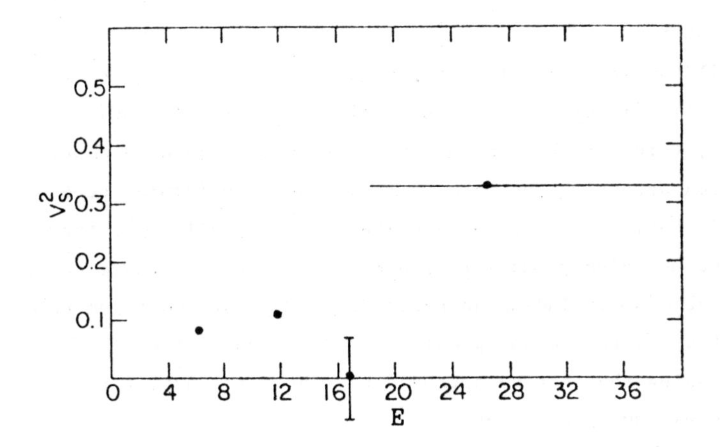

Fig. 7 The velocity of sound squared versus ε for the quenched QCD on an $8^3 \times 4$ lattice.

obtains a similar picture. Since the effective m_H is then expected to decrease, one should see the height of the maximum in the confined phase increase appreciably, which is what one finds in the Monte Carlo simulations[15].

4. CONCLUSIONS

While we are still rather far away from obtaining the phase diagram in fig. 1 completely, especially the critical (end) points in the diagram, I feel rather optimistic that it will soon be done. We now have good approximation schemes to include the fermion determinant, which do satisfy some necessary checks. In particular, the results obtained by using pseudo-fermion method agree with those obtained by other methods, including an exact one. One has now begun to obtain quantities of phenomenological interest, such as the velocity of sound, using the lattice approach and the first set of results in this area appear quite encouraging.

This work was supported by the U. S. Department of Energy under contract No. DE-AC-02-76CH00016.

REFERENCES

1) For a review of lattice QCD: M. Creutz, L. Jacobs, and C. Rebbi, Phys. Rep. 93, 201 (1983);
For finite temperature lattice QCD: J. Cleymans, R. V. Gavai, and E. Suhonen, Phys. Rep. 130, 217 (1986);
B. Svetitsky, Phys. Rep. 132, 1 (1986).
2) Creutz, M., Phys. Rev. Lett. 43, 553 (1979);
__________ Phys. Rev. D21, 2308 (1980).
3) Kajantie, K., Montonen, C., and Pietarinen, E., Z. Phys. C9, 253 (1981);
Kogut, J. et al., Phys. Rev. Lett. 50, 393 (1983);
Çelik, T., Engels, J., and Satz, H., Phys. Lett. 125B, 411 (1983).
4) Tomboulis, E.T. and Yaffe, L.G., Phys. Rev. Lett. 52, 2115 (1984).
5) Fucito, F., Marinari, E., Parisi, G., and Rebbi, C., Nucl. Phys. B180, [FS2] 360 (1981).
6) Gavai, R.V., and Karsch, F., Nucl. Phys. B261, 273 (1985).
7) Fucito, F., Solomon, S., and Rebbi, C., Nucl. Phys. B248, 397 (1984).
8) Fucito, F., Moriarty, K.J.M., Rebbi, C., and Solomon, S., "The hadronic spectrum with dynamical fermions", Brookhaven Preprint, BNL 37546.
9) Gavai, R.V., Nucl. Phys. B269, 530 (1986).
10) Polonyi, J., Wyld, H.W., Kogut, J.B., Shigemitsu, J., and Sinclair, D.K., Phys. Rev. Lett. 53, 644 (1984).
11) Fukugita, M. and Ukawa, A., "Deconfining and Chiral Transitions of finite temperature Quantum Chromodynamics in the presence of dynamical quark loops", Kyoto preprint, RIFP-642.
12) Scalpino, D.J. and Sugar, R.L., Phys. Rev. Lett. 46, 519 (1981).
13) Gavai, R.V. and Gocksch, A., "Comparing an exact fermion Monte Carlo algorithm with the pseudo-fermion method using the staggered fermions", Brookhaven preprint 37773.
14) Gavai, R.V. and Gocksch, A., Phys. Rev. D33, 614 (1986).
15) Redlich, K. and Satz, H., Phys. Rev. D (1986), in press.

Detection of the Charmonium Mass Shift near the Deconfinement Temperature

Takaaki Hashimoto and Osamu Miyamura
Department of Applied Mathematics, Faculty of Engineering Science, Osaka University, Toyonaka, Osaka 560, Japan
Kikuji Hirose
Teikoku Women's University, Moriguchi, Osaka 570, Japan
Takeshi Kanki
College of General Education, Osaka University, Toyonaka, Osaka 560, Japan
(presented by T.Hashimoto)

Abstract

Mass shift of charmonium near critical temperature of deconfining transition is investigated by $c-\bar{c}$ potential model through change of string tension. Based on the results, detection of the shift by lepton pair production is discussed by the use of hydrodynamical model of ultra-relativistic nucleus-nucleus collision. It is shown that the mass shift is detectable if the critical temperature is not lower than 250 MeV.

Introduction

It seems hard to discriminate the signals of the transition and the new phase from ordinary hadronic processes without ambiguity. It is believed that the best approach is to observe simultaneous occurence or correlation of the various proposed signals such as hydrodynamical effects,[1,2] excessive strange particle production,[3] critical phenomena[4] etc.. In this situation further efforts to find unambiguous signals of the transition are required. Hadronic mass shift related with restoration of chiral symmetry has been discussed recently.[5,6] In spite of uniqueness of the phenomenon, detectability of the effect in actual collision is still uncertain. In this report, we present some results for mass shift of charmonium and change of its leptonic width based on c-$\bar{c}$ potential model with variation of string tension which is expected to be temperature dependent. Recent works in Monte Calro simulation[7-9] and model investigation[10,11] suggest critical behaviour of the string tension near deconfining transition. Further, we discuss detection of the mass shift by lepton pair production in heavy ion collision assuming hydrodynamical development.[12-14] We see that the effect could be detectable if the deconfining temperature is not lower than 250 MeV.

Mass Shift of Charmonium

Computation of the mass shift adopted here is based on c-$\bar{c}$ potential model

$$V(r) = -\frac{4}{3}\frac{\alpha_s(r)}{r} + kr \ , \qquad (1)$$

where the first term is gluon exchange Coulombic term given by Buchmüller and Tye,[15] i.e.,

$$\alpha_s(r) = \frac{8}{b_0}\int_0^\infty \frac{\sin(\Lambda tr)}{t}\left(\frac{1}{\ln(1+t^2)} - \frac{1}{t^2}\right) , \qquad (2)$$

with Λ=250 MeV and b_0=9 (3 flavours).

Since we consider the string tension k at finite temperature, our present interest is to solve c-$\bar{c}$ states under variation of k. Generally speaking, the Coloumbic term might be also temperature dependent. Here we restrect ourselves to variation of the string tension because the long range part, kr, would primarily receive finite temperature effect below the critical temperature. Mass of charm quark in potential model may also be temperature dependent. In the case of heavy quarks, this would be minor effect and is omitted in the present calculation. We shall make a comment in later discussion.

Calculations are made by variational method with variation of string tension. In the calculation of the leptonic width, reliability of the value of variational wave function at the origin is rather poor. So we adopt the values from sum rule for potential, i.e.,

$$|\psi(0)|^2 = \frac{m_c}{2\pi} < \frac{dV}{dr} > . \qquad (3)$$

In Fig.1, the dependence of masses and widths on the string tension are shown. We fix the charm quark mass m_c to be 1.41 GeV and we can reproduce the results of Buchmüller and Tye at k=0.18 GeV^2. We see that the both quantities decrease gradually as the string tension decreases. The observed feature is understood as an effect due to broadening of wave function. Our simple calculations suggest that mass shift of a few hundred MeV is expected for charmonium near critical temperature of deconfining transition.

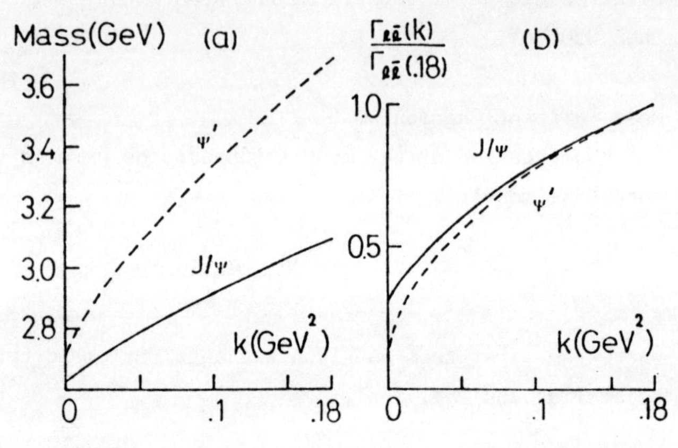

Figure 1 (a) : Mass shift of J/ψ and ψ'(3.7) with respect to variation of string tension k.
(b) : Change of leptonic pair decay width of J/ψ and ψ'(3.7) versus string tension.

Lepton Pair Production

Let us consider actual heavy ion collision. We imagine that well thermalized hadronic fire ball is formed in the collision. We assume that initial temperature is close to but <u>below</u> the critical temperature and subsequent developement is described by hydrodynamics. In order to incorporate the mass shift discussed above to this thermalized system, we need to specify dependence of the string tension on temperature. Here we use a form parametrized by critical exponent b as

$$k(T) = k(0) \left(\frac{T_{dec} - T}{T_{dec}} \right)^{b}. \tag{4}$$

As is well known, the expression (4) is suitable for the second order phase transition. At present, the order of deconfining transition in a system including dynamical quark freedom is still an open question. In case of weak first order transition, the critical behavior of string tension can be effectively expressed by appropriate choice of b. It is noted that critical behavior of the string tension has been observed by Monte Carlo simulation in SU(3) gauge system in which the transition is first order.[7)-9)] On the other hand , in string model, the transition is second order and the critical exponent is predicted to be 0.5.[10,11)] In Fig. 2, dependence of the mass on temperature is shown for several values of b.

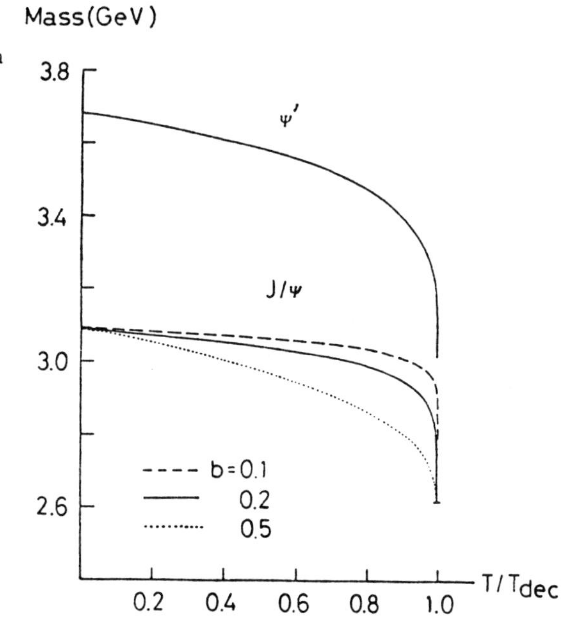

Figure 2 : Mass shift of J/ψ and ψ'(3.7) with respect to temperature.

As for a model of hydrodynamical developement in ultra-relativistic heavy ion collision we choose the one dimensional scaling expansion as an example. The temperature profile function is in this case given[12-14] by

$$\Phi(T) = \int dx^4\ \delta(T-T(x)) = 6\pi\ (R\tau_o)^2\ \frac{Y_{cm}}{T_o}\ \left(\frac{T}{T_o}\right)^{-7}, \quad (5)$$

where R is a radius of fire disk; τ_0 and T_0 is starting time of the evolution and starting temperature, respectively, while T_f is temperature of freeze-out; Ycm is center of mass rapidity of incidents. Emission rate of leptonic pair from four dimensional volume element dx^4 is given[14] by

$$\frac{dw}{dx^4} = n(M(T),T)\ \Gamma_{\ell\bar{\ell}}(T)\ , \quad (6)$$

with

$$n(M,T) = 3\ \left(\frac{MT}{2\pi}\right)^{3/2}\ \exp\left(-\frac{M}{T}\right)\ , \quad (7)$$

for $T<M$. Combining equations (5), (6) and the results of mass shift and leptonic width of J/ψ , invariant mass distribution of lepton pairs is given by

$$\frac{1}{\sigma_{FD}}\frac{d\sigma_{FD}}{dM} = \int_{T_f}^{T_o} dT\ \Phi(T)\ \frac{dw}{dx^4}(M(T),T)\ \delta(M-M(T))$$

$$= 3\ \left|\frac{\partial T}{\partial M(T)}\right|\ \Phi(T)\ \left(\frac{MT}{2\pi}\right)^{3/2}\ \exp\left(-\frac{M}{T}\right)\ \Gamma_{\ell\bar{\ell}}(T)$$

$$\times\ \theta(M(T_f)-M)\ \theta(M-M(T_o))\ , \quad (8)$$

where T is such that $M(T)=M$ and σ_{FD} is a formation cross-section of the fire disk. In the above expression, we have assumed that the total width of J/ψ does not change drastically and narrow width approximation has been used.

Results for the case of T_{dec}=250-350 MeV, T_f=100 MeV and b=0.1-0.5 at $T_0=T_{dec}$ are presented in Fig.3. As for the values of parameters of the fire disk, we take R=3 fm, τ_0=1 fm/c and Y_{cm}=3 as an example. For the sake of comparison, the figure includes data of proton (225 GeV/c) - Carbon collision[16] multiplied by a factors 3 and 12 as a simple estimaion for Carbon-Carbon and Tungsten-Carbon collisions in a superposition picture.[17] As shown in the figure, the yield of lepton pair from shifted J/ψ is well separated from original position for $b \geq 0.2$. This separation is due to the fact that Boltzmann factor in eq.(8) prefers high temperature at which the mass shift is significant. Strength of the yield strongly depends on the deconfinement

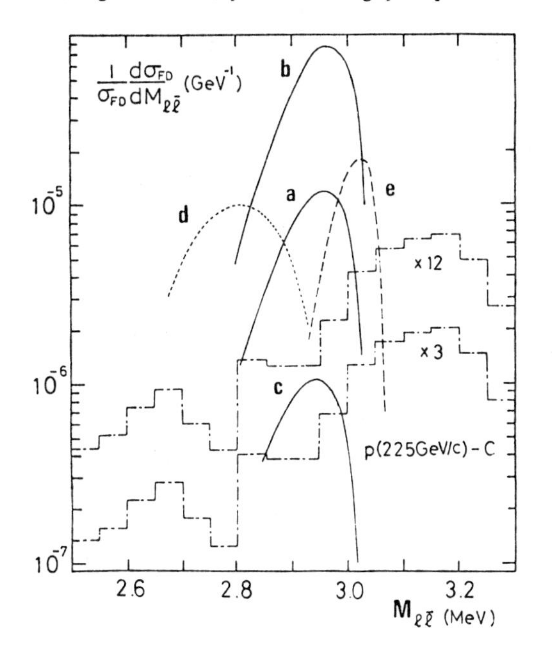

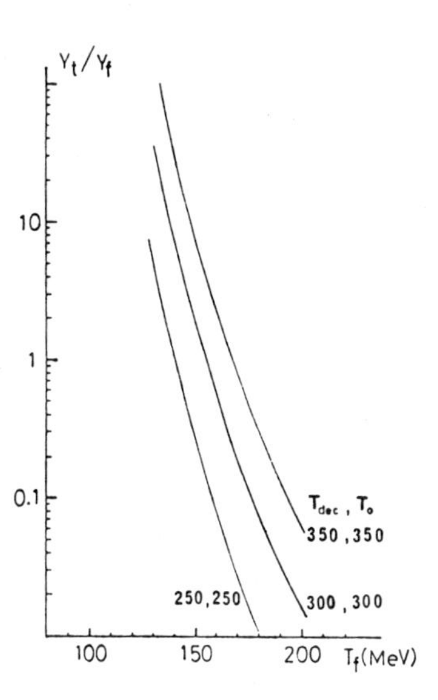

Figure 3 : Lepton pair yield from mass sfifted J/ψ for different values of deconfinement temperature T_{dec} and critical exponent b of string tension.

Figure 4 : The ratio between thermal yield Y_t and freeze-out yield Y_f.

The dotted dashed curves correspond to the data of proton-Carbon collision at 225 GeV/c (see Ref.16) multiplied by 3 and 12.

a) T_{dec}=300 MeV, b=0.2 b) T_{dec}=350 MeV, b=0.2 c) T_{dec}=250 MeV, b=0.2
d) T_{dec}=300 MeV, b=0.5 e) T_{dec}=300 MeV, b=0.1

temperature. For T_{dec} greater than 300 MeV, the clear evidence of shifted J/ψ peak will be expected as shown in Fig.3. In the case of lower T_{dec} we need better resoluton in lepton pair invariant mass to observe the peak. If T_{dec} is smaller than 200 MeV, the yield is severely suppressed.

Dependence on T_f is related with thermalization of J/ψ. If the total of cross-sections of $\pi + J/\psi \to D + \bar{D}$, $\eta_c + \pi$, etc. is larger than 10 mb, the thermalization of J/ψ is maintained in the region $T \geq 150$ MeV. On the other hand, if it is about a few mb, J/ψ decouples even at $T = 200$ MeV. In Fig.4, the ratio between lepton pair yield from thermalized J/ψ and that from freeze-out J/ψ is shown. The former strongly depends on T_f, while the latter is is almost insensitive to T_f. Present scenario works if T_f is smaller than 150 MeV, otherwise the yield from decoupled J/ψ dominates. In Fig.5, dependence on the starting temperature is shown. As shown in the figure, the mass shift reveals only when T_0 becomes close to T_{dec} and would be observed as a signal of the phase transition if we can control T_0 as a function of observation of the mass shift of charmonium by lepton pair production in heavy ion experiment at energies 10^2 Gev/nucleon to TeV/nucleon is one of new possibilities of detecting the phase transition.

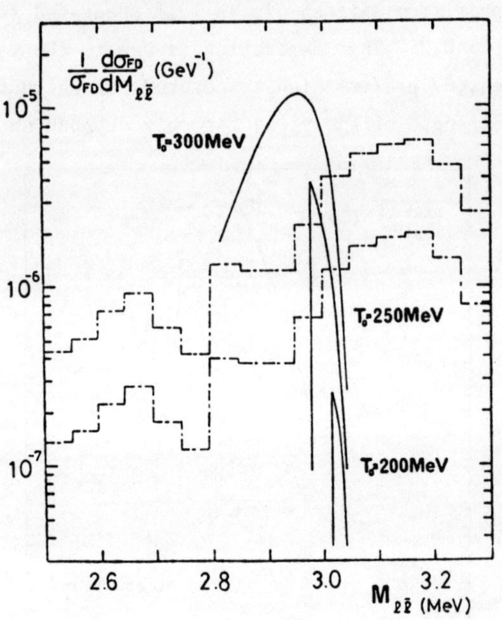

Figure 5 : Lepton pair yields for different values of starting temperature T_0. Deconfinement temperature is fixed at 300 MeV.

Discussions

So far we have assumed that charm quark mass is independent of temperature. However, if chiral restoration occurs and light quark condensation $\langle \bar{q}q \rangle$ decreases, dressed mass also may decrease. This means that further reduction of mass of charmonium is expected. As for temperature dependence of Coulombic term (1), finite temperature effect would be important in deconfinement phase. In fact, screening of color charge has been observed in Monte Carlo analyses.[8,9] A related question is whether charmonium like cluster still exists in quark gluon plasma. We have made tentative caluculation by screened Coulombic potential and found that possibility is little. Thus contribution to lepton pair in J/ψ mass region from deconfinement phase would be mainly thermal quark-antiquark annihilation.[18] In connection with this point, we make comments on our calculation for the case $T_0 \geq T_{dec}$. If there is considerable period of mixed phase of hadron and quark-gluon plasma, this gives aditional δ-function to the temperature profile function and we expect a spike in lepton pair invariant mass spectrum.

Observation of mass shift of ρ[5] or ϕ[19] meson might be also interesting and realistic evaluation of mass shifts in lepton pair production should be examined.

References

1) L. Van Hove, Z. Phys. C21, 93 (1983).
2) M. Gyulassy, Nucl. Phys. A418, 59c (1984).
3) J. Rafelski and B. Müller, Phys. Rev. Lett. 48, 1066 (1982).
4) T. Hatsuda and T. Kunihiro, Phys. Rev. Lett. 55, 158 (1985).
5) R.D. Pisarski, Phys. Lett. 110B, 155 (1982).
6) R.D. Pisarski and F. Wilczek, Phys. Rev. D29, 338 (1984).
7) L. McLerran and B. Svetitsky, Phys. Rev. D24, 450 (1981).
8) M. Fukugita, T. Kaneko and A. Ukawa, Phys. Lett. 154B, 185 (1985).
9) C. Dorninger, H. Leeb and H. Markum, Z. Phys. C29, 531 (1985).
10) R.D. Pisarski and O. Alvarez, Phys. Rev. D26, 3735 (1982).
11) F. Takagi, preprint Tohoku Unov. TU/85/290 (1985).

12) J.D. Bjorken, Phys. Rev. D27, 140 (1983).

13) K. Kajantie, R. Raito and P.V. Ruuskanen, Nucl. Phys. B222, 152 (1983).

14) E.V. Shuryak, Phys. Lett. 78B, 150 (1978).

15) W. Buchmüller and S.-H.H. Tye, Phys. Rev. D24, 132 (1981).

16) Mass number dependence of cross-section of lepton pair production in nucleus(A)-nucleus(B) and proton-nucleus(A) collisions are AB dσ(pp) / dM and A dσ(pp) / dM in superposition picture. As for total inelastic cross-sections, we have used emperical formulae given in A.S.Carrol et al., Phys. Lett. B80, 319 (1979) and J.Jaros et al., Phys. Rev. C18, 2273 (1978) to obtain $\sigma_{FD}^{-1}\, d\sigma_{FD} / dM$.

17) J.G. Branson et al., Phys. Rev. Lett. 38, 1331 (1977).

18) G. Domokos and J. Goldman, Phys. Rev. D23, 203 (1981); S.A. Chin, Phys. Lett. 119B, 51 (1982); K. Kajantie and H. Miettinen, Z. Phys. C14, 356 (1982).

19) Y. Takahashi and P.B. Eby, in proceedings of the 19th Cosmic Ray Conference at La Jolla, HE 1.3-13 (1985).

Charmonium Sum Rule at Finite Temperature

Takaaki Hashimoto, Kikuji Hirose[#], Takeshi Kanki[¶] and Osamu Miyamura

Department of Applied Mathematics, Faculty of Engineering Science, Osaka University, Toyonaka, Osaka 560, Japan

Teikoku Women's University, Moriguchi, Osaka 570, Japan

¶ Institute of Physics, College of General Education, Osaka University, Toyonaka, Osaka 560, Japan

(Presented by K.Hirose)

Abstract

New QCD sum rule which is available for finite temperature is proposed and applied to the charmonium system. We find some temperature dependence of the gluon condensation.

§1. Introduction

Since pioneering works by Shifman,Vainstein and Zakharov (SVZ)[1],QCD sum rules have been used extensively for determining low energy parameters of QCD, i.e., quark masses m_q, strong coupling constant α_s and some vacuum condensations $\langle G_{\mu\nu}G^{\mu\nu}\rangle$, $\langle\bar{q}q\rangle$, $\cdots$. These parameters can not be measured directly by experiment because of the confinement. QCD sum rule, which is essentially a dispersion relation, correlates these parameters with hadronic data, i.e., masses and widths of resonances in appropriate channel. Important point is that $\langle G_{\mu\nu}G^{\mu\nu}\rangle$ and $\langle\bar{q}q\rangle$ $\cdots$ are the nonperturbative quantities. So that,the obtained numerical values provide valuable information on the nonperturbative physics of QCD.

One then expects that a similar situation[2] occurs for the physics of hadronic matter at finite temperature ($0<T<T_{dec}$). The gluon and quark condensations would be T-dependent. There may be mass shifts of hadronic states. QCD sum rule at $T\neq 0$ would correlate these phenomena.

In this talk, we report on our two recent investigations:

(1) Derivation of a new sum rule which is appropriate for $T\neq 0$ case.

(2) Application of this sum rule to charmonium system and determination of the T-dependence of $\langle G_{\mu\nu}G^{\mu\nu}\rangle$ (for $T\lesssim 100$ MeV).

§2. Proposal of new sum rule at $T\neq 0$

Retarded current correlator $\Pi^{QCD}_{(Ret)}(q)$ at $T\neq 0$ (See Fig.1.);

$$q^2\, \Pi^{QCD}_{(Ret)}(q) \equiv \Pi^{QCD}_{(Ret)}{}^{\mu}{}_{\mu}(q)$$

$$= \int_{-\infty}^{\infty} d^4x\ \theta(x_0)\ \exp(iqx)\ \langle\langle \left[J_\mu(x),\ J^\mu(0) \right] \rangle\rangle, \tag{1}$$

is connected with Matsubara Green-function $\Pi^{QCD}_{(M)}$ as

$$\Pi^{QCD}_{(M)}(q_0{=}2\pi i n T) = \frac{1}{\pi}\int_0^\infty ds\, \frac{\mathrm{Im}\ \Pi^{QCD}_{(Ret)}(s)}{s + (2\pi nT)^2}\ , \tag{2}$$

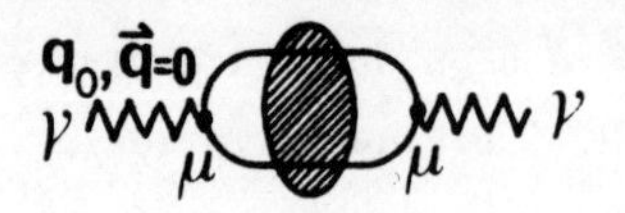

Figure 1

where n is an integer and we have chosen the $\vec{q} = 0$ frame. In the r.h.s. of (2), $\mathrm{Im}\ \Pi^{QCD}_{(Ret)}(s)$ is saturated by resonances and continuum at finit temperature. Hereafter, we denote $\mathrm{Im}\ \Pi^{QCD}_{(Ret)}$ in this case by $\mathrm{Im}\ \Pi^{had}$. It is noted that their parameters such as masses are temperature dependent.

For the l.h.s. of (2), we use the operator product expansion (OPE);

$$\Pi^{QCD}_{(M)}(T,q_0) = \left[C^0_I(T,q_0) + C^1_I(T,q_0) \right] I + C_G(T,q_0)\ \langle G_{\mu\nu}G^{\mu\nu}\rangle^{*)} + \cdots \tag{3}$$

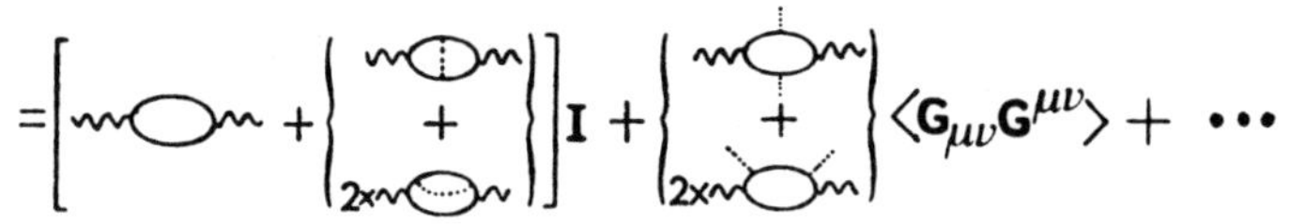

(4)

Wilson coefficients $C(T,q_0)$ are calculated diagramatically by the use of Matsubara method.

The coefficient C_G of the second term gives very small values compared with values of C_I. In order to extract the information on $\langle G_{\mu\nu}G^{\mu\nu}\rangle$, SVZ use the technique (higher moment method or Borel transformation) that leads the cancellation of large terms in the sum rule in one hand and on the other hand, enhances the small C_G term by taking the higher derivative of q^2. In the case of $T\neq 0$, we obtain the sum rules only for discrete values of q_0. Analytic continuation to all values of q_0 is extremely difficult for C_G. So we can not use the higher derivative method such as Borel transformation. Then we propose the following new technique.

Using only discrete q_0 values, we obtain the following sum rule from Eq.(2):

$$\sum_n g_n \, \Pi^{QCD}_{(M)}(2\pi i n T) = \frac{1}{\pi}\int_0^\infty ds \; \mathrm{Im}\,\Pi^{had}(s) \; \Big(\sum_n \frac{g_n}{s + (2\pi n T)^2} \Big) \; . \quad (5)$$

Here g_n is an "oscillating" weight and we require that the sum in the r.h.s. gives a certain "exponential" decreasing factor on s. For this purpose, here we take

$$g_n = \frac{2\pi}{\beta} \left(\cos\left(\frac{2\pi n}{\beta} x\right) - (-1)^n \right) , \quad (6)$$

where $\beta = 1/T$ and x is a parameter which corresponds to the 1/M in SVZ sum rule. Then the final form of the sum rule for $T\neq 0$ is,

$$\frac{2\pi}{\beta} \sum_{n=1}^{\infty} \left(\cos\left(\frac{2\pi n}{\beta} x\right) - (-1)^n \right) \Pi^{QCD}_{(M)}\left(\frac{2\pi i n}{\beta}\right)$$

$$= \int_0^\infty dw \ \mathrm{Im}\ \Pi^{had}(w^2) \left(\frac{ch(\beta/2-x)w - 1}{sh(\beta w/2)} \right) , \qquad (7)$$

where $w = \sqrt{s}$.

Note that Eq.(7) is formally derived. Actually we should take into account the convergence condition for interchange of the sum and the integral. For large W, one finds the exponential factor e^{-xw} in the r.h.s. of (7). Since g_n is a periodic function of x, the range of x is limited to $0 \leq x \leq \beta/2$.

§3. Charmonium sum rule

Now we consider the $c\bar{c}$ channel and charmonium problem, which is successful in SVZ sum rule. The calculation of $C(T,q_0)$'s is straightforward but tedious. The result for C_I^0 was already published[2]. We finished the calculation of C_G at $T \neq 0$, but the final expression is very lengthy and exceed one page. So that, here, we show only the numerical result for typical temperature $T = 100$ MeV in Fig.2; $\Delta C_G \equiv C_G(T,q_0) - C_G(0,q_0)$ is the correction to the SVZ value (T=0) as the finite temperature effect. In magnitude, ΔC_G is very small compared with $C_G(T{=}0,q_0)$. One, however, sees that ΔC_G has large (negative) slope for $n \to 0$. Because of this property, ΔC_G term gives a sensible contribution to the sum rule as seen later. At present, we have not yet completed the calculation of ΔC_I^1 (two loop correction). We suppose that ΔC_I^1 gives

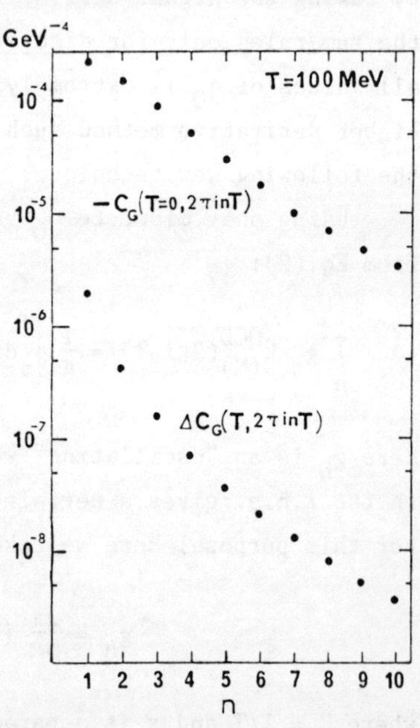

Figure 2 : Coefficient for gluon condensation as function of n.

only negligible contribution, so that our present result remains unchanged.

If the sum rule (7) works well in the $c\bar{c}$ channel, the l.h.s. should coincide with the r.h.s. in the wide range of the parameter x, i.e.,the left-right (L/R) ratio should be <u>unity</u>.

First, we show that new sum rule (7) reproduces the result of SVZ at $T = 0$. At $T \to 0$ limit, (7) becomes

$$\int_0^\infty dQ_0 \cos(Q_0 x)\, \Pi^{QCD}(iQ_0) = \int_0^\infty dw \exp(-xw)\, \mathrm{Im}\, \Pi^{had}(w^2) \ . \quad (8)$$

The L/R ratio of this sum rule is shown in Fig.3. Here we take the same values of the parameters as SVZ;

$$\left.\begin{array}{l} m_c = 1.26 \text{ GeV}^{**)} ,\ \alpha_s = 0.24 \ , \\ \langle G^2\rangle \equiv \langle G_{\mu\nu}G^{\mu\nu}\rangle = 0.1884 \text{ GeV}^4 , \\ \text{and other standard exp. data } (J/\psi \text{ and } \psi'). \end{array}\right\} \quad (9)$$

One clearly sees how the finite gluon condensation improves the sum rule at T=0.

Let us consider the T≠0 case. Shown in Fig.4a is the result of the sum rule (7) for $T \lesssim$ 130 MeV, where we keep all parameters to be the same values as those of the T=0 case (9). For $T \lesssim$ 70 MeV, this sum rule still works and therefore, no appreciable changes of

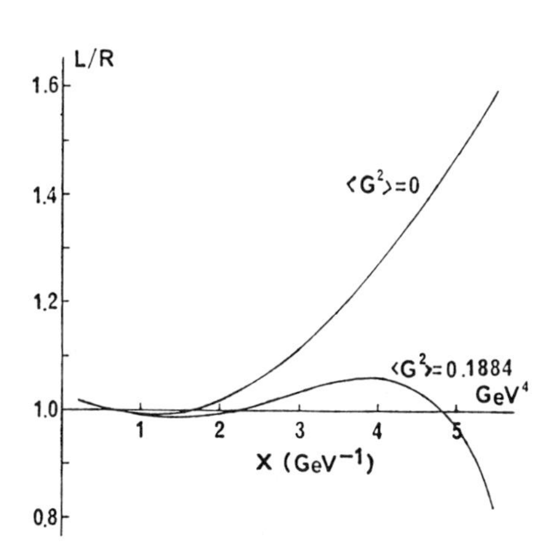

Figure 3 : L/R-ratio versus x at T = 0 .

$\langle G^2\rangle$ and also of masses and widths of J/ψ and ψ' would be expected. However, for $T \gtrsim 100$ MeV, we find the breakdown of the sum rule. This is indeed the effect of ΔC_G term.

In order to remedy the breakdown of the sum rule, we have to take the gluon condensation $\langle G^2 \rangle$ as temperature dependent. At the same time, we may expect that the hadronic parameters are also temperature dependent. At present we have no experimental information on this interesting phenomena.

To proceed further, we have to employ a certain theoretical model of hadrons as giving the bridge between $\langle G^2 \rangle$ and resonance parameters. Here we take the Buchmüller-Tye model by $c-\bar{c}$ potential[3];

$$V(r) = -\frac{4}{3}\frac{\overline{\alpha_s}(r)}{r} + kr . \tag{10}$$

The relation between the string tension k and $\langle G^2 \rangle$ is given by flux-tube model[4];

$$k = \sqrt{3}\,\alpha_s\,(\langle G^2 \rangle)^{1/2} . \tag{11}$$

Using the values (9), we find $k = 0.18\ \mathrm{GeV}^2$. This value gives the standard Regge slope and thus the relation(11) is successful at T=0. With the same k value, the potential (10) reproduces the very good charmonium parameters.

We calculate the masses and the leptonic widths of J/ψ and ψ' under the variation of the string tension k (See Ref.5). The charm quark mass m_c(on-shell) = 1.41 GeV and $\alpha_s = 0.24$ are fixed as temperature independent. Combining these results with the sum rule, we obtain the curves for T = 100 MeV in Fig.4b. For smaller value of $\langle G^2 \rangle$, the sum rule gives the better result.

In conclusion, we propose the sum rule available for $T \neq 0$ case. This sum rule works well for $T \lesssim 100$ MeV. It is suggested that no appreciable change of the charmonium parameters and the gluon condensation $\langle G^2 \rangle$ is expected for $T \lesssim 70$ MeV but smaller $\langle G^2 \rangle$ (0.16 GeV^4) is favorable at T = 100 MeV .

Since the allowed range of x is limited to $0 \leq x \leq \beta/2$, it is not clear whether the sum rule still works sensitively for $T \gg 100$ MeV . The reason for this limitation is due to the fact that we take information only at the discrete q_0 values.

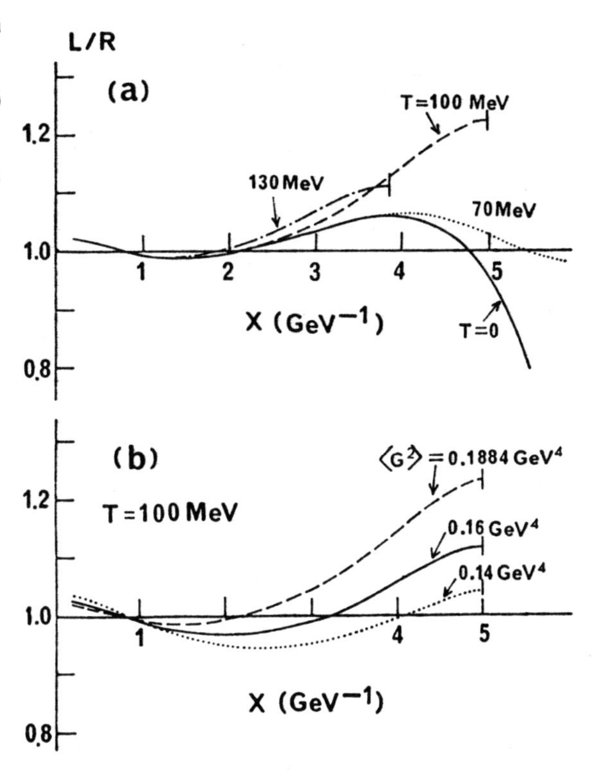

Figure 4 : L/R-ratio versus x at finite temperature.

References

1) M.A.Shifman, A.I.Vainshtein and V.I.Zakharov,Nucl.Phys. B147, 385, 448, 519 (1979).
2) A.I. Bochkarev and M.E. Shaposhnikov, Phys.Lett. 145B, 276(1984).
3) W.Buchmüller and S.-H.H. Tye, Phys.Rev. D24, 132 (1981).
4) C.Quigg, FERMILAB-Conf-81/78-THY, November (1981).
5) T.Hashimoto, 'Detection of the charmonium mass shift near the deconfinement temperature', talk given at this conference.

*) For $T \neq 0$, we have no longer Lorentz invariance.The separate

condensations $\langle E^2 \rangle$ and $\langle H^2 \rangle$ might generally be provided. We take here simpler form $\langle G_{\mu\nu} G^{\mu\nu} \rangle$.

**) m_c is defined at $p^2 = -m_c^2$ (on-shell) (See Ref.1).). Corresponding on-shell mass is m_c(on-shell)=1.41GeV^2.

THE QUARK-HADRON PHASE TRANSITION IN THE EARLY UNIVERSE

K. Kajantie

Department of Theoretical Physics and Academy of Finland

Siltavuorenpenger 20C

00170 Helsinki, Finland

ABSTRACT

When the temperature of the universe was about 200 MeV and its age about 10 microseconds, it underwent a phase transition in which the quarks and gluons of the cosmic fluid condensed into pions, nucleons and other hadrons. Assuming that the phase transition is of first order, various scenarios of going through the phase transition and their consequences were studied. Characteristic of all, unless the initial supercooling is very large, is the existence of a mixed phase, in which the expansion takes place by converting more and more matter from the dense quark phase to a dilute hadron phase. The dynamics of initial nucleation leads to a new scale in the problem: an average initial distance R_i between the hadron bubbles. R_i is on the average 10^{-4} times the Hubble distance 1/H; observable consequences, for example, energy density inhomogeneities, are poportional to R_iH. Enrichment of baryon number in quark droplets and the stability of quark nuggets is discussed. The results are contained in refs. (1-6).

1. E. Suhonen, Phys. Lett. 119B, 81 (1982).
2. E. Witten, Phys. Rev. D30, 272 (1984)
3. T. De Grand and K. Kajantie, Phys. Lett. 147B, 273 ('1984)
4. J. Applegate and C. Hogan, Phys. Rev. D31, 3037 (1985)
5. K. Kajantie and H. Kurki-Suonio, Univ. of Texas, Austin, preprint, 1986.
6. J. Madsen and K. Riisager, Phys. Lett. 158B, 208 (1985).

List of Participants

Jorg Aichelin
Inst. f. Theor. Physik
Univ. of Heidelberg
Philosophenweg 19
D-6900 Heidelberg, W. Germany

Johann Bartel
T-9, MS B279
Los Alamos National Laboratory
Los Alamos, NM 87545

Jerzy Bartke
Physics Division, Bldg. 50, 149
Lawrence Berkeley Laboratory
#1 Cyclotron Road
Berkeley, CA 94720

Gordon Baym
T-DO, MS B210
Los Alamos National Laboratory
Los Alamos, NM 87545

David H. Boal
Department of Physics
University of Illinois
1110 W. Green Street
Urbana, IL 61801

Peter BraunMunzinger
Physics Department
SUNY
Stony Brook, NY 11794

Harold C. Britt
P-DO, MS D456
Los Alamos National Laboratory
Los Alamos, NM 87545

Renato Campanini
Dipartimento di Fisica
Universita di Bologna
Via Irncrio 46
Bologna 40126 Italy

Peter A. Carruthers
T-8, MS B285
Los Alamos National Laboratory
Los Alamos, NM 87545

Laszlo P. Csernai
Physics Department
University of Minnesota
Minneapolis, MN 55455

M. Derrick
High Energy Physics Division
Argonne National Laboratory
Argonne, IL 60439

Bernice Durand
Department of Physics
University of Wisconsin
Madison, WI 53706

Gosta A. Ekspong
Institute of Physics
Vanadisv. 9
Stockholm Sweden S-11346

H.-Th. Elze
Nuclear Science Division, 70-A
Lawrence Berkeley Laboratory
Berkeley, Ca. 94720

Erwin M. Friedlander
Nuclear Science Division, 70-A
Lawrence Berkeley Laboratory
Berkeley, CA 94720

Goro Fujioka
Fermilab, MS 221, E653
PO Box 500
Batavia, IL 60510

Rajiv V. Gavai
Tata Inst. of Fundamental
Research
Homi Bhabha Rd.
Bombay 400 005 India

Avigdor Gavron
P-2, MS D456
Los Alamos National Laboratory
Los Alamos, NM 87545

Claus-Konrad Gelbke
Cyclotron Laboratory
Michigan State University
East Lansing, MI 48824

Gerson Goldhaber
Physics Division
Bldg. 50A/2160
Lawrence Berkeley Laboratory
Berkeley, CA 94530

A. Goldhaber
Department of Physics
SUNY
Stony Brook, NY 11974

Cheryl Grant
Physics and Astronomy
University of Minnesota
Minneapolis, MN 55455

Clive Halliwell
Physics Department
University of Illinois
PO Box 4348
Chicago, IL 60626

Yogiro Hama
Instituto de Fisica USP
C. Postal 20 516
Sao Paulo, SP
CEP 01498 Brazil

Takeaki Hashimoto
Department of Applied
Mathematics
Osaka University
Toyonaka, Osaka 560 Japan

Ulrich W. Heinz
Physics Department
Brookhaven National Laboratory
Upton, NY 11973

Kikuji Hirose
Teikoku Women's University
Moriguchi
Osaka 570 Japan

Hafeez R. Hoorani
Dept. of Physics
Simon Fraser U.
Burnaby, BC
V5A 1S6 Canada

Rudolph C. Hwa
Inst. of Theoretical Science
Univ. of Oregon
Eugene, OR 97403

Barbara V. Jacak
P-2, MS D456
Los Alamos National Laboratory
Los Alamos, NM 87545

List of Participants

Mikkel B. Johnson
MP-DO, MS H-850
Los Alamos National Laboratory
Los Alamos, NM 87545

K. Kajantie
Department of
Theoretical Physics
Siltanvorenpenger 20C
00170 Helsinki, Finland

Steve Koonin
Physics Department
California Institute of
Technology
Pasadena, CA 91125

Robert J. Ledoux
Department of Physics
Massachusetts Institute
of Technology
Cambridge, MA 02139

T. Ludlam
Physics Department
Brookhaven National Laboratory
Upton, NY 11786

David G. Madland
T-2, MS B243
Los Alamos National Laboratory
Los Alamos, NM 87545

Gabriel Mamane
P-2, MS D456
Los Alamos National Laboratory
Los Alamos, NM 87545

Larry McLerran
Physics Section
Fermilab
Batavia, IL 60510

Richard E. Mischke
MP-4, MS H846
Los Alamos National Laboratory
Los Alamos, NM 87545

J. R. Nix
T-9, MS B279
Los Alamos National Laboratory
Los Alamos, NM 87545

Eivind Osnes
Institute of Physics
Blindern
Oslo 3, Norway

Amalio Pacheco
Dpto. de Fisica Teorica
Universidad de Zaragoza
Zaragoza, Spain

Chris Pethick
Loomis Laboratory of Physics
University of Illinois
1110 W. Green Street
Urbana, IL 61801

Fred W. Pottag
Physics Department
University of Marburg
Marburg 3550 West Germany

Scott Pratt
Physics Department
Texas A&M
College Station, TX 77843

Howel G. Pugh
Physics Division, 70A-3307
Lawrence Berkeley Laboratory
Berkeley, CA 94720

J. Rafelski
Department of Physics
University of Cape Town
Rondebosch 7700 Cape
South Africa

Mark J. Rhoades-Brown
Physics Department
SUNY
Stony Brook, NY 11794

Louis Rosen
LAMPF
Los Alamos National Laboratory
Los Alamos, NM 87545

Ina Sarcevic
School of Physics & Astronomy
Minneapolis, Minn. 55455

Bernd Schurmann
Theoretische Physik
Technische Universitat
Munich, W. Germany

David K. Scott
Office of the Provost
Admin. Bldg.
Michigan State University
East Lansing, Michigan 48824

Chia C. Shih
Department of Physics
University of Tennessee
Knoxville, TN 37996-1200

Ed Siciliano
T-2, MS B243
Los Alamos National Laboratory
Los Alamos, NM 87545

Madeleine L. Soyeur
CEN de Saclay
Science de Physique Theorique
F-91191 Gif-sur-Yvette
Cedex, France

Johanna Stachel
Physics Department
SUNY
Stony Brook, NY 11794

Gerry Stephenson
P-DO, MS D434
Los Alamos National Laboratory
Los Alamos, NM 87545

Bruno J. Strack
GSI, Planckstr. 1
D-6100 Darmstadt 11
FRG

Daniel D. Strottman
T-9, MS B279
Los Alamos National Laboratory
Los Alamos, NM 87545

Esko E. Suhonen
Department of Physics
University of Oulu
Oulu 90570 Finland

List of Participants

Thomas Throwe
Physics Department
SUNY
Stony Brook, NY 11794

Wolfgang Trautmann
GSI-1 Planckstrasse
D-6100 Darmstadt, W. Germany

T. L. Trueman
Brookhaven National Laboratory
Upton, NY 11973

Leon Van Hove
Theoretical Division
CERN
1211 Geneva 23
Switzerland

Lou Voyvodic
Fermilab
Batavia, IL 60510

W. D. Walker
Physics Department
Duke University
Durham, NC 27706

Klaus Wehrberger
Inst. fur Kernphysik,
Schlossgartenstrasse 9
6100 Darmstadt, W. Germany

Richard M. Weiner
University of Marburg
Marburg 3550 West Germany

Geert Wenes
T-5, MS B283
Los Alamos National Laboratory
Los Alamos, NM 87545

Klaus Werner
Inst. f. Theor. Physik
Univ. of Heidelberg
Philosophenweg 19
D-6900 Heidelberg, W. Germany

Geof West
T-8, MS B285
Los Alamos National Laboratory
Los Alamos, NM 87545

G. Wilk
Institute for Nuclear Studies
ul. Hoza 69
00681 Warsaw, Poland

William A. Zajc
Physics Department
University of Pennsylvania
Philadelphia, PA 19104

RANDALL LIBRARY-UNCW
3 0490 0090925 +